AF324926

ADVANCED INEQUALITIES

SERIES ON CONCRETE AND APPLICABLE MATHEMATICS

ISSN: 1793-1142

Series Editor: Professor George A. Anastassiou
Department of Mathematical Sciences
The University of Memphis
Memphis, TN 38152, USA

Series on Concrete and Applicable Mathematics – Vol. 11

ADVANCED INEQUALITIES

George A Anastassiou

University of Memphis, USA

NEW JERSEY · LONDON · SINGAPORE · BEIJING · SHANGHAI · HONG KONG · TAIPEI · CHENNAI

Published by

World Scientific Publishing Co. Pte. Ltd.

5 Toh Tuck Link, Singapore 596224

USA office: 27 Warren Street, Suite 401-402, Hackensack, NJ 07601

UK office: 57 Shelton Street, Covent Garden, London WC2H 9HE

British Library Cataloguing-in-Publication Data
A catalogue record for this book is available from the British Library.

ADVANCED INEQUALITIES
Series on Concrete and Applicable Mathematics — Vol. 11

ISBN-13 978-981-4317-62-7
ISBN-10 981-4317-62-4

Printed in Singapore by B & Jo Enterprise Pte Ltd

Dedicated to my daughter Peggy

Preface

In this monograph we present univariate and multivariate classical analysis advanced inequalities. This treatise relies on author's last thirteen years related research work, more precisely see [15]-[61], and it is a natural outgrowth of it. Chapters are self-contained and several advanced courses can be taught out of this book. Extensive background and motivations are given per chapter. A very extensive list of references is given at the end.

The topics covered are very diverse. We present recent advances on Ostrowski type inequalities, see Chapters 2-9; on Opial type inequalities, see Chapters 10-16; Poincare and Sobolev type inequalities, see Chapters 17-20. Also we present Hardy-Opial type inequalities, as well as Ostrowski like ones, see Chapters 21-22. Then we present some work on ordinary and distributional Taylor formulae with estimates for their remainders and applications, see Chapters 23-24. Finally in Chapters 25-29 we study Chebyshev-Gruss, Gruss, Comparison of Means inequalities.

Our book's results appeared for the first time in our published articles mentioned above, and they are mostly optimal, that is usually our inequalities are sharp and attained. They are expected to find applications in many areas of pure and applied mathematics, such as mathematical analysis, probability, ordinary and partial differential equations, numerical analysis, information theory, etc. As such this monograph is suitable for researchers, graduate students, and seminars of the above subjects, also to be in all science libraries.

The preparation of book took place during 2008-2009 in Memphis, Tennessee, USA.

I would like to thank my family for their dedication and love to me, which was the strongest support during the writing of this monograph. Also many thanks go to my typist Rodica Gal for an excellent and on time technical job.

January 1, 2010
George A. Anastassiou
Department of Mathematical Sciences
The University of Memphis, TN
U.S.A.

Contents

Chapter 1

Introduction

Here we describe the material contained in this monograph.

Our results are mostly optimal, i.e. sharp, attained inequalities. We give also applications.

The exposed results are brought for the first time in a book form.

CHAPTER 2: Very general univariate Ostrowski type inequalities are given, involving the $\|\cdot\|_\infty$ and $\|\cdot\|_p, p \geq 1$ norms of the engaged nth order derivative, $n \geq 1$. In establishing them, several important univariate identities of Montgomery type are developed.

CHAPTER 3: We generalize Ostrowski inequality for higher order derivatives, by using a generalized Euler-type identity. Some of the produced inequalities are sharp, namely attained by basic functions. The rest of the estimates are tight. We give applications to trapezoidal and midpoint rules. Estimates are given with respect to L_∞ norm.

CHAPTER 4: We present a general multivariate Euler type identity. Using it we derive general tight multivariate high order Ostrowski type inequalities for the estimate on the error of a multivariate function f evaluated at a point from its average. The estimates are involving only the single partial derivatives of f and are with respect to $\|\cdot\|_p, 1 \leq p \leq \infty$. We give specific applications of our main results to the multivariate trapezoid and midpoint rules for functions f differentiable up to order 6. We prove sharpness of our inequalities for differentiation orders $m = 1, 2, 4$ and with respect to $\|\cdot\|_\infty$.

CHAPTER 5: Very general multidimensional Ostrowski type inequalities are presented, some of them are proved to be sharp. They involve the $\|\cdot\|_\infty$ and $\|\cdot\|_p$ norms of the engaged mixed partial of nth order $n \geq 1$. In establishing them, other important multivariate results of Montgomery type identity are developed.

CHAPTER 6: The classical Ostrowski inequality for functions on intervals is extended to functions on general domains in Euclidean space. For radial functions on balls the inequality is sharp.

CHAPTER 7: The classical Ostrowski inequality for functions on intervals estimates the value of the function minus its average in terms of the maximum of its first derivative. This result is extended to functions on general domains using the

L_∞ norm of its nth partial derivatives. For radial functions on balls the inequality is sharp.

CHAPTER 8: Here are presented Ostrowski type inequalities over spherical shells. These regard sharp or close to sharp estimates to the difference of the average of a multivariate function from its value at a point.

CHAPTER 9: The classical Ostrowski inequality for functions on intervals estimates the value of the function minus its average in terms of the maximum of its first derivative. This result is extended to higher order over shells and balls of $\mathbb{R}^N$, $N \geq 1$, with respect to an extended complete Chebyshev system and the generalized radial derivatives of Widder type. We treat radial and non-radial functions.

CHAPTER 10: Here we generalize Opial inequalities in the multidimensional case over balls. The inequalities carry weights and are proved to be sharp. The functions under consideration vanish at the center of the ball.

CHAPTER 11: Here we give Opial type weighted multidimensional inequalities over balls and arbitrary smooth bounded domains. The inequalities are mostly sharp. The functions under consideration vanish on the boundary.

CHAPTER 12: Various L_p form Opial type inequalities are given for Widder derivatives.

CHAPTER 13: Various L_p form Opial type inequalities are presented for a linear differential operator L, involving its related initial value problem solution y, Ly, the associated Green's function H and initial conditions point $x_0 \in \mathbb{R}$.

CHAPTER 14: Various L_p form Opial type inequalities are shown for functions valued in a Banach vector space. We give applications in $\mathbb{C}$.

CHAPTER 15: Various L_p form Opial type inequalities are given for semigroups with applications.

CHAPTER 16: Various L_p form Opial type inequalities are shown for cosine and sine operator functions with applications.

CHAPTER 17: Various L_p form Poincaré like inequalities, forward and reverse, are given for a linear differential operator L, involving its related initial value problem solution y, Ly, the associated Green's function H and initial conditions at point $x_0 \in \mathbb{R}$.

CHAPTER 18: Various L_p form Poincaré and Sobolev like inequalities, forward and reverse, are presented involving Widder derivative.

CHAPTER 19: Various L_p form Poincaré and Sobolev like inequalities are given for functions valued in a Banach vector space. We give applications on $\mathbb{C}$.

CHAPTER 20: Here we present Poincaré type general L_p inequalities regarding semigroups, cosine and sine operator functions. We apply inequalities to specific cases of them.

CHAPTER 21: Various L_p form Hardy–Opial type sharp integral inequalities are derived involving two functions.

CHAPTER 22: A sharp multidimensional integral type inequality is presented involving nth order ($n \in \mathbb{N}$) mixed partial derivatives. This is subject to some basic boundary condition satisfied by the involved multivariate function.

CHAPTER 23: Important estimates of the remainder in Taylor's formula are given. We treat both univariate and multivariate cases.

CHAPTER 24:We derive a distributional Taylor formula with precise integral remainder. We give applications of it and estimates for the remainder.

CHAPTER 25: Here we present Chebyshev–Grüss type univariate inequalities by using the generalized Euler type and Fink identities. The results involve the functions $f, g, f^{(n)}, g^{(n)}, n \in \mathbb{N}$, and are with respect to $\|\cdot\|_p, 1 \le p \le \infty$.

CHAPTER 26: Here are presented Grüss type multivariate integral inequalities.

CHAPTER 27: Here we give Chebyshev–Grüss type inequalities on $\mathbb{R}^N$ over spherical shells and balls by extending some basic univariate results of Pachpatte [206].

CHAPTER 28: We present tight multivariate Chebyshev–Grüss and comparison of integral means inequalities by using a general multivariate Euler type identity. Our results involve the functions f, g and their high order single partials and are with respect to $\|\cdot\|_p, 1 \le p \le \infty$.

CHAPTER 29: We describe a general multivariate Fink type identity which is a representation formula for a multivariate function. Using it we derive general tight multivariate high order Ostrowski type, comparison of means and Grüss type inequalities. The estimates involve L_p norms, any $1 \le p \le \infty$.

Our presented results are expected to find applications in many areas of pure and applied mathematics, such as mathematical analysis, probability, ordinary and partial differential equations, numerical analysis, information theory, etc.

Chapter 2

Advanced Univariate Ostrowski Type Inequalities

Very general univariate Ostrowski type inequalities are presented, involving the $\|\cdot\|_\infty$ and $\|\cdot\|_p$, $p \geq 1$ norms of the engaged nth order derivative, $n \geq 1$. In proving them, several important univariate identities of Montgomery type are given. This chapter follows [24].

2.1 Introduction

In 1938, A. Ostrowski [196] proved the following important inequality:

Theorem 2.1. *Let $f\colon [a,b] \to \mathbb{R}$ be continuous on $[a,b]$ and differentiable on (a,b) whose derivative $f'\colon (a,b) \to \mathbb{R}$ is bounded on (a,b), i.e., $\|f'\|_\infty :=$ $\sup_{t\in(a,b)} |f'(t)| < +\infty$. Then*

$$\left| \frac{1}{b-a} \int_a^b f(t)dt - f(x) \right| \leq \left[\frac{1}{4} + \frac{(x - \frac{a+b}{2})^2}{(b-a)^2} \right] \cdot (b-a)\|f'\|_\infty, \qquad (2.1)$$

for any $x \in [a,b]$. The constant $\frac{1}{4}$ is the best possible.

Since then there has been a lot of activity around these inequalities with important applications to Numerical Analysis and Probability.

This chapter is greatly motivated and inspired also by the following results.

Theorem 2.2 (see [16]). *Let $f \in C^{n+1}([a,b])$, $n \in \mathbb{N}$ and $x \in [a,b]$ be fixed, such that $f^{(k)}(x) = 0$, $k = 1, \ldots, n$. Then it holds*

$$\left| \frac{1}{b-a} \int_a^b f(y)dy - f(x) \right| \leq \frac{\|f^{(n+1)}\|_\infty}{(n+2)!} \cdot \left(\frac{(x-a)^{n+2} + (b-x)^{n+2}}{b-a} \right). \qquad (2.2)$$

Inequality (2.2) is sharp. In particular, when n is odd is attained by $f^*(y) :=$ $(y-x)^{n+1} \cdot (b-a)$, while when n is even the optimal function is

$$\tilde{f}(y) := |y - x|^{n+\alpha} \cdot (b-a), \quad \alpha > 1.$$

Clearly inequality (2.2) generalizes inequality (2.1) for higher order derivatives of f.

5

One of the goals of this chapter is to give a generalization Theorem 4.17, p. 191 of [85], see here Theorem 2.13 later. We would like to mention that result.

Theorem 2.3. *Let $f\colon [a,b] \to \mathbb{R}$ be continuous on $[a,b]$ and twice differentiable function on (a,b), whose second derivative $f''\colon (a,b) \to \mathbb{R}$ is bounded on (a,b). Then*

$$\left| f(x) - \frac{1}{b-a} \int_a^b f(t)dt - \frac{f(b)-f(a)}{b-a}\left(x - \frac{a+b}{2}\right) \right|$$

$$\leq \frac{1}{2}\left\{ \left[\frac{(x-\frac{a+b}{2})^2}{(b-a)^2} + \frac{1}{4}\right]^2 + \frac{1}{12}\right\}$$

$$\cdot (b-a)^2 \cdot \|f''\|_\infty \leq \frac{\|f''\|_\infty}{6}(b-a)^2, \quad \text{for all } x \in [a,b]. \tag{2.3}$$

In this chapter we present very general Ostrowski type inequalities in the one dimensional case. Involved integrals here are also of Riemann–Stieltjes form. Along the way to establishing these results, we show several important side univariate related results, generalizing Montgomery's identity ([188], Ch. XVII, p. 565), see Section 2.2.

2.2   Auxilliary Results

We present the following results we need, which by themselves are of independent interest.

Theorem 2.4. *Let $f\colon [a,b] \to \mathbb{R}$ be n-times differentiable on $[a,b]$, $n \in \mathbb{N}$. The nth derivative $f^{(n)}\colon [a,b] \to \mathbb{R}$ is integrable on $[a,b]$. Let $x \in [a,b]$. Define the kernel*

$$P(r,s) := \begin{cases} \dfrac{s-a}{b-a}, & a \leq s \leq r, \\[2mm] \dfrac{s-b}{b-a}, & r < s \leq b, \end{cases}$$

where $r,s \in [a,b]$. Then

$$\theta_{1,n} := f(x) - \frac{1}{(b-a)} \int_a^b f(s_1)ds_1$$

$$-\sum_{k=0}^{n-2}\left(\frac{f^{(k)}(b) - f^{(k)}(a)}{b-a}\right) \cdot \underbrace{\int_a^b \cdots \int_a^b}_{(k+1)\text{th}-\text{integral}} P(x,s_1)\prod_{i=1}^{k} P(s_i,s_{i+1}) \cdot ds_1 ds_2$$

$$\cdots ds_{k+1} = \int_a^b \cdots \int_a^b P(x,s_1)\prod_{i=1}^{n-1} P(s_i,s_{i+1})f^{(n)}(s_n)ds_1 ds_2 \cdots ds_n$$

$$=: \theta_{2,n}. \tag{2.4}$$

Here and later we make the conventions that $\prod_{k=0}^{-1} \bullet = 0,\ \prod_{i=1}^{0} \bullet = 1.$

Proof.   Applying integration by parts twice we obtain Montgomery identity ([188], Ch. XVII, p. 565)

$$f(x) = \frac{1}{b-a} \int_a^b f(s_1)ds_1 + \int_a^b P(x,s_1)f'(s_1)ds_1.$$

Doing the same for the derivative of f we have

$$f'(s_1) = \frac{1}{b-a} \int_a^b f'(s_2)ds_2 + \int_a^b P(s_1,s_2)f''(s_2)ds_2.$$

That is

$$f'(s_1) = \frac{f(b)-f(a)}{b-a} + \int_a^b P(s_1,s_2)f''(s_2)ds_2.$$

Similarly for f'' we get

$$f''(s_2) = \frac{f'(b)-f'(a)}{b-a} + \int_a^b P(s_2,s_3)f'''(s_3)ds_3.$$

And in general we obtain

$$f^{(n-1)}(s_{n-1}) = \frac{f^{(n-2)}(b)-f^{(n-2)}(a)}{b-a} + \int_a^b P(s_{n-1},s_n)f^{(n)}(s_n)ds_n.$$

We observe that

$$\int_a^b P(x,s_1)f'(s_1)ds_1 = \left(\frac{f(b)-f(a)}{b-a}\right)\int_a^b P(x,s_1)ds_1$$
$$+ \int_a^b \int_a^b P(x,s_1)P(s_1,s_2)f''(s_2)ds_2 ds_1.$$

That is

$$f(x) = \frac{1}{b-a}\int_a^b f(s_1)ds_1 + \left(\frac{f(b)-f(a)}{b-a}\right)\int_a^b P(x,s_1)ds_1$$
$$+ \int_a^b \int_a^b P(x,s_1)P(s_1,s_2)f''(s_2)ds_2 ds_1.$$

Next we see that

$$\int_a^b \int_a^b P(x,s_1)P(s_1,s_2)f''(s_2)ds_2 ds_1$$
$$= \left(\frac{f'(b)-f'(a)}{b-a}\right)\int_a^b \int_a^b P(x,s_1)P(s_1,s_2)ds_2 ds_1$$
$$+ \int_a^b \int_a^b \int_a^b P(x,s_1)P(s_1,s_2)P(s_2,s_3)f'''(s_3)ds_3 ds_2 ds_1.$$

Hence we get

$$f(x) = \frac{1}{b-a}\int_a^b f(s_1)ds_1 + \left(\frac{f(b)-f(a)}{b-a}\right)\int_a^b P(x,s_1)ds_1$$
$$+ \left(\frac{f'(b)-f'(a)}{b-a}\right)\int_a^b\int_a^b P(x,s_1)P(s_1,s_2)ds_2ds_1$$
$$+ \int_a^b\int_a^b\int_a^b P(x,s_1)P(s_1,s_2)P(s_2,s_3)f'''(s_3)ds_3ds_2ds_1.$$

Therefore in general we derive

$$f(x)$$
$$= \frac{1}{b-a}\int_a^b f(s_1)ds_1 + \left(\frac{f(b)-f(a)}{b-a}\right)\int_a^b P(x,s_1)ds_1$$
$$+ \left(\frac{f'(b)-f'(a)}{b-a}\right)\int_a^b\int_a^b P(x,s_1)P(s_1,s_2)ds_2ds_1$$
$$+ \left(\frac{f''(b)-f''(a)}{b-a}\right)\int_a^b\int_a^b\int_a^b P(x,s_1)P(s_1,s_2)P(s_2,s_3)ds_3ds_2ds_1$$
$$+ \left(\frac{f'''(b)-f'''(a)}{b-a}\right)\int_a^b\int_a^b\int_a^b\int_a^b P(x,s_1)P(s_1,s_2)P(s_2,s_3)P(s_3,s_4)ds_4\cdots ds_1$$
$$+ \cdots + \left(\frac{f^{(n-2)}(b)-f^{(n-2)}(a)}{b-a}\right)\underbrace{\int_a^b\cdots\int_a^b}\ P(x,s_1)\prod_{i=1}^{n-2}P(s_i,s_{i+1})ds_{n-1}\cdots ds_1$$
$$+ \underbrace{\int_a^b\cdots\int_a^b}_{n\text{th}-\text{integral}} P(x,s_1)\prod_{i=1}^{n-1}P(s_i,s_{i+1})f^{(n)}(s_n)ds_n\cdots ds_1.$$

The last equality is written briefly as follows

$$f(x) = \frac{1}{b-a}\int_a^b f(s_1)ds_1 + \sum_{k=0}^{n-2}\left(\frac{f^{(k)}(b)-f^{(k)}(a)}{b-a}\right)$$
$$\cdot\left(\underbrace{\int_a^b\cdots\int_a^b}_{(k+1)\text{th}-\text{integral}} P(x,s_1)\prod_{i=1}^{k}P(s_i,s_{i+1})ds_1ds_2\cdots ds_{k+1}\right)$$
$$+ \int_a^b\cdots\int_a^b P(x,s_1)\prod_{i=1}^{n-1}P(s_i,s_{i+1})f^{(n)}(s_n)ds_1ds_2\cdots ds_n. \qquad (2.5)$$

So we have proved (2.4). $\square$

A special very common case follows.

Corollary 2.5. *Under the assumptions and notations of Theorem 2.4, additionally assume that*

$$f^{(k)}(a) = f^{(k)}(b), \qquad k = 0,1,\ldots,n-2; \quad \text{when } n \geq 2.$$

Then we get

$$\theta_{1,n} = f(x) - \frac{1}{b-a} \int_a^b f(s_1)ds_1$$

$$= \int_a^b \cdots \int_a^b P(x,s_1) \cdot \prod_{i=1}^{n-1} P(s_i, s_{i+1}) \cdot f^{(n)}(s_n)ds_1 \cdots ds_n = \theta_{2,n}, \quad (2.6)$$

for $n \in \mathbb{N}$, $x \in [a,b]$.

Proof.   Directly from (2.4). $\qquad\qquad\qquad\qquad\qquad\qquad\qquad\qquad\square$

Next we generalize Montgomery's identity for Riemann–Stieltjes integrals (see also in [102] a related but different result).

Proposition 2.6.   *Let $f\colon [a,b] \to \mathbb{R}$ be differentiable on $[a,b]$. The derivative $f'\colon [a,b] \to \mathbb{R}$ is integrable on $[a,b]$. Let also $g\colon [a,b] \to \mathbb{R}$ be of bounded variation, such that $g(a) \neq g(b)$. Let $x \in [a,b]$. We define*

$$P(g(x),g(t)) = \begin{cases} \dfrac{g(t) - g(a)}{g(b) - g(a)}, & a \leq t \leq x, \\[3mm] \dfrac{g(t) - g(b)}{g(b) - g(a)}, & x < t \leq b. \end{cases}$$

Then

$$f(x) = \frac{1}{(g(b) - g(a))} \int_a^b f(t)dg(t) + \int_a^b P(g(x),g(t))f'(t)dt. \quad (2.7)$$

Proof.   Here f is continuous on $[a,b]$, and g of bounded variation on $[a,b]$. Then f is Riemann–Stieltjes integrable with respect to g, and it holds

$$\int_a^b f(x)dg(x) + \int_a^b g(x)df(x) = f(b)g(b) - f(a)g(a).$$

Also it holds that g is Riemann–Stieltjes integrable with respect to f. Here f is differentiable on $[a,b]$.

By the above we obtain

$$\int_a^x (g(t) - g(a))f'(t)dt = \int_a^x (g(t) - g(a))df(t)$$

$$= f(t)(g(t) - g(a))\Big|_a^x - \int_a^x f(t)dg(t)$$

$$= f(x)(g(x) - g(a)) - \int_a^x f(t)dg(t).$$

Also we have

$$\int_x^b (g(t) - g(b))f'(t)dt = \int_x^b (g(t) - g(b))df(t)$$

$$= (g(t) - g(b))f(t)\Big|_x^b - \int_x^b f(t)dg(t)$$

$$= (g(b) - g(x))f(x) - \int_x^b f(t)dg(t).$$

Adding the above equalities we get

$$\int_a^x (g(t) - g(a))f'(t)dt + \int_x^b (g(t) - g(b))f'(t)dt$$

$$= f(x)(g(b) - g(a)) - \int_a^b f(t)dg(t).$$

Hence we derive that

$$f(x) = \frac{1}{(g(b) - g(a))} \int_a^b f(t)dg(t) + \frac{1}{(g(b) - g(a))}$$

$$\cdot \left[\int_a^x (g(t) - g(a))f'(t)dt + \int_x^b (g(t) - g(b))f'(t)dt \right].$$

So we have established (2.7). $\qquad\qquad\qquad\qquad\qquad\qquad\qquad\square$

For twice differentiable functions we obtain the following general Montgomery type identity.

Proposition 2.7. *Let $f\colon [a,b] \to \mathbb{R}$ be twice differentiable on $[a,b]$. The second derivative $f''\colon [a,b] \to \mathbb{R}$ is integrable on $[a,b]$. Let also $g\colon [a,b] \to \mathbb{R}$ be of bounded variation, such that $g(a) \neq g(b)$. Let $x \in [a,b]$. The kernel P is defined as in Proposition 2.6. Then*

$$f(x) = \frac{1}{(g(b) - g(a))} \int_a^b f(t)dg(t) + \frac{1}{(g(b) - g(a))}$$

$$\cdot \left(\int_a^b P(g(x), g(t)) \cdot dt \right) \cdot \left(\int_a^b f'(t_1)dg(t_1) \right)$$

$$+ \int_a^b \int_a^b P(g(x), g(t)) \cdot P(g(t), g(t_1)) \cdot f''(t_1)dt_1 dt. \qquad (2.8)$$

Proof.   Apply (2.7) for f', we get

$$f'(t) = \frac{1}{(g(b) - g(a))} \int_a^b f'(t_1)dg(t_1) + \int_a^b P(g(t), g(t_1))f''(t_1)dt_1.$$

Therefore it holds

$$\int_a^b P(g(x), g(t)) \cdot f'(t)dt = \int_a^b P(g(x), g(t))$$

$$\cdot \left[\frac{1}{(g(b) - g(a))} \int_a^b f'(t_1)dg(t_1) + \int_a^b P(g(t), g(t_1))f''(t_1)dt_1 \right] dt$$

$$= \frac{1}{(g(b) - g(a))} \cdot \int_a^b P(g(x), g(t)) \cdot \left(\int_a^b f'(t_1)dg(t_1) \right) dt$$

$$+ \int_a^b P(g(x), g(t)) \cdot \left(\int_a^b P(g(t), g(t_1))f''(t_1)dt_1 \right) dt$$

$$= \frac{1}{(g(b) - g(a))} \cdot \left(\int_a^b P(g(x), g(t)) \cdot dt \right) \cdot \left(\int_a^b f'(t_1)dg(t_1) \right)$$

$$+ \int_a^b \int_a^b P(g(x), g(t)) \cdot P(g(t), g(t_1))f''(t_1)dt_1 dt.$$

Using the last equality along with (2.7), we have proved (2.8). $\square$

At the greatest generality it holds

Theorem 2.8. *Let* $f \colon [a, b] \to \mathbb{R}$ *be n-times differentiable on* $[a, b]$, $n \in \mathbb{N}$. *The nth derivative* $f^{(n)} \colon [a, b] \to \mathbb{R}$ *is integrable on* $[a, b]$. *Let also* $g \colon [a, b] \to \mathbb{R}$ *be of bounded variation, such that* $g(a) \neq g(b)$. *Let* $x \in [a, b]$. *The kernel* P *is defined as in Proposition 2.6. Then*

$$f(x) = \frac{1}{(g(b) - g(a))} \cdot \int_a^b f(s_1)dg(s_1) + \frac{1}{(g(b) - g(a))}$$

$$\cdot \sum_{k=0}^{n-2} \left(\int_a^b f^{(k+1)}(s_1)dg(s_1) \right)$$

$$\cdot \left(\underbrace{\int_a^b \cdots \int_a^b}_{(k+1)\text{th}-\text{integral}} P(g(x), g(s_1)) \prod_{i=1}^{k} P(g(s_i), g(s_{i+1})) \cdot ds_1 ds_2 \cdots ds_{k+1} \right)$$

$$+ \int_a^b \cdots \int_a^b P(g(x), g(s_1)) \prod_{i=1}^{n-1} P(g(s_i), g(s_{i+1}))f^{(n)}(s_n)ds_1 \cdots ds_n. \quad (2.9)$$

Proof. Similar to Proposition 2.7. $\square$

Another important special case follows.

Corollary 2.9. *All as in Theorem 2.8. Additionally suppose that*

$$\int_a^b f^{(k+1)}(s_1)dg(s_1) = 0, \qquad k = 0, 1, \ldots, n - 2; \text{ when } n \geq 2.$$

Then

$$M_{1,n} := f(x) - \frac{1}{(g(b) - g(a))} \cdot \int_a^b f(s_1)dg(s_1)$$

$$= \int_a^b \cdots \int_a^b P(g(x), g(s_1)) \prod_{i=1}^{n-1} P(g(s_i), g(s_{i+1})) f^{(n)}(s_n) \cdot ds_1 \cdots ds_n$$

$$=: M_{2,n}, \tag{2.10}$$

for $n \in \mathbb{N}$, $x \in [a,b]$.

Proof.   From Theorem 2.8. $\square$

A similar multivariate result comes next.

Theorem 2.10. *Let Q be a compact convex subset of $\mathbb{R}^k$, $k \geq 2$; $x := (x_1, \ldots, x_k)$ and $O := (0, \ldots, 0) \in Q$. Let $f \in C^1(Q)$ and suppose that all partial derivatives of f of order one are coordinatewise absolutely continuous functions. Then*

$$f(x) = \int_0^1 f(t_1 x)dt_1 + \sum_{j=1}^k x_j \int_0^1 \int_0^1 t_1 \frac{\partial f}{\partial x_j}(t_1 t_2 x)dt_1 dt_2$$

$$+ \sum_{i=1}^k \sum_{j=1}^k x_i x_j \int_0^1 \int_0^1 t_1^2 t_2 \frac{\partial^2 f(t_1 t_2 x)}{\partial x_i \partial x_j} dt_1 dt_2. \tag{2.11}$$

Note. Parameterized integrals as above are met in Radon Transform Theory and in Computerized Tomography.

Proof of Theorem 2.10. Set $g_x(t) := f(tx)$, $x := (x_1, \ldots, x_k)$, $x \in \mathbb{R}^k$, $k \geq 2$, $0 \leq t \leq 1$. Then $g_x(0) = f(0)$, $g_x(1) = f(x)$. Integrating by parts we obtain that

$$g_x(1) = \int_0^1 g_x(t)dt + \int_0^1 t g_x'(t)dt.$$

Here we have

$$g_x'(t) = \sum_{i=1}^k x_i \frac{\partial f}{\partial x_i}(tx).$$

Therefore

$$f(x) = \int_0^1 f(tx)dt + \int_0^1 t \left(\sum_{i=1}^k x_i \frac{\partial f(tx)}{\partial x_i} \right) dt.$$

So that

$$f(x) = \int_0^1 f(t_1 x)dt_1 + \sum_{i=1}^k x_i \int_0^1 t_1 \left(\frac{\partial f(t_1 x)}{\partial x_i} \right) dt_1. \tag{2.12}$$

Next apply (2.12) for $\frac{\partial f}{\partial x_j}$ to obtain

$$\frac{\partial f}{\partial x_j}(x) = \int_0^1 \frac{\partial f}{\partial x_j}(t_2 x)dt_2 + \sum_{i=1}^k x_i \int_0^1 t_2 \frac{\partial^2 f(t_2 x)}{\partial x_i \partial x_j} dt_2.$$

Hence

$$t_1\frac{\partial f}{\partial x_j}(t_1 x) = t_1 \int_0^1 \frac{\partial f}{\partial x_j}(t_2 t_1 x)dt_2 + \sum_{i=1}^k t_1^2 x_i \int_0^1 t_2 \frac{\partial^2 f(t_2 t_1 x)}{\partial x_i \partial x_j}dt_2.$$

Furthermore it holds

$$\int_0^1 t_1 \frac{\partial f(t_1 x)}{\partial x_j}dt_1 = \int_0^1 t_1 \left(\int_0^1 \frac{\partial f}{\partial x_j}(t_2 t_1 x)dt_2 \right)dt_1$$

$$+ \int_0^1 \sum_{i=1}^k t_1^2 x_i \left(\int_0^1 t_2 \frac{\partial^2 f(t_1 t_2 x)}{\partial x_i \partial x_j}dt_2 \right)dt_1. \qquad (2.13)$$

Consequently by (2.13) we obtain

$$\sum_{j=1}^k x_j \int_0^1 t_1 \left(\frac{\partial f(t_1 x)}{\partial x_j} \right)dt_1 = \sum_{j=1}^k x_j \int_0^1 \int_0^1 t_1 \frac{\partial f}{\partial x_j}(t_1 t_2 x)dt_2 dt_1$$

$$+ \sum_{j=1}^k x_j \int_0^1 \sum_{i=1}^k t_1^2 x_i \left(\int_0^1 t_2 \frac{\partial^2 f(t_1 t_2 x)}{\partial x_i \partial x_j}dt_2 \right)dt_1$$

$$= \sum_{j=1}^k x_j \int_0^1 \int_0^1 t_1 \frac{\partial f}{\partial x_j}(t_1 t_2 x)dt_2 dt_1$$

$$+ \sum_{j=1}^k \sum_{i=1}^k x_j x_i \int_0^1 \int_0^1 t_1^2 t_2 \frac{\partial^2 f(t_1 t_2 x)}{\partial x_i \partial x_j}dt_2 dt_1. \qquad (2.14)$$

At last combining (2.12) and (2.14) we have established (2.11). $\qquad\square$

Corollary 2.11. *All as in Theorem 2.10. Additionally suppose that*

$$\int_0^1 \int_0^1 t_1 \frac{\partial f}{\partial x_j}(t_1 t_2 x)dt_1 dt_2 = 0, \quad all \; j = 1, \ldots, k.$$

Then

$$f(x) - \int_0^1 f(t_1 x)dt_1 = \sum_{i=1}^k \sum_{j=1}^k x_i x_j \int_0^1 \int_0^1 t_1^2 t_2 \frac{\partial^2 f(t_1 t_2 x)}{\partial x_i \partial x_j}dt_1 dt_2, \qquad (2.15)$$

for any $x \in Q$; $0 \in Q$.

We will use the next

Lemma 2.12. *Let $f\colon [a,b] \to \mathbb{R}$ be 3-times differentiable on $[a,b]$. Assume that f''' is integrable on $[a,b]$. Let $x \in [a,b]$.*

Define

$$P(r,s) := \begin{cases} \dfrac{s-a}{b-a}, & a \le s \le r, \\[2em] \dfrac{s-b}{b-a}, & r < s \le b; \; r,s \in [a,b]. \end{cases}$$

Then

$$\theta_{1,3} := f(x) - \frac{1}{(b-a)} \int_a^b f(s_1)ds_1 - \left(\frac{f(b)-f(a)}{b-a}\right) \int_a^b P(x,s_1)ds_1$$

$$- \left(\frac{f'(b)-f'(a)}{b-a}\right) \int_a^b \int_a^b P(x,s_1)P(s_1,s_2)ds_1 ds_2$$

$$= \int_a^b \int_a^b \int_a^b P(x,s_1)P(s_1,s_2)P(s_2,s_3)f'''(s_3)ds_1 ds_2 ds_3 =: \theta_{2,3}. \quad (2.16)$$

Proof.   By Theorem 2.4. $\square$

2.3   Main Results

We present an Ostrowski type inequality.

Theorem 2.13. *Let $f\colon [a,b] \to \mathbb{R}$ be 3-times differentiable on $[a,b]$. Assume that f''' is bounded on $[a,b]$. Let $x \in [a,b]$. Then*

$$\left| f(x) - \frac{1}{(b-a)} \int_a^b f(s_1)ds_1 - \left(\frac{f(b)-f(a)}{b-a}\right)\left(x - \left(\frac{a+b}{2}\right)\right) \right.$$

$$\left. - \left(\frac{f'(b)-f'(a)}{2(b-a)}\right) \cdot \left[x^2 - (a+b)x + \left(\frac{a^2+b^2+4ab}{6}\right) \right] \right|$$

$$\leq \frac{\|f'''\|_\infty}{(b-a)^3} \cdot A, \quad (2.17)$$

where

$$A := \left[abx^4 - \frac{1}{3}a^2b^3x + \frac{1}{3}a^3bx^2 - ab^2x^3 - \frac{1}{3}a^3b^2x \right.$$

$$+ \frac{1}{3}ab^3x^2 + a^2b^2x^2 - a^2bx^3 - \frac{1}{2}ax^5 - \frac{1}{2}bx^5$$

$$+ \frac{1}{6}x^6 + \frac{3}{4}a^2x^4 + \frac{3}{4}b^2x^4 + \frac{1}{3}b^2a^4 - \frac{2}{3}a^3x^3$$

$$- \frac{2}{3}b^3x^3 - \frac{1}{3}b^3a^3 + \frac{5}{12}a^4x^2 + \frac{5}{12}b^4x^2$$

$$\left. + \frac{1}{3}b^4a^2 - \frac{2}{15}ba^5 - \frac{2}{15}ab^5 - \frac{1}{6}a^5x - \frac{1}{6}b^5x + \frac{a^6}{20} + \frac{b^6}{20} \right]. \quad (2.18)$$

Inequality (2.17) is attained by

$$f(x) = (x-a)^3 + (b-x)^3, \quad (2.19)$$

in that case both sides of the inequality equal zero.

Proof. Since $\|f'''\|_\infty < +\infty$, by (2.16) we obtain

$$|\theta_{1,3}| = |\theta_{2,3}| \le \|f'''\|_\infty \cdot \int_a^b \int_a^b \int_a^b |P(x,s_1)| \cdot |P(s_1,s_2)| \cdot |P(s_2,s_3)| ds_1 ds_2 ds_3$$

$$= \frac{\|f'''\|_\infty}{(b-a)^3} \cdot \int_a^b |p(x,s_1)| \left(\int_a^b |p(s_1,s_2)| \left(\int_a^b |p(s_2,s_3)| ds_3 \right) ds_2 \right) ds_1,$$

$$(2.20)$$

where

$$p(r,s) := \begin{cases} s-a, & a \le s \le r, \\ s-b, & r < s \le b; \ r,s \in [a,b]. \end{cases}$$

We see that

$$\int_a^b |p(s_2,s_3)| ds_3 = \frac{(s_2-a)^2 + (b-s_2)^2}{2}.$$

Furthermore we find with some calculations that

$$\int_a^b |p(s_1,s_2)| \left(\frac{(s_2-a)^2 + (b-s_2)^2}{2} \right) ds_2$$

$$= \frac{1}{12} \left[6 \left(s_1 - \left(\frac{a+b}{2} \right) \right)^4 + 3(b-a)^2 \left(s_1 - \left(\frac{a+b}{2} \right) \right)^2 + \frac{7}{8}(b-a)^4 \right].$$

To finish calculating the integral in (2.20) we find

$$\frac{1}{12} \int_a^b |p(x,s_1)| \left[6 \left(s_1 - \left(\frac{a+b}{2} \right) \right)^4 \right.$$

$$\left. +3(b-a)^2 \left(s_1 - \left(\frac{a+b}{2} \right) \right)^2 + \frac{7}{8}(b-a)^4 \right] ds_1$$

$$= \frac{1}{12} \int_a^x (s_1 - a) \left[6 \left(s_1 - \left(\frac{a+b}{2} \right) \right)^4 \right.$$

$$\left. +3(b-a)^2 \left(s_1 - \left(\frac{a+b}{2} \right) \right)^2 + \frac{7}{8}(b-a)^4 \right] ds_1$$

$$+ \frac{1}{12} \int_x^b (b - s_1) \left[6 \left(s_1 - \left(\frac{a+b}{2} \right) \right)^4 \right.$$

$$\left. +3(b-a)^2 \left(s_1 - \left(\frac{a+b}{2} \right) \right)^2 + \frac{7}{8}(b-a)^4 \right] ds_1$$

$$= A.$$

Consequently we have

$$|\theta_{1,3}| \le \frac{\|f'''\|_\infty}{(b-a)^3} \cdot A.$$

$$(2.21)$$

Next we find that

$$\int_a^b P(x,s_1)ds_1 = \left(x - \left(\frac{a+b}{2}\right)\right). \tag{2.22}$$

Furthermore we obtain

$$\int_a^b \int_a^b P(x,s_1)P(s_1,s_2)ds_1 ds_2$$
$$= \frac{1}{2}\left[(x^2 + (a+b)^2) - (a+b)x - \frac{(5a^2 + 5b^2 + 8ab)}{6}\right]. \tag{2.23}$$

Therefore we derive

$$\theta_{1,3} = f(x) - \frac{1}{(b-a)}\int_a^b f(s_1)ds_1$$
$$- \left(\frac{f(b)-f(a)}{b-a}\right)\left(x - \left(\frac{a+b}{2}\right)\right) - \left(\frac{f'(b)-f'(a)}{2(b-a)}\right)$$
$$\cdot \left[(x^2 + (a+b)^2) - (a+b)x - \frac{(5a^2 + 5b^2 + 8ab)}{6}\right]. \tag{2.24}$$

So we have established inequality (2.17). $\qquad\qquad\qquad\qquad\square$

We have an interesting special case.

Corollary 2.14. *All as in Theorem 2.13. Additionally suppose that $f(a) = f(b)$, and $f'(a) = f'(b)$. Then*

$$\left|\frac{1}{(b-a)}\int_a^b f(s_1)ds_1 - f(x)\right| \le \frac{\|f'''\|_\infty}{(b-a)^3} \cdot A. \tag{2.25}$$

Other Ostrowski type results follow:

Theorem 2.15. *All as in Theorem 2.4. Additionally assume that $\|f^{(n)}\|_\infty < +\infty$. Then*

$$|\theta_{1,n}| \le \|f^{(n)}\|_\infty \cdot \int_a^b \cdots \int_a^b |P(x,s_1)| \cdot \prod_{i=1}^{n-1} |P(s_i, s_{i+1})|ds_1 \cdots ds_n. \tag{2.26}$$

Theorem 2.16. *All as in Proposition 2.6. Additionally suppose that $\|f'\|_\infty < +\infty$. Then*

$$\left|\frac{1}{(g(b)-g(a))}\int_a^b f(t)dg(t) - f(x)\right| \le \|f'\|_\infty \cdot \int_a^b |P(g(x),g(t))|dt. \tag{2.27}$$

Theorem 2.17. *All as in Proposition 2.7. Additionally assume that $\|f''\|_\infty < +\infty$. Then*

$$\left|f(x) - \frac{1}{(g(b)-g(a))}\int_a^b f(t)dg(t) - \frac{1}{(g(b)-g(a))}\right.$$
$$\left. \cdot \left(\int_a^b P(g(x),g(t))dt\right) \cdot \left(\int_a^b f'(t_1)dg(t_1)\right)\right|$$
$$\le \|f''\|_\infty \cdot \left(\int_a^b \int_a^b |P(g(x),g(t))| \cdot |P(g(t),g(t_1))| \cdot dt_1 dt\right). \tag{2.28}$$

Theorem 2.18. *All as in Theorem 2.8. Additionally suppose that* $\|f^{(n)}\|_\infty < +\infty$. *Then*

$$\left| f(x) - \frac{1}{(g(b) - g(a))} \int_a^b f(s_1)dg(s_1) - \frac{1}{(g(b) - g(a))} \cdot \sum_{k=0}^{n-2} \left(\int_a^b f^{(k+1)}(s_1)dg(s_1) \right) \right.$$

$$\left. \cdot \left(\int_a^b \cdots \int_a^b P(g(x), g(s_1)) \cdot \prod_{i=1}^k P(g(s_i), g(s_{i+1}))ds_1 \cdots ds_{k+1} \right) \right|$$

$$\leq \|f^{(n)}\|_\infty \cdot \left(\int_a^b \cdots \int_a^b |P(g(x), g(s_1))| \cdot \prod_{i=1}^{n-1} |P(g(s_i), g(s_{i+1}))|ds_1 \cdots ds_n \right).$$

$$(2.29)$$

Theorem 2.19. *All as in Theorem 2.10. Additionally assume that*

$$\gamma_{ij} := \left\| \frac{\partial^2 f}{\partial x_i \partial x_j} \right\|_\infty < +\infty, \qquad for\ all\ i, j = 1, \ldots, k.$$

Then

$$\left| f(x) - \int_0^1 f(t_1 x)dt_1 - \sum_{j=1}^k x_j \int_0^1 \int_0^1 t_1 \frac{\partial f}{\partial x_j}(t_1 t_2 x)dt_1 dt_2 \right|$$

$$\leq \frac{1}{6} \left(\sum_{i=1}^k \sum_{j=1}^k |x_i| |x_j| \cdot \gamma_{ij} \right).$$

$$(2.30)$$

Next we present L_p, $p > 1$, Ostrowski type results.

Theorem 2.20. *All as in Theorem 2.4. Additionally assume that* $\|f^{(n)}\|_p < +\infty$, $p > 1$. *Here* p, q: $\frac{1}{p} + \frac{1}{q} = 1$. *Then*

$$|\theta_{1,n}| \leq \|f^{(n)}\|_p \cdot \int_a^b \cdots \int_a^b |P(x, s_1)| \cdot \left(\prod_{i=1}^{n-2} |P(s_i, s_{i+1})| \right)$$

$$\cdot \|P(s_{n-1}, \bullet)\|_q ds_1 ds_2 \cdots ds_{n-1}.$$

$$(2.31)$$

Theorem 2.21. *All as in Proposition 2.6. Additionally assume that* $\|f'\|_p < +\infty$, $p > 1$. *Here* p, q: $\frac{1}{p} + \frac{1}{q} = 1$. *Then*

$$\left| f(x) - \frac{1}{(g(b) - g(a))} \int_a^b f(t)dg(t) \right| \leq \|f'\|_p \cdot \|P(g(x), g(t))\|_{q,t}.$$

$$(2.32)$$

Here $\| \cdot \|_{q,t}$ *means integration with respect to t.*

Theorem 2.22. *All as in Proposition 2.8. Additionally assume that* $\|f''\|_p < +\infty$, $p > 1$. *Here* p, q: $\frac{1}{p} + \frac{1}{q} = 1$. *Then*

$$\left| f(x) - \frac{1}{(g(b) - g(a))} \int_a^b f(t)dg(t) - \frac{1}{(g(b) - g(a))} \right.$$

$$\left. \cdot \left(\int_a^b P(g(x), g(t))dt \right) \left(\int_a^b f'(t_1)dg(t_1) \right) \right|$$

$$\leq \|f''\|_p \cdot \int_a^b |P(g(x), g(t))| \cdot \|P(g(t), g(t_1)\|_{q,t_1} \ dt.$$

$$(2.33)$$

Theorem 2.23. *All as in Theorem 2.8. Additionally suppose that $\|f^{(n)}\|_p < +\infty$, $p > 1$. Here $p, q: \frac{1}{p} + \frac{1}{q} = 1$. Then*

$$\left| f(x) - \frac{1}{(g(b) - g(a))} \int_a^b f(s_1) dg(s_1) - \frac{1}{(g(b) - g(a))} \right.$$
$$\cdot \sum_{k=0}^{n-2} \left(\int_a^b f^{(k+1)}(s_1) dg(s_1) \right) \cdot \left(\int_a^b \cdots \int_a^b P(g(x), g(s_1)) \right.$$
$$\left. \left. \cdot \prod_{i=1}^{k} P(g(s_i), g(s_{i+1})) \cdot ds_1 \cdots ds_{k+1} \right) \right|$$
$$\leq \|f^{(n)}\|_p \cdot \left[\int_a^b \cdots \int_a^b |P(g(x), g(s_1))| \cdot \left(\prod_{i=1}^{n-2} |P(g(s_i), g(s_{i+1}))| \right) \right.$$
$$\left. \cdot \|P(g(s_{n-1}), g(s_n))\|_{q,s_n} \cdot ds_1 ds_2 \cdots ds_{n-1} \right]. \tag{2.34}$$

Finally we present L_1 Ostrowski type results.

Theorem 2.24. *All as in Theorem 2.4. Additionally suppose that $\|f^{(n)}\|_1 < +\infty$. Then*

$$|\theta_{1,n}| \leq \|f^{(n)}\|_1 \cdot \left(\int_a^b \cdots \int_a^b |P(x, s_1)| \cdot \left(\prod_{i=1}^{n-2} |P(s_i, s_{i+1})| \right) \right.$$
$$\left. \cdot \|P(s_{n-1}, \bullet)\|_\infty \cdot ds_1 ds_2 \cdots ds_{n-1} \right). \tag{2.35}$$

The interested reader can consult with [10], which is a further study on this basic chapter.

Theorem 2.25. *All as in Proposition 2.6. Additionally assume that $\|f'\|_1 < +\infty$. Then*

$$\left| f(x) - \frac{1}{(g(b) - g(a))} \int_a^b f(t) dg(t) \right| \leq \|f'\|_1 \cdot \|P(g(x), g(t))\|_{\infty, t}. \tag{2.36}$$

Here $\| \cdot \|_{\infty, t}$ is taken with respect to t.

Theorem 2.26. *All as in Proposition 2.8. Additionally assume that $\|f''\|_1 < +\infty$. Then*

$$\left| f(x) - \frac{1}{(g(b) - g(a))} \int_a^b f(t) dg(t) - \frac{1}{(g(b) - g(a))} \right.$$
$$\left. \cdot \left(\int_a^b P(g(x), g(t)) dt \right) \cdot \left(\int_a^b f'(t_1) dg(t_1) \right) \right|$$
$$\leq \|f''\|_1 \cdot \left(\int_a^b |P(g(x), g(t))| \cdot \|P(g(t), g(t_1))\|_{\infty, t_1} \cdot dt \right). \tag{2.37}$$

Finally we give

Theorem 2.27. *All as in Theorem 2.8. Additionally suppose that $\|f^{(n)}\|_1 < +\infty$.*
Then

$$
\left| f(x) - \frac{1}{(g(b) - g(a))} \cdot \int_a^b f(s_1)dg(s_1) - \frac{1}{(g(b) - g(a))} \right.
$$

$$
\cdot \sum_{k=0}^{n-2} \left(\int_a^b f^{(k+1)}(s_1)dg(s_1) \right) \cdot \left(\int_a^b \cdots \int_a^b P(g(x), g(s_1)) \right.
$$

$$
\left. \left. \cdot \prod_{i=1}^{k} P(g(s_i), g(s_{i+1}))ds_1 ds_2 \cdots ds_{k+1} \right) \right|
$$

$$
\leq \|f^{(n)}\|_1 \cdot \left(\int_a^b \cdots \int_a^b |P(g(x), g(s_1))| \cdot \left(\prod_{i=1}^{n-2} |P(g(s_i), g(s_{i+1}))| \right) \right.
$$

$$
\left. \cdot \|P(g(s_{n-1}), g(s_n))\|_{\infty, s_n} \cdot ds_1 \cdots ds_{n-1} \right). \tag{2.38}
$$

Chapter 3

Higher Order Ostrowski Inequalities

We present a generalization of Ostrowski inequality for higher order derivatives, by using a generalized Euler-type identity. Some of the produced inequalities are sharp, namely attained by basic functions. The rest of the estimates are tight. We give applications to trapezoidal and midpoint rules. Estimates are given with respect to L_∞ norm. This treatement is based on [35].

3.1 Introduction

We mention as motivation to our work the great Ostrowski inequality [196]:

$$\left| f(x) - \frac{1}{b-a} \int_a^b f(t)dt \right| \leq \left[\frac{1}{4} + \frac{\left(x - \frac{a+b}{2}\right)^2}{(b-a)^2} \right] (b-a)\|f'\|_\infty, \qquad (3.1)$$

where $f \in C'([a,b])$, $x \in [a,b]$, which is a sharp inequality, see [16]. Other motivations come from [10], [24], [16], [98] and [121].

We use here the sequence $\{B_k(t), k \geq 0\}$ of Bernoulli polynomials which is uniquely determined by the following identities:

$$B_k'(t) = kB_{k-1}(t), \quad k \geq 1, \quad B_0(t) = 1$$

and

$$B_k(t+1) - B_k(t) = kt^{k-1}, \quad k \geq 0.$$

The values $B_k = B_k(0)$, $k \geq 0$ are the known Bernoulli numbers. We need to mention

$$B_0(t) = 1, \quad B_1(t) = t - \frac{1}{2}, \quad B_2(t) = t^2 - t + \frac{1}{6},$$

$$B_3(t) = t^3 - \frac{3}{2}t^2 + \frac{1}{2}t, \quad B_4(t) = t^4 - 2t^3 + t^2 - \frac{1}{30},$$

$$B_5(t) = t^5 - \frac{5}{2}t^4 + \frac{5}{3}t^3 - \frac{1}{6}t, \quad \text{and} \quad B_6(t) = t^6 - 3t^5 + \frac{5}{2}t^4 - \frac{t^2}{2} + \frac{1}{42}.$$

Let $\{B_k^*(t), k \geq 0\}$ be a sequence of periodic functions of period 1, related to Bernoulli polynomials as

$$B_k^*(t) = B_k(t), \quad 0 \leq t < 1, \quad B_k^*(t+1) - B_k^*(t), \quad t \in \mathbb{R}.$$

21

We have that $B_0^*(t) = 1$, B_1^* is a discontinuous function with a jump of -1 at each integer, while B_k^*, $k \geq 2$ are continuous functions. Notice that $B_k(0) = B_k(1) = B_k$, $k \geq 2$.

We use the following result.

Theorem 3.1. *Let $f \colon [a, b] \to \mathbb{R}$ be such that $f^{(n-1)}$, $n \geq 1$, is a continuous function and $f^{(n)}(x)$ exists and is finite for all but a countable set of x in (a, b) and that $f^{(n)} \in L_1([a, b])$. Then for every $x \in [a, b]$ we have*

$$f(x) = \frac{1}{b-a}\int_a^b f(t)dt + \sum_{k=1}^{n-1} \frac{(b-a)^{k-1}}{k!} B_k\left(\frac{x-a}{b-a}\right)[f^{(k-1)}(b) - f^{(k-1)}(a)]$$

$$+ \frac{(b-a)^{n-1}}{n!}\int_{[a,b]}\left[B_n\left(\frac{x-a}{b-a}\right) - B_n^*\left(\frac{x-t}{b-a}\right)\right]f^{(n)}(t)dt. \qquad (3.2)$$

The sum in (3.2) when $n = 1$ is zero.

Proof. By using Theorem 2 of [98], Exercise 18.41(d), p. 299 in [158] and Problem 14(c), p. 264 in [224]. And that $f^{(n-1)}$ as implied absolutely continuous it is also of bounded variation. $\qquad \square$

If $f^{(n-1)}$ is just absolutely continuous then (3.2) is valid again. Formula (3.2) is a generalized *Euler type identity*, see also [171]. We set

$$\Delta_n(x) := f(x) - \frac{1}{b-a}\int_a^b f(t)dt$$

$$- \sum_{k=1}^{n-1} \frac{(b-a)^{k-1}}{k!} B_k\left(\frac{x-a}{b-a}\right)[f^{(k-1)}(b) - f^{(k-1)}(a)], \quad x \in [a, b]. \quad (3.3)$$

We have by (3.2) that

$$\Delta_n(x) = \frac{(b-a)^{n-1}}{n!}\int_{[a,b]}\left[B_n\left(\frac{x-a}{b-a}\right) - B_n^*\left(\frac{x-t}{b-a}\right)\right]f^{(n)}(t)dt. \qquad (3.4)$$

In this chapter we give sharp, namely attained, upper bounds for $|\Delta_4(x)|$ and tight upper bounds for $|\Delta_n(x)|$, $n \geq 5$, $x \in [a, b]$, with respect to L_∞ norm. That is generalizing (3.1) for higher order derivatives. High computational difficulties in this direction prevent us for shoming sharpness for $n \geq 5$ cases.

3.2   Main Results

We give

Theorem 3.2. *Let $f \colon [a, b] \to \mathbb{R}$ be such that $f^{(n-1)}$, $n \geq 1$, is a continuous function and $f^{(n)}(x)$ exists and is finite for all but a countable set of x in (a, b) and that $f^{(n)} \in L_\infty([a, b])$. Then for every $x \in [a, b]$ we have*

$$|\Delta_n(x)| \leq \frac{(b-a)^{n-1}}{n!}\left(\int_a^b \left|B_n\left(\frac{x-a}{b-a}\right) - B_n^*\left(\frac{x-t}{b-a}\right)\right|dt\right)\|f^{(n)}\|_\infty. \qquad (3.5)$$

Proof. By Theorem 3.1. $\qquad \square$

Performing the change of variable method in the integral of (3.5) we derive

Corollary 3.3. *All assumptions as in Theorem 3.2. Then for every $x \in [a, b]$ we have*

$$|\Delta_n(x)| \leq \frac{(b-a)^n}{n!} \left(\int_0^1 \left| B_n(t) - B_n\left(\frac{x-a}{b-a}\right) \right| dt \right) \|f^{(n)}\|_\infty, \quad n \geq 1. \qquad (3.6)$$

Note. Inequality (3.6) appeared first as Theorem 7, p. 350, in [98], wrongly under the sole assumption of $f^{(n)} \in L_\infty([a, b])$. Also in the rest of [98] the complete assumptions of our Theorem 3.2 are missing, whenever it applies.

Using Cauchy–Schwartz inequality we find that

$$\int_0^1 \left| B_n(t) - B_n\left(\frac{x-a}{b-a}\right) \right| dt \leq \left(\int_0^1 \left(B_n(t) - B_n\left(\frac{x-a}{b-a}\right) \right)^2 dt \right)^{1/2}$$

$$= \sqrt{\frac{(n!)^2}{(2n)!} |B_{2n}| + B_n^2\left(\frac{x-a}{b-a}\right)}, \quad n \geq 1, \qquad (3.7)$$

the last comes by [98], p. 352.

We give

Corollary 3.4. *All assumptions as in Theorem 3.2. Then for every $x \in [a, b]$ we have*

$$|\Delta_n(x)| \leq \frac{(b-a)^n}{n!} \left(\sqrt{\frac{(n!)^2}{(2n)!} |B_{2n}| + B_n^2\left(\frac{x-a}{b-a}\right)} \right) \|f^{(n)}\|_\infty, \quad n \geq 1. \qquad (3.8)$$

Here we elaborate on (3.6) and (3.8). We introduce the parameter

$$\lambda := \frac{x-a}{b-a}, \quad a \leq x \leq b. \qquad (3.9)$$

We see that

$$\lambda = 0 \quad \text{iff} \quad x = a,$$

$$\lambda = 1 \quad \text{iff} \quad x = b,$$

and

$$\lambda = \frac{1}{2} \quad \text{iff} \quad x = \frac{a+b}{2}.$$

Consider

$$p_4(t) := B_4(t) - B_4(\lambda) = t^4 - 2t^3 + t^2 - \lambda^4 + 2\lambda^3 - \lambda^2. \qquad (3.10)$$

We need to compute

$$I_4(\lambda) := \int_0^1 |p_4(t)| dt, \quad 0 < \lambda < 1. \qquad (3.11)$$

Lemma 3.5. *We find*

$$
I_4(\lambda) =
\begin{cases}
\dfrac{16\lambda^5}{5} - 7\lambda^4 + \dfrac{14}{3}\lambda^3 - \lambda^2 + \dfrac{1}{30}, & 0 \le \lambda \le 1/2, \\[3mm]
-\dfrac{16\lambda^5}{5} + 9\lambda^4 - \dfrac{26\lambda^3}{3} + 3\lambda^2 - \dfrac{1}{10}, & 1/2 \le \lambda \le 1,
\end{cases}
\tag{3.12}
$$

which is continuous in $\lambda \in [0,1]$. *We obtain*

$$
I_4(0) = I_4(1) = \frac{1}{30},
$$

$$
I_4\left(\frac{1}{2}\right) = \frac{7}{240}.
\tag{3.13}
$$

Proof. Here all calculations are done by Mathematica 4. The equation $p_4(t) = 0$ has four real roots,

$$
r_1 = 1 - \lambda, \quad r_2 = \lambda, \quad r_3 = \frac{1}{2}\left(1 - \sqrt{1 + 4\lambda - 4\lambda^2}\right)
$$

and

$$
r_4 = \frac{1}{2}\left(1 + \sqrt{1 + 4\lambda - 4\lambda^2}\right).
$$

We find the following orders:

i) if $0 \le \lambda \le 1/2$, then

$$
r_3 \le 0 \le r_2 \le r_1 \le 1 \le r_4,
$$

and
ii) if $1/2 \le \lambda \le 1$, then

$$
r_3 \le 0 \le r_1 \le r_2 \le 1 \le r_4.
$$

So we have $p_4(t) = (t - r_1)(t - r_2)(t - r_3)(t - r_4)$, $t \in [0,1]$. We easily derive that when $0 \le \lambda \le 1/2$, $p_4(t)$ is greater equal zero over $[\lambda, 1 - \lambda]$ and smaller equal zero over $[0, \lambda]$ and $[1 - \lambda, 1]$.

Similarly when $1/2 \le \lambda \le 1$ we get that $p_4(t) \ge 0$ over $[1 - \lambda, \lambda]$ and $p_4(t) \le 0$ over $[0, 1 - \lambda]$ and $[\lambda, 1]$. Therefore when $0 \le \lambda \le 1/2$ we have

$$
I_4(\lambda) = \int_\lambda^0 p_4(t)dt + \int_\lambda^{1-\lambda} p_4(t)dt + \int_1^{1-\lambda} p_4(t)dt,
$$

while when $1/2 \le \lambda \le 1$ we have

$$
I_4(\lambda) = \int_{1-\lambda}^0 p_4(t)dt + \int_{1-\lambda}^{\lambda} p_4(t)dt + \int_1^{\lambda} p_4(t)dt,
$$

proving (3.12) after the computations are done. $\square$

Using basic calculus we obtain

Lemma 3.6.

$$\min_{\lambda \in [0,1]} I_4(\lambda) = I_4\left(\frac{1}{4}\right) = I_4\left(\frac{3}{4}\right) = \frac{5}{256} = 0.01953125, \tag{3.14}$$

and

$$\max_{\lambda \in [0,1]} I_4(\lambda) = I_4(0) = I_4(1) = \frac{1}{30} = 0.033333\,. \tag{3.15}$$

Consequently by Lemmas 3.5 and 3.6 we get

Theorem 3.7. *Assumptions as in Theorem 3.2, case of $n = 4$. For every $x \in [a, b]$ it holds*

$$|\Delta_4(x)| \le \frac{(b-a)^4}{24} I_4(\lambda) \|f^{(4)}\|_\infty, \tag{3.16}$$

where $I_4(\lambda)$ is given by (3.12) with $\lambda = \frac{x-a}{b-a}$. Furthermore we have that

$$|\Delta_4(x)| \le \frac{(b-a)^4}{720} \|f^{(4)}\|_\infty, \quad \forall x \in [a, b]. \tag{3.17}$$

Optimality is achieved in

Theorem 3.8. *Assumptions as in Theorem 3.2, case of $n = 4$. Inequality (3.17) is sharp, namely it is attained when $x = a, b$ by the functions $(t - a)^4$ and $(t - b)^4$.*

Proof. We have

$$\Delta_4(a) = \Delta_4(b) = \left(\frac{f(a) + f(b)}{2}\right) - \frac{(b-a)}{12}(f'(b) - f'(a)) - \frac{1}{b-a}\int_a^b f(t)dt. \tag{3.18}$$

So by (3.17) we have

$$|\Delta_4(a)| = |\Delta_4(b)| \le \frac{(b-a)^4}{720} \|f^{(4)}\|_\infty. \tag{3.19}$$

Let $f(t) = (t - a)^4$ or $f(t) = (t - b)^4$, then

$$|\Delta_4(a)| = |\Delta_4(b)| = \frac{(b-a)^4}{30} = \text{R.H.S.}(3.19),$$

proving that (3.19) is attained. That is proving (3.17) sharp. $\qquad\square$

The trapezoid and midpoint inequalities follow.

Corollary 3.9. *Assumptions as in Theorem 3.2, case of $n = 4$. It holds*

$$\left|\left(\frac{f(a) + f(b)}{2}\right) - \frac{(b-a)}{12}(f'(b) - f'(a)) - \frac{1}{b-a}\int_a^b f(t)dt\right|$$

$$\le \frac{(b-a)^4}{720} \|f^{(4)}\|_\infty, \tag{3.20}$$

the last inequality is attained by $(t - a)^4$ and $(t - b)^4$, that is sharp.

Furthermore we obtain

$$\left| f\left(\frac{a+b}{2}\right) + \frac{(b-a)}{24}(f'(b) - f'(a)) - \frac{1}{b-a}\int_a^b f(t)dt \right|$$

$$\leq \frac{7}{5760}(b-a)^4\|f^{(4)}\|_\infty. \tag{3.21}$$

Remark 3.10. We do obtain the trapezoidal formula

$$\Delta_5(a) = \Delta_5(b) = \Delta_6(a) = \Delta_6(b) = \frac{(f(a)+f(b))}{2} - \frac{(b-a)}{12}(f'(b)-f'(a))$$

$$+ \frac{(b-a)^3}{720}(f'''(b) - f'''(a)) - \frac{1}{b-a}\int_a^b f(t)dt. \tag{3.22}$$

We also find the midpoint formula

$$\Delta_5\left(\frac{a+b}{2}\right) = \Delta_6\left(\frac{a+b}{2}\right) = f\left(\frac{a+b}{2}\right) + \frac{(b-a)}{24}(f'(b) - f'(a))$$

$$- \frac{7(b-a)^3}{5760}(f'''(b) - f'''(a)) - \frac{1}{b-a}\int_a^b f(t)dt. \tag{3.23}$$

Using (3.8) and Mathematica 7 we get

Theorem 3.11. *Assumptions as in Theorem 3.2, cases of $n = 5, 6$. It holds*

$$\left\{\begin{array}{l} |\Delta_5(a)|, \\ \left|\Delta_5\left(\dfrac{a+b}{2}\right)\right| \end{array}\right. \leq \frac{(b-a)^5}{144\sqrt{2310}}\|f^{(5)}\|_\infty, \tag{3.24}$$

and

$$|\Delta_6(a)| \leq \frac{1}{1440}\sqrt{\frac{101}{30030}}(b-a)^6\|f^{(6)}\|_\infty, \tag{3.25}$$

with

$$\left|\Delta_6\left(\frac{a+b}{2}\right)\right| \leq \frac{1}{46080}\sqrt{\frac{7081}{2145}}(b-a)^6\|f^{(6)}\|_\infty. \tag{3.26}$$

Chapter 4

Multidimensional Euler Identity and Optimal Multidimensional Ostrowski Inequalities

We develop and establish a general multivariate Euler type identity. Using it we derive general tight multivariate high order Ostrowski type inequalities for the estimate on the error of a multivariate function f evaluated at a point from its average. The estimates are involving only the single partial derivatives of f and are with respect to $\|\cdot\|_p$, $1 \le p \le \infty$. We give specific applications of the main results to the multidimensional trapezoid and midpoint rules for functions f differentiable up to order 6. We show sharpness of the inequalities for differentiation orders $m = 1, 2, 4$ and with respect to $\|\cdot\|_\infty$. This treatment relies on [38].

4.1 Introduction

We mention as motivation to this chapter the great Ostrowski inequality, see [196], [16], [21], [24].

Theorem 4.1. *It holds*

$$\left| f(x) - \frac{1}{b-a} \int_a^b f(t)dt \right| \le \left[\frac{1}{4} + \frac{\left(x - \frac{a+b}{2}\right)^2}{(b-a)^2} \right] (b-a)\|f'\|_\infty,$$

where $f \in C^1([a,b])$, $x \in [a,b]$, which is a sharp inequality.

The author proved also in [17], [21] the following result:

Theorem 4.2. *Let $f \in C^{n+1}([a,b])$, $n \in \mathbb{N}$ and $x \in [a,b]$ be fixed, such that $f^{(k)}(x) = 0$, $k = 1, \ldots, n$. Then*

$$\left| \frac{1}{b-a} \int_a^b f(y)dy - f(x) \right| \le \frac{\|f^{(n+1)}\|_\infty}{(n+2)!} \cdot \left(\frac{(x-a)^{n+2} + (b-x)^{n+2}}{b-a} \right).$$

The last inequality is sharp. In particular, when n is odd it is attained by $f^*(y) := (y-x)^{n+1} \cdot (b-a)$, while when n is even the optimal function is

$$\tilde{f}(y) := |y-x|^{n+\alpha} \cdot (b-a), \quad \alpha > 1.$$

Recently the author in [35], proved

27

Theorem 4.3. *Let $f \colon [a,b] \to \mathbb{R}$ be such that $f^{(n-1)}$, $n \geq 1$, is a continuous function and $f^{(n)}(x)$ exists and is finite for all but a countable set of x in (a,b) and that $f^{(n)} \in L_\infty([a,b])$. Then for every $x \in [a,b]$ we have*

$$|\Delta_n(x)| \leq \frac{(b-a)^{n-1}}{n!} \left(\int_a^b \left| B_n\left(\frac{x-a}{b-a}\right) - B_n^*\left(\frac{x-t}{b-a}\right) \right| dt \right) \|f^{(n)}\|_\infty,$$

where

$$\Delta_n(x) := f(x) - \frac{1}{b-a} \int_a^b f(t)dt$$
$$- \sum_{k=1}^{n-1} \frac{(b-a)^{k-1}}{k!} B_k\left(\frac{x-a}{b-a}\right)[f^{(k-1)}(b) - f^{(k-1)}(a)], x \in [a,b],$$

with $B_k(t)$ the Bernoulli polynomial of degree $k \geq 1$. Here $\{B_k^*(t), \ k \geq 0\}$ is the sequence of periodic functions of period 1, related to Bernoulli polynomials as $B_k^*(t) = B_k(t)$, $0 \leq t < 1$, $B_k^*(t+1) = B_k^*(t)$, $t \in \mathbb{R}$.
    We give also

Corollary 4.4. *All assumptions as in Theorem 4.3. Then for every $x \in [a,b]$ we have*

$$|\Delta_n(x)| \leq \frac{(b-a)^n}{n!} \left(\int_0^1 \left| B_n(t) - B_n\left(\frac{x-a}{b-a}\right) \right| dt \right) \|f^{(n)}\|_\infty, \quad n \geq 1.$$

Note. The last inequality appeared first as Theorem 7, p. 350, in [98], wrongly under the sole assumption of $f^{(n)} \in L_\infty([a,b])$.
    Theorem 4.3 and Corollary 4.4 along with [35], [98] are the greatest motivations for the Euler identity method we use in this chapter. Further motivation is given by (see also [17], [21])

Theorem 4.5. *Let $f \in C^1\left(\prod_{i=1}^n [a_i, b_i] \right)$, where $a_i < b_i$; $a_i, b_i \in \mathbb{R}$, $i = 1, \ldots, k$, and let $\overrightarrow{x}_0 := (x_{01}, \ldots, x_{0k}) \in \prod_{i=1}^k [a_i, b_i]$ be fixed. Then*

$$\left| \frac{1}{\prod\limits_{i=1}^k (b_i - a_i)} \int_{a_1}^{b_1} \cdots \int_{a_i}^{b_i} \cdots \int_{a_k}^{b_k} f(z_1, \ldots, z_k) dz_1 \cdots dz_k - f(\overrightarrow{x}_0) \right|$$

$$\leq \sum_{i=1}^k \left(\frac{(x_{0i} - a_i)^2 + (b_i - x_{0i})^2}{2(b_i - a_i)} \right) \left\| \frac{\partial f}{\partial z_i} \right\|_\infty.$$

The last inequality is sharp, here the optimal function is

$$f^*(z_1, \ldots, z_k) := \sum_{i=1}^k |z_i - x_{0i}|^{\alpha_i}, \ \alpha_i > 1.$$

Theorem 4.5 generalizes Theorem 4.1 to multidimension.

We also mention (see [17], [21])

Theorem 4.6. *Let Q be a compact and convex subset of $\mathbb{R}^k$, $k \geq 1$. Let $f \in C^{n+1}(Q)$, $n \in \mathbb{N}$ and $\overrightarrow{x}_0 \in Q$ be fixed such that all partial derivatives $f_\alpha := \frac{\partial^\alpha f}{\partial z^\alpha}$, where $\alpha = (\alpha_1, \ldots, \alpha_k)$, $\alpha_i \in \mathbb{Z}^+$, $i = 1, \ldots, k$, $|\alpha| = \sum_{i=1}^{k} \alpha_i = j$, $j = 1, \ldots, n$ fulfill $f_\alpha(\overrightarrow{x}_0) = 0$. Then*

$$\left| \frac{1}{\mathrm{Vol}(Q)} \int_Q f(\overrightarrow{z}) d\overrightarrow{z} - f(\overrightarrow{x}_0) \right| \leq \frac{D_{n+1}(f)}{(n+1)!\,\mathrm{Vol}(Q)} \int_Q (\|\overrightarrow{z} - \overrightarrow{x}_0\|_{\ell_1})^{n+1} d\overrightarrow{z},$$

where

$$D_{n+1}(f) := \max_{\alpha:\,|\alpha|=n+1} \|f_\alpha\|_\infty$$

and

$$\|\overrightarrow{z} - \overrightarrow{x}_0\|_{\ell_1} := \sum_{i=1}^{k} |z_i - x_{0i}|.$$

As a related result we give

Corollary 4.7. *Under the assumptions of Theorem 4.6 we find that*

$$\left| \frac{1}{\mathrm{Vol}(Q)} \int_Q f(\overrightarrow{z}) d\overrightarrow{z} - f(\overrightarrow{x}_0) \right|$$

$$\leq \frac{1}{(n+1)!\,\mathrm{Vol}(Q)} \int_Q \left[\left(\sum_{i=1}^{k} |z_i - x_{0i}| \cdot \left\| \frac{\partial}{\partial z_i} \right\|_\infty \right)^{n+1} f \right] d\overrightarrow{z}.$$

Furthermore, the last inequality is sharp: when n is odd it is attained by

$$f^*(z_1, \ldots, z_k) := \sum_{i=1}^{k} (z_i - x_{0i})^{n+1},$$

while when n is even the optimal function is

$$\tilde{f}(z_1, \ldots, z_k) := \sum_{i=1}^{k} |z_i - x_{0i}|^{n+\alpha_i}, \quad \alpha_i > 1.$$

Theorems 4.5, 4.6 and Corollary 4.7 motivate this chapter in the multivariate setting. Here we develop a very general Multivariate Euler type identity, see Theorems 4.14, 4.15, 4.18, 4.21. Based on this identity we prove some general tight multivariate Ostrowski type inequalities, see Theorems 4.32, 4.33, 4.34, 4.35, 4.60. Then we give lots of specific and important applications of Theorems 4.32, 4.33, 4.34, see Corollaries 4.36–4.59. There are produced many multivariate Ostrowski type inequalities for differentiation orders $m = 1, \ldots, 6$, mostly related to multivariate trapezoid and midpoint rules. When we impose some basic and natural boundary conditions, then inequalities become very simple and elegant, see Corollaries 4.56 – 4.59. The surprising fact there is, that only a very small number of sets of boundary conditions is needed comparely to the higher order of differentiation of the involved functions.

At the end we establish sharpness of the inequalities with respect to $\|\cdot\|_\infty$ and for differentiation orders $m = 1, 2, 4$, see Corollaries 4.61–4.69.

4.2   Background

We will use

Theorem 4.8 ([171, p. 17]). *Let $f \in C^n([a,b])$, $n \in \mathbb{N}$, $x \in [a,b]$. Then*

$$f(x) = \frac{1}{b-a}\int_a^b f(t)dt + \phi_{n-1}(x) + \mathcal{R}_n(x), \tag{4.1}$$

where for $m \in \mathbb{N}$ we call

$$\phi_m(x) := \sum_{k=1}^{m} \frac{(b-a)^{k-1}}{k!} B_k\left(\frac{x-a}{b-a}\right)[f^{(k-1)}(b) - f^{(k-1)}(a)]$$

with the convention $\phi_0(x) = 0$, and

$$R_n(x) := -\frac{(b-a)^{n-1}}{n!}\int_a^b \left[B_n^*\left(\frac{x-t}{b-a}\right) - B_n\left(\frac{x-a}{b-a}\right)\right]f^{(n)}(t)dt. \tag{4.2}$$

Here $B_k(x)$, $k \geq 0$, are the Bernoulli polynomials, $B_k = B_k(0)$, $k \geq 0$, the Bernoulli numbers, and $B_k^*(x)$, $k \geq 0$, are periodic functions of period one, related to the Bernoulli polynomials as

$$B_k^*(x) = B_k(x), \quad 0 \leq x < 1, \tag{4.3}$$

and

$$B_k^*(x+1) = B_k^*(x), \quad x \in \mathbb{R}. \tag{4.4}$$

Some basic properties of Bernoulli polynomials follow (see [1, 23.1]). We have

$$B_0(x) = 1, \quad B_1(x) = x - \frac{1}{2}, \quad B_2(x) = x^2 - x + \frac{1}{6},$$

and

$$B_k'(x) = kB_{k-1}(x), \quad k \in \mathbb{N}, \tag{4.5}$$

$$B_k(x+1) - B_k(x) = kx^{k-1}, \quad k \geq 0. \tag{4.6}$$

Clearly $B_0^* = 1$, B_1^* is a discontinuous function with a jump of -1 at each integer, and B_k^*, $k \geq 2$, is a continuous function. Notice that $B_k(0) = B_k(1) = B_k$, $k \geq 2$.
  We need also the more general

Theorem 4.9 (see [35]). *Let $f\colon [a,b] \to \mathbb{R}$ be such that $f^{(n-1)}$, $n \geq 1$, is a continuous function and $f^{(n)}(x)$ exists and is finite for all but a countable set of x in (a,b) and that $f^{(n)} \in L_1([a,b])$. Then for every $x \in [a,b]$ we have*

$$f(x) = \frac{1}{b-a}\int_a^b f(t)dt + \sum_{k=1}^{n-1} \frac{(b-a)^{k-1}}{k!} B_k\left(\frac{x-a}{b-a}\right)[f^{(k-1)}(b) - f^{(k-1)}(a)]$$

$$+ \frac{(b-a)^{n-1}}{n!}\int_{[a,b]}\left[B_n\left(\frac{x-a}{b-a}\right) - B_n^*\left(\frac{x-t}{b-a}\right)\right]f^{(n)}(t)dt. \tag{4.7}$$

The sum in (4.7) when $n = 1$ is zero. If $f^{(n-1)}$ is just absolutely continuous then (4.7) is valid again. Formula (4.7) is a *generalized Euler type identity*, see also [98].

4.3 Main Results

We make

Assumption 4.10. *Let f and the existing $\frac{\partial^\ell f}{\partial x_j^\ell}$, all $\ell = 1, \ldots, m$; $j = 1, \ldots, n$, be continuous real valued functions on $\prod_{i=1}^{n}[a_i, b_i]$; $m, n \in \mathbb{N}$, $a_i, b_i \in \mathbb{R}$.*

We give

Proposition 4.11. *All as in Assumption 4.10 for $m = n = 2$, $x_i \in [a_i, b_i]$, $i = 1, 2$. Then*

$$f(x_1, x_2) = \frac{1}{(b_1 - a_1)(b_2 - a_2)} \left(\int_{a_1}^{b_1} \int_{a_2}^{b_2} f(s_1, s_2) ds_1 ds_2 \right) + T_2(x_2) + T_1(x_1, x_2),$$
(4.8)

where

$$T_2 := T_2(x_2) := \frac{1}{(b_1 - a_1)} B_1\left(\frac{x_2 - a_2}{b_2 - a_2}\right) \left(\int_{a_1}^{b_1} (f(s_1, b_2) - f(s_1, a_2)) ds_1 \right)$$
$$+ \frac{(b_2 - a_2)}{2(b_1 - a_1)} \int_{a_1}^{b_1} \int_{a_2}^{b_2} \left[\left(B_2\left(\frac{x_2 - a_2}{b_2 - a_2}\right) - B_2^*\left(\frac{x_2 - s_2}{b_2 - a_2}\right) \right) \frac{\partial^2 f}{\partial x_2^2}(s_1, s_2) \right] ds_1 ds_2,$$
(4.9)

and

$$T_1 := T_1(x_1, x_2) := B_1\left(\frac{x_1 - a_1}{b_1 - a_1}\right) (f(b_1, x_2) - f(a_1, x_2))$$
$$+ \frac{(b_1 - a_1)}{2} \int_{a_1}^{b_1} \left(B_2\left(\frac{x_1 - a_1}{b_1 - a_1}\right) - B_2^*\left(\frac{x_1 - s_1}{b_1 - a_1}\right) \right) \frac{\partial^2 f}{\partial x_1^2}(s_1, x_2) ds_1.$$
(4.10)

Proof. By Theorem 4.8 we have

$$f(x_1, x_2) = \frac{1}{b_1 - a_1} \int_{a_1}^{b_1} f(s_1, x_2) ds_1 + B_1\left(\frac{x_1 - a_1}{b_1 - a_1}\right)(f(b_1, x_2) - f(a_1, x_2))$$
$$+ \frac{(b_1 - a_1)}{2} \int_{a_1}^{b_1} \left(B_2\left(\frac{x_1 - a_1}{b_1 - a_1}\right) - B_2^*\left(\frac{x_1 - s_1}{b_1 - a_1}\right) \right) \frac{\partial^2 f}{\partial x_1^2}(s_1, x_2) ds_1$$
$$= \frac{1}{b_1 - a_1} \int_{a_1}^{b_1} f(s_1, x_2) ds_1 + T_1(x_1, x_2).$$
(4.11)

And also we obtain

$$f(s_1, x_2) = \frac{1}{b_2 - a_2} \int_{a_2}^{b_2} f(s_1, s_2) ds_2 + B_1\left(\frac{x_2 - a_2}{b_2 - a_2}\right)(f(s_1, b_2) - f(s_1, a_2))$$
$$+ \frac{(b_2 - a_2)}{2} \int_{a_2}^{b_2} \left(B_2\left(\frac{x_2 - a_2}{b_2 - a_2}\right) - B_2^*\left(\frac{x_2 - s_2}{b_2 - a_2}\right) \right) \frac{\partial^2 f}{\partial x_2^2}(s_1, s_2) ds_2.$$
(4.12)

Thus we derive

$$f(x_1, x_2) = \frac{1}{b_1 - a_1} \int_{a_1}^{b_1} \left\{ \frac{1}{b_2 - a_2} \int_{a_2}^{b_2} f(s_1, s_2) ds_2 \right.$$

$$+ B_1 \left(\frac{x_2 - a_2}{b_2 - a_2} \right) (f(s_1, b_2) - f(s_1, a_2))$$

$$+ \frac{(b_2 - a_2)}{2} \int_{a_2}^{b_2} \left(B_2 \left(\frac{x_2 - a_2}{b_2 - a_2} \right) \right.$$

$$\left. \left. - B_2^* \left(\frac{x_2 - s_2}{b_2 - a_2} \right) \right) \frac{\partial^2 f}{\partial x_2^2}(s_1, s_2) ds_2 \right\} ds_1 + T_1(x_1, x_2), \qquad (4.13)$$

proving the claim. $\square$

We continue with

Proposition 4.12. *All as in Assumption 4.10, case $n = 3$ and $m \in \mathbb{N}$, $x_i \in [a_i, b_i]$, $i = 1, 2, 3$. Then*

$$f(x_1, x_2, x_3) = \frac{1}{\prod\limits_{i=1}^{3}(b_i - a_i)} \int_{a_1}^{b_1} \int_{a_2}^{b_2} \int_{a_3}^{b_3} f\left(s_1, s_2, s_3\right) ds_1 ds_2 ds_3$$

$$+ \ T_3(x_3) + T_2(x_2, x_3) + T_1(x_1, x_2, x_3), \qquad (4.14)$$

where

$$T_3 := T_3(x_3) := \frac{1}{(b_1 - a_1)(b_2 - a_2)} \sum_{k=1}^{m-1} \frac{(b_3 - a_3)^{k-1}}{k!} B_k \left(\frac{x_3 - a_3}{b_3 - a_3} \right)$$

$$\times \left(\int_{a_1}^{b_1} \int_{a_2}^{b_2} \left(\frac{\partial^{k-1} f}{\partial x_3^{k-1}}(s_1, s_2, b_3) - \frac{\partial^{k-1} f}{\partial x_3^{k-1}}\left(s_1, s_2, a_3\right) \right) ds_1 ds_2 \right)$$

$$+ \frac{(b_3 - a_3)^{m-1}}{m!(b_1 - a_1)(b_2 - a_2)} \int_{a_1}^{b_1} \int_{a_2}^{b_2} \int_{a_3}^{b_3} \left(B_m \left(\frac{x_3 - a_3}{b_3 - a_3} \right) \right.$$

$$\left. - B_m^* \left(\frac{x_3 - s_3}{b_3 - a_3} \right) \right) \frac{\partial^m f}{\partial x_3^m}(s_1, s_2, s_3) ds_1 ds_2 ds_3, \qquad (4.15)$$

$$T_2 :=$$

$$T_2(x_2, x_3) := \frac{1}{(b_1 - a_1)} \sum_{k=1}^{m-1} \frac{(b_2 - a_2)^{k-1}}{k!} B_k \left(\frac{x_2 - a_2}{b_2 - a_2} \right) \left(\int_{a_1}^{b_1} \left(\frac{\partial^{k-1} f}{\partial x_2^{k-1}}(s_1, b_2, x_3) \right. \right.$$

$$\left. \left. - \frac{\partial^{k-1} f}{\partial x_2^{k-1}}(s_1, a_2, x_3) \right) ds_1 \right) + \frac{(b_2 - a_2)^{m-1}}{m!(b_1 - a_1)} \int_{a_1}^{b_1} \left(\int_{a_2}^{b_2} \left(B_m \left(\frac{x_2 - a_2}{b_2 - a_2} \right) \right. \right.$$

$$\left. \left. - B_m^* \left(\frac{x_2 - s_2}{b_2 - a_2} \right) \right) \frac{\partial^m f}{\partial x_2^m}(s_1, s_2, x_3) ds_2 \right) ds_1, \qquad (4.16)$$

and

$$T_1 := T_1(x_1, x_2, x_3) := \sum_{k=1}^{m-1} \frac{(b_1 - a_1)^{k-1}}{k!} B_k\left(\frac{x_1 - a_1}{b_1 - a_1}\right)$$

$$\left(\frac{\partial^{k-1} f}{\partial x_1^{k-1}}(b_1, x_2, x_3) - \frac{\partial^{k-1} f}{\partial x_1^{k-1}}(a_1, x_2, x_3)\right)$$

$$+ \frac{(b_1 - a_1)^{m-1}}{m!} \int_{a_1}^{b_1} \left(B_m\left(\frac{x_1 - a_1}{b_1 - a_1}\right) - B_m^*\left(\frac{x_1 - s_1}{b_1 - a_1}\right)\right) \frac{\partial^m f}{\partial x_1^m}(s_1, x_2, x_3) ds_1.$$

$$(4.17)$$

When $m = 1$ then the sums in (4.15)–(4.17) are zero.

Proof.   Applying Theorem 4.8 we have

$$f(x_1, x_2, x_3) = \frac{1}{b_1 - a_1} \int_{a_1}^{b_1} f(s_1, x_2, x_3) ds_1 + T_1(x_1, x_2, x_3). \qquad (4.18)$$

Furthermore we find

$$f(s_1, x_2, x_3) = \frac{1}{b_2 - a_2} \int_{a_2}^{b_2} f(s_1, s_2, x_3) ds_2$$

$$+ \sum_{k=1}^{m-1} \frac{(b_2 - a_2)^{k-1}}{k!} B_k\left(\frac{x_2 - a_2}{b_2 - a_2}\right) \left(\frac{\partial^{k-1} f}{\partial x_2^{k-1}}(s_1, b_2, x_3) - \frac{\partial^{k-1} f}{\partial x_2^{k-1}}(s_1, a_2, x_3)\right)$$

$$+ \frac{(b_2 - a_2)^{m-1}}{m!} \int_{a_2}^{b_2} \left(B_m\left(\frac{x_2 - a_2}{b_2 - a_2}\right) - B_m^*\left(\frac{x_2 - s_2}{b_2 - a_2}\right)\right) \frac{\partial^m f}{\partial x_2^m}(s_1, s_2, x_3) ds_2$$

$$(4.19)$$

and

$$f(s_1, s_2, x_3) = \frac{1}{b_3 - a_3} \int_{a_3}^{b_3} f(s_1, s_2, s_3) ds_3$$

$$+ \sum_{k=1}^{m-1} \frac{(b_3 - a_3)^{k-1}}{k!} B_k\left(\frac{x_3 - a_3}{b_3 - a_3}\right) \left(\frac{\partial^{k-1} f}{\partial x_3^{k-1}}(s_1, s_2, b_3) - \frac{\partial^{k-1} f}{\partial x_3^{k-1}}(s_1, s_2, a_3)\right)$$

$$+ \frac{(b_3 - a_3)^{m-1}}{m!} \int_{a_3}^{b_3} \left(B_m\left(\frac{x_3 - a_3}{b_3 - a_3}\right) - B_m^*\left(\frac{x_3 - s_3}{b_3 - a_3}\right)\right) \frac{\partial^m f}{\partial x_3^m}(s_1, s_2, s_3) ds_3.$$

$$(4.20)$$

Putting the above together we derive

$$
f(x_1, x_2, x_3) = \frac{1}{b_1 - a_1} \int_{a_1}^{b_1} \left\{ \frac{1}{b_2 - a_2} \int_{a_2}^{b_2} f(s_1, s_2, x_3) ds_2 \right.
$$

$$
+ \sum_{k=1}^{m-1} \frac{(b_2 - a_2)^{k-1}}{k!} B_k\left(\frac{x_2 - a_2}{b_2 - a_2}\right) \left(\frac{\partial^{k-1} f}{\partial x_2^{k-1}}(s_1, b_2, x_3) - \frac{\partial^{k-1} f}{\partial x_2^{k-1}}(s_1, a_2, x_3)\right)
$$

$$
+ \frac{(b_2 - a_2)^{m-1}}{m!} \int_{a_2}^{b_2} \left(B_m\left(\frac{x_2 - a_2}{b_2 - a_2}\right)\right.
$$

$$
\left.\left. - B_m^*\left(\frac{x_2 - s_2}{b_2 - a_2}\right)\right) \frac{\partial^m f}{\partial x_2^m}(s_1, s_2, x_3) ds_2 \right\} ds_1 + T_1 \tag{4.21}
$$

$$
= \frac{1}{(b_1 - a_1)(b_2 - a_2)} \int_{a_1}^{b_1} \int_{a_2}^{b_2} f(s_1, s_2, x_3) ds_2 ds_1
$$

$$
+ \frac{1}{(b_1 - a_1)} \sum_{k=1}^{m-1} \frac{(b_2 - a_2)^{k-1}}{k!} B_k\left(\frac{x_2 - a_2}{b_2 - a_2}\right) \left(\int_{a_1}^{b_1} \frac{\partial^{k-1} f}{\partial x_2^{k-1}}(s_1, b_2, x_3) ds_1\right.
$$

$$
\left. - \int_{a_1}^{b_1} \frac{\partial^{k-1} f}{\partial x_2^{k-1}}(s_1, a_2, x_3) ds_1\right) + \frac{(b_2 - a_2)^{m-1}}{m!(b_1 - a_1)} \int_{a_1}^{b_1} \left(\int_{a_2}^{b_2} \left(B_m\left(\frac{x_2 - a_2}{b_2 - a_2}\right)\right.\right.
$$

$$
\left.\left. - B_m^*\left(\frac{x_2 - s_2}{b_2 - a_2}\right)\right) \frac{\partial^m f}{\partial x_2^m}(s_1, s_2, x_3) ds_2 \right) ds_1 + T_1. \tag{4.22}
$$

So far we have

$$
f(x_1, x_2, x_3)
$$

$$
= \frac{1}{(b_1 - a_1)(b_2 - a_2)} \int_{a_1}^{b_1} \int_{a_2}^{b_2} f(s_1, s_2, x_3) ds_2 ds_1 + T_2(x_2, x_3) + T_1(x_1, x_2, x_3). \tag{4.23}
$$

Finally we conclude that

$$
f(x_1, x_2, x_3)
$$

$$
= \frac{1}{(b_1 - a_1)(b_2 - a_2)} \int_{a_1}^{b_1} \int_{a_2}^{b_2} \left[\frac{1}{b_3 - a_3} \int_{a_3}^{b_3} f(s_1, s_2, s_3) ds_3 \right.
$$

$$
+ \sum_{k=1}^{m-1} \frac{(b_3 - a_3)^{k-1}}{k!} B_k\left(\frac{x_3 - a_3}{b_3 - a_3}\right) \left(\frac{\partial^{k-1} f}{\partial x_3^{k-1}}(s_1, s_2, b_3) - \frac{\partial^{k-1} f}{\partial x_3^{k-1}}(s_1, s_2, a_3)\right)
$$

$$
+ \frac{(b_3 - a_3)^{m-1}}{m!} \int_{a_3}^{b_3} \left(B_m\left(\frac{x_3 - a_3}{b_3 - a_3}\right)\right.
$$

$$
\left.\left. - B_m^*\left(\frac{x_3 - s_3}{b_3 - a_3}\right)\right) \frac{\partial^m f}{\partial x_3^m}(s_1, s_2, s_3) ds_3 \right] ds_2 ds_1 + T_2 + T_1 \tag{4.24}
$$

$$= \frac{1}{\prod\limits_{i=1}^{3}(b_i - a_i)} \int_{a_1}^{b_1} \int_{a_2}^{b_2} \int_{a_3}^{b_3} f(s_1, s_2, s_3) ds_1 ds_2 ds_3 + T_3 + T_2 + T_1, \qquad (4.25)$$

proving the claim. $\qquad\square$

We further present

Proposition 4.13. *All as in Assumption 4.10, case of $n = 4$ and $m \in \mathbb{N}$, $x_i \in [a_i, b_i]$, $i = 1, \ldots, 4$. Then*

$$f(x_1, x_2, x_3, x_4) = \frac{1}{\prod\limits_{i=1}^{4}(b_i - a_i)} \int_{\prod\limits_{i=1}^{4}[a_i,b_i]} f(s_1, s_2, s_3, s_4) ds_1 ds_2 ds_3 ds_4$$

$$+ \sum_{j=1}^{4} T_j. \qquad (4.26)$$

Here

$$T_4 := T_4(x_4) := \frac{1}{\prod\limits_{i=1}^{3}(b_i - a_i)} \sum_{k=1}^{m-1} \frac{(b_4 - a_4)^{k-1}}{k!} B_k \left(\frac{x_4 - a_4}{b_3 - a_4} \right)$$

$$\left(\int_{\prod\limits_{i=1}^{3}[a_i,b_i]} \left(\frac{\partial^{k-1} f}{\partial x_4^{k-1}}(s_1, s_2, s_3, b_4) - \frac{\partial^{k-1} f}{\partial x_4^{k-1}}(s_1, s_2, s_3, a_4) \right) ds_1 ds_2 ds_3 \right)$$

$$+ \frac{(b_4 - a_4)^{m-1}}{\prod\limits_{i=1}^{3}(b_i - a_i)m!} \int_{\prod\limits_{i=1}^{4}[a_i,b_i]} \left(\left(B_m \left(\frac{x_4 - a_4}{b_4 - a_4} \right) \right.\right.$$

$$\left.\left. - B_m^* \left(\frac{x_4 - s_4}{b_4 - a_4} \right) \right) \frac{\partial^m f}{\partial x_4^m}(s_1, s_2, s_3, s_4) \right) ds_1 ds_2 ds_3 ds_4, \qquad (4.27)$$

$$T_3 := T_3(x_3, x_4) := \frac{1}{(b_1 - a_1)(b_2 - a_2)} \sum_{k=1}^{m-1} \frac{(b_3 - a_3)^{k-1}}{k!} B_k \left(\frac{x_3 - a_3}{b_3 - a_3} \right)$$

$$\times \left(\int_{a_1}^{b_1} \int_{a_2}^{b_2} \left(\frac{\partial^{k-1} f}{\partial x_3^{k-1}}(s_1, s_2, b_3, x_4) - \frac{\partial^{k-1} f}{\partial x_3^{k-1}}(s_1, s_2, a_3, x_4) \right) ds_1 ds_2 \right)$$

$$+ \frac{(b_3 - a_3)^{m-1}}{m!(b_1 - a_1)(b_2 - a_2)} \left(\int_{a_1}^{b_1} \int_{a_2}^{b_2} \int_{a_3}^{b_3} \left(\left(B_m \left(\frac{x_3 - a_3}{b_3 - a_3} \right) \right.\right.\right.$$

$$\left.\left.\left. - B_m^* \left(\frac{x_3 - s_3}{b_3 - a_3} \right) \right) \frac{\partial^m f}{\partial x_3^m}(s_1, s_2, s_3, x_4) \right) ds_1 ds_2 ds_3 \right), \qquad (4.28)$$

$$T_2 := T_2(x_2, x_3, x_4)$$

$$:= \frac{1}{(b_1 - a_1)}\left(\sum_{k=1}^{m-1}\frac{(b_2 - a_2)^{k-1}}{k!}B_k\left(\frac{x_2 - a_2}{b_2 - a_2}\right)\right.$$

$$\times\left(\int_{a_1}^{b_1}\left(\frac{\partial^{k-1}f}{\partial x_2^{k-1}}(s_1, b_2, x_3, x_4) - \frac{\partial^{k-1}}{\partial x_2^{k-1}}f(s_1, a_2, x_3, x_4)\right)ds_1\right)\right)$$

$$+\frac{(b_2 - a_2)^{m-1}}{m!(b_1 - a_1)}\left(\int_{a_1}^{b_1}\int_{a_2}^{b_2}\left(\left(B_m\left(\frac{x_2 - a_2}{b_2 - a_2}\right)\right.\right.\right.$$

$$\left.\left.\left.- B_m^*\left(\frac{x_2 - s_2}{b_2 - a_2}\right)\right)\frac{\partial^m f}{\partial x_2^m}(s_1, s_2, x_3, x_4)\right)ds_1 ds_2\right), \tag{4.29}$$

and

$$T_1 := T_1(x_1, x_2, x_3, x_4)$$

$$:= \sum_{k=1}^{m-1}\frac{(b_1 - a_1)^{k-1}}{k!}B_k\left(\frac{x_1 - a_1}{b_1 - a_1}\right)\left(\frac{\partial^{k-1}f}{\partial x_1^{k-1}}(b_1, x_2, x_3, x_4)\right.$$

$$\left.-\frac{\partial^{k-1}f}{\partial x_1^{k-1}}(a_1, x_2, x_3, x_4)\right) + \frac{(b_1 - a_1)^{m-1}}{m!}\int_{a_1}^{b_1}\left(B_m\left(\frac{x_1 - a_1}{b_1 - a_1}\right)\right.$$

$$\left.- B_m^*\left(\frac{x_1 - s_1}{b_1 - a_1}\right)\right)\frac{\partial^m f}{\partial x_1^m}(s_1, x_2, x_3, x_4)ds_1. \tag{4.30}$$

When $m = 1$ then the sums in (4.27)–(4.30) are zero.

Proof.　By Theorem 4.8 we have

$$f(x_1, x_2, x_3, x_4) = \frac{1}{b_1 - a_1}\int_{a_1}^{b_1} f(s_1, x_2, x_3, x_4)ds_1$$

$$+ \sum_{k=1}^{m-1}\frac{(b_1 - a_1)^{k-1}}{k!}B_k\left(\frac{x_1 - a_1}{b_1 - a_1}\right)\left(\frac{\partial^{k-1}f}{\partial x_1^{k-1}}(b_1, x_2, x_3, x_4)\right.$$

$$\left.-\frac{\partial^{k-1}f}{\partial x_1^{k-1}}(a_1, x_2, x_3, x_4)\right) + \frac{(b_1 - a_1)^{m-1}}{m!}\int_{a_1}^{b_1}\left(B_m\left(\frac{x_1 - a_1}{b_1 - a_1}\right)\right.$$

$$\left.- B_m^*\left(\frac{x_1 - s_1}{b_1 - a_1}\right)\right)\frac{\partial^m f}{\partial x_1^m}(s_1, x_2, x_3, x_4)ds_1 \tag{4.31}$$

$$= \frac{1}{(b_1 - a_1)}\int_{a_1}^{b_1} f(s_1, x_2, x_3, x_4)ds_1 + T_1(x_1, x_2, x_3, x_4). \tag{4.32}$$

Similarly we obtain

$$f(s_1, x_2, x_3, x_4) = \frac{1}{b_2 - a_2}\int_{a_2}^{b_2} f(s_1, s_2, x_3, x_4)ds_2$$

$$+ \sum_{k=1}^{m-1}\frac{(b_2 - a_2)^{k-1}}{k!}B_k\left(\frac{x_2 - a_2}{b_2 - a_2}\right)\left(\frac{\partial^{k-1}f}{\partial x_2^{k-1}}(s_1, b_2, x_3, x_4)\right.$$

$$-\frac{\partial^{k-1}f}{\partial x_2^{k-1}}(s_1,a_2,x_3,x_4)\Bigg) + \frac{(b_2-a_2)^{m-1}}{m!}\int_{a_2}^{b_2}\Bigg(B_m\left(\frac{x_2-a_2}{b_2-a_2}\right)$$

$$- B_m^*\left(\frac{x_2-s_2}{b_2-a_2}\right)\Bigg)\frac{\partial^m f}{\partial x_2^m}(s_1,s_2,x_3,x_4)ds_2, \tag{4.33}$$

$$f(s_1,s_2,x_3,x_4) = \frac{1}{b_3-a_3}\int_{a_3}^{b_3}f(s_1,s_2,s_3,x_4)ds_3$$

$$+\sum_{k=1}^{m-1}\frac{(b_3-a_3)^{k-1}}{k!}B_k\left(\frac{x_3-a_3}{b_3-a_3}\right)\left(\frac{\partial^{k-1}f}{\partial x_3^{k-1}}(s_1,s_2,b_3,x_4)\right.$$

$$\left.-\frac{\partial^{k-1}f}{\partial x_3^{k-1}}(s_1,s_2,a_3,x_4)\right) + \frac{(b_3-a_3)^{m-1}}{m!}\int_{a_3}^{b_3}\Bigg(B_m\left(\frac{x_3-a_3}{b_3-a_3}\right)$$

$$- B_m^*\left(\frac{x_3-s_3}{b_3-a_3}\right)\Bigg)\frac{\partial^m f}{\partial x_3^m}(s_1,s_2,s_3,x_4)ds_3, \tag{4.34}$$

and finally

$$f(s_1,s_2,s_3,x_4) = \frac{1}{b_4-a_4}\int_{a_4}^{b_4}f(s_1,s_2,s_3,s_4)ds_4$$

$$+\sum_{k=1}^{m-1}\frac{(b_4-a_4)^{k-1}}{k!}B_k\left(\frac{x_4-a_4}{b_4-a_4}\right)\left(\frac{\partial^{k-1}f}{\partial x_4^{k-1}}(s_1,s_2,s_3,b_4)\right.$$

$$\left.-\frac{\partial^{k-1}f}{\partial x_4^{k-1}}(s_1,s_2,s_3,a_4)\right) + \frac{(b_4-a_4)^{m-1}}{m!}\int_{a_4}^{b_4}\Bigg(B_m\left(\frac{x_4-a_4}{b_4-a_4}\right)$$

$$- B_m^*\left(\frac{x_4-s_4}{b_4-a_4}\right)\Bigg)\frac{\partial^m f}{\partial x_4^m}(s_1,s_2,s_3,s_4)ds_4. \tag{4.35}$$

Consequently, by using the above we derive

$$f(x_1,x_2,x_3,x_4)$$

$$= \frac{1}{(b_1-a_1)}\int_{a_1}^{b_1}\Bigg[\frac{1}{(b_2-a_2)}\int_{a_2}^{b_2}f(s_1,s_2,x_3,x_4)ds_2$$

$$+\sum_{k=1}^{m-1}\frac{(b_2-a_2)^{k-1}}{k!}B_k\left(\frac{x_2-a_2}{b_2-a_2}\right)\left(\frac{\partial^{k-1}f}{\partial x_2^{k-1}}(s_1,b_2,x_3,x_4)\right.$$

$$\left.-\frac{\partial^{k-1}f}{\partial x_2^{k-1}}(s_1,a_2,x_3,x_4)\right) + \frac{(b_2-a_2)^{m-1}}{m!}\int_{a_2}^{b_2}\Bigg(B_m\left(\frac{x_2-a_2}{b_2-a_2}\right)$$

$$-B_m^*\left(\frac{x_2-s_2}{b_2-a_2}\right)\Bigg)\frac{\partial^m f}{\partial x_2^m}(s_1,s_2,x_3,x_4)ds_2\Bigg]ds_1 + T_1 \tag{4.36}$$

$$
= \frac{1}{(b_1 - a_1)(b_2 - a_2)} \left(\int_{a_1}^{b_1} \int_{a_2}^{b_2} f(s_1, s_2, x_3, x_4) ds_1 ds_2 \right)
$$

$$
+ \frac{1}{(b_1 - a_1)} \sum_{k=1}^{m-1} \frac{(b_2 - a_2)^{k-1}}{k!} B_k \left(\frac{x_2 - a_2}{b_2 - a_2} \right) \left(\int_{a_1}^{b_1} \left(\frac{\partial^{k-1} f}{\partial x_2^{k-1}} (s_1, b_2, x_3, x_4) \right. \right.
$$

$$
\left. - \frac{\partial^{k-1} f}{\partial x_2^{k-1}} (s_1, a_2, x_3, x_4) ds_1 \right) + \frac{(b_2 - a_2)^{m-1}}{m!(b_1 - a_1)} \int_{a_1}^{b_1} \int_{a_2}^{b_2} \left(\left(B_m \left(\frac{x_2 - a_2}{b_2 - a_2} \right) \right. \right.
$$

$$
\left. - B_m^* \left(\frac{x_2 - s_2}{b_2 - a_2} \right) \right) \frac{\partial^m f}{\partial x_2^m} (s_1, s_2, x_3, x_4) ds_1 ds_2 + T_1 \tag{4.37}
$$

$$
= \frac{1}{(b_1 - a_1)(b_2 - a_2)} \left(\int_{a_1}^{b_1} \int_{a_2}^{b_2} \left\{ \frac{1}{b_3 - a_3} \int_{a_3}^{b_3} f(s_1, s_2, s_3, x_4) ds_3 \right. \right.
$$

$$
+ \sum_{k=1}^{m-1} \frac{(b_3 - a_3)^{k-1}}{k!} B_k \left(\frac{x_3 - a_3}{b_3 - a_3} \right) \left(\frac{\partial^{k-1} f}{\partial x_3^{k-1}} (s_1, s_2, b_3, x_4) \right.
$$

$$
\left. - \frac{\partial^{k-1} f}{\partial x_3^{k-1}} (s_1, s_2, a_3, x_3) \right) + \frac{(b_3 - a_3)^{m-1}}{m!} \left(\int_{a_3}^{b_3} \left(B_m \left(\frac{x_3 - a_3}{b_3 - a_3} \right) \right. \right.
$$

$$
\left. - B_m^* \left(\frac{x_3 - s_3}{b_3 - a_3} \right) \right) \frac{\partial^m f}{\partial x_3^m} (s_1, s_2, s_3, x_4) ds_3 \right) ds_1 ds_2 \right) + T_2 + T_1. \tag{4.38}
$$

That is, we found that

$$
f(x_1, x_2, x_3, x_4) = \frac{1}{\prod\limits_{i=1}^{3} (b_i - a_i)} \int_{\prod\limits_{i=1}^{3} [a_i, b_i]} f(s_1, s_2, s_3, x_4) ds_1 ds_2 ds_3 + \sum_{j=1}^{3} T_j. \tag{4.39}
$$

At last we observe that

$$
f(x_1, x_2, x_3, x_4)
$$

$$
= \frac{1}{\prod\limits_{i=1}^{3} (b_i - a_i)} \int_{\prod\limits_{i=1}^{3} [a_i, b_i]} \left(\frac{1}{b_4 - a_4} \int_{a_4}^{b_4} f(s_1, s_2, s_3, s_4) ds_4 \right.
$$

$$
+ \sum_{k=1}^{m-1} \frac{(b_4 - a_4)^{k-1}}{k!} B_k \left(\frac{x_4 - a_4}{b_4 - a_4} \right) \left(\frac{\partial^{k-1} f}{\partial x_4^{k-1}} (s_1, s_2, s_3, b_4) - \frac{\partial^{k-1} f}{\partial x_4^{k-1}} (s_1, s_2, s_3, a_4) \right)
$$

$$
+ \frac{(b_4 - a_4)^{m-1}}{m!} \int_{a_4}^{b_4} \left(B_m \left(\frac{x_4 - a_4}{b_4 - a_4} \right) \right.
$$

$$
\left. - B_m^* \left(\frac{x_4 - s_4}{b_4 - a_4} \right) \right) \frac{\partial^m f}{\partial x_4^m} (s_1, s_2, s_3, s_4) ds_4 \right) ds_1 ds_2 ds_3 + \sum_{i=1}^{3} T_i \tag{4.40}
$$

$$
= \frac{1}{\prod\limits_{i=1}^{4} (b_i - a_i)} \int_{\prod\limits_{i=1}^{4} [a_i, b_i]} f(s_1, s_2, s_3, s_4) ds_1 ds_2 ds_3 ds_4 + \sum_{j=1}^{4} T_j, \tag{4.41}
$$

proving the claim. $\qquad\square$

We finally give the following general Multivariate Euler-type identity in

Theorem 4.14. *All as in Assumption 4.10 for $m, n \in \mathbb{N}$, $x_i \in [a_i, b_i]$, $i = 1, 2, \ldots, n$. Then*

$$f(x_1, x_2, \ldots, x_n) = \frac{1}{\prod\limits_{i=1}^{n} (b_i - a_i)} \int_{\prod\limits_{i=1}^{n} [a_i, b_i]} f(s_1, s_2, \ldots, s_n) ds_1 ds_2 \cdots ds_n + \sum_{j=1}^{n} T_j,$$

$$(4.42)$$

where for $j = 1, \ldots, n$ we have

$$T_j := T_j(x_j, x_{j+1}, \ldots, x_n) := \frac{1}{\left(\prod\limits_{i=1}^{j-1} (b_i - a_i)\right)} \left\{ \sum_{k=1}^{m-1} \frac{(b_j - a_j)^{k-1}}{k!} B_k\left(\frac{x_j - a_j}{b_j - a_j}\right) \right.$$

$$\times \left(\int_{\prod\limits_{i=1}^{j-1} [a_i, b_i]} \left(\frac{\partial^{k-1} f}{\partial x_j^{k-1}} (s_1, s_2, \ldots, s_{j-1}, b_j, x_{j+1}, \ldots, x_n) \right.\right.$$

$$\left.\left.\left. - \frac{\partial^{k-1} f}{\partial x_j^{k-1}} (s_1, s_2, \ldots, s_{j-1}, a_j, x_{j+1}, \ldots, x_n) \right) ds_1 \cdots ds_{j-1} \right) \right\}$$

$$+ \frac{(b_j - a_j)^{m-1}}{m! \left(\prod\limits_{i=1}^{j-1} (b_i - a_i)\right)} \left\{ \int_{\prod\limits_{i=1}^{j} [a_i, b_i]} \left(\left(B_m\left(\frac{x_j - a_j}{b_j - a_j}\right) \right.\right.\right.$$

$$\left.\left.\left. - B_m^*\left(\frac{x_j - s_j}{b_j - a_j}\right) \right) \frac{\partial^m f}{\partial x_j^m} (s_1, s_2, \ldots, s_j, x_{j+1}, \ldots, x_n) \right) ds_1 ds_2 \cdots ds_j \right\}.$$

$$(4.43)$$

When $m = 1$ then the sum in (4.43) is zero.

Proof. Theorem 4.14 is true for $n = 3, 4$, see Propositions 4.12, 4.13. Assume that Theorem 4.14, is true for n, see (4.42), (4.43). Then we prove it for $n+1$, $n \in \mathbb{N}$.

We have

$$f(x_1, x_2, \ldots, x_n, x_{n+1}) = \frac{1}{\prod\limits_{i=1}^{n} (b_i - a_i)} \int_{\prod\limits_{i=1}^{n} [a_i, b_i]} f(s_1, \ldots, s_n, x_{n+1}) ds_1, \ldots ds_n$$

$$+ \sum_{j=1}^{n} T_j, \quad \text{for} \quad j = 1, \ldots, n,$$

where

$$
T_j := T_j\big(x_j, x_{j+1}, \ldots, x_n, x_{n+1}\big) = \frac{1}{\big(\prod\limits_{i=1}^{(j-1)} (b_i - a_i)\big)} \left\{ \sum_{k=1}^{m-1} \frac{(b_j - a_j)^{k-1}}{k!} \right.
$$

$$
B_k\left(\frac{x_j - a_j}{b_j - a_j}\right)\left(\int_{\prod\limits_{i=1}^{j-1}[a_i,b_i]} \left(\frac{\partial^{k-1} f}{\partial x_j^{k-1}}(s_1, \ldots, s_{j-1}, b_j, x_{j+1}, \ldots, x_n, x_{n+1})\right.\right.
$$

$$
\left.\left. - \frac{\partial^{k-1} f}{\partial x_j^{k-1}}\Big(s_1, \ldots, s_{j-1}, a_j, x_{j+1}, \ldots, x_n, x_{n+1}\Big)\right) ds_1 \ldots ds_{j-1}\right)\right\}
$$

$$
+ \frac{(b_j - a_j)^{m-1}}{m!\big(\prod\limits_{i=1}^{j-1} (b_i - a_i)\big)} \left\{ \int_{\prod\limits_{i=1}^{j}[a_i,b_i]} \left(\left(B_m\left(\frac{x_j - a_j}{b_j - a_j}\right)\right.\right.\right.
$$

$$
\left.\left.\left. - B_m^*\left(\frac{x_j - s_j}{b_j - a_j}\right)\right) \frac{\partial^m f}{\partial x_j^m}\Big(s_1, \ldots, s_j, x_{j+1}, \ldots, x_n, x_{n+1}\Big)\right) ds_1 \ldots ds_j\right\}.
$$

But it holds

$$
f(s_1 \ldots, s_n, x_{n+1}) = \frac{1}{(b_{n+1} - a_{n+1})} \int_{a_{n+1}}^{b_{n+1}} f(s_1, \ldots, s_n, s_{n+1}) ds_{n+1}
$$

$$
+ \sum_{k=1}^{m-1} \frac{(b_{n+1} - a_{n+1})^{k-1}}{k!} B_k\left(\frac{x_{n+1} - a_{n+1}}{b_{n+1} - a_{n+1}}\right)
$$

$$
\left[\frac{\partial^{k-1} f(s_1, \ldots s_n, b_{n+1})}{\partial x_{n+1}^{k-1}} - \frac{\partial^{k-1} f}{\partial x_{n+1}^{k-1}}\right]\Big(s_1, \ldots, s_n, a_{n+1}\Big)
$$

$$
+ \frac{(b_{n+1} - a_{n+1})^{m-1}}{m!} \int_{a_{n+1}}^{b_{n+1}} \left[B_m\left(\frac{x_{n+1} - a_{n+1}}{b_{n+1} - a_{n+1}}\right)\right.
$$

$$
\left. - B_m^*\left(\frac{x_{n+1} - s_{n+1}}{b_{n+1} - a_{n+1}}\right)\right] \frac{\partial^m f(s_1, \ldots, s_n, s_{n+1})}{\partial x_{n+1}^m} ds_{n+1}.
$$

Thus we get

$$
f(x_1, x_2, \ldots, x_n, x_{n+1})
$$

$$
= \frac{1}{\prod\limits_{i=1}^{n} (b_i - a_i)} \int_{\prod\limits_{i=1}^{n}[a_i,b_i]} \left\{ \frac{1}{(b_{n+1} - a_{n+1})} \int_{a_{n+1}}^{b_{n+1}} f(s_1, \ldots, s_n, s_{n+1}) ds_{n+1}\right.
$$

$$
+ \sum_{k=1}^{m-1} \frac{(b_{n+1} - a_{n+1})^{k-1}}{k!} B_k\left(\frac{x_{n+1} - a_{n+1}}{b_{n+1} - a_{n+1}}\right)
$$

$$
\left[\frac{\partial^{k-1} f(s_1, \ldots, s_n, b_{n+1})}{\partial x_{n+1}^{k-1}} - \frac{\partial^{k-1} f(s_1, \ldots, s_n, a_{n+1})}{\partial x_{n+1}^{k-1}}\right]
$$

$$+\frac{(b_{n+1}-a_{n+1})^{m-1}}{m!}\int_{a_{n+1}}^{b_{n+1}}\left[B_m\left(\frac{x_{n+1}-a_{n+1}}{b_{n+1}-a_{n+1}}\right)\right.$$

$$\left.-B_m^*\left(\frac{x_{n+1}-s_{n+1}}{b_{n+1}-a_{n+1}}\right)\right]\frac{\partial^m f(s_1,\ldots,s_n,s_{n+1})}{\partial x_{n+1}^m}ds_{n+1}\Bigg\}ds_1\ldots,ds_n+\sum_{j=1}^{n}T_j.$$

Hence

$$f(x_1,\ldots,x_{n+1})=\frac{1}{\prod_{i=1}^{n+1}(b_i-a_i)}\int_{\prod_{i=1}^{n+1}[a_i,b_i]}f(s_1,\ldots,s_{n+1})\,ds_1\ldots ds_{n+1}+\sum_{j=1}^{n+1}T_j,$$

where

$$T_{n+1}(x_{n+1}):=T_{n+1}:=\frac{1}{\prod_{i=1}^{n}(b_i-a_i)}\left\{\sum_{k=1}^{m-1}\frac{(b_{n+1}-a_{n+1})^{k-1}}{k!}B_k\left(\frac{x_{n+1}-a_{n+1}}{b_{n+1}-a_{n+1}}\right)\right.$$

$$\left\{\int_{\prod_{i=1}^{n}[a_i,b_i]}\left[\frac{\partial^{k-1}f(s_1,\ldots,s_n,b_{n+1})}{\partial x_{n+1}^{k-1}}-\frac{\partial^{k-1}f}{\partial x_{n+1}^{k-1}}\left(s_1,\ldots,s_n,a_{n+1}\right)\right]ds_1\ldots ds_n\right\}\right\}$$

$$+\frac{(b_{n+1}-a_{n+1})^{m-1}}{m!\left(\prod_{i=1}^{n}(b_i-a_i)\right)}\left\{\int_{\prod_{i=1}^{n+1}[a_i,b_i]}\left[\left[B_m\left(\frac{x_{n+1}-a_{n+1}}{b_{n+1}-a_{n+1}}\right)\right.\right.\right.$$

$$\left.\left.\left.-B_m^*\left(\frac{x_{n+1}-s_{n+1}}{b_{n+1}-a_{n+1}}\right)\right]\frac{\partial^m f}{\partial x_{n+1}^m}\left(s_1,\ldots,s_n,s_{n+1}\right)\right]ds_1\ldots ds_{n+1}\right\}.$$

Thus is proving the claim. $\qquad\square$

Next we rewrite the last theorem.

Theorem 4.15. *All as in Assumption 4.10 for $m,n\in\mathbb{N}$, $x_i\in[a_i,b_i]$, $i=1,2,\ldots,n$. Then*

$$E_m^f(x_1,x_2,\ldots,x_n):=f(x_1,x_2,\ldots,x_n)$$

$$-\frac{1}{\prod_{i=1}^{n}(b_i-a_i)}\int_{\prod_{i=1}^{n}[a_i,b_i]}f(s_1,\ldots,s_n)ds_1\cdots ds_n$$

$$-\sum_{j=1}^{n}A_j=\sum_{j=1}^{n}B_j,\tag{4.44}$$

where for $j = 1, \ldots, n$ we have

$$A_j := A_j(x_j, x_{j+1}, \ldots, x_n) = \frac{1}{\left(\prod_{i=1}^{j-1}(b_i - a_i)\right)} \left\{ \sum_{k=1}^{m-1} \frac{(b_j - a_j)^{k-1}}{k!} B_k\left(\frac{x_j - a_j}{b_j - a_j}\right) \right.$$

$$\times \left(\int_{\prod_{i=1}^{j-1}[a_i,b_i]} \left(\frac{\partial^{k-1} f}{\partial x_j^{k-1}}(s_1, s_2, \ldots, s_{j-1}, b_j, x_{j+1}, \ldots, x_n) \right.\right.$$

$$\left.\left.\left. - \frac{\partial^{k-1} f}{\partial x_j^{k-1}}(s_1, s_2, \ldots, s_{j-1}, a_j, x_{j+1}, \ldots, x_n) \right) ds_1 \cdots ds_{j-1} \right) \right\}, \qquad (4.45)$$

and

$$B_j := B_j(x_j, x_{j+1}, \ldots, x_n) := \frac{(b_j - a_j)^{m-1}}{m!\left(\prod_{i=1}^{j-1}(b_i - a_i)\right)} \left\{ \int_{\prod_{i=1}^{j}[a_i,b_i]} \left(\left(B_m\left(\frac{x_j - a_j}{b_j - a_j}\right) \right.\right.\right.$$

$$\left.\left.\left. - B_m^*\left(\frac{x_j - s_j}{b_j - a_j}\right) \right) \frac{\partial^m f}{\partial x_j^m}(s_1, s_2, \ldots, s_j, x_{j+1}, \ldots, x_n) \right) ds_1 ds_2 \cdots ds_j \right\}.$$

$$(4.46)$$

When $m = 1$ then $A_j = 0$, and $T_j = B_j$, $j = 1, \ldots, n$.

Remark 4.16. *Notice above that $T_j = A_j + B_j$, $j = 1, \ldots, n$. Also we have that*

$$|E_m^f(x_1, x_2, \ldots, x_n)| \le \sum_{j=1}^{n} |B_j|. \qquad (4.47)$$

Also by denoting

$$\Delta := f(x_1, \ldots, x_n) - \frac{1}{\prod_{i=1}^{n}(b_i - a_i)} \int_{\prod_{i=1}^{n}[a_i,b_i]} f(s_1, \ldots, s_n) ds_1 \cdots ds_n \qquad (4.48)$$

we get

$$|\Delta| \le \sum_{j=1}^{n} (|A_j| + |B_j|). \qquad (4.49)$$

Later we will estimate A_j, B_j.

A general set of suppositions follow

Assumption 4.17. *Here $m \in \mathbb{N}$, $j = 1, \ldots, n$. We suppose*

1) *$f \colon \prod_{i=1}^{n}[a_i, b_i] \to \mathbb{R}$ is continuous.*

2) *$\frac{\partial^\ell f}{\partial x_j^\ell}$ are existing real valued functions for all $j = 1, \ldots, n$; $\ell = 1, \ldots, m - 2$.*

3) *For each $j = 1, \ldots, n$ we assume that $\frac{\partial^{m-1} f}{\partial x_j^{m-1}}(x_1, \ldots, x_{j-1}, \cdot, x_{j+1}, \ldots, x_n)$ is a continuous real valued function.*

4) *For each $j = 1, \ldots, n$ we assume that $g_j(\cdot) := \frac{\partial^m f}{\partial x_j^m}(x_1, \ldots, x_{j-1}, \cdot, x_{j+1}, \ldots, x_n)$ exists and is real valued with the possibility of being infinite only over an at most countable subset of (a_j, b_j).*

5) *Parts #3, #4 are true for all*

$$(x_1, \ldots, x_{j-1}, x_{j+1}, \ldots, x_n) \in \prod_{\substack{i=1 \\ i \neq j}}^{n} [a_i, b_i].$$

6) *The functions for $j = 2, \ldots, n;\ \ell = 1, \ldots, m - 2,$*

$$q_j\left(\overbrace{\cdot, \cdot, \ldots, \cdot}^{j-1}\right) := \frac{\partial^\ell f}{\partial x_j^\ell}\left(\overbrace{\cdot, \cdot, \cdot, \ldots, \cdot}^{j-1}, x_j, x_{j+1}, \ldots, x_n\right)$$

are continuous on $\prod_{i=1}^{j-1}[a_i, b_i]$, for each $(x_j, x_{j+1}, \ldots, x_n) \in \prod_{i=j}^{n}[a_i, b_i]$.

7) *The functions for each $j = 1, \ldots, n,$*

$$\varphi_j\left(\overbrace{\cdot, \cdot, \cdot, \ldots, \cdot, \cdot, \cdot}^{j}\right) := \frac{\partial^m f}{\partial x_j^m}\left(\overbrace{\cdot, \cdot, \cdot, \ldots, \cdot, \cdot}^{j}, x_{j+1}, \ldots, x_n\right) \in L_1\left(\prod_{i=1}^{j}[a_i, b_i]\right),$$

for any $(x_{j+1}, \ldots, x_n) \in \prod_{i=j+1}^{n}[a_i, b_i].$

We present

Theorem 4.18. *All as in Assumption* 4.17. *Then (4.42) and (4.44) are true again.*

Proof.   Use of Assumption 4.17, Theorem 4.9 and it is similar to the proof of the so far results of this chapter. $\square$

Remark 4.19. *The results of Propositions 4.11, 4.12, 4.13 are still valid under Assumption 4.17 for their cases.*

Some weaker general suppositions follow.

Assumption 4.20. *Here $m \in \mathbb{N}$, $j = 1, \ldots, n$ and only the Parts #1, #2, #6, #7 of Assumption 4.17 remain the same. We further suppose that for each $j = 1, \ldots, n$ and over $[a_j, b_j]$, the function*

$$\frac{\partial^{m-1}}{\partial x_j^{m-1}} f(x_1, \ldots, x_{j-1}, \cdot, x_{j+1}, \ldots, x_n)$$

is absolutely continuous, and this is true for all

$$(x_1, \ldots, x_{j-1}, x_{j+1}, \ldots, x_n) \in \prod_{\substack{i=1 \\ i \neq j}}^{n} [a_i, b_i].$$

We give

Theorem 4.21. *All as in Assumption 4.20. Then (4.42) and (4.44) are still true.*

Remark 4.22. *The results of Propositions 4.11, 4.12, 4.13 are again valid under Assumption 4.20 for their cases.*

We proceed with the estimates of our remainders in (4.42) and (4.44).

We need to make

Remark 4.23. *We are operating under the Assumptions 4.10 or 4.17 or 4.20. We observe for $j = 1, \ldots, n$ that*

$$|B_j| \leq \Gamma_j, \tag{4.50}$$

where

$$\Gamma_j := \frac{(b_j - a_j)^{m-1}}{m! \left(\prod_{i=1}^{j-1} (b_i - a_i) \right)} \left(\int_{\prod_{i=1}^{j} [a_i, b_i]} \left| B_m \left(\frac{x_j - a_j}{b_j - a_j} \right) - B_m^* \left(\frac{x_j - s_j}{b_j - a_j} \right) \right| \right.$$

$$\left. \times \left| \frac{\partial^m f}{\partial x_j^m} (s_1, s_2, \ldots, s_j, x_{j+1}, \ldots, x_n) \right| ds_1 \cdots ds_j \right). \tag{4.51}$$

Here we assume

$$\frac{\partial^m f}{\partial x_j^m} \left(\overbrace{\cdots}^{j}, x_{j+1}, \ldots, x_n \right) \in L_\infty \left(\prod_{i=1}^{j} [a_i, b_i] \right)$$

for any $(x_{j+1}, \ldots, x_n) \in \prod_{i=j+1}^{n} [a_i, b_i]$, any $x_j \in [a_j, b_j]$.

Thus we obtain

$$\Gamma_j \leq \frac{(b_j - a_j)^{m-1}}{m! \left(\prod_{i=1}^{j-1} (b_i - a_i) \right)} \left(\int_{\prod_{i=1}^{j} [a_i, b_i]} \left| B_n \left(\frac{x_j - a_j}{b_j - a_j} \right) - B_m^* \left(\frac{x_j - s_j}{b_j - a_j} \right) \right| ds_1 \cdots ds_j \right)$$

$$\times \left\| \frac{\partial^m f}{\partial x_j^m} \left(\overbrace{\cdot, \cdot, \cdot, \cdots, \cdot}^{j}, x_{j+1}, \ldots, x_n \right) \right\|_{\infty, \prod_{i=1}^{j} [a_i, b_i]} \tag{4.52}$$

$$= \frac{(b_j - a_j)^{m-1}}{m!} \left(\int_{a_j}^{b_j} \left| B_m \left(\frac{x_j - a_j}{b_j - a_j} \right) - B_m^* \left(\frac{x_j - s_j}{b_j - a_j} \right) \right| ds_j \right)$$

$$\times \left\| \frac{\partial^m f}{\partial x_j^m} \left(\overbrace{\cdots, \cdots}^{j}, x_{j+1}, \ldots, x_n \right) \right\|_{\infty, \prod_{i=1}^{j} [a_i, b_i]} \tag{4.53}$$

(by letting

$$\lambda_j := \lambda_j(x_j) := \frac{x_j - a_j}{b_j - a_j} \tag{4.54}$$

we have)

$$= \frac{(b_j - a_j)^m}{m!}$$

$$\cdot \left(\int_0^1 |B_m(\lambda_j) - B_m(t_j)| dt_j \right) \left\| \frac{\partial^m f}{\partial x_j^m} \left(\overbrace{\cdots}^{j}, x_{j+1}, \ldots, x_n \right) \right\|_{\infty, \prod\limits_{i=1}^{j} [a_i, b_i]} \tag{4.55}$$

$$\leq \frac{(b_j - a_j)^m}{m!}$$

$$\cdot \left(\sqrt{\int_0^1 (B_m(\lambda_j) - B_m(t_j))^2 dt_j} \right) \left\| \frac{\partial^m f}{\partial x_j^m} \left(\overbrace{\cdots}^{j}, x_{j+1}, \ldots, x_n \right) \right\|_{\infty, \prod\limits_{i=1}^{j} [a_i, b_i]} \tag{4.56}$$

(using [98], p. 352 we get)

$$= \frac{(b_j - a_j)^m}{m!}$$

$$\cdot \left(\sqrt{\frac{(m!)^2}{(2m)!} |B_{2m}| + B_m^2 \left(\frac{x_j - a_j}{b_j - a_j} \right)} \right) \left\| \frac{\partial^m f}{\partial x_j^m} \left(\overbrace{\cdots}^{j}, x_{j+1}, \ldots, x_n \right) \right\|_{\infty, \prod\limits_{i=1}^{j} [a_i, b_i]} .$$

$$\tag{4.57}$$

We have established

Lemma 4.24. *Suppose Assumptions 4.10 or 4.17 or 4.20. Let B_j as in (4.46). In particular suppose that*

$$\frac{\partial^m f}{\partial x_j^m} \left(\overbrace{\cdots}^{j}, x_{j+1}, \ldots, x_n \right) \in L_\infty \left(\prod_{i=1}^{j} [a_i, b_i] \right),$$

for any $(x_{j+1}, \ldots, x_n) \in \prod\limits_{i=j+1}^{n} [a_i, b_i]$, *all* $j = 1, \ldots, n$. *Then for any* $(x_j, x_{j+1}, \ldots, x_n) \in \prod\limits_{i=j}^{n} [a_i, b_i]$ *we have*

$$|B_j| = |B_j(x_j, x_{j+1}, \ldots, x_n)| \leq \frac{(b_j - a_j)^m}{m!} \left(\sqrt{\frac{(m!)^2}{(2m)!} |B_{2m}| + B_m^2 \left(\frac{x_j - a_j}{b_j - a_j} \right)} \right)$$

$$\times \left\| \frac{\partial^m f}{\partial x_j^m} \left(\overbrace{\cdots}^{j}, x_{j+1}, \ldots, x_n \right) \right\|_{\infty, \prod\limits_{i=1}^{j} [a_i, b_i]}, \tag{4.58}$$

for all $j = 1, \ldots, n$.

We continue with

Remark 4.25. *Continuing from Remark 4.23. Let $p_j, q_j > 1$: $\frac{1}{p_j} + \frac{1}{q_j} = 1$; $j = 1, \ldots, n$, with the assumption that*

$$\frac{\partial^m f}{\partial x_j^m} (\cdots, x_{j+1}, \ldots, x_n) \in L_{q_j} \left(\prod_{i-1}^{j} [a_i, b_i] \right),$$

for any $(x_{j+1}, \ldots, x_n) \in \prod_{i=j+1}^{n} [a_i, b_i]$, $x_j \in [a_j, b_j]$. *From (4.50) and (4.51) we get*

$$|B_j| \leq \frac{(b_j - a_j)^{m-1}}{m! \left(\prod_{i=1}^{j-1} (b_i - a_i) \right)} \left(\int_{\prod_{i=1}^{j} [a_i, b_i]} \left| B_m \left(\frac{x_j - a_j}{b_j - a_j} \right) \right. \right.$$

$$\left. \left. - B_m^* \left(\frac{x_j - s_j}{b_j - a_j} \right) \right|^{p_j} ds_1 \cdots ds_j \right)^{1/p_j}$$

$$\times \left(\int_{\prod_{i=1}^{j} [a_i, b_i]} \left| \frac{\partial^m f}{\partial x_j^m} (s_1, \ldots, s_j, x_{j+1}, \ldots, x_n) \right|^{q_j} ds_1 \cdots ds_j \right)^{1/q_j} \tag{4.59}$$

$$= \frac{(b_j - a_j)^{m-1}}{m! \left(\prod_{i=1}^{j-1} (b_i - a_i) \right)} \left(\prod_{i=1}^{j-1} (b_i - a_i) \right)^{1/p_j} \left(\int_{a_j}^{b_j} \left| B_m(\lambda_j) \right. \right.$$

$$\left. \left. - B_m^* \left(\frac{x_j - s_j}{b_j - a_j} \right) \right|^{p_j} ds_j \right)^{1/p_j} \left\| \frac{\partial^m f}{\partial x_j^m} (\cdots, x_{j+1}, \ldots, x_n) \right\|_{q_j, \prod_{i=1}^{j} [a_i, b_i]} \tag{4.60}$$

$$= \frac{(b_j - a_j)^{m - \frac{1}{q_j}}}{m!} \left(\prod_{i=1}^{j-1} (b_i - a_i) \right)^{-1/q_j} \left(\int_0^1 \left| B_m(\lambda_j) \right. \right.$$

$$\left. \left. - B_m(t_j) \right|^{p_j} dt_j \right)^{1/p_j} \left\| \frac{\partial^m f}{\partial x_j^m} (\cdots, x_{j+1}, \ldots, x_n) \right\|_{q_j, \prod_{i=1}^{j} [a_i, b_i]}. \tag{4.61}$$

We have derived

Lemma 4.26. *Suppose Assumptions 4.10 or 4.17 or 4.20. Let B_j as in (4.46) and $p_j, q_j > 1$: $\frac{1}{p_j} + \frac{1}{q_j} = 1$; $j = 1, \ldots, n$. In particular we suppose that*

$$\frac{\partial^m f}{\partial x_j^m} (\cdots, x_{j+1}, \ldots, x_n) \in L_{q_j} \left(\prod_{i=1}^{j} [a_i, b_i] \right),$$

for any $(x_{j+1}, \ldots, x_n) \in \prod_{i=j+1}^{n} [a_i, b_i]$, *for all* $j = 1, \ldots, n$. *Then for any* $(x_j, x_{j+1}, \ldots, x_n) \in \prod_{i=j}^{n} [a_i, b_i]$ *we have*

$$|B_j| = |B_j(x_j, x_{j+1}, \ldots, x_n)|$$

$$\leq \frac{(b_j - a_j)^{m - \frac{1}{q_j}}}{m!} \left(\prod_{i=1}^{j-1} (b_i - a_i) \right)^{-1/q_j} \left(\int_0^1 \left| B_m \left(\frac{x_j - a_j}{b_j - a_j} \right) \right. \right.$$

$$\left. \left. - B_m(t_j) \right|^{p_j} dt_j \right)^{1/p_j} \left\| \frac{\partial^m f}{\partial x_j^m} (\cdots, x_{j+1}, \ldots, x_n) \right\|_{q_j, \prod_{i=1}^{j} [a_i, b_i]}. \tag{4.62}$$

When $p_j = q_j = 2$, then we obtain

$$|B_j| \leq \frac{(b_j - a_j)^{m-\frac{1}{2}}}{m!} \left(\prod_{i=1}^{j-1} (b_i - a_i) \right)^{-1/2} \left(\sqrt{\frac{(m!)^2}{(2m)!} |B_{2m}| + B_m^2 \left(\frac{x_j - a_j}{b_j - a_j} \right)} \right)$$
$$\times \left\| \frac{\partial^m f}{\partial x_j^m} (\cdots, x_{j+1}, \ldots, x_n) \right\|_{2, \prod\limits_{i=1}^{j} [a_i, b_i]}. \tag{4.63}$$

The inequalities (4.62), (4.63) are true for all $j = 1, \ldots, n$.

At last we make

Remark 4.27. *Continuing from Remark 4.25. Assume for $j = 1, \ldots, n$ that*

$$\frac{\partial^m f}{\partial x_j^m} (\cdots, x_{j+1}, \ldots, x_n) \in L_1 \left(\prod_{i=1}^{j} [a_i, b_i] \right),$$

for any $(x_{j+1}, \ldots, x_n) \in \prod_{i=j+1}^{n} [a_i, b_i]$, any $x_j \in [a_j, b_j]$. Then from (4.50) and (4.51) we obtain

$$|B_j| \leq \frac{(b_j - a_j)^{m-1}}{m! \left(\prod\limits_{i=1}^{j-1} (b_i - a_i) \right)} \left(\int_{\prod\limits_{i=1}^{j} [a_i, b_i]} \right.$$
$$\left. \left| \frac{\partial^m f}{\partial x_j^m} (s_1, \ldots, s_j, x_{j+1}, \ldots, x_n) \right| ds_1 \cdots ds_j \right)$$
$$\cdot \left\| B_m \left(\frac{x_j - a_j}{b_j - a_j} \right) - B_m^* \left(\frac{x_j - s_j}{b_j - a_j} \right) \right\|_{\infty, [a_j, b_j]} \tag{4.64}$$
$$= \frac{(b_j - a_j)^{m-1}}{m! \left(\prod_{i=1}^{j-1} (b_i - a_i) \right)} \left\| B_m \left(\frac{x_j - a_j}{b_j - a_j} \right) - B_m^* \left(\frac{x_j - s_j}{b_j - a_j} \right) \right\|_{\infty, [a_j, b_j]}$$
$$\times \left\| \frac{\partial^m f}{\partial x_j^m} (\cdots, x_{j+1}, \ldots, x_n) \right\|_{1, \prod\limits_{i=1}^{j} [a_i, b_i]} \tag{4.65}$$
$$\overset{(by\ [98],\ p.\ 347)}{=} \frac{(b_j - a_j)^{m-1}}{m! \left(\prod\limits_{i=1}^{j-1} (b_i - a_i) \right)} \left\| B_m(t) - B_m \left(\frac{x_j - a_j}{b_j - a_j} \right) \right\|_{\infty, [0,1]}$$
$$\times \left\| \frac{\partial^m f}{\partial x_j^m} (\cdots, x_{j+1}, \ldots, x_n) \right\|_{1, \prod\limits_{i=1}^{j} [a_i, b_i]}. \tag{4.66}$$

From [98], pp. 347–348 we have:

i) case $m - 2r$, $r \in \mathbb{N}$, then

$$\left\| B_m(t) - B_m\left(\frac{x_j - a_j}{b_j - a_j}\right) \right\|_{\infty,[0,1]} = \left\| B_{2r}(t) - B_{2r}\left(\frac{x_j - a_j}{b_j - a_j}\right) \right\|_{\infty,[0,1]}$$

$$= (1 - 2^{-2r})|B_{2r}| + \left| 2^{-2r} B_{2r} - B_{2r}\left(\frac{x_j - a_j}{b_j - a_j}\right) \right|; \tag{4.67}$$

ii) case $m = 2r + 1$, $r \in \mathbb{N}$, *then*

$$\left\| B_m(t) - B_m\left(\frac{x_j - a_j}{b_j - a_j}\right) \right\|_{\infty,[0,1]} = \left\| B_{2r+1}(t) - B_{2r+1}\left(\frac{x_j - a_j}{b_j - a_j}\right) \right\|_{\infty,[0,1]}$$

$$\leq \frac{2(2r+1)!}{(2\pi)^{2r+1}(1 - 2^{-2r})} + \left| B_{2r+1}\left(\frac{x_j - a_j}{b_j - a_j}\right) \right|; \tag{4.68}$$

iii) special case of $m = 1$, *then*

$$\left\| B_m(t) - B_m\left(\frac{x_j - a_j}{b_j - a_j}\right) \right\|_{\infty,[0,1]} = \left\| B_1(t) - B_1\left(\frac{x_j - a_j}{b_j - a_j}\right) \right\|_{\infty,[0,1]}$$

$$= \frac{1}{2} + \left| x_j - \left(\frac{a_j + b_j}{2}\right) \right|. \tag{4.69}$$

We have proved

Lemma 4.28. *Suppose Assumptions 4.10 or 4.17 or 4.20. Let B_j as in (4.46), $j = 1, \ldots, n$. In particular we suppose for $j = 1, \ldots, n$ that*

$$\frac{\partial^m f}{\partial x_j^m}(\ldots, x_{j+1}, \ldots, x_n) \in L_1\left(\prod_{i=1}^{j}[a_i, b_i]\right),$$

for any $(x_{j+1}, \ldots, x_n) \in \prod_{i=j+1}^{n}[a_i, b_i]$. *Then for any* $(x_j, x_{j+1}, \ldots, x_n) \in \prod_{i=j}^{n}[a_i, b_i]$ *we have*

$$|B_j| = |B_j(x_j, x_{j+1}, \ldots, x_n)| \leq \frac{(b_j - a_j)^{m-1}}{m!\left(\prod_{i=1}^{j-1}(b_i - a_i)\right)} \left\| \frac{\partial^m f}{\partial x_j^m}(\ldots, x_{j+1}, \ldots, x_n) \right\|_{1, \prod_{i=1}^{j}[a_i, b_i]}$$

$$\times \left\| B_m(t) - B_m\left(\frac{x_j - a_j}{b_j - a_j}\right) \right\|_{\infty,[0,1]}. \tag{4.70}$$

The special cases follow:
1) When $m = 2r$, $r \in \mathbb{N}$ *we have*

$$|B_j| \leq \frac{(b_j - a_j)^{2r-1}}{(2r)!\left(\prod_{i=1}^{j-1}(b_i - a_i)\right)} \left\| \frac{\partial^{2r} f}{\partial x_j^{2r}}(\ldots, x_{j+1}, \ldots, x_n) \right\|_{1, \prod_{i=1}^{j}[a_i, b_i]}$$

$$\times \left[(1 - 2^{-2r})|B_{2r}| + \left| 2^{-2r} B_{2r} - B_{2r}\left(\frac{x_j - a_j}{b_j - a_j}\right) \right| \right]. \tag{4.71}$$

2) *When* $m = 2r + 1$, $r \in \mathbb{N}$ *we obtain*

$$|B_j| \leq \frac{(b_j - a_j)^{2r}}{(2r+1)!\left(\prod\limits_{i=1}^{j-1}(b_i - a_i)\right)} \left\| \frac{\partial^{2r+1} f}{\partial x_j^{2r+1}}(\ldots, x_{j+1}, \ldots, x_n) \right\|_{1, \prod\limits_{i=1}^{j}[a_i, b_i]}$$

$$\times \left[\frac{2(2r+1)!}{(2\pi)^{2r+1}(1 - 2^{-2r})} + \left| B_{2r+1}\left(\frac{x_j - a_j}{b_j - a_j} \right) \right| \right]. \tag{4.72}$$

3) *When* $m = 1$ *we get*

$$|B_j| \leq \frac{1}{\prod\limits_{i=1}^{j-1}(b_i - a_i)} \left\| \frac{\partial f}{\partial x_j}(\ldots, x_{j+1}, \ldots, x_n) \right\|_{1, \prod\limits_{i=1}^{j}[a_i, b_i]} \left[\frac{1}{2} + \left| x_j - \left(\frac{a_j + b_j}{2} \right) \right| \right].$$

$$\tag{4.73}$$

We need

Definition 4.29. *Let* $\overrightarrow{x} = (x_1, \ldots, x_\theta) \in \prod\limits_{i=1}^{\theta}[a_i, b_i]$, $\theta \in \mathbb{N}$, *where* $\|\overrightarrow{x}\| := \sqrt{x_1^2 + \cdots + x_\theta^2}$. *Let* $F \colon \prod\limits_{i=1}^{\theta}[a_i, b_i] \to \mathbb{R}$ *be continuous. We define the (first) modulus of continuity of* F *by*

$$\omega_1(F, \delta) := \sup_{\substack{\text{all } \overrightarrow{x}, \overrightarrow{y} \in \prod\limits_{i=1}^{\theta}[a_i, b_i], \\ \text{with } \|\overrightarrow{x} - \overrightarrow{y}\| \leq \delta}} |F(\overrightarrow{x}) - F(\overrightarrow{y})|, \tag{4.74}$$

for all $\delta > 0$.

Notice 4.30. Under Assumption 4.10 we have valid that

$$\left| \frac{\partial^{k-1} f}{\partial x_j^{k-1}}(s_1, \ldots, s_{j-1}, b_j, x_{j+1}, \ldots, x_n) - \frac{\partial^{k-1} f}{\partial x_j^{k-1}}(s_1, \ldots, s_{j-1}, a_j, x_{j+1}, \ldots, x_n) \right|$$

$$\leq \omega_1\left(\frac{\partial^{k-1} f}{\partial x_j^{k-1}}\left(\overbrace{\cdots}^{j}, x_{j+1}, \ldots, x_n \right), b_j - a_j \right), \text{ all } j = 1, \ldots, n; \ k = 1, \ldots, m-1.$$

$$\tag{4.75}$$

We give

Lemma 4.31. *Suppose Assumption 4.10. Let* A_j *as in (4.45),* $j = 1, \ldots, n$. *Then for any*

$$(x_j, x_{j+1}, \ldots, x_n) \in \prod\limits_{i=j}^{n}[a_i, b_i]$$

we have

$$|A_j| = |A_j(x_j, x_{j+1}, \ldots, x_n)| \leq \sum_{k=1}^{m-1} \frac{(b_j - a_j)^{k-1}}{k!} \left| B_k \left(\frac{x_j - a_j}{b_j - a_j} \right) \right|$$

$$\cdot \, \omega_1 \left(\frac{\partial^{k-1} f}{\partial x_j^{k-1}} (\overset{j}{\overbrace{\cdots}}, x_{j+1}, \ldots, x_n), b_j - a_j \right), \text{ for all } j = 1, \ldots, n. \quad (4.76)$$

Putting together all these above auxilliary results, we derive the following multivariate Ostrowski type inequalities.

Theorem 4.32. *Suppose Assumptions 4.10 or 4.17 or 4.20. Let $E_m^f(x_1, x_2, \ldots, x_n)$ as in (4.44) and A_j for $j = 1, \ldots, n$ as in (4.45), $m \in \mathbb{N}$. In particular we suppose that*

$$\frac{\partial^m f}{\partial x_j^m} (\overset{j}{\overbrace{\cdots}}, x_{j+1}, \ldots, x_n) \in L_\infty \left(\prod_{i=1}^{j} [a_i, b_i] \right),$$

for any $(x_{j+1}, \ldots, x_n) \in \prod_{i=j+1}^{n} [a_i, b_i]$, all $j = 1, \ldots, n$. Then

$$|E_m^f(x_1, \ldots, x_n)|$$

$$= \left| f(x_1, \ldots, x_n) - \frac{1}{\prod_{i=1}^{n}(b_i - a_i)} \int_{\prod_{i=1}^{n} [a_i, b_i]} f(s_1, \ldots, s_n) ds_1 \cdots ds_n - \sum_{j=1}^{n} A_j \right|$$

$$\leq \frac{1}{m!} \sum_{j=1}^{n} \left[(b_j - a_j)^m \left(\sqrt{\frac{(m!)^2}{(2m)!} |B_{2m}| + B_m^2 \left(\frac{x_j - a_j}{b_j - a_j} \right)} \right) \right.$$

$$\left. \times \left\| \frac{\partial^m f}{\partial x_j^m} (\overset{j}{\overbrace{\cdots}}, x_{j+1}, \ldots, x_n) \right\|_{\infty, \prod_{i=1}^{j} [a_i, b_i]} \right]. \quad (4.77)$$

Proof. By Theorems 4.15, 4.18, 4.21 and Lemma 4.24. $\qquad\qquad\square$

Next comes

Theorem 4.33. *Suppose Assumptions 4.10 or 4.17 or 4.20. Let $E_m^f(x_1, \ldots, x_n)$ as in (4.44), $m \in \mathbb{N}$. Let $p_j, q_j > 1: \frac{1}{p_j} + \frac{1}{q_j} = 1; j = 1, \ldots, n$. In particular we assume that*

$$\frac{\partial^m f}{\partial x_j^m} (\ldots, x_{j+1}, \ldots, x_n) \in L_{q_j} \left(\prod_{i=1}^{j} [a_i, b_i] \right),$$

for any $(x_{j+1}, \ldots, x_n) \in \prod_{i=j+1}^{n} [a_i, b_i]$, for all $j = 1, \ldots, n$. Then

$$|E_m^f(x_1,\ldots,x_n)| \le \frac{1}{m!}\sum_{j=1}^n \left[(b_j-a_j)^{m-\frac{1}{q_j}}\left(\prod_{i=1}^{j-1}(b_i-a_i)\right)^{-\frac{1}{q_j}}\left(\int_0^1\left|B_m\left(\frac{x_j-a_j}{b_j-a_j}\right)\right.\right.\right.$$

$$\left.\left.\left.-B_m(t_j)\right|^{p_j}dt_j\right)^{1/p_j}\left\|\frac{\partial^m f}{\partial x_j^m}(\ldots,x_{j+1},\ldots,x_n)\right\|_{q_j,\,\prod\limits_{i=1}^{j}[a_i,b_i]}\right].$$

$$(4.78)$$

When $p_j = q_j = 2$, all $j = 1,\ldots,n$, then

$$|E_m^f(x_1,\ldots,x_n)| \le \frac{1}{m!}\sum_{j=1}^n\left[(b_j-a_j)^{m-\frac{1}{2}}\left(\prod_{i=1}^{j-1}(b_i-a_i)\right)^{-1/2}\right.$$

$$\times\left(\sqrt{\frac{(m!)^2}{(2m)!}|B_{2m}|+B_m^2\left(\frac{x_j-a_j}{b_j-a_j}\right)}\right)$$

$$\left.\times\left\|\frac{\partial^m f}{\partial x_j^m}(\ldots,x_{j+1},\ldots,x_n)\right\|_{2,\,\prod\limits_{i=1}^{j}[a_i,b_i]}\right].$$

$$(4.79)$$

Proof.   By Theorems 4.15, 4.18, 4.21 and Lemma 4.26. $\square$

Finally we have

Theorem 4.34. *Suppose Assumptions 4.10 or 4.17 or 4.20. Let $E_m^f(x_1,\ldots,x_n)$ as in (4.44), $m \in \mathbb{N}$. In particular we assume for $j = 1,\ldots,n$ that*

$$\frac{\partial^m f}{\partial x_j^m}(\ldots,x_{j+1},\ldots,x_n) \in L_1\left(\prod_{i=1}^{j}[a_i,b_i]\right),$$

for any $(x_{j+1},\ldots,x_n) \in \prod_{i=j+1}^n[a_i,b_i]$. Then

$$|E_m^f(x_1,\ldots,x_n)| \le \frac{1}{m!}\sum_{j=1}^n\left[\frac{(b_j-a_j)^{m-1}}{\left(\prod\limits_{i=1}^{j-1}(b_i-a_i)\right)}\right.$$

$$\left(\left\|\frac{\partial^m f}{\partial x_j^m}(\ldots,x_{j+1},\ldots,x_n)\right\|_{1,\,\prod\limits_{i=1}^{j}[a_i,b_i]}\right)\left.\left\|B_m(t)-B_m\left(\frac{x_j-a_j}{b_j-a_j}\right)\right\|_{\infty,[0,1]}\right].$$

$$(4.80)$$

The special cases are calculated and estimated further as follows:
1) *When $m = 2r$, $r \in \mathbb{N}$, then*

$$|E_{2r}^f(x_1,\ldots,x_n)|$$

$$\leq \frac{1}{(2r)!}\sum_{j=1}^n\left\{\frac{(b_j-a_j)^{2r-1}}{\left(\prod\limits_{i=1}^{j-1}(b_i-a_i)\right)}\left(\left\|\frac{\partial^{2r}f}{\partial x_j^{2r}}(\ldots,x_{j+1},\ldots,x_n)\right\|_{1,\prod\limits_{i=1}^{j}[a_i,b_i]}\right)\right.$$

$$\left.\times\left[(1-2^{-2r})|B_{2r}|+\left|2^{-2r}B_{2r}-B_{2r}\left(\frac{x_j-a_j}{b_j-a_j}\right)\right|\right]\right\}. \tag{4.81}$$

2) *When $m = 2r+1$, $r \in \mathbb{N}$, then*

$$|E_{2r+1}^f(x_1,\ldots,x_n)|$$

$$\leq \frac{1}{(2r+1)!}\sum_{j=1}^n\left\{\frac{(b_j-a_j)^{2r}}{\left(\prod\limits_{i=1}^{j-1}(b_i-a_i)\right)}\left(\left\|\frac{\partial^{2r+1}f}{\partial x_j^{2r+1}}(\ldots,x_{j+1},\ldots,x_n)\right\|_{1,\prod\limits_{i=1}^{j}[a_i,b_i]}\right)\right.$$

$$\left.\times\left[\frac{2(2r+1)!}{(2\pi)^{2r+1}(1-2^{-2r})}+\left|B_{2r+1}\left(\frac{x_j-a_j}{b_j-a_j}\right)\right|\right]\right\}. \tag{4.82}$$

And at last

3) *When $m = 1$, then*

$$|E_1^f(x_1,\ldots,x_n)| \leq \sum_{j=1}^n\left\{\frac{1}{\left(\prod\limits_{i=1}^{j-1}(b_i-a_i)\right)}\left[\left\|\frac{\partial f}{\partial x_j}(\ldots,x_{j+1},\ldots,x_n)\right\|_{1,\prod\limits_{i=1}^{j}[a_i,b_i]}\right.\right.$$

$$\left.\left.\times\left[\frac{1}{2}+\left|x_j-\left(\frac{a_j+b_j}{2}\right)\right|\right]\right\}. \tag{4.83}$$

Proof. By Theorems 4.15, 4.18, 4.21 and Lemma 4.28. $\square$

The final general Ostrowski type estimate follows:

Theorem 4.35. *All as in Assumption 4.10, $m,n \in \mathbb{N}$, $x_i \in [a_i,b_i]$, $i = 1,2,\ldots,n$. Here B_j is as in (4.46), for $j = 1,\ldots,n$. Then*

$$\left|f(x_1,\ldots,x_n)-\frac{1}{\prod\limits_{i=1}^{n}(b_i-a_i)}\int_{\prod\limits_{i=1}^{n}[a_i,b_i]}f(s_1,s_2,\ldots,s_n)ds_1 ds_2\cdots ds_n\right|$$

$$\leq \sum_{j=1}^n|B_j|+\sum_{j=1}^n\left[\sum_{k=1}^{m-1}\frac{(b_j-a_j)^{k-1}}{k!}\left|B_k\right.\right.$$

$$\left.\left.\left(\frac{x_j-a_j}{b_j-a_j}\right)\right|\omega_1\left(\frac{\partial^{k-1}f}{\partial x_j^{k-1}}(\ldots,x_{j+1},\ldots,x_n),b_j-a_j\right)\right]. \tag{4.84}$$

Estimates for B_j are given by Lemmas 4.24, 4.26, 4.28.

Proof. By Theorems 4.14, 4.15 and Lemma 4.31, see also Remark 4.16, (4.48), (4.49). $\qquad\square$

4.4 Applications

Here we apply Theorems 4.32, 4.33, 4.34. We give

Corollary 4.36. *Suppose Assumptions 4.10 or 4.17 or 4.20, case $m = 1$.*

i) Assume $\frac{\partial f}{\partial x_j}(\ldots, x_{j+1}, \ldots, x_n) \in L_\infty\left(\prod_{i=1}^{j}([a_i, b_i])\right)$, for any $(x_{j+1}, \ldots, x_n) \in \prod_{i=j+1}^{n}[a_i, b_i]$, all $j = 1, \ldots, n$. Then

$$\left| f(x_1, x_2, \ldots, x_n) - \frac{1}{\prod_{i=1}^{n}(b_i - a_i)} \int_{\prod_{i=1}^{n}[a_i, b_i]} f(s_1, \ldots, s_n) ds_1 \cdots ds_n \right|$$

$$\leq \sum_{j=1}^{n} \left[(b_j - a_j) \left(\sqrt{\frac{1}{12} + \left(\frac{x_j - a_j}{b_j - a_j} - \frac{1}{2} \right)^2} \right) \right\|$$

$$\times \left. \frac{\partial f}{\partial x_j}(\ldots, x_{j+1}, \ldots, x_n) \right\|_{\infty, \prod_{i=1}^{j}[a_i, b_i]} \right]. \tag{4.85}$$

ii) Let $p_j, q_j > 1$: $\frac{1}{p_j} + \frac{1}{q_j} = 1$; $j = 1, \ldots, n$. In particular we suppose that

$$\frac{\partial f}{\partial x_j}(\ldots, x_{j+1}, \ldots, x_n) \in L_{q_j}\left(\prod_{i=1}^{j}[a_i, b_i]\right),$$

for any $(x_{j+1}, \ldots, x_n) \in \prod_{i=j+1}^{n}[a_i, b_i]$, for all $j = 1, \ldots, n$. Then

$$\left| f(x_1, x_2, \ldots, x_n) - \frac{1}{\prod_{i=1}^{n}(b_i - a_i)} \int_{\prod_{i=1}^{n}[a_i, b_i]} f(s_1, \ldots, s_n) ds_1 \cdots ds_n \right|$$

$$\leq \sum_{j=1}^{n} \left[(b_j - a_j)^{1 - \frac{1}{q_j}} \left(\prod_{i=1}^{j-1}(b_i - a_i) \right)^{-\frac{1}{q_j}} \left(\int_0^1 \left| \left(\frac{x_j - a_j}{b_j - a_j} \right) - t_j \right|^{p_j} dt_j \right)^{1/p_j} \right.$$

$$\times \left. \left\| \frac{\partial f}{\partial x_j}(\ldots, x_{j+1}, \ldots, x_n) \right\|_{q_j, \prod_{i=1}^{j}[a_i, b_i]} \right]. \tag{4.86}$$

When $p_j = q_j = 2$, all $j = 1, \ldots, n$, then

$$\left| f(x_1,\ldots,x_n) - \frac{1}{\prod_{i=1}^{n}(b_i - a_i)} \int_{\prod_{i=1}^{n}[a_i,b_i]} f(s_1,\ldots,s_n)ds_1\cdots ds_n \right|$$

$$\leq \sum_{j=1}^{n}\left[\left(\sqrt{(b_j-a_j)}\Big/\sqrt{\prod_{i=1}^{j-1}(b_i-a_i)}\right)\left(\sqrt{\frac{1}{12}+\left(\frac{x_j-a_j}{b_j-a_j}-\frac{1}{2}\right)^2}\right)\right.$$

$$\left. \times\left\|\frac{\partial f}{\partial x_j}(\ldots,x_{j+1},\ldots,x_n)\right\|_{2,\prod_{i=1}^{j}[a_i,b_i]}\right]. \tag{4.87}$$

iii) *Here assume for $j = 1,\ldots,n$ that*

$$\frac{\partial f}{\partial x_j}(\ldots,x_{j+1},\ldots,x_n) \in L_1\left(\prod_{i=1}^{j}[a_i,b_i]\right),$$

for any $(x_{j+1},\ldots,x_n) \in \prod_{i=j+1}^{n}[a_i,b_i]$. *Then*

$$\left| f(x_1,\ldots,x_n) - \frac{1}{\prod_{i=1}^{n}(b_i - a_i)} \int_{\prod_{i=1}^{n}[a_i,b_i]} f(s_1,\ldots,s_n)ds_1\cdots ds_n \right|$$

$$\leq \sum_{j=1}^{n}\left\{\frac{1}{\left(\prod_{i=1}^{j-1}(b_i-a_i)\right)}\left[\left\|\frac{\partial f}{\partial x_j}(\ldots,x_{j+1},\ldots,x_n)\right\|_{1,\prod_{i=1}^{j}[a_i,b_i]}\right]\right.$$

$$\left. \times\left[\frac{1}{2}+\left|x_j-\left(\frac{a_j+b_j}{2}\right)\right|\right]\right\}. \tag{4.88}$$

Notice 4.37. We have for $j = 1,\ldots,n$:

$$\begin{aligned}
\lambda_j &:= \frac{x_j-a_j}{b_j-a_j} = 0 \quad &&\text{iff}\quad x_j = a_j,\\
\lambda_j &= \qquad\qquad 1 \quad &&\text{iff}\quad x_j = b_j,\\
\lambda_j &= \qquad\qquad \frac{1}{2} \quad &&\text{iff}\quad x_j = \frac{a_j+b_j}{2}.
\end{aligned} \tag{4.89}$$

We continue with Corollaries to Corollary 4.36.

Corollary 4.38. *Suppose Assumptions 4.10 or 4.17 or 4.20, Case $m = 1$, all $x_j = a_j$, $j = 1,\ldots,n$.*

i) *Assume*

$$\frac{\partial f}{\partial x_j}(\ldots,a_{j+1},\ldots,a_n) \in L_\infty\left(\prod_{i=1}^{j}[a_i,b_i]\right), \quad \text{all } j = 1,\ldots,n.$$

Then

$$\left| f(a_1, a_2, \ldots, a_n) - \frac{1}{\prod\limits_{i=1}^{n} (b_i - a_i)} \int_{\prod\limits_{i=1}^{n} [a_i, b_i]} f(s_1, \ldots, s_n) ds_1 \cdots ds_n \right|$$

$$\leq \frac{\sqrt{3}}{3} \left\{ \sum_{j=1}^{n} \left[(b_j - a_j) \left\| \frac{\partial f}{\partial x_j} (\ldots, a_{j+1}, \ldots, a_n) \right\|_{\infty, \prod\limits_{i=1}^{j} [a_i, b_i]} \right] \right\}. \tag{4.90}$$

ii) *Let* $p_j, q_j > 1 \colon \frac{1}{p_j} + \frac{1}{q_j} = 1;\ j = 1, \ldots, n.$ *In particular we suppose that*

$$\frac{\partial f}{\partial x_j} (\ldots, a_{j+1}, \ldots, a_n) \in L_{q_j} \left(\prod_{i=1}^{j} [a_i, b_i] \right),$$

for all $j = 1, \ldots, n.$ *Then*

$$\left| f(a_1, \ldots, a_n) - \frac{1}{\prod\limits_{i=1}^{n} (b_i - a_i)} \int_{\prod\limits_{i=1}^{n} [a_i, b_i]} f(s_1, \ldots, s_n) ds_1 \cdots ds_n \right|$$

$$\leq \sum_{j=1}^{n} \left[\frac{(b_j - a_j)^{1 - \frac{1}{q_j}} \left(\prod\limits_{i=1}^{j-1} (b_i - a_i) \right)^{-1/q_j}}{(p_j + 1)^{1/p_j}} \right.$$

$$\left. \times \left\| \frac{\partial f}{\partial x_j} (\ldots, a_{j+1}, \ldots, a_n) \right\|_{q_j, \prod\limits_{i=1}^{j} [a_i, b_i]} \right]. \tag{4.91}$$

When $p_j = q_j = 2,$ *all* $j = 1, \ldots, n,$ *then*

$$\left| f(a_1, \ldots, a_n) - \frac{1}{\prod\limits_{i=1}^{n} (b_i - a_i)} \int_{\prod\limits_{i=1}^{n} [a_i, b_i]} f(s_1, \ldots, s_n) ds_1 \cdots ds_n \right|$$

$$\leq \frac{\sqrt{3}}{3} \left(\sum_{j=1}^{n} \left[\left(\sqrt{(b_j - a_j)} \middle/ \sqrt{\prod_{i=1}^{j-1} (b_i - a_i)} \right) \right.\right.$$

$$\left.\left. \times \left\| \frac{\partial f}{\partial x_j} (\ldots, a_{j+1}, \ldots, a_n) \right\|_{2, \prod\limits_{i=1}^{j} [a_i, b_i]} \right] \right). \tag{4.92}$$

iii) *Here assume for* $j = 1, \ldots, n,$ *that*

$$\frac{\partial f}{\partial x_j} (\ldots, a_{j+1}, \ldots, a_n) \in L_1 \left(\prod_{i=1}^{j} [a_i, b_i] \right).$$

Then

$$\left| f(a_1,\ldots,a_n) - \frac{1}{\prod\limits_{i=1}^{n}(b_i-a_i)} \int_{\prod\limits_{i=1}^{n}[a_i,b_i]} f(s_1,\ldots,s_n)ds_1\cdots ds_n \right|$$

$$\leq \frac{1}{2}\sum_{j=1}^{n}\left\{ \frac{(b_j-a_j+1)}{\left(\prod\limits_{i=1}^{j-1}(b_i-a_i)\right)} \left\| \frac{\partial f}{\partial x_j}(\ldots,a_{j+1},\ldots,a_n) \right\|_{1,\prod\limits_{i=1}^{j}[a_i,b_i]} \right\}. \qquad (4.93)$$

Corollary 4.39. *Suppose Assumptions 4.10 or 4.17 or 4.20, case $m = 1$, all $x_j = b_j$, $j = 1,\ldots,n$.*
   i) *Assume*

$$\frac{\partial f}{\partial x_j}(\ldots,b_{j+1},\ldots,b_n) \in L_\infty\left(\prod_{i=1}^{j}[a_i,b_i]\right),$$

for all $j = 1,\ldots,n$. Then

$$\left| f(b_1,\ldots,b_n) - \frac{1}{\prod\limits_{i=1}^{n}(b_i-a_i)} \int_{\prod\limits_{i=1}^{n}[a_i,b_i]} f(s_1,\ldots,s_n)ds_1\cdots ds_n \right|$$

$$\leq \frac{\sqrt{3}}{3}\sum_{j=1}^{n}\left[(b_j-a_j)\left\| \frac{\partial f}{\partial x_j}(\ldots,b_{j+1},\ldots,b_n) \right\|_{\infty,\prod\limits_{i=1}^{j}[a_i,b_i]} \right]. \qquad (4.94)$$

   ii) *Let $p_j,q_j > 1$: $\frac{1}{p_j} + \frac{1}{q_j} = 1$; $j = 1,\ldots,n$. In particular we assume that*

$$\frac{\partial f}{\partial x_j}(\ldots,b_{j+1},\ldots,b_n) \in L_{q_j}\left(\prod_{i=1}^{j}[a_i,b_i]\right),$$

for all $j = 1,\ldots,n$. Then

$$\left| f(b_1,\ldots,b_n) - \frac{1}{\prod\limits_{i=1}^{n}(b_i-a_i)} \int_{\prod\limits_{i=1}^{n}[a_i,b_i]} f(s_1,\ldots,s_n)ds_1\cdots ds_n \right|$$

$$\leq \sum_{j=1}^{n}\left[(p_j+1)^{-1/p_j}(b_j-a_j)^{1-\frac{1}{q_j}}\left(\prod_{i=1}^{j-1}(b_i-a_i)\right)^{-\frac{1}{q_j}} \right.$$

$$\left. \left\| \frac{\partial f}{\partial x_j}(\ldots,b_{j+1},\ldots,b_n) \right\|_{q_j,\prod\limits_{i=1}^{j}[a_i,b_i]} \right]. \qquad (4.95)$$

When $p_j = q_j = 2$, all $j = 1, \ldots, n$, then

$$\left| f(b_1, \ldots, b_n) - \frac{1}{\prod_{i=1}^{n}(b_i - a_i)} \int_{\prod_{i=1}^{n}[a_i,b_i]} f(s_1, \ldots, s_n) ds_1 \cdots ds_n \right|$$

$$\leq \frac{\sqrt{3}}{3} \sum_{j=1}^{n} \left[\left(\sqrt{b_j - a_j} \Big/ \sqrt{\prod_{i=1}^{j-1}(b_i - a_i)} \right) \left\| \frac{\partial f}{\partial x_j}(\ldots, b_{j+1}, \ldots, b_n) \right\|_{2, \prod_{i=1}^{j}[a_i,b_i]} \right].$$

$$(4.96)$$

iii) *Here suppose for $j = 1, \ldots, n$ that*

$$\frac{\partial f}{\partial x_j}(\ldots, b_{j+1}, \ldots, b_n) \in L_1\left(\prod_{i=1}^{j}[a_i, b_i] \right).$$

Then

$$\left| f(b_1, \ldots, b_n) - \frac{1}{\prod_{i=1}^{n}(b_i - a_i)} \int_{\prod_{i=1}^{n}[a_i,b_i]} f(s_1, \ldots, s_n) ds_1 \cdots ds_n \right|$$

$$\leq \frac{1}{2} \sum_{j=1}^{n} \left\{ \frac{(1 + b_j - a_j)}{\left(\prod_{i=1}^{j-1}(b_i - a_i) \right)} \left\| \frac{\partial f}{\partial x_j}(\ldots, b_{j+1}, \ldots, b_n) \right\|_{1, \prod_{i=1}^{j}[a_i,b_i]} \right\}. \qquad (4.97)$$

Next come the multivariate midpoint rule inequalities.

Corollary 4.40. *Suppose Assumptions 4.10 or 4.17 or 4.20, case $m = 1$, all $x_j = \frac{a_j + b_j}{2}$, $j = 1, \ldots, n$.*
  i) *Assume*

$$\frac{\partial f}{\partial x_j}\left(\ldots, \frac{a_{j+1} + b_{j+1}}{2}, \ldots, \frac{a_n + b_n}{2} \right) \in L_\infty\left(\prod_{i=1}^{j}[a_i, b_i] \right), \quad \text{all } j = 1, \ldots, n.$$

Then

$$\left| f\left(\frac{a_1 + b_1}{2}, \frac{a_2 + b_2}{2}, \ldots, \frac{a_n + b_n}{2} \right) - \frac{1}{\prod_{i=1}^{n}(b_i - a_i)} \int_{\prod_{i=1}^{n}[a_i,b_i]} f(s_1, \ldots, s_n) ds_1 \cdots ds_n \right|$$

$$\leq \frac{1}{2\sqrt{3}} \sum_{j=1}^{n} \left[(b_j - a_j) \left\| \frac{\partial f}{\partial x_j}\left(\ldots, \frac{a_{j+1} + b_{j+1}}{2}, \ldots, \frac{a_n + b_n}{2} \right) \right\|_{\infty, \prod_{i=1}^{j}[a_i,b_i]} \right] \quad (4.98)$$

ii) *Let $p_j, q_j > 1$: $\frac{1}{p_j} + \frac{1}{q_j} = 1$; $j = 1, \ldots, n$. In particular we suppose that*

$$\frac{\partial f}{\partial x_j}\left(\ldots, \frac{a_{j+1} + b_{j+1}}{2}, \ldots, \frac{a_n + b_n}{2} \right) \in L_{q_j}\left(\prod_{i=1}^{j}[a_i, b_i] \right),$$

for all $j = 1, \ldots, n$. Then

$$\left| f\left(\frac{a_1 + b_1}{2}, \ldots, \frac{a_n + b_n}{2}\right) - \frac{1}{\prod_{i=1}^{n}(b_i - a_i)} \int_{\prod_{i=1}^{n}[a_i,b_i]} f(s_1, \ldots, s_n) ds_1 \cdots ds_n \right|$$

$$\leq \frac{1}{2} \sum_{j=1}^{n} \left[(p_{j+1})^{-\frac{1}{p_j}} (b_j - a_j)^{1 - \frac{1}{q_j}} \left(\prod_{i=1}^{j-1}(b_i - a_i) \right)^{-\frac{1}{q_j}} \right.$$

$$\left. \times \left\| \frac{\partial f}{\partial x_j} \left(\ldots, \frac{a_{j+1} + b_{j+1}}{2}, \ldots, \frac{a_n + b_n}{2} \right) \right\|_{q_j, \prod_{i=1}^{j}[a_i,b_i]} \right]. \tag{4.99}$$

When $p_j = q_j = 2$, all $j = 1, \ldots, n$, then

$$\left| f\left(\frac{a_1 + b_1}{2}, \ldots, \frac{a_n + b_n}{2}\right) - \frac{1}{\prod_{i=1}^{n}(b_i - a_i)} \int_{\prod_{i=1}^{n}[a_i,b_i]} f(s_1, \ldots, s_n) ds_1 \cdots ds_n \right|$$

$$\leq \frac{1}{2\sqrt{3}} \sum_{j=1}^{n} \left[\left(\sqrt{b_j - a_j} \Big/ \sqrt{\prod_{i=1}^{j-1}(b_i - a_i)} \right) \right.$$

$$\left. \times \left\| \frac{\partial f}{\partial x_j} \left(\ldots, \frac{a_{j+1} + b_{j+1}}{2}, \ldots, \frac{a_n + b_n}{2} \right) \right\|_{2, \prod_{i=1}^{j}[a_i,b_i]} \right]. \tag{4.100}$$

iii) *Here assume for $j = 1, \ldots, n$ that*

$$\frac{\partial f}{\partial x_j} \left(\ldots, \frac{a_{j+1} + b_{j+1}}{2}, \ldots, \frac{a_n + b_n}{2} \right) \in L_1 \left(\prod_{i=1}^{j}[a_i, b_i] \right).$$

Then

$$\left| f\left(\frac{a_1 + b_1}{2}, \ldots, \frac{a_n + b_n}{2}\right) - \frac{1}{\prod_{i=1}^{n}(b_i - a_i)} \int_{\prod_{i=1}^{n}[a_i,b_i]} f(s_1, \ldots, s_n) ds_1 \cdots ds_n \right|$$

$$\leq \frac{1}{2} \sum_{j=1}^{n} \left\{ \frac{1}{\left(\prod_{i=1}^{j-1}(b_i - a_i) \right)} \left\| \frac{\partial f}{\partial x_j} \left(\ldots, \frac{a_{j+1} + b_{j+1}}{2}, \ldots, \frac{a_n + b_n}{2} \right) \right\|_{1, \prod_{i=1}^{j}[a_i,b_i]} \right\}.$$

$$\tag{4.101}$$

Next we treat the case of $m = 2$ and only for the norms $\|\cdot\|_\infty$, $\|\cdot\|_2$, and specifically for $\lambda_j = 0, 1, \frac{1}{2}$, $j = 1, \ldots, n$. The multivariate trapezoid rule estimates follow immediately.

Corollary 4.41. *Suppose Assumptions 4.10 or 4.17 or 4.20, case $m = 2$, all $x_j = a_j$, $j = 1, \ldots, n$.*

i) *Assume*

$$\frac{\partial^2 f}{\partial x_j^2}(\ldots, a_{j+1}, \ldots, a_n) \in L_\infty\left(\prod_{i=1}^{j}[a_i, b_i]\right),$$

all $j = 1, \ldots, n$. *Then*

$$K_2 := \left|\left(\frac{f(a_1, a_2, \ldots, a_n) + f(b_1, a_2, \ldots, a_n)}{2}\right)\right.$$

$$-\frac{1}{\prod_{i=1}^{n}(b_i - a_i)}\int_{\prod_{i=1}^{n}[a_i,b_i]} f(s_1, \ldots, s_n)ds_1 \cdots ds_n$$

$$+\frac{1}{2}\left\{\sum_{j=2}^{n}\left[\frac{1}{\left(\prod_{i=1}^{j-1}(b_i - a_i)\right)}\left\{\int_{\prod_{i=1}^{j-1}[a_i,b_i]}\left(f(s_1, s_2, \ldots, s_{j-1}, b_j, a_{j+1}, \ldots, a_n)\right.\right.\right.$$

$$\left.\left.\left.\left.- f(s_1, \ldots, s_{j-1}, a_j, a_{j+1}, \ldots, a_n))ds_1 \cdots ds_{j-1}\right\}\right]\right\}\right|$$

$$\leq \frac{1}{2\sqrt{30}}\left\{\sum_{j=1}^{n}\left[(b_i - a_j)^2\left\|\frac{\partial^2 f}{\partial x_j^2}(\ldots, a_{j+1}, \ldots, a_n)\right\|_{\infty, \prod_{i=1}^{j}[a_i,b_i]}\right]\right\}. \qquad (4.102)$$

ii) *Suppose*

$$\frac{\partial^2 f}{\partial x_j^2}(\ldots, a_{j+1}, \ldots, a_n) \in L_2\left(\prod_{i=1}^{j}[a_i, b_i]\right),$$

all $j = 1, \ldots, n$. *Then*

$$K_2$$

$$\leq \frac{1}{2\sqrt{30}}\left\{\sum_{j=1}^{n}\left[(b_j - a_j)^{3/2}\left(\prod_{i=1}^{j-1}(b_i - a_i)\right)^{-1/2}\left\|\frac{\partial^2 f}{\partial x_j^2}(\ldots, a_{j+1}, \ldots, a_n)\right\|_{2, \prod_{i=1}^{j}[a_i,b_i]}\right]\right\}.$$

$$(4.103)$$

We continue with trapezoid rule estimates.

Corollary 4.42. *Suppose Assumptions 4.10 or 4.17 or 4.20, case* $m = 2$*, all* $x_j = b_j$, $j = 1, \ldots, n$.
  i) *Assume*

$$\frac{\partial^2 f}{\partial x_j^2}(\ldots, b_{j+1}, \ldots, b_n) \in L_\infty\left(\prod_{i=1}^{j}[a_i, b_i]\right),$$

all $j = 1, \ldots, n$. Then

$$\Lambda_2$$

$$:= \left| \frac{f(b_1, \ldots, b_n) + f(a_1, b_2, \ldots, b_n)}{2} - \frac{1}{\prod\limits_{i=1}^{n} (b_i - a_i)} \int_{\prod\limits_{i=1}^{n} [a_i, b_i]} f(s_1, \ldots, s_n) ds_1 \cdots ds_n \right.$$

$$- \frac{1}{2} \left\{ \sum_{j=2}^{n} \left[\frac{1}{\left(\prod\limits_{i=1}^{j-1} (b_i - a_i) \right)} \left\{ \int_{\prod\limits_{i=1}^{j-1} [a_i, b_i]} (f(s_1, \ldots, s_{j-1}, b_j, b_{j+1}, \ldots, b_n) \right. \right. \right.$$

$$\left. \left. \left. \left. - f(s_1, \ldots, s_{j-1}, a_j, b_{j+1}, \ldots, b_n)) ds_1 \cdots ds_{j-1} \right\} \right] \right\} \right|$$

$$\leq \frac{1}{2\sqrt{30}} \left\{ \sum_{j=1}^{n} (b_j - a_j)^2 \left\| \frac{\partial^2 f}{\partial x_j^2}(\ldots, b_{j+1}, \ldots, b_n) \right\|_{\infty, \prod\limits_{i=1}^{j} [a_i, b_i]} \right\}. \tag{4.104}$$

ii) Assume

$$\frac{\partial^2 f}{\partial x_j^2}(\ldots, b_{j+1}, \ldots, b_n) \in L_2 \left(\prod_{i=1}^{j} [a_i, b_i] \right),$$

all $j = 1, \ldots, n$. Then

$$\Lambda_2 \leq \frac{1}{2\sqrt{30}} \left\{ \sum_{j=1}^{n} \left[(b_j - a_j)^{3/2} \left(\prod_{i=1}^{j-1} (b_i - a_i) \right)^{-1/2} \right. \right.$$

$$\left. \left. \times \left\| \frac{\partial^2 f}{\partial x_j^2}(\ldots, b_{j+1}, \ldots, b_n) \right\|_{2, \prod\limits_{i=1}^{j} [a_i, b_i]} \right] \right\}. \tag{4.105}$$

The multivariate midpoint rule estimates follow.

Corollary 4.43. *Suppose Assumptions 4.10 or 4.17 or 4.20, case $m = 2$, all $x_j = \frac{a_j + b_j}{2}$, $j = 1, \ldots, n$.*

i) Assume

$$\frac{\partial^2 f}{\partial x_j^2} \left(\ldots, \frac{a_{j+1} + b_{j+1}}{2}, \ldots, \frac{a_n + b_n}{2} \right) \in L_\infty \left(\prod_{i=1}^{j} [a_i, b_i] \right),$$

all $j = 1, \ldots, n$. Then

$$M_2 := \left| f\left(\frac{a_1 + b_1}{2}, \ldots, \frac{a_n + b_n}{2} \right) \right.$$

$$\left. - \frac{1}{\prod\limits_{i=1}^{n} (b_i - a_i)} \int_{\prod\limits_{i=1}^{n} [a_i, b_i]} f(s_1, \ldots, s_n) ds_1 \cdots ds_n \right|$$

$$\leq \frac{1}{8\sqrt{5}} \left\{ \sum_{j=1}^{n} \left[(b_j - a_j)^2 \left\| \frac{\partial^2 f}{\partial x_j^2} \left(\ldots, \frac{a_{j+1} + b_{j+1}}{2}, \ldots, \frac{a_n + b_n}{2} \right) \right\|_{\infty, \prod\limits_{i=1}^{j} [a_i, b_i]} \right] \right\}. \tag{4.106}$$

ii) *Suppose*

$$\frac{\partial^2 f}{\partial x_j^2}\left(\ldots,\frac{a_{j+1}+b_{j+1}}{2},\ldots,\frac{a_n+b_n}{2}\right) \in L_2\left(\prod_{i=1}^{j}[a_i,b_i]\right),$$

all $j = 1,\ldots,n$. *Then*

$$M_2 \leq \frac{1}{8\sqrt{5}}\left\{\sum_{j=1}^{n}\left[(b_j-a_j)^{3/2}\left(\prod_{i=1}^{j-1}(b_i-a_i)\right)^{-1/2}\left\|\frac{\partial^2 f}{\partial x_j^2}\right.\right.\right.$$

$$\left.\left.\left.\left(\ldots,\frac{a_{j+1}+b_{j+1}}{2},\ldots,\frac{a_n+b_n}{2}\right)\right\|_{2,\prod\limits_{i=1}^{j}[a_i,b_i]}\right]\right\}. \tag{4.107}$$

We continue with trapezoid and midpoint rules inequalities for $m = 3$.

Corollary 4.44. *Suppose Assumptions 4.10 or 4.17 or 4.20, case* $m = 3$, *all* $x_j = a_j$, $j = 1,\ldots,n$.

i) *Assume*

$$\frac{\partial^3 f}{\partial x_j^3}(\ldots,a_{j+1},\ldots,a_n) \in L_\infty\left(\prod_{i=1}^{j}[a_i,b_i]\right),$$

all $j = 1,\ldots,n$. *Then*

$$K_3 := \left|\left(\frac{f(a_1,\ldots,a_n)+f(b_1,a_2,\ldots,a_n)}{2}\right)\right.$$

$$-\frac{1}{\prod\limits_{i=1}^{n}(b_i-a_i)}\int_{\prod\limits_{i=1}^{n}[a_i,b_i]}f(s_1,\ldots,s_n)ds_1\cdots ds_n$$

$$-\sum_{\substack{j=1\\ \text{with } j=k\neq 1}}^{n}\left\{\frac{1}{\prod\limits_{i=1}^{j-1}(b_i-a_i)}\left\{\sum_{k=1}^{2}\frac{(b_j-a_j)^{k-1}}{k!}B_k(0)\right.\right.$$

$$\times\left(\int_{\prod\limits_{i=1}^{j-1}[a_i,b_i]}\left(\frac{\partial^{k-1}f}{\partial x_j^{k-1}}(s_1,\ldots,s_{j-1},b_j,a_{j+1},\ldots,a_n)\right.\right.$$

$$\left.\left.\left.\left.-\frac{\partial^{k-1}f}{\partial x_j^{k-1}}(s_1,\ldots,s_{j-1},a_j,a_{j+1},\ldots,a_n)\right)ds_1\cdots ds_{j-1}\right)\right\}\right\}$$

$$\leq \frac{1}{12\sqrt{210}}\left\{\sum_{j=1}^{n}\left[(b_j-a_j)^3\left\|\frac{\partial^3 f}{\partial x_j^3}(\ldots,a_{n+1},\ldots,a_n)\right\|_{\infty,\prod\limits_{i=1}^{j}[a_i,b_i]}\right]\right\}. \tag{4.108}$$

ii) *Assume*

$$\frac{\partial^3 f}{\partial x_j^3}(\ldots,a_{j+1},\ldots,a_n) \in L_2\left(\prod_{i=1}^{j}[a_i,b_i]\right),$$

all $j = 1, \ldots, n$. Then

$$K_3 \leq \frac{1}{12\sqrt{210}} \sum_{j=1}^{n} \left[(b_j - a_j)^{5/2} \left(\prod_{i=1}^{j-1} (b_i - a_i) \right)^{-1/2} \left\| \frac{\partial^3 f}{\partial x_j^3} \right. \right.$$

$$\left. \left. (\ldots, a_{j+1}, \ldots, a_n) \right\|_{2, \prod_{i=1}^{j} [a_i, b_i]} \right]. \tag{4.109}$$

Corollary 4.45. *Suppose Assumptions 4.10 or 4.17 or 4.20, case $m = 3$, all $x_j = b_j$, $j = 1, \ldots, n$.*

  i) *Assume*

$$\frac{\partial^3 f}{\partial x_j^3}(\ldots, b_{j+1}, \ldots, b_n) \in L_\infty \left(\prod_{i=1}^{j} [a_i, b_i] \right),$$

all $j = 1, \ldots, n$. Then

$$\Lambda_3 := \left| \left(\frac{f(b_1, \ldots, b_n) + f(a_1, b_2, \ldots, b_n)}{2} \right) \right.$$

$$- \frac{1}{\prod\limits_{i=1}^{n} (b_i - a_i)} \int_{\prod\limits_{i=1}^{n} [a_i, b_i]} f(s_1, \ldots, s_n) ds_1 \cdots ds_n$$

$$- \sum_{\substack{j=1 \\ \text{with } j=k\neq 1}}^{n} \left\{ \frac{1}{\prod\limits_{i=1}^{j-1} (b_i - a_i)} \left\{ \sum_{k=1}^{2} \frac{(b_j - a_j)^{k-1}}{k!} B_k(1) \right. \right.$$

$$\times \left(\int_{\prod\limits_{i=1}^{j-1} [a_i, b_i]} \left(\frac{\partial^{k-1} f}{\partial x_j^{k-1}}(s_1, s_2, \ldots, s_{j-1}, b_j, b_{j+1}, \ldots, b_n) \right. \right.$$

$$\left. \left. \left. \left. - \frac{\partial^{k-1} f}{\partial x_j^{k-1}}(s_1, s_2, \ldots, s_{j-1}, a_j, b_{j+1}, \ldots, b_n) \right) ds_1 \cdots ds_{j-1} \right) \right\} \right\} \right|$$

$$\leq \frac{1}{12\sqrt{210}} \sum_{j=1}^{n} \left[(b_j - a_j)^3 \left\| \frac{\partial^3 f}{\partial x_j^3}(\ldots, b_{j+1}, \ldots, b_n) \right\|_{\infty, \prod_{i=1}^{j} [a_i, b_i]} \right]. \tag{4.110}$$

  ii) *Suppose*

$$\frac{\partial^3 f}{\partial x_j^3}(\ldots, b_{j+1}, \ldots, b_n) \in L_2 \left(\prod_{i=1}^{j} [a_i, b_i] \right),$$

all $j = 1, \ldots, n$. Then

$$\Lambda_3 \leq \frac{1}{12\sqrt{210}} \sum_{j=1}^{n} \left[(b_j - a_j)^{5/2} \left(\prod_{i=1}^{j-1} (b_i - a_i) \right)^{-1/2} \left\| \frac{\partial^3 f}{\partial x_j^3} \right. \right.$$

$$\left. \left. (\ldots, b_{j+1}, \ldots, b_n) \right\|_{2, \prod_{i=1}^{j} [a_i, b_i]} \right]. \tag{4.111}$$

Corollary 4.46. *Suppose Assumptions 4.10 or 4.17 or 4.20, case* $m = 3$, *all* $x_j = \frac{a_j + b_j}{2}$, $j = 1, \ldots, n$.

i) *Assume*

$$\frac{\partial^3 f}{\partial x_j^3}\left(\ldots, \frac{a_{j+1} + b_{j+1}}{2}, \ldots, \frac{a_n + b_n}{2}\right) \in L_\infty\left(\prod_{i=1}^{j}[a_i, b_i]\right),$$

all $j = 1, \ldots, n$. *Then*

$$M_3 := \left| f\left(\frac{a_1 + b_1}{2}, \ldots, \frac{a_n + b_n}{2}\right) \right.$$

$$- \frac{1}{\prod\limits_{i=1}^{n}(b_i - a_i)} \int_{\prod\limits_{i=1}^{n}[a_i, b_i]} f(s_1, \ldots, s_n) ds_1 \cdots ds_n$$

$$+ \frac{1}{24} \sum_{j=1}^{n} \left\{ \frac{(b_j - a_j)}{\left(\prod\limits_{i=1}^{j-1}(b_i - a_i)\right)} \left(\int_{\prod\limits_{i=1}^{j-1}[a_i, b_i]} \left(\frac{\partial f}{\partial x_j} \right. \right. \right.$$

$$(s_1, s_2, \ldots, s_{j-1}, b_j, \frac{a_{j+1} + b_{j+1}}{2}, \ldots, \frac{a_n + b_n}{2})$$

$$\left. \left. \left. - \frac{\partial f}{\partial x_j}\left(s_1, \ldots, s_{j-1}, a_j, \frac{a_{j+1} + b_{j+1}}{2}, \ldots, \frac{a_n + b_n}{2}\right)\right) ds_1 \cdots ds_{j-1} \right\} \right|$$

$$\leq \frac{1}{12\sqrt{210}} \sum_{j=1}^{n} \left[(b_j - a_j)^3 \left\| \frac{\partial^3 f}{\partial x_j^3}\left(\ldots, \frac{a_{j+1} + b_{j+1}}{2}, \ldots, \frac{a_n + b_n}{2}\right) \right\|_{\infty, \prod\limits_{i=1}^{j}[a_i, b_i]} \right].$$

$$(4.112)$$

ii) *Suppose*

$$\frac{\partial^3 f}{\partial x_j^3}\left(\ldots, \frac{a_{j+1} + b_{j+1}}{2}, \ldots, \frac{a_n + b_n}{2}\right) \in L_2\left(\prod_{i=1}^{j}[a_i, b_i]\right),$$

all $j = 1, \ldots, n$. *Then*

$$M_3 \leq \frac{1}{12\sqrt{210}} \sum_{j=1}^{n} \left[(b_j - a_j)^{5/2} \left(\prod_{i=1}^{j-1}(b_i - a_i)\right)^{-1/2} \right.$$

$$\left. \times \left\| \frac{\partial^3 f}{\partial x_j^3}\left(\ldots, \frac{a_{j+1} + b_{j+1}}{2}, \ldots, \frac{a_n + b_n}{2}\right) \right\|_{2, \prod\limits_{i=1}^{j}[a_i, b_i]} \right]. \quad (4.113)$$

Next we present trapezoid and midpoint rules inequalities for $m = 4$.

Corollary 4.47. *Suppose Assumptions 4.10 or 4.17 or 4.20, case* $m = 4$, *all* $x_j = a_j$, $j = 1, \ldots, n$.

i) *Assume*

$$\frac{\partial^4 f}{\partial x_j^4}(\ldots, a_{j+1}, \ldots, a_n) \in L_\infty\left(\prod_{i=1}^{j}[a_i, b_i]\right),$$

all $j = 1, \ldots, n$. Then

$$K_3 = \left| \left(\frac{f(a_1, \ldots, a_n) + f(b_1, a_2, \ldots, a_n)}{2} \right) \right.$$

$$- \frac{1}{\prod\limits_{i=1}^{n}(b_i - a_i)} \int_{\prod\limits_{i=1}^{n}[a_i, b_i]} f(s_1, \ldots, s_n) ds_1 \cdots ds_n$$

$$- \sum_{\substack{j=1 \\ (\text{with } j=k \neq 1)}}^{n} \left\{ \frac{1}{\prod\limits_{i=1}^{j-1}(b_i - a_i)} \left\{ \sum_{k=1}^{2} \frac{(b_j - a_j)^{k-1}}{k!} \right.\right.$$

$$\times B_k(0) \left(\int_{\prod\limits_{i=1}^{j-1}[a_i, b_i]} \left(\frac{\partial^{k-1} f}{\partial x_j^{k-1}}(s_1, \ldots, s_{j-1}, b_j, a_{j+1}, \ldots, a_n) \right. \right.$$

$$\left. \left. \left. \left. - \frac{\partial^{k-1} f}{\partial x_j^{k-1}}(s_1, s_2, \ldots, s_{j-1}, a_j, a_{j+1}, \ldots, a_n) \right) ds_1 \cdots ds_{j-1} \right) \right\} \right\} \right|$$

$$\leq \frac{1}{24\sqrt{630}} \sum_{j=1}^{n} \left[(b_j - a_j)^4 \left\| \frac{\partial^4 f}{\partial x_j^4}(\ldots, a_{j+1}, \ldots, a_n) \right\|_{\infty, \prod\limits_{i=1}^{j}[a_i, b_i]} \right]. \tag{4.114}$$

ii) *Assume*

$$\frac{\partial^4 f}{\partial x_j^4}(\ldots, a_{j+1}, \ldots, a_n) \in L_2 \left(\prod_{i=1}^{j}[a_i, b_i] \right),$$

all $j = 1, \ldots, n$. Then

$$K_3 \leq \frac{1}{24\sqrt{630}} \sum_{j=1}^{n} \left[(b_j - a_j)^{7/2} \left(\prod_{i=1}^{j-1}(b_i - a_i) \right)^{-1/2} \left\| \frac{\partial^4 f}{\partial x_j^4} \right.\right.$$

$$\left. \left. (\ldots, a_{j+1}, \ldots, a_n) \right\|_{2, \prod\limits_{i=1}^{j}[a_i, b_i]} \right]. \tag{4.115}$$

Corollary 4.48. *Suppose Assumptions 4.10 or 4.17 or 4.20, case $m = 4$, all $x_j = b_j$, $j = 1, \ldots, n$.*

i) *Assume*

$$\frac{\partial^4 f}{\partial x_j^4}(\ldots, b_{j+1}, \ldots, b_n) \in L_\infty \left(\prod_{i=1}^{j}[a_i, b_i] \right),$$

all $j = 1, \ldots, n$. Then

$$\Lambda_3 = \left| \left(\frac{f(b_1, \ldots, b_n) + f(a_1, b_2, \ldots, b_n)}{2} \right) \right.$$

$$- \frac{1}{\prod\limits_{i=1}^{n} (b_i - a_i)} \int_{\prod\limits_{i=1}^{n} [a_i, b_i]} f(s_1, \ldots, s_n) ds_1 \cdots ds_n$$

$$- \sum_{\substack{j=1 \\ (\text{with } j=k\neq 1)}}^{n} \left\{ \frac{1}{\left(\prod\limits_{i=1}^{j-1} (b_i - a_i) \right)} \left\{ \sum_{k=1}^{2} \frac{(b_j - a_j)^{k-1}}{k!} \right. \right.$$

$$\times B_k(1) \left(\int_{\prod_{i=1}^{j-1}[a_i,b_i]} \left(\frac{\partial^{k-1} f}{\partial x_j^{k-1}}(s_1, s_2, \ldots, s_{j-1}, b_j, b_{j+1}, \ldots, b_n) \right. \right.$$

$$\left. \left. \left. - \frac{\partial^{k-1} f}{\partial x_j^{k-1}}(s_1, s_2, \ldots, s_{j-1}, a_j, b_{j+1}, \ldots, b_n) \right) ds_1 \cdots ds_{j-1} \right) \right\} \right\} \right|$$

$$\leq \frac{1}{24\sqrt{630}} \left\{ \sum_{j=1}^{n} \left[(b_j - a_j)^4 \left\| \frac{\partial^4 f}{\partial x_j^4}(\ldots, b_{j+1}, \ldots, b_n) \right\|_{\infty, \prod\limits_{i=1}^{j} [a_i,b_i]} \right] \right\}. \qquad (4.116)$$

ii) *Suppose*

$$\frac{\partial^4 f}{\partial x_j^4}(\ldots, b_{j+1}, \ldots, b_n) \in L_2 \left(\prod_{i=1}^{j} [a_i, b_i] \right),$$

all $j = 1, \ldots, n$. Then

$$\Lambda_3 \leq \frac{1}{24\sqrt{630}} \sum_{j=1}^{n} \left[(b_j - a_j)^{7/2} \left(\prod_{i=1}^{j-1} (b_i - a_i) \right)^{-1/2} \left\| \frac{\partial^4 f}{\partial x_j^4} \right. \right.$$

$$\left. \left. (\ldots, b_{j+1}, \ldots, b_n) \right\|_{2, \prod\limits_{i=1}^{j} [a_i,b_i]} \right]. \qquad (4.117)$$

Corollary 4.49. *Suppose Assumptions 4.10 or 4.17 or 4.20, case $m = 4$, all $x_j = \frac{a_j + b_j}{2}$, $j = 1, \ldots, n$.*
   i) *Assume*

$$\frac{\partial^4 f}{\partial x_j^4} \left(\ldots, \frac{a_{j+1} + b_{j+1}}{2}, \ldots, \frac{a_n + b_n}{2} \right) \in L_\infty \left(\prod_{i=1}^{j} [a_i, b_i] \right),$$

all $j = 1, \ldots, n$. Then

$$M_3 = \left| f\left(\frac{a_1 + b_1}{2}, \ldots, \frac{a_n + b_n}{2}\right) \right.$$

$$- \frac{1}{\prod\limits_{i=1}^{n}(b_i - a_i)} \int_{\prod\limits_{i=1}^{n}[a_i,b_i]} f(s_1, \ldots, s_n)ds_1 \cdots ds_n$$

$$+ \frac{1}{24}\left\{\sum_{j=1}^{n}\left\{\frac{(b_j - a_j)}{\left(\prod\limits_{i=1}^{j-1}(b_i - a_i)\right)}\left(\int_{\prod\limits_{i=1}^{j-1}[a_i,b_i]}\left(\frac{\partial f}{\partial x_j}\left(s_1, s_2, \ldots, s_{j-1},\right.\right.\right.\right.\right.$$

$$\left. b_j, \frac{a_{j+1} + b_{j+1}}{2}, \ldots, \frac{a_n + b_n}{2}\right)$$

$$\left.\left.\left.\left.\left. - \frac{\partial f}{\partial x_j}\left(s_1, s_2, \ldots, s_{j-1}, a_j, \frac{a_{j+1} + b_{j+1}}{2}, \ldots, \frac{a_n + b_n}{2}\right)\right)ds_1 \cdots ds_{j-1}\right)\right\}\right\}\right|$$

$$\leq \frac{1}{152}\sqrt{\frac{107}{35}}\left\{\sum_{j=1}^{n}\left[(b_j - a_j)^4\left\|\frac{\partial^4 f}{\partial x_j^4}\right.\right.\right.$$

$$\left.\left.\left.\left(\ldots, \frac{a_{j+1} + b_{j+1}}{2}, \ldots, \frac{a_n + b_n}{2}\right)\right\|_{\infty, \prod\limits_{i=1}^{j}[a_i,b_i]}\right]\right\}. \tag{4.118}$$

ii) *Assume*

$$\frac{\partial^4 f}{\partial x_j^4}\left(\ldots, \frac{a_{j+1} + b_{j+1}}{2}, \ldots, \frac{a_n + b_n}{2}\right) \in L_2\left(\prod_{i=1}^{j}[a_i, b_i]\right),$$

all $j = 1, \ldots, n$. Then

$$M_3 \leq \frac{1}{1152}\sqrt{\frac{107}{35}}\left\{\sum_{j=1}^{n}\left[(b_j - a_j)^{7/2}\left(\prod_{i=1}^{j-1}(b_i - a_i)\right)^{-1/2}\right.\right.$$

$$\left.\left. \times \left\|\frac{\partial^4 f}{\partial x_j^4}\left(\ldots, \frac{a_{j+1} + b_{j+1}}{2}, \ldots, \frac{a_n + b_n}{2}\right)\right\|_{2, \prod\limits_{i=1}^{j}[a_i,b_i]}\right]\right\}. \tag{4.119}$$

Also we present trapezoid and midpoint rules inequalities for $m = 5$.

Corollary 4.50. *Suppose Assumptions 4.10 or 4.17 or 4.20, case $m = 5$, all $x_j = a_j$, $j = 1, \ldots, n$.*
    i) *Assume*

$$\frac{\partial^5 f}{\partial x_j^5}(\ldots, a_{j+1}, \ldots, a_n) \in L_\infty\left(\prod_{i=1}^{j}[a_i, b_i]\right),$$

all $j = 1, \ldots, n$. Then

$$K_5 := \left| \left(\frac{f(a_1, \ldots, a_n) + f(b_1, a_2, \ldots, a_n)}{2} \right) \right.$$

$$- \frac{1}{\prod\limits_{i=1}^{n}(b_i - a_i)} \int_{\prod\limits_{i=1}^{n}[a_i, b_i]} f(s_1, \ldots, s_n) ds_1 \cdots ds_n$$

$$- \sum_{\substack{j=1 \\ (\text{with } j=k\neq 1)}}^{n} \left\{ \frac{1}{\left(\prod\limits_{i=1}^{j-1}(b_i - a_i)\right)} \left\{ \sum_{k \in \{1,2,4\}} \frac{(b_j - a_j)^{k-1}}{k!} \right. \right.$$

$$\times B_k(0) \left(\int_{\prod\limits_{i=1}^{j-1}[a_i,b_i]} \left(\frac{\partial^{k-1} f}{\partial x_j^{k-1}}(s_1, \ldots, s_{j-1}, b_j, a_{j+1}, \ldots, a_n) \right. \right.$$

$$\left. \left. \left. - \frac{\partial^{k-1} f}{\partial x_j^{k-1}}(s_1, \ldots, s_{j-1}, a_j, a_{j+1}, \ldots, a_n) \right) ds_1 \cdots ds_{j-1} \right) \right\} \right\} \right|$$

$$\leq \frac{1}{720} \sqrt{\frac{5}{462}} \left\{ \sum_{j=1}^{n} \left[(b_j - a_j)^5 \left\| \frac{\partial^5 f}{\partial x_j^5}(\ldots, a_{j+1}, \ldots, a_n) \right\|_{\infty, \prod\limits_{i=1}^{j}[a_i,b_i]} \right] \right\}.$$

$$(4.120)$$

ii) *Suppose*

$$\frac{\partial^5 f}{\partial x_j^5}(\ldots, a_{j+1}, \ldots, a_n) \in L_2 \left(\prod_{i=1}^{j}[a_i, b_i] \right),$$

all $j = 1, \ldots, n$. Then

$$K_5 \leq \frac{1}{720} \sqrt{\frac{5}{462}} \left\{ \sum_{j=1}^{n} \left[(b_j - a_j)^{9/2} \left(\prod_{i=1}^{j-1}(b_i - a_i) \right)^{-1/2} \right. \right.$$

$$\left. \left. \times \left\| \frac{\partial^5 f}{\partial x_j^5}(\ldots, a_{j+1}, \ldots, a_n) \right\|_{2, \prod\limits_{i=1}^{j}[a_i,b_i]} \right] \right\}.$$

$$(4.121)$$

Corollary 4.51. *Suppose Assumptions 4.10 or 4.17 or 4.20, case $m = 5$, all $x_j = b_j$, $j = 1, \ldots, n$.*
   i) *Assume*

$$\frac{\partial^5 f}{\partial x_j^5}(\ldots, b_{j+1}, \ldots, b_n) \in L_\infty \left(\prod_{i=1}^{j}[a_i, b_i] \right),$$

all $j = 1, \ldots, n$. Then

$$\Lambda_5 := \left| \left(\frac{f(b_1, \ldots, b_n) + f(a_1, b_2, \ldots, b_n)}{2} \right) \right.$$

$$- \frac{1}{\prod\limits_{i=1}^{n} (b_i - a_i)} \int_{\prod\limits_{i=1}^{n} [a_i, b_i]} f(s_1, \ldots, s_n) ds_1 \cdots ds_n$$

$$- \sum_{\substack{j=1 \\ (\text{with } j=k\neq 1)}}^{n} \left\{ \frac{1}{\left(\prod\limits_{i=1}^{j-1} (b_i - a_i) \right)} \left\{ \sum_{k \in \{1,2,4\}} \frac{(b_j - a_j)^{k-1}}{k!} \right. \right.$$

$$\times B_k(1) \left(\int_{\prod\limits_{i=1}^{j-1} [a_i, b_i]} \left(\frac{\partial^{k-1} f}{\partial x_j^{k-1}}(s_1, \ldots, s_{j-1}, b_j, b_{j+1}, \ldots, b_n) \right.\right.$$

$$\left.\left.\left.\left. - \frac{\partial^{k-1} f}{\partial x_j^{k-1}}(s_1, \ldots, s_{j-1}, a_j, b_{j+1}, \ldots, b_n) \right) ds_1 \cdots ds_{j-1} \right) \right\} \right\} \right|$$

$$\leq \frac{1}{720} \sqrt{\frac{5}{462}} \left\{ \sum_{j=1}^{n} \left[(b_j - a_j)^5 \left\| \frac{\partial^5 f}{\partial x_j^5}(\ldots, b_{j+1}, \ldots, b_n) \right\|_{\infty, \prod\limits_{i=1}^{j} [a_i, b_i]} \right] \right\}.$$

$$(4.122)$$

ii) *Suppose*

$$\frac{\partial^5 f}{\partial x_j^5}(\ldots, b_{j+1}, \ldots, b_n) \in L_2 \left(\prod_{i=1}^{j} [a_i, b_i] \right),$$

all $j = 1, \ldots, n$. Then

$$\Lambda_5 \leq \frac{1}{720} \sqrt{\frac{5}{462}} \left\{ \sum_{j=1}^{n} \left[(b_j - a_j)^{9/2} \left(\prod_{i=1}^{j-1} (b_i - a_i) \right)^{-1/2} \right.\right.$$

$$\left.\left. \times \left\| \frac{\partial^5 f}{\partial x_j^5}(\ldots, b_{j+1}, \ldots, b_n) \right\|_{2, \prod\limits_{i=1}^{j} [a_i, b_i]} \right] \right\}.$$

$$(4.123)$$

Corollary 4.52. *Suppose Assumptions 4.10 or 4.17 or 4.20, case $m = 5$, all* $x_j = \frac{a_j + b_j}{2}$, $j = 1, \ldots, n$.

i) *Assume*

$$\frac{\partial^5 f}{\partial x_j^5} \left(\ldots, \frac{a_{j+1} + b_{j+1}}{2}, \ldots, \frac{a_n + b_n}{2} \right) \in L_\infty \left(\prod_{i=1}^{j} [a_i, b_i] \right),$$

all $j = 1, \ldots, n.$ *Then*

$$M_5 := \left| f\left(\frac{a_1 + b_1}{2}, \ldots, \frac{a_n + b_n}{2}\right) \right.$$

$$- \frac{1}{\prod\limits_{i=1}^{n}(b_i - a_i)} \int_{\prod\limits_{i=1}^{n}[a_i,b_i]} f(s_1, \ldots, s_n)ds_1 \cdots ds_n$$

$$- \sum_{j=1}^{n}\left\{\frac{1}{\left(\prod\limits_{i=1}^{j-1}(b_i - a_i)\right)}\left\{\sum_{k\in\{2,4\}}\frac{(b_j - a_j)^{k-1}}{k!}\right.\right.$$

$$\times B_k\left(\frac{1}{2}\right)\left(\int_{\prod\limits_{i=1}^{j-1}[a_i,b_i]}\left(\frac{\partial^{k-1}f}{\partial x_j^{k-1}}\left(s_1, \ldots, s_{j-1}, b_j, \frac{a_{j+1} + b_{j+1}}{2}, \ldots, \frac{a_n + b_n}{2}\right)\right.\right.$$

$$\left.\left.\left.\left.- \frac{\partial^{k-1}f}{\partial x_j^{k-1}}\left(s_1, \ldots, s_{j-1}, a_j, \frac{a_{j+1} + b_{j+1}}{2}, \ldots, \frac{a_n + b_n}{2}\right)\right)ds_1 \cdots ds_{j-1}\right)\right\}\right\}\right|$$

$$\leq \frac{1}{720}\sqrt{\frac{5}{462}}\left\{\sum_{j=1}^{n}\left[(b_j - a_j)^5\left\|\frac{\partial^5 f}{\partial x_j^5}\left(\ldots,\right.\right.\right.\right.$$

$$\left.\left.\left.\left.\frac{a_{j+1} + b_{j+1}}{2}, \ldots, \frac{a_n + b_n}{2}\right)\right\|_{\infty,\prod\limits_{i=1}^{j}[a_i,b_i]}\right]\right\}. \tag{4.124}$$

ii) *Assume*

$$\frac{\partial^5 f}{\partial x_j^5}\left(\ldots, \frac{a_{j+1} + b_{j+1}}{2}, \ldots, \frac{a_n + b_n}{2}\right) \in L_2\left(\prod_{i=1}^{j}[a_i, b_i]\right),$$

all $j = 1, \ldots, n.$ *Then*

$$M_5 \leq \frac{1}{720}\sqrt{\frac{5}{462}}\left\{\sum_{j=1}^{n}\left[(b_j - a_j)^{9/2}\left(\prod_{i=1}^{j-1}(b_i - a_i)\right)^{-1/2}\right.\right.$$

$$\left.\left.\times\left\|\frac{\partial^5 f}{\partial x_j^5}\left(\ldots, \frac{a_{j+1} + b_{j+1}}{2}, \ldots, \frac{a_n + b_n}{2}\right)\right\|_{2,\prod\limits_{i=1}^{j}[a_i,b_i]}\right]\right\}. \tag{4.125}$$

Finally we present trapezoid and midpoint rules inequalities for $m = 6$.

Corollary 4.53. *Suppose Assumptions 4.10 or 4.17 or 4.20, case* $m = 6$, *all* $x_j = a_j,\ j = 1, \ldots, n.$
  i) *Assume*

$$\frac{\partial^6 f}{\partial x_j^6}(\ldots, a_{j+1}, \ldots, a_n) \in L_\infty\left(\prod_{i=1}^{j}[a_i, b_i]\right),$$

all $j = 1, \ldots, n$. Then

$$K_5 = \left| \left(\frac{f(a_1, \ldots, a_n) + f(b_1, a_2, \ldots, a_n)}{2} \right) \right.$$

$$- \frac{1}{\prod_{i=1}^{n} (b_i - a_i)} \int_{\prod_{i=1}^{n} [a_i, b_i]} f(s_1, \ldots, s_n) ds_1 \cdots ds_n$$

$$- \sum_{\substack{j=1 \\ (\text{with } j=k \neq 1)}}^{n} \left\{ \frac{1}{\left(\prod_{i=1}^{j-1} (b_i - a_i) \right)} \left\{ \sum_{k \in \{1,2,4\}} \frac{(b_j - a_j)^{k-1}}{k!} \right. \right.$$

$$\times B_k(0) \left(\int_{\prod_{i=1}^{j-1} [a_i, b_i]} \left(\frac{\partial^{k-1} f}{\partial x_j^{k-1}} (s_1, \ldots, s_{j-1}, b_j, a_{j+1}, \ldots, a_n) \right. \right.$$

$$\left. \left. \left. \left. - \frac{\partial^{k-1} f}{\partial x_j^{k-1}} (s_1, \ldots, s_{j-1}, a_j, a_{j+1}, \ldots, a_n) \right) ds_1 \cdots ds_{j-1} \right) \right\} \right\} \right|$$

$$\leq \frac{1}{1440} \sqrt{\frac{101}{30030}} \left\{ \sum_{j=1}^{n} \left[(b_j - a_j)^6 \left\| \frac{\partial^6 f}{\partial x_j^6} (\ldots, a_{j+1}, \ldots, a_n) \right\|_{\infty, \prod_{i=1}^{j} [a_i, b_i]} \right] \right\}.$$

$$(4.126)$$

ii) *Suppose*

$$\frac{\partial^6 f}{\partial x_j^6} (\ldots, a_{j+1}, \ldots, a_n) \in L_2 \left(\prod_{i=1}^{j} [a_i, b_i] \right),$$

all $j = 1, \ldots, n$. Then

$$K_5 \leq \frac{1}{1440} \sqrt{\frac{101}{30030}} \left\{ \sum_{j=1}^{n} \left[(b_j - a_j)^{11/2} \left(\prod_{i=1}^{j-1} (b_i - a_i) \right)^{-1/2} \right. \right.$$

$$\left. \left. \times \left\| \frac{\partial^6 f}{\partial x_j^6} (\ldots, a_{j+1}, \ldots, a_n) \right\|_{2, \prod_{i=1}^{j} [a_i, b_i]} \right] \right\}. \qquad (4.127)$$

Corollary 4.54. *Suppose Assumptions 4.10 or 4.17 or 4.20, case $m = 6$, all $x_j = b_j$, $j = 1, \ldots, n$.*

     i) *Assume*

$$\frac{\partial^6 f}{\partial x_j^6} (\ldots, b_{j+1}, \ldots, b_n) \in L_\infty \left(\prod_{i=1}^{j} [a_i, b_i] \right),$$

all $j = 1, \ldots, n$. Then

$$\Lambda_5 = \left| \left(\frac{f(b_1, \ldots, b_n) + f(a_1, b_2, \ldots, b_n)}{2} \right) \right.$$

$$- \frac{1}{\prod\limits_{i=1}^{n}(b_i - a_i)} \int_{\prod\limits_{i=1}^{n}[a_i,b_i]} f(s_1, \ldots, s_n)ds_1 \cdots ds_n$$

$$- \sum_{\substack{j=1 \\ (\text{with } j=k \neq 1)}} \left\{ \frac{1}{\left(\prod\limits_{i=1}^{j-1}(b_i - a_i) \right)} \left\{ \sum_{k \in \{1,2,4\}} \frac{(b_j - a_j)^{k-1}}{k!} \right. \right.$$

$$\times B_k(1) \left(\int_{\prod\limits_{i=1}^{j-1}[a_i,b_i]} \left(\frac{\partial^{k-1} f}{\partial x_j^{k-1}}(s_1, \ldots, s_{j-1}, b_j, b_{j+1}, \ldots, b_n) \right. \right.$$

$$\left. \left. \left. \left. - \frac{\partial^{k-1} f}{\partial x_j^{k-1}}(s_1, \ldots, s_{j-1}, a_j, b_{j+1}, \ldots, b_n) \right) ds_1 \cdots ds_{j-1} \right) \right\} \right\} \right|$$

$$\leq \frac{1}{1440} \sqrt{\frac{101}{30030}} \left\{ \sum_{j=1}^{n} \left[(b_j - a_j)^6 \left\| \frac{\partial^6 f}{\partial x_j^6}(\ldots, b_{j+1}, \ldots, b_n) \right\|_{\infty, \prod\limits_{i=1}^{j}[a_i,b_i]} \right] \right\}.$$

$$(4.128)$$

ii) *Assume*

$$\frac{\partial^6 f}{\partial x_j^6}(\ldots, b_{j+1}, \ldots, b_n) \in L_2 \left(\prod_{i=1}^{j}[a_i, b_i] \right),$$

all $j = 1, \ldots, n.$ *Then*

$$\Lambda_5 \leq \frac{1}{1440} \sqrt{\frac{101}{30030}} \left\{ \sum_{j=1}^{n} \left[(b_j - a_j)^{11/2} \left(\prod_{i=1}^{j-1}(b_i - a_i) \right)^{-1/2} \right. \right.$$

$$\left. \left. \times \left\| \frac{\partial^6 f}{\partial x_j^6}(\ldots, b_{j+1}, \ldots, b_n) \right\|_{2, \prod\limits_{i=1}^{j}[a_i,b_i]} \right] \right\}.$$

$$(4.129)$$

Corollary 4.55. *Suppose Assumptions 4.10 or 4.17 or 4.20, case $m = 6$, all* $x_j = \frac{a_j + b_j}{2}$, $j = 1, \ldots, n.$

   i) *Assume*

$$\frac{\partial^6 f}{\partial x_j^6}\left(\ldots, \frac{a_{j+1} + b_{j+1}}{2}, \ldots, \frac{a_n + b_n}{2} \right) \in L_\infty \left(\prod_{i=1}^{j}[a_i, b_i] \right),$$

all $j = 1, \ldots, n$. Then

$$M_5 = \left| f\left(\frac{a_1 + b_1}{2}, \ldots, \frac{a_n + b_n}{2}\right)\right.$$

$$- \frac{1}{\prod\limits_{i=1}^{n}(b_i - a_i)} \int_{\prod\limits_{i=1}^{n}[a_i,b_i]} f(s_1, \ldots, s_n)ds_1 \cdots ds_n$$

$$- \sum_{j=1}^{n}\left\{\frac{1}{\left(\prod\limits_{i=1}^{j-1}(b_i - a_i)\right)}\left\{\sum_{k\in\{2,4\}} \frac{(b_j - a_j)^{k-1}}{k!}\right.\right.$$

$$\times B_k\left(\frac{1}{2}\right)\left(\int_{\prod\limits_{i=1}^{j-1}[a_i,b_i]}\left(\frac{\partial^{k-1}f}{\partial x_j^{k-1}}\left(s_1, \ldots, s_{j-1}, b_j, \frac{a_{j+1} + b_{j+1}}{2}, \ldots, \frac{a_n + b_n}{2}\right)\right.\right.$$

$$\left.\left.\left.\left. - \frac{\partial^{k-1}f}{\partial x_j^{k-1}}\left(s_1, \ldots, s_{j-1}, a_j, \frac{a_{j+1} + b_{j+1}}{2}, \ldots, \frac{a_n + b_n}{2}\right)\right)ds_1 \cdots ds_{j-1}\right)\right\}\right\}\right|$$

$$\leq \frac{1}{46080}\sqrt{\frac{7081}{2145}}\left\{\sum_{j=1}^{n}\left[(b_j - a_j)^6\right.\right.$$

$$\left.\left.\times \left\|\frac{\partial^6 f}{\partial x_j^6}\left(\ldots, \frac{a_{j+1} + b_{j+1}}{2}, \ldots, \frac{a_n + b_n}{2}\right)\right\|_{\infty, \prod\limits_{i=1}^{j}[a_i,b_i]}\right]\right\}. \tag{4.130}$$

ii) *Suppose*

$$\frac{\partial^6 f}{\partial x_j^6}\left(\ldots, \frac{a_{j+1} + b_{j+1}}{2}, \ldots, \frac{a_n + b_n}{2}\right) \in L_2\left(\prod_{i=1}^{j}[a_i, b_i]\right),$$

all $j = 1, \ldots, n$. Then

$$M_5 \leq \frac{1}{46080}\sqrt{\frac{7081}{2145}}\left\{\sum_{j=1}^{n}\left[(b_j - a_j)^{11/2}\left(\prod_{i=1}^{j-1}(b_i - a_i)\right)^{-1/2}\right.\right.$$

$$\left.\left.\times \left\|\frac{\partial^6 f}{\partial x_j^6}\left(\ldots, \frac{a_{j+1} + b_{j+1}}{2}, \ldots, \frac{a_n + b_n}{2}\right)\right\|_{2, \prod\limits_{i=1}^{j}[a_i,b_i]}\right]\right\}. \tag{4.131}$$

At the end we give a simplified special case of Theorems 4.32 and 4.33.

Corollary 4.56. *Suppose Assumptions 4.10 or 4.17 or 4.20. We further assume that*

$$\frac{\partial^\ell f}{\partial x_j^\ell}(\ldots, b_j, \ldots) = \frac{\partial^\ell f}{\partial x_j^\ell}(\ldots, a_j, \ldots), \tag{4.132}$$

for all $j = 1, \ldots, n$ and all $\ell = 0, 1, \ldots, m - 2$. Here $m, n \in \mathbb{N}$, $x_i \in [a_i, b_i]$, $i = 1, 2, \ldots, n$.

i) *In particular we assume that*

$$\frac{\partial^m f}{\partial x_j^m}(\ldots, x_{j+1}, \ldots, x_n) \in L_\infty \left(\prod_{i=1}^{j} [a_i, b_i] \right),$$

for any $(x_{j+1}, \ldots, x_n) \in \prod_{i=j+1}^{n} [a_i, b_i]$, *all* $j = 1, \ldots, n$. *Then*

$$\left| f(x_1, \ldots, x_n) - \frac{1}{\prod_{i=1}^{n}(b_i - a_i)} \int_{\prod_{i=1}^{n} [a_i, b_i]} f(s_1, \ldots, s_n) ds_1 \cdots ds_n \right|$$

$$\leq \frac{1}{m!} \left\{ \sum_{j=1}^{n} \left[(b_j - a_j)^m \left(\sqrt{\frac{(m!)^2}{(2m)!}|B_{2m}| + B_m^2\left(\frac{x_j - a_j}{b_j - a_j}\right)} \right) \right. \right.$$

$$\left. \left. \times \left\| \frac{\partial^m f}{\partial x_j^m}(\ldots, x_{j+1}, \ldots, x_n) \right\|_{\infty, \prod_{i=1}^{j} [a_i, b_i]} \right] \right\}, \tag{4.133}$$

true $\forall (x_1, \ldots, x_n) \in \prod_{i=1}^{n} [a_i, b_i]$.

ii) *Suppose*

$$\frac{\partial^m f}{\partial x_j^m}(\ldots, x_{j+1}, \ldots, x_n) \in L_2 \left(\prod_{i=1}^{j} [a_i, b_i] \right),$$

for any $(x_{j+1}, \ldots, x_n) \in \prod_{i=j+1}^{n} [a_i, b_i]$, *for all* $j = 1, \ldots, n$. *Then*

$$\left| f(x_1, \ldots, x_n) - \frac{1}{\prod_{i=1}^{n}(b_i - a_i)} \int_{\prod_{i=1}^{n} [a_i, b_i]} f(s_1, \ldots, s_n) ds_1 \cdots ds_n \right|$$

$$\leq \frac{1}{m!} \left\{ \sum_{j=1}^{n} \left[(b_j - a_j)^{m-\frac{1}{2}} \left(\prod_{i=1}^{j-1}(b_i - a_i) \right)^{-1/2} \left(\sqrt{\frac{(m!)^2}{(2m)!}|B_{2m}| + B_m^2\left(\frac{x_j - a_j}{b_j - a_j}\right)} \right) \right. \right.$$

$$\left. \left. \times \left\| \frac{\partial^m f}{\partial x_j^m}(\ldots, x_{j+1}, \ldots, x_n) \right\|_{2, \prod_{i=1}^{j} [a_i, b_i]} \right] \right\}, \tag{4.134}$$

true $\forall (x_1, \ldots, x_n) \in \prod_{i=1}^{n} [a_i, b_i]$.

Proof.   Clearly here $A_j = 0$, $j = 1, \ldots, n$. Then proof is obvious. $\square$

Similarly as in Corollary 4.56 we obtain

Corollary 4.57. *Suppose Assumptions 4.10 or 4.17 or 4.20, case* $m = 6$, *all* $x_j = a_j$, $j = 1, \ldots, n$. *Also assume*

$$\frac{\partial^\ell f}{\partial x_j^\ell}(\ldots, b_j, a_{j+1}, \ldots, a_n) = \frac{\partial^\ell f}{\partial x_j^\ell}(\ldots, a_j, a_{j+1}, \ldots, a_n), \tag{4.135}$$

for all $j = 1, \ldots, n$, and all $\ell = 0, 1, 3$.

    i) *Assume*

$$\frac{\partial^6 f}{\partial x_j^6}(\ldots, a_{j+1}, \ldots, a_n) \in L_\infty \left(\prod_{i=1}^{j} [a_i, b_i] \right), \quad \text{all } j = 1, \ldots, n.$$

Then

$$\left| f(a_1, \ldots, a_n) - \frac{1}{\prod_{i=1}^{n}(b_i - a_i)} \int_{\prod_{i=1}^{n}[a_i, b_i]} f(s_1, \ldots, s_n) ds_1 \cdots ds_n \right|$$

$$\leq \frac{1}{1440} \sqrt{\frac{101}{30030}} \left\{ \sum_{j=1}^{n} \left[(b_j - a_j)^6 \left\| \frac{\partial^6 f}{\partial x_j^6}(\ldots, a_{j+1}, \ldots, a_n) \right\|_{\infty, \prod_{i=1}^{j}[a_i, b_i]} \right] \right\}.$$

$$\tag{4.136}$$

    ii) *Suppose*

$$\frac{\partial^6 f}{\partial x_j^6}(\ldots, a_{j+1}, \ldots, a_n) \in L_2 \left(\prod_{i=1}^{j} [a_i, b_i] \right), \quad \text{all } j = 1, \ldots, n.$$

Then

$$\left| f(a_1, \ldots, a_n) - \frac{1}{\prod_{i=1}^{n}(b_i - a_i)} \int_{\prod_{i=1}^{n}[a_i, b_i]} f(s_1, \ldots, s_n) ds_1 \cdots ds_n \right|$$

$$\leq \frac{1}{1440} \sqrt{\frac{101}{30030}} \left\{ \sum_{j=1}^{n} \left[(b_j - a_j)^{11/2} \left(\prod_{i=1}^{j-1}(b_i - a_i) \right)^{-1/2} \right. \right.$$

$$\left. \left. \times \left\| \frac{\partial^6 f}{\partial x_j^6}(\ldots, a_{j+1}, \ldots, a_n) \right\|_{2, \prod_{i=1}^{j}[a_i, b_i]} \right] \right\}. \tag{4.137}$$

Proof.   By Corollary 4.53.   □

Corollary 4.58. *Suppose Assumptions 4.10 or 4.17 or 4.20, case $m = 6$, all $x_j = b_j$, $j = 1, \ldots, n$. Also assume*

$$\frac{\partial^\ell f}{\partial x_j^\ell}(\ldots, b_j, b_{j+1}, \ldots, b_n) = \frac{\partial^\ell f}{\partial x_j^\ell}(\ldots, a_j, b_{j+1}, \ldots, b_n), \tag{4.138}$$

for all $j = 1, \ldots, n$ and all $\ell = 0, 1, 3$.

    i) *Assume*

$$\frac{\partial^6 f}{\partial x_j^6}(\ldots, b_{j+1}, \ldots, b_n) \in L_\infty \left(\prod_{i=1}^{j} [a_i, b_i] \right), \quad \text{all } j = 1, \ldots, n.$$

Then

$$\left| f(b_1,\ldots,b_n) - \frac{1}{\prod\limits_{i=1}^{n}(b_i - a_i)} \int_{\prod\limits_{i=1}^{n}[a_i,b_i]} f(s_1,\ldots,s_n)ds_1\cdots ds_n \right|$$

$$\leq \frac{1}{1440}\sqrt{\frac{101}{30030}}\left\{\sum_{j=1}^{n}\left[(b_j - a_j)^6\left\|\frac{\partial^6 f}{\partial x_j^6}(\ldots,b_{j+1},\ldots,b_n)\right\|_{\infty,\prod\limits_{i=1}^{j}[a_i,b_i]}\right]\right\}.$$

$$(4.139)$$

ii) *Suppose*

$$\frac{\partial^6 f}{\partial x_j^6}(\ldots,b_{j+1},\ldots,b_n) \in L_2\left(\prod_{i=1}^{j}[a_i,b_i]\right), \quad all\ j = 1,\ldots,n.$$

Then

$$\left| f(b_1,\ldots,b_n) - \frac{1}{\prod\limits_{i=1}^{n}(b_i - a_i)} \int_{\prod\limits_{i=1}^{n}[a_i,b_i]} f(s_1,\ldots,s_n)ds_1\cdots ds_n \right|$$

$$\leq \frac{1}{1440}\sqrt{\frac{101}{30030}}\left\{\sum_{j=1}^{n}\left[(b_j - a_j)^{11/2}\left(\prod_{i=1}^{j-1}(b_i - a_i)\right)^{-1/2}\right.\right.$$

$$\left.\left.\times \left\|\frac{\partial^6 f}{\partial x_j^6}(\ldots,b_{j+1},\ldots,b_n)\right\|_{2,\prod\limits_{i=1}^{j}[a_i,b_i]}\right]\right\}. \qquad (4.140)$$

Proof.   By Corollary 4.54. $\square$

Corollary 4.59.   *Suppose Assumptions 4.10 or 4.17 or 4.20, case $m = 6$, all* $x_j = \frac{a_j+b_j}{2}$, $j = 1,\ldots,n$. *Also assume*

$$\frac{\partial^\ell f}{\partial x_j^\ell}\left(\ldots,b_j,\frac{a_{j+1}+b_{j+1}}{2},\ldots,\frac{a_n+b_n}{2}\right) = \frac{\partial^\ell f}{\partial x_j^\ell}\left(\ldots,a_j,\frac{a_{j+1}+b_{j+1}}{2},\ldots,\frac{a_n+b_n}{2}\right),$$

$$(4.141)$$

for all $j = 1,\ldots,n$ *and* $\ell = 1,3$.

   i) *Assume*

$$\frac{\partial^6 f}{\partial x_j^6}\left(\ldots,\frac{a_{j+1}+b_{j+1}}{2},\ldots,\frac{a_n+b_n}{2}\right) \in L_\infty\left(\prod_{i=1}^{j}[a_i,b_i]\right),$$

all $j = 1, \ldots, n$. Then

$$\left| f\left(\frac{a_1 + b_1}{2}, \ldots, \frac{a_n + b_n}{2} \right) - \frac{1}{\prod\limits_{i=1}^{n} (b_i - a_i)} \int_{\prod\limits_{i=1}^{n} [a_i, b_i]} f(s_1, \ldots, s_n) ds_1 \cdots ds_n \right|$$

$$\leq \frac{1}{46080} \sqrt{\frac{7081}{2145}} \left\{ \sum_{j=1}^{n} \left[(b_j - a_j)^6 \right.\right.$$

$$\left.\left. \times \left\| \frac{\partial^6 f}{\partial x_j^6} \left(\ldots, \frac{a_{j+1} + b_{j+1}}{2}, \ldots, \frac{a_n + b_n}{2} \right) \right\|_{\infty, \prod\limits_{i=1}^{j} [a_i, b_i]} \right] \right\}. \tag{4.142}$$

ii) *Suppose*

$$\frac{\partial^6 f}{\partial x_j^6} \left(\ldots, \frac{a_{j+1} + b_{j+1}}{2}, \ldots, \frac{a_n + b_n}{2} \right) \in L_2 \left(\prod_{i=1}^{j} [a_i, b_i] \right),$$

all $j = 1, \ldots, n$. Then

$$\left| f\left(\frac{a_1 + b_1}{2}, \ldots, \frac{a_n + b_n}{2} \right) - \frac{1}{\prod\limits_{i=1}^{n} (b_i - a_i)} \int_{\prod\limits_{i=1}^{n} [a_i, b_i]} f(s_1, \ldots, s_n) ds_1 \cdots ds_n \right|$$

$$\leq \frac{1}{46080} \sqrt{\frac{7081}{2145}} \left\{ \sum_{j=1}^{n} \left[(b_j - a_j)^{11/2} \left(\prod_{i=1}^{j-1} (b_i - a_i) \right)^{-1/2} \right.\right.$$

$$\left.\left. \times \left\| \frac{\partial^6 f}{\partial x_j^6} \left(\ldots, \frac{a_{j+1} + b_{j+1}}{2}, \ldots, \frac{a_j + b_n}{2} \right) \right\|_{2, \prod\limits_{i=1}^{j} [a_i, b_i]} \right] \right\}. \tag{4.143}$$

Proof.   By Corollary 4.55. $\square$

One can apply similar conditions to (4.132) for the cases of $m = 2, 3, 4, 5$ and simplify a lot the results of Corollaries 4.41, 4.42 and of Corollaries 4.44 – 4.52, exactly as we did in Corollaries 4.56–4.59 for general $m \in \mathbb{N}$ and $m = 6$, etc.

4.5   Sharpness

We need to include

Theorem 4.60. *Suppose Assumptions 4.10 or 4.17 or 4.20. Let $E_m^f(x_1, \ldots, x_n)$ as in (4.44), $m \in \mathbb{N}$. In particular we assume that*

$$\frac{\partial^m f}{\partial x_j^m} \left(\overbrace{\cdots}^{j}, x_{j+1}, \ldots, x_n \right) \in L_\infty \left(\prod_{i=1}^{j} [a_i, b_i] \right),$$

for any $(x_{j+1}, \ldots, x_n) \in \prod_{i=j+1}^{n} [a_i, b_i]$, *all* $j = 1, \ldots, n$. *Then*

$$|E_m^f(x_1, \ldots, x_n)| \leq \frac{1}{m!} \sum_{j=1}^{n} \left[(b_j - a_j)^m \left(\int_0^1 \left| B_m \left(\frac{x_j - a_j}{b_j - a_j} \right) - B_m(t_j) \right| dt_j \right) \right.$$
$$\left. \times \left\| \frac{\partial^m f}{\partial x_j^m} (\ldots, x_{j+1}, \ldots, x_n) \right\|_{\infty, \prod_{i=1}^{j} [a_i, b_i]} \right]. \tag{4.144}$$

Proof.   By Remark 4.23, see (4.55). $\square$

We give the important

Corollary 4.61. *Suppose Assumptions 4.10 or 4.17 or 4.20. Let* $E_m^f(x_1, \ldots, x_n)$ *as in (4.44),* $m \in \mathbb{N}$. *In particular we assume that*

$$\frac{\partial^m f}{\partial x_j^m} \left(\overset{j}{\overbrace{\ldots}}, x_{j+1}, \ldots, x_n \right) \in L_\infty \left(\prod_{i=1}^{j} [a_i, b_i] \right),$$

for any $(x_{j+1}, \ldots, x_n) \in \prod_{i=j+1}^{n} [a_i, b_i]$, *all* $j = 1, \ldots, n$. *And also suppose* $\frac{\partial^m f}{\partial x_j^m} \in L_\infty \left(\prod_{i=1}^{n} [a_i, b_i] \right)$, $j = 1, \ldots, n-1$. *Call*

$$D_m(f) := \max_{1 \leq j \leq n} \left\{ \left\| \frac{\partial^m f}{\partial x_j^m} \right\|_\infty \right\}. \tag{4.145}$$

Then

$$|E_m^f(x_1, \ldots, x_n)| \leq \frac{D_m(f)}{m!} \sum_{j=1}^{n} \left[(b_j - a_j)^m \left(\int_0^1 \left| B_m \left(\frac{x_j - a_j}{b_j - a_j} \right) - B_m(t_j) \right| dt_j \right) \right]. \tag{4.146}$$

Comment 4.62. We observe that (see also [98])

$$I_1(\lambda_j) := \int_0^1 |B_1(\lambda_j) - B_1(t_j)| dt_j = \frac{1}{4} + \frac{\left(x_j - \left(\frac{a_j + b_j}{2} \right) \right)^2}{(b_j - a_j)^2}, \tag{4.147}$$

where $\lambda_j = \frac{x_j - a_j}{b_j - a_j}$, $j = 1, \ldots, n$. Notice that

$$\max_{\lambda_j \in [0,1]} I_1(\lambda_j) = I_1(0) = I_1(1) = \frac{1}{2}, \tag{4.148}$$

i.e. when $x_j = a_j$ or b_j.

Thus we have

Corollary 4.63. *All here assumed as in Corollary 4.61 when* $m = 1$. *Then*

$$\left| f(x_1, \ldots, x_n) - \frac{1}{\prod_{i=1}^{n} (b_i - a_i)} \int_{\prod_{i=1}^{n} [a_i, b_i]} f(s_1, \ldots, s_n) ds_1 \cdots ds_n \right|$$

$$\leq \frac{D_1(f)}{2}\left(\sum_{j=1}^{n}(b_j - a_j)\right). \tag{4.149}$$

Inequality (4.149) is sharp, that is attained by $f_1(s_1,\ldots,s_n) := \sum_{j=1}^{n}(s_j - a_j)$ *when*

$x_j = a_j$, $j = 1,\ldots,n$, *and by* $f_2(s_1,\ldots,s_n) := \sum_{j=1}^{n}(b_j - s_j)$ *when* $x_j = b_j$, $j = 1,\ldots,n$.

Proof.   i) Case of $x_j = a_j$, $j = 1,\ldots,n$. Then $\frac{\partial f_1}{\partial x_j} = 1$, $j = 1,\ldots,n$, i.e. $\left\|\frac{\partial f_1}{\partial x_j}\right\|_\infty = 1$ and $D_1(f_1) = 1$. Clearly then we have

$$\text{L.H.S.}(4.149) = \text{R.H.S.}(4.149) = \frac{1}{2}\sum_{j=1}^{n}(b_j - a_j),$$

proving sharpness.

ii) Case of $x_j = b_j$, $j = 1,\ldots,n$. Then $\frac{\partial f_2}{\partial x_j} = -1$, $j = 1,\ldots,n$, i.e. $\left\|\frac{\partial f_2}{\partial x_j}\right\|_\infty = 1$ and $D_1(f_2) = 1$. Clearly we have

$$\text{L.H.S.}(4.149) = \text{R.H.S.}(4.149) = \frac{1}{2}\sum_{j=1}^{n}(b_j - a_j),$$

proving again sharpness. $\square$

Comment 4.64. We observe that ([98])

$$I_2(\lambda_j) := \int_0^1 |B_2(\lambda_j) - B_2(t_j)|dt_j = \frac{8}{3}\delta_j^3(x) - \delta_j^2(x) + \frac{1}{12}, \quad j = 1,\ldots,n, \tag{4.150}$$

where

$$\delta_j(x_j) := \frac{\left|x_j - \frac{a_j+b_j}{2}\right|}{b_j - a_j}, \quad x_j \in [a_j, b_j].$$

Also from [98] we have that

$$\max_{0\leq\lambda_j\leq1} I_2(\lambda_j) = I_2(0) = I_2(1) = \frac{1}{6}, \tag{4.151}$$

i.e. when $x_j = a_j$ or b_j.

We continue with

Corollary 4.65. *All here assumed as in Corollary 4.61 when $m = 2$. Then*

$$|E_2^f(x_1,\ldots,x_n)| \leq \frac{D_2(f)}{12}\sum_{j=1}^{n}(b_j - a_j)^2. \tag{4.152}$$

Inequality (4.152) is sharp, that is attained by $f_1(s_1,\ldots,s_n) := \sum_{j=1}^{n}(s_j - a_j)^2$ *when*

$x_j = a_j$, $j = 1,\ldots,n$ *and by* $f_2(s_1,\ldots,s_n) := \sum_{j=1}^{n}(s_j - b_j)^2$ *when* $x_j = b_j$, $j = 1,\ldots,n$.

Proof.   i) Case of $x_j = a_j$, $j = 1, \ldots, n$. Then $\frac{\partial f_1}{\partial x_j} = 2(s_j - a_j)$, $\frac{\partial^2 f_1}{\partial x_j^2} = 2$, and $\left\| \frac{\partial^2 f_1}{\partial x_j^2} \right\|_\infty = 2$, with $D_2(f) = 2$. Clearly then we have

$$\text{L.H.S.}(4.152) = \text{R.H.S.}(4.152) = \frac{1}{6} \sum_{j=1}^{n} (b_i - a_j)^2,$$

proving sharpness.

ii) Case of $x_j = b_j$, $j = 1, \ldots, n$. Then $\frac{\partial f_2}{\partial x_j} = 2(s_j - b_j)$, $\frac{\partial^2 f_2}{\partial x_j^2} = 2$, and $\left\| \frac{\partial^2 f_2}{\partial x_j^2} \right\|_\infty = 2$, with $D_2(f) = 2$. Clearly again we have

$$\text{L.H.S.}(4.152) = \text{R.H.S.}(4.152) = \frac{1}{6} \sum_{j=1}^{n} (b_i - a_j)^2,$$

proving again sharpness. $\square$

Comment 4.66. By [10] we have that

$$\max_{0 \leq \lambda_j \leq 1} I_3(\lambda_j) = I_3\left(\frac{3 - \sqrt{3}}{6} \right) = I_3\left(\frac{3 + \sqrt{3}}{6} \right) = \frac{\sqrt{3}}{36}, \tag{4.153}$$

where

$$I_3(\lambda_j) := \int_0^1 |B_3(\lambda_j) - B_3(t_j)| dt_j, \quad j = 1, \ldots, n. \tag{4.154}$$

Consequently we have

Corollary 4.67. *All here assumed as in Corollary 4.61 when* $m = 3$. *Then*

$$|E_3^f(x_1, \ldots, x_n)| \leq \frac{\sqrt{3} D_3(f)}{216} \sum_{j=1}^{n} (b_j - a_j)^3. \tag{4.155}$$

Comment 4.68. We call

$$I_m(\lambda_j) := \int_0^1 |B_m(\lambda_j) - B_m(t_j)| dt_j, \tag{4.156}$$

where $\lambda_j := \frac{x_j - a_j}{b_j - a_j}$, $j = 1, \ldots, n, m \in \mathbb{N}$. In [35] we found that

$$\max_{\lambda_j \in [0,1]} I_4(\lambda_j) = I_4(0) = I_4(1) = \frac{1}{30}. \tag{4.157}$$

So we give

Corollary 4.69. *All here supposed as in Corollary 4.61 when* $m = 4$. *Then*

$$|E_4^f(x_1, \ldots, x_n)| \leq \frac{D_4(f)}{720} \left(\sum_{j=1}^{n} (b_j - a_j)^4 \right). \tag{4.158}$$

Inequality (4.158) is sharp, that is attained by $f_1(s_1, \ldots, s_n) := \sum_{j=1}^{n} (s_j - a_j)^4$ *when*

$x_j = a_j$, $j = 1, \ldots, n$ *and by* $f_2(s_1, \ldots, s_n) := \sum_{j=1}^{n} (s_j - b_j)^4$ *when* $x_j = b_j$, $j =$

$1, \ldots, n.$

Proof.   Case of $x_j = a_j$, $j = 1, \ldots, n$. Then

$$\frac{\partial f_1}{\partial x_j} = 4(s_j - a_j)^3, \quad \frac{\partial^2 f_1}{\partial x_j^2} = 12(s_j - a_j)^2, \quad \frac{\partial^3 f_1}{\partial x_j^3} = 24(s_j - a_j), \quad \frac{\partial^4 f_1}{\partial x_j^4} = 24,$$

with $\left\| \frac{\partial^4 f_1}{\partial x_j^4} \right\|_{\infty} = 24$ and $D_4(f) = 24$. Clearly then we have

$$\text{L.H.S.}(4.158) = \text{R.H.S.}(4.158) = \frac{1}{30} \left(\sum_{j=1}^{n} (b_j - a_j)^4 \right),$$

proving sharpness.

   ii) Case $x_j = b_j$, $j = 1, \ldots, n$. Then

$$\frac{\partial f_2}{\partial x_j} = 4(s_j - b_j)^3, \quad \frac{\partial^2 f_2}{\partial x_j^2} = 12(s_j - b_j)^2, \quad \frac{\partial^3 f_2}{\partial x_j^3} = 24(s_j - b_j), \quad \frac{\partial^4 f_2}{\partial x_j^4} = 24,$$

with $D_4(f_2) = 24$. Clearly again we have

$$\text{L.H.S.}(4.158) = \text{R.H.S.}(4.158) = \frac{1}{30} \left(\sum_{j=1}^{n} (b_j - a_j)^4 \right),$$

proving again sharpness. $\square$

Comment 4.70. Inequality (4.144) is sharper than (4.77), however the integral $I_m(\lambda_j)$ (see (4.156)) in its right hand side, is difficult to compute and find its maximum value for $m \geq 5$. That is why (4.77) is more practical, also less restrictive, and we used it extensively here in the applications.

Chapter 5

More on Multidimensional Ostrowski Type Inequalities

Very general multidimensional Ostrowski type inequalities are established, some of them prove to be sharp. They involve the $\| \cdot \|_\infty$ and $\| \cdot \|_p$ norms of the engaged mixed partial of nth order $n \geq 1$. In establishing them, other important multivariate results of Montgomery type identity are developed and presented. This chapter relies on [23].

5.1 Introduction

In 1938, A. Ostrowski [196] proved the following inequality:

Theorem 5.1. *Let* $f \colon [a,b] \to \mathbb{R}$ *be continuous on* $[a,b]$ *and differentiable on* (a,b) *whose derivative* $f' \colon (a,b) \to \mathbb{R}$ *is bounded on* (a,b), *i.e.,* $\|f'\|_\infty = \sup\limits_{t \in (a,b)} |f'(t)| < +\infty$. *Then*

$$\left| \frac{1}{b-a} \int_a^b f(t)dt - f(x) \right| \leq \left[\frac{1}{4} + \frac{\left(x - \frac{a+b}{2}\right)^2}{(b-a)^2} \right] \cdot (b-a)\|f'\|_\infty, \qquad (5.1)$$

for all $x \in [a,b]$. *The constant* $\frac{1}{4}$ *is the best possible.*

Since then there has been a lot of activity around these inequalities with important applications to Numerical Analysis.

This chapter is greatly motivated by the following results:

Theorem 5.2 (see [17]). *Let* $f \in C^1\left(\prod\limits_{i=1}^{k} [a_i, b_i] \right)$, *where* $a_i < b_i$; $a_i, b_i \in \mathbb{R}$, $i = 1, \ldots, k$, *and let* $\overrightarrow{x}_0 := (x_{01}, \ldots, x_{0k}) \in \prod\limits_{i=1}^{k} [a_i, b_i]$ *be fixed. Then*

$$\left| \frac{1}{\prod\limits_{i=1}^{k}(b_i - a_i)} \int_{a_1}^{b_1} \cdots \int_{a_i}^{b_i} \cdots \int_{a_k}^{b_k} f(z_1, \ldots, z_k)dz_1 \cdots dz_k - f(\overrightarrow{x}_0) \right|$$

$$\leq \sum_{i=1}^{k} \left(\frac{(x_{0i} - a_i)^2 + (b_i - x_{0i})^2}{2(b_i - a_i)} \right) \left\| \frac{\partial f}{\partial z_i} \right\|_\infty. \qquad (5.2)$$

81

Inequality (5.2) is sharp, here the optimal function is

$$f^*(z_1, \ldots, z_k) := \sum_{i=1}^{k} |z_i - x_{0i}|^{\alpha_i}, \quad \alpha_i > 1.$$

Clearly inequality (5.2) generalizes inequality (5.1) to multidimension.

We also would like to mention

Theorem 5.3 (see [122]). *Let* $f: [a,b] \times [c,d] \to \mathbb{R}$ *be a continuous mapping on* $[a,b] \times [c,d]$, $f''_{x,y} := \frac{\partial^2 f}{\partial x \partial y}$ *exists on* $(a,b) \times (c,d)$ *and is in* $L_p((a,b) \times (c,d))$, *i.e.,*

$$\|f''_{s,t}\|_p := \left(\int_a^b \int_c^d \left| \frac{\partial^2 f(x,y)}{\partial x \partial y} \right|^p dx dy \right)^{1/p} < +\infty, \quad p > 1.$$

Then

$$\left| \int_a^b \int_c^d f(s,t) ds dt \right.$$

$$\left. - \left[(b-a) \int_c^d f(x,t) dt + (d-c) \int_a^b f(s,y) ds - (d-c)(b-a) f(x,y) \right] \right|$$

$$\leq \left[\frac{(x-a)^{q+1} + (b-x)^{q+1}}{q+1} \right]^{1/q} \cdot \left[\frac{(y-c)^{q+1} + (d-y)^{q+1}}{q+1} \right]^{1/q} \cdot \|f''_{s,t}\|_p, \quad (5.3)$$

for all $(x,y) \in [a,b] \times [c,d]$, *where* $\frac{1}{p} + \frac{1}{q} = 1$.

In this chapter we develop more Ostrowski type inequalities in the multidimension, some of them are sharp. Along the way to establishing them we produce important side related results. Among others, one of the purposes of the chapter is to generalize Theorem 5.3, as far as possible, over $\times_{i=n}^n [a_i, b_i]$, $n \in \mathbb{N}$, where we have involved $\frac{\partial^n f(x_1, \ldots, x_n)}{\partial x_1 \cdots \partial x_n}$. The main results are Ostrowski type inequalities involving the norms $\| \cdot \|_\infty$ and $\| \cdot \|_p$. Reference [62] provided us with important analytical tools.

5.2   Auxilliary Results

The following results are also by themselves of interest.

We give

Theorem 5.4. *Let* $f: [a,A] \times [b,B] \times [c,C] \to \mathbb{R}$ *be a continuous mapping on* $[a,A] \times [b,B] \times [c,C]$, *and* $f'''_{x,y,z} := \frac{\partial^3 f}{\partial x \partial y \partial z}$ *exists on* $[a,A] \times [b,B] \times [c,C]$ *and is integrable. Let also* $(x,y,z) \in [a,A] \times [b,B] \times [c,C]$ *be fixed. We define the kernels* $p: [a,A]^2 \to \mathbb{R}$, $q: [b,B]^2 \to \mathbb{R}$, *and* $\theta: [c,C]^2 \to \mathbb{R}$:

$$p(x,s) := \begin{cases} s-a, s \in [a,x] \\ s-A, s \in (x,A], \end{cases}$$

$$q(y,t) := \begin{cases} t - b, t \in [b, y], \\ t - B, t \in (y, B] \end{cases}$$

and

$$\theta(z,r) := \begin{cases} r - c, r \in [c, z], \\ r - C, r \in (z, C]. \end{cases}$$

Then

$$\theta_{1,3} := \int_a^A \int_b^B \int_c^C p(x,s)q(y,t)\theta(z,r)f'''_{s,t,r}(s,t,r)\,dsdtdr$$

$$= \{(A-a)(B-b)(C-c) \cdot f(x,y,z)\} - \left[(B-b)(C-c)\int_a^A f(s,y,z)ds \right.$$

$$+ (A-a)(C-c)\int_b^B f(x,t,z)dt + (A-a)(B-b)\int_c^C f(x,y,r)dr\Bigg]$$

$$+ \left[(C-c)\int_a^A \int_b^B f(s,t,z)dsdt + (B-b)\int_a^A \int_c^C f(s,y,r)dsdr \right.$$

$$+ (A-a)\int_b^B \int_c^C f(x,t,r)dtdr\Bigg]$$

$$- \int_a^A \int_b^B \int_c^C f(s,t,r)dsdtdr =: \theta_{2,3}. \tag{5.4}$$

Proof. Integrating by parts repeatedly we obtain the following eight equalities:

$$\int_a^x \int_b^y \int_c^z (s-a)(t-b)(r-c)f'''_{s,t,r}(s,t,r)\,drdtds$$

$$= (x-a)(y-b)(z-c)f(x,y,z) - (y-b)(z-c)\int_a^x f(s,y,z)ds$$

$$- (x-a)(z-c)\int_b^y f(x,t,z)dt - (x-a)(y-b)\int_c^z f(x,y,r)dr$$

$$+ (z-c)\int_a^x \int_b^y f(s,t,z)dtds + (y-b)\int_a^x \int_c^z f(s,y,r)drds$$

$$+ (x-a)\int_b^y \int_c^z f(x,t,r)drdt - \int_a^x \int_b^y \int_c^z f(s,t,r)drdtds. \tag{5.5}$$

Similarly, we have

$$\int_x^A \int_y^B \int_z^C (s-A)(t-B)(r-C)f'''_{s,t,r}(s,t,r)\,drdtds$$

$$= (A-x)(B-y)(C-z)f(x,y,z) - (B-y)(C-z)\int_x^A f(s,y,z)ds$$

$$- (A-x)(C-z)\int_y^B f(x,t,z)dt - (A-x)(B-y)\int_z^C f(x,y,r)dr$$

$$+ (C-z)\int_x^A \int_y^B f(s,t,z)dtds + (B-y)\int_x^A \int_z^C f(s,y,r)drds$$

$$+ (A-x)\int_y^B \int_z^C f(x,t,r)drdt - \int_x^A \int_y^B \int_z^C f(s,t,r)drdtds. \tag{5.6}$$

Furthermore,

$$\int_a^x \int_b^y \int_z^C (s-a)(t-b)(r-C)f'''_{s,t,r}(s,t,r)drdtds$$

$$= (x-a)(y-b)(C-z)f(x,y,z) - (y-b)(C-z)\int_a^x f(s,y,z)ds$$

$$- (x-a)(C-z)\int_b^y f(x,t,z)dt - (x-a)(y-b)\int_z^C f(x,y,r)dr$$

$$+ (C-z)\int_a^x \int_b^y f(s,t,z)dtds + (y-b)\int_a^x \int_z^C f(s,y,r)drds$$

$$+ (x-a)\int_b^y \int_z^C f(x,t,r)drdt - \int_a^x \int_b^y \int_z^C f(s,t,r)drdtds. \qquad (5.7)$$

Also,

$$\int_A^x \int_B^y \int_c^z (s-A)(t-B)(r-c)f'''_{s,t,r}(s,t,r)drdtds$$

$$= (A-x)(B-y)(z-c)f(x,y,z) - (B-y)(z-c)\int_x^A f(s,y,z)ds$$

$$- (A-x)(z-c)\int_y^B f(x,t,z)dt - (A-x)(B-y)\int_c^z f(x,y,r)dr$$

$$+ (z-c)\int_x^A \int_y^B f(s,t,z)dtds + (B-y)\int_x^A \int_c^z f(s,y,r)drds$$

$$+ (A-x)\int_y^B \int_c^z f(x,t,r)drdt - \int_x^A \int_y^B \int_c^z f(s,t,r)drdtds. \qquad (5.8)$$

Next we find,

$$\int_a^x \int_y^B \int_z^C (s-a)(t-B)(r-C)f'''_{s,t,r}(s,t,r)drdtds$$

$$= (x-a)(B-y)(C-z)f(x,y,z) - (B-y)(C-z)\int_a^x f(s,y,z)ds$$

$$- (x-a)(C-z)\int_y^B f(x,t,z)dt - (x-a)(B-y)\int_z^C f(x,y,r)dr$$

$$+ (C-z)\int_a^x \int_y^B f(s,t,z)dtds + (B-y)\int_a^x \int_z^C f(s,y,r)drds$$

$$+ (x-a)\int_y^B \int_z^C f(x,t,r)drdt - \int_a^x \int_y^B \int_z^C f(s,t,r)drdtds. \qquad (5.9)$$

Also,

$$\int_x^A \int_b^y \int_c^z (s-A)(t-b)(r-c)f'''_{s,t,r}(s,t,r)drdtds$$

$$= (A-x)(y-b)(z-c)f(x,y,z) - (y-b)(z-c)\int_x^A f(s,y,z)ds$$

$$- (A-x)(z-c)\int_b^y f(x,t,z)dt - (A-x)(y-b)\int_c^z f(x,y,r)dr$$

$$+ (y-b)\int_x^A \int_c^z f(s,y,r)drds + (z-c)\int_x^A \int_b^y f(s,t,z)dtds$$

$$+ (A-x)\int_b^y \int_c^z f(x,t,r)drdt - \int_x^A \int_b^y \int_c^z f(s,t,r)drdtds. \qquad (5.10)$$

And,

$$\int_a^x \int_y^B \int_c^z (s-a)(t-B)(r-c)f'''_{s,t,r}(s,t,r)drdtds$$

$$= (x-a)(B-y)(z-c)f(x,y,z) - (B-y)(z-c)\int_a^x f(s,y,z)ds$$

$$- (x-a)(z-c)\int_y^B f(x,t,z)dt - (x-a)(B-y)\int_c^z f(x,y,r)dr$$

$$+ (z-c)\int_a^x \int_y^B f(s,t,z)dtds + (B-y)\int_a^x \int_c^z f(s,y,r)drds$$

$$+ (x-a)\int_y^B \int_c^z f(x,t,r)drdt - \int_a^x \int_y^B \int_c^z f(s,t,r)drdtds. \qquad (5.11)$$

Finally we have,

$$\int_x^A \int_b^y \int_z^C (s-A)(t-b)(r-C)f'''_{s,t,r}(s,t,r)drdtds$$

$$= (A-x)(y-b)(C-z)f(x,y,z) - (y-b)(C-z)\int_x^A f(s,y,z)ds$$

$$- (A-x)(C-z)\int_b^y f(x,t,z)dt - (A-x)(y-b)\int_z^C f(x,y,r)dr$$

$$+ (C-z)\int_x^A \int_b^y f(s,t,z)dtds + (y-b)\int_x^A \int_z^C f(s,y,r)drds$$

$$+ (A-x)\int_b^y \int_z^C f(x,t,r)drdt - \int_x^A \int_b^y \int_z^C f(s,t,r)drdtds \qquad (5.12)$$

Adding all the right-hand sides of (5.5) – (5.12), we derive:

$$\text{Big R.H.S.} = (A-a)(B-b)(C-c)f(x,y,z) - (B-b)(C-c)\int_a^A f(s,y,z)ds$$

$$- (A-a)(C-c)\int_b^B f(x,t,z)dt - (A-a)(B-b)\int_c^C f(x,y,r)dr$$

$$+ (C-c)\int_a^A \int_b^B f(s,t,z)dtds + (B-b)\int_a^A \int_c^C f(s,y,r)dsdr$$

$$+ (A-a)\int_b^B \int_c^C f(x,t,r)dtdr - \int_a^A \int_b^B \int_c^C f(s,t,r)drdtds.$$

$$(5.13)$$

Adding all the left-hand sides of (5.5) – (5.12), we find

$$\text{Big L.H.S.} = \int_a^A \int_b^B \int_c^C p(x,s)q(y,t)\theta(z,r)f'''_{s,t,r}(s,t,r)drdtds. \qquad (5.14)$$

Clearly we have

$$\text{Big L.H.S.} = \text{Big R.H.S.} \qquad (5.15)$$

We have established (5.4). $\qquad\qquad\qquad\qquad\qquad\qquad\qquad\qquad\qquad\qquad\square$

In general we state

Theorem 5.5. *Let* $f\colon \times_{i=1}^n [a_i,b_i] \to \mathbb{R}$ *be a continuous mapping on* $\times_{i=1}^n [a_i,b_i]$, *and* $\frac{\partial^n f(x_1,\ldots,x_n)}{\partial x_1 \cdots \partial x_n}$ *exists on* $\times_{i=1}^n [a_i,b_i]$ *and is integrable. Let also* $(x_1,\ldots,x_n) \in \times_{i=1}^n [a_i,b_i]$ *be fixed. We define the kernels* $p_i\colon [a_i,b_i]^2 \to \mathbb{R}$:

$$p_i(x_i, s_i) := \begin{cases} s_i - a_i, s_i \in [a_i, x_i] \\ s_i - b_i, s_i \in (x_i, b_i], \end{cases}$$

for all $i = 1,\ldots,n$.
   Then

$$\theta_{1,n} := \int_{\times_{i=1}^n [a_i,b_i]} \prod_{i=1}^n p_i(x_i, s_i) \frac{\partial^n f(s_1,\ldots,s_n)}{\partial s_1 \cdots \partial s_n} ds_1 \cdots ds_n$$

$$= \left\{ \left(\prod_{i=1}^n (b_i - a_i) \right) \cdot f(x_1,\ldots,x_n) \right\}$$

$$-\left[\sum_{i=1}^{\binom{n}{1}}\left(\prod_{\substack{j=1\\j\neq i}}^{n}(b_j-a_j)\int_{a_i}^{b_i}f(x_1,\ldots,s_i,\ldots,x_n)ds_i\right)\right]$$

$$+\left[\sum_{\ell=1}^{\binom{n}{2}}\left(\prod_{\substack{k=1\\k\neq i,j}}^{n}(b_k-a_k)\left(\int_{a_i}^{b_i}\int_{a_j}^{b_j}f(x_1,\ldots,s_i,\ldots,s_j,\ldots,x_n)ds_ids_j\right)\right)_{(\ell)}\right]$$

$$-+\cdots-+\cdots+(-1)^{n-1}$$

$$\cdot\left[\sum_{j=1}^{\binom{n}{n-1}}(b_j-a_j)\int_{\times_{\substack{i=1\\i\neq j}}^{n}[a_i,b_i]}f(s_1,\ldots,x_j,\ldots,s_n)ds_1\cdots\widehat{ds_j}\cdots ds_n\right]$$

$$+(-1)^n\int_{\times_{i=1}^{n}[a_i,b_i]}f(s_1,\ldots,s_n)ds_1\cdots ds_n=:\theta_{2,n}. \tag{5.16}$$

The above ℓ counts all the (i,j)'s, $i<j$ and $i,j=1,\ldots,n$. Also $\widehat{ds_j}$ means ds_j is missing.

Proof. Similar to Theorem 5.4. $\qquad\qquad\Box$

5.3 Main Results

Here we present the consequences of Section 5.2. We have

Theorem 5.6. *Under the notations and assumptions of Theorem 5.4, additionally we suppose that* $\|f'''_{s,t,r}\|_\infty<+\infty$. *Then*

$$|\theta_{2,3}|\leq\frac{\|f'''_{s,t,r}\|_\infty}{8}\cdot\left\{((x-a)^2+(A-x)^2)\cdot((y-b)^2+(B-y)^2)\right.$$

$$\left.\cdot((z-c)^2+(C-z)^2)\right\},\text{for all }(x,y,z)\in[a,A]\times[b,B]\times[c,C]. \tag{5.17}$$

Proof. Notice that

$$|\theta_{2,3}|=|\theta_{1,3}|\leq\|f'''_{s,t,r}\|_\infty\left(\int_a^A|p(x,s)|ds\right)\left(\int_b^B|q(y,t)|dt\right)\left(\int_c^C|\theta(z,r)|dr\right).$$

Also observe that

$$\int_a^A|p(x,s)|ds=\frac{1}{2}\{(x-a)^2+(A-x)^2\},$$

etc. $\qquad\qquad\Box$

The counterpart of the last theorem is

Theorem 5.7. *Under the notations and assumptions of Theorem 5.5, additionally we suppose that*

$$\left\|\frac{\partial^n f(s_1,\ldots,s_n)}{\partial s_1\cdots\partial s_n}\right\|_\infty<+\infty.$$

Then

$$|\theta_{2,n}| \le \frac{\left\| \frac{\partial^n f(s_1,\ldots,s_n)}{\partial s_1 \cdots \partial s_n} \right\|_\infty}{2^n} \cdot \left\{ \prod_{i=1}^{n} [(x_i - a_i)^2 + (b_i - x_i)^2] \right\},$$

$$\text{for all } (x_1,\ldots,x_n) \in \times_{i=1}^{n} [a_i, b_i]. \tag{5.18}$$

Proof.   As in Theorem 5.6. $\square$

The L_p-analogues follow.

Theorem 5.8. *Under the notations and assumptions of Theorem 5.4, additionally, we assume that $f'''_{s,t,r} \in L_p([a,A] \times [b,B] \times [c,C])$, i.e.,*

$$\|f'''_{s,t,r}\|_p := \left(\int_a^A \int_b^B \int_c^C \left| \frac{\partial^3 f(x,y,z)}{\partial x \partial y \partial z} \right|^p dx\,dy\,dz \right)^{1/p} < +\infty,$$

where $p > 1$. Then

$$|\theta_{2,3}| \le \frac{\|f'''_{s,t,r}\|_p}{(q+1)^{3/q}} \cdot \left\{ [((x-a)^{q+1} + (A-x)^{q+1}) \cdot ((y-b)^{q+1} + (B-y)^{q+1}) \right.$$

$$\left. \cdot ((z-c)^{q+1} + (C-z)^{q+1})]^{1/q} \right\}, \quad \text{for all } (x,y,z) \in [a,A] \times [b,B] \times [c,C],$$

$$\tag{5.19}$$

where q is such that $\frac{1}{p} + \frac{1}{q} = 1$.

Proof.   Notice that

$$|\theta_{2,3}| = |\theta_{1,3}| = \left| \int_a^A \int_b^B \int_c^C p(x,s)q(y,t)\theta(z,r) f'''_{s,t,r}(s,t,r)\,ds\,dt\,dr \right|$$

$$\le \int_a^A \int_a^B \int_c^C |p(x,s)|\,|q(y,t)|\,|\theta(z,r)|\,|f'''_{s,t,r}(s,t,r)|\,ds\,dt\,dr$$

$$\text{(by Hölder's inequality)}$$

$$\le \left(\int_a^A \int_b^B \int_c^C |f'''_{s,t,r}(s,t,r)|^p\,ds\,dt\,dr \right)^{1/p}$$

$$\cdot \left(\int_a^A \int_b^B \int_c^C (|p(x,s)|\,|q(y,t)|\,|\theta(z,r)|)^q\,ds\,dt\,dr \right)^{1/q}$$

$$= \|f'''_{s,t,r}\|_p \cdot \left\{ \left(\int_a^A |p(x,s)|^q\,ds \right) \cdot \left(\int_b^B |q(y,t)|^q\,dt \right) \right.$$

$$\left. \cdot \left(\int_c^C |\theta(z,r)|^q\,dr \right) \right\}^{1/q}$$

$$= \|f'''_{s,t,r}\|_p \cdot \left\{ \left(\frac{(x-a)^{q+1} + (A-x)^{q+1}}{q+1} \right) \right.$$

$$\left. \cdot \left(\frac{(y-b)^{q+1} + (B-y)^{q+1}}{q+1} \right) \cdot \left(\frac{(z-c)^{q+1} + (C-z)^{q+1}}{q+1} \right) \right\}^{1/q}. \quad \square$$

The corresponding general L_p-case follows.

Theorem 5.9. *Under the notations and assumptions of Theorem 5.5, additionally we assume that* $\frac{\partial^n f}{\partial x_1 \cdots \partial x_n} \in L_p\left(\times_{i=1}^n [a_i, b_i]\right)$, *i.e.*

$$\left\| \frac{\partial^n f(x_1, \ldots, x_n)}{\partial x_1 \cdots \partial x_n} \right\|_p < +\infty, \qquad \text{where } p > 1.$$

Then

$$|\theta_{2,n}| \leq \frac{\left\| \frac{\partial^n f(x_1,\ldots,x_n)}{\partial x_1 \cdots \partial x_n} \right\|_p}{(q+1)^{n/q}} \cdot \left\{ \prod_{i=1}^n [(x_i - a_i)^{q+1} + (b_i - x_i)^{q+1}] \right\}^{1/q}, \tag{5.20}$$

for any $(x_1, \ldots, x_n) \in \times_{i=1}^n [a_i, b_i]$, *and* $q\colon \frac{1}{p} + \frac{1}{q} = 1$.

Proof. Same as in Theorem 5.8. $\qquad\qquad\qquad\qquad\qquad\qquad\qquad\square$

Remark 5.10. Equalities (5.4) and (5.16) can simplify dramatically, if for instance we suppose in Theorem 5.4 that there exists an $(x_0, y_0, z_0) \in [a, A] \times [b, B] \times [c, C]$ such that

$$f(x_0, \cdot, \cdot) = f(\cdot, y_0, \cdot) = f(\cdot, \cdot, z_0) = 0.$$

Also in Theorem 5.5 we may assume that there exists an $(x_1^0, x_2^0, \ldots, x_n^0) \in \times_{i=1}^n [a_i, b_i]$ such that

$$f(x_1^0, x_2, \ldots, x_n) = f(x_1, x_2^0, x_3, \ldots, x_n) = \cdots = f(x_1, \ldots, x_{n-1}, x_n^0) = 0,$$

for any $(x_1, \ldots, x_n) \in \times_{i=1}^n [a_i, b_i]$. So in these particular cases we derive that

$$\theta_{2,3}(x_0, y_0, z_0) = -\int_a^A \int_b^B \int_c^C f(s,t,r)\,dsdtdr, \tag{5.21}$$

and

$$\theta_{2,n}(x_1^0, x_2^0, \ldots, x_n^0) = (-1)^n \int_{\times_{i=1}^n [a_i, b_i]} f(s_1, \ldots, s_n)\,ds_1 \cdots ds_n. \tag{5.22}$$

Hence in these cases we have

$$|\theta_{2,3}(x_0, y_0, z_0)| = \left| \int_a^A \int_b^B \int_c^C f(s,t,r)\,dsdtdr \right|, \tag{5.23}$$

and

$$|\theta_{2,n}(x_1^0, x_2^0, \ldots, x_n^0)| = \left| \int_{\times_{i=1}^n [a_i, b_i]} f(s_1, \ldots, s_n)\,ds_1 \cdots ds_n \right|.$$

So according to Theorems $5.6 - 5.9$ we get respectively:

$$\left| \int_a^A \int_b^B \int_c^C f(s,t,r)\,dsdtdr \right| \leq \frac{\| f_{s,t,r}''' \|_\infty}{8}$$

$$\cdot \{ ((x_0 - a)^2 + (A - x_0)^2) \cdot ((y_0 - b)^2 + (B - y_0)^2) \cdot ((z_0 - c)^2 + (C - z_0)^2) \}; \tag{5.24}$$

and

$$\left| \int_{\times_{i=1}^n [a_i,b_i]} f(s_1,\ldots,s_n) ds_1 \cdots ds_n \right| \leq \frac{\left\| \frac{\partial^n f(s_1,\ldots,s_n)}{\partial s_1 \cdots \partial s_n} \right\|_\infty}{2^n}$$
$$\cdot \left\{ \prod_{i=1}^n [(x_i^0 - a_i)^2 + (b_i - x_i^0)^2] \right\}. \qquad (5.25)$$

Also it holds

$$\left| \int_a^A \int_b^B \int_c^C f(s,t,r) ds dt dr \right|$$
$$\leq \frac{\| f_{s,t,r}''' \|_p}{(q+1)^{3/q}} \cdot \left\{ [((x_0 - a)^{q+1} + (A - x_0)^{q+1}) \cdot ((y_0 - b)^{q+1} + (B - y_0)^{q+1}) \right.$$
$$\left. \cdot ((z_0 - c)^{q+1} + (C - z_0)^{q+1})]^{1/q} \right\}; \qquad (5.26)$$

and

$$\left| \int_{\times_{i=1}^n [a_i,b_i]} f(s_1,\ldots,s_n) ds_1 \cdots ds_n \right| \leq \frac{\left\| \frac{\partial^n f(x_1,\ldots,x_n)}{\partial x_1 \cdots \partial x_n} \right\|_p}{(q+1)^{n/q}}$$
$$\cdot \left\{ \prod_{i=1}^n [(x_i^0 - a_i)^{q+1} + (b_i - x_i^0)^{q+1}] \right\}^{1/q}. \qquad (5.27)$$

Finally we present the optimality of inequalities in Theorems 5.6 and 5.7.

Theorem 5.11. *Inequalities (5.17) and (5.18) are sharp.*

Proof.   It is enough to prove that (5.18) is sharp. Here the optimal function will be

$$f^*(s_1,\ldots,s_n) := \prod_{i=1}^n |s_i - x_i^0|^\alpha (b_i - \alpha_i), \qquad \alpha > 1, \qquad (5.28)$$

where $(x_1^0, x_2^0, \ldots, x_n^0)$ is fixed in $\times_{i=1}^n [a_i, b_i]$. Notice here that

$$f^*(s_1,\ldots,x_j^0,\ldots,s_n) = 0, \quad \text{for all } j = 1,\ldots,n, \text{ and any } (s_1,\ldots,s_n) \in \times_{i=1}^n [a_i,b_i].$$

Therefore by Remark 5.10, inequality (5.18) collapses to inequality (5.25). We see that

$$\frac{\partial^n f^*(s_1,\ldots,s_n)}{\partial s_1 \cdots \partial s_n} = \alpha^n \cdot \left(\prod_{i=1}^n (b_i - a_i) \right) \cdot \left(\prod_{i=1}^n |s_i - x_i^0|^{\alpha-1} \text{sign}(s_i - x_i^0) \right), \qquad (5.29)$$

and

$$\left| \frac{\partial^n f^*(s_1,\ldots,s_n)}{\partial s_1 \cdots \partial s_n} \right| = \alpha^n \cdot \left(\prod_{i=1}^n (b_i - a_i) \right) \cdot \left(\prod_{i=1}^n |s_i - x_i^0|^{\alpha-1} \right). \qquad (5.30)$$

Consequently we find that

$$\left\|\frac{\partial^n f^*(s_1,\ldots,s_n)}{\partial s_1 \cdots \partial s_n}\right\|_\infty = \alpha^n \left(\prod_{i=1}^n (b_i - a_i)\right)$$
$$\cdot \left(\prod_{i=1}^n (\max(b_i - x_i^0, x_i^0 - a_i))^{\alpha-1}\right). \qquad (5.31)$$

First we calculate the left-hand side of corresponding inequality (5.25). We have

$$\left|\int_{\times_{i=1}^n [a_i,b_i]} f^*(s_1,\ldots,s_n)ds_1\cdots ds_n\right|$$

$$= \int_{\times_{i=1}^n [a_i,b_i]} \left(\prod_{i=1}^n (b_i - a_i)\right) \cdot \left(\prod_{i=1}^n |s_i - x_i^0|^\alpha\right) ds_1 \cdots ds_n$$

$$= \left(\prod_{i=1}^n (b_i - a_i)\right) \int_{\times_{i=1}^n [a_i,b_i]} \left(\prod_{i=1}^n |s_i - x_i^0|^\alpha\right) ds_1 \cdots ds_n$$

$$= \left(\prod_{i=1}^n (b_i - a_i)\right) \cdot \left(\prod_{i=1}^n \left(\int_{a_i}^{b_i} |s_i - x_i^0|^\alpha ds_i\right)\right)$$

$$= \left(\prod_{i=1}^n (b_i - a_i)\right) \cdot \left(\prod_{i=1}^n \left(\frac{(x_i^0 - a_i)^{\alpha+1} + (b_i - x_i^0)^{\alpha+1}}{\alpha+1}\right)\right).$$

That is,

$$\text{L.H.S.}(5.25) = \frac{\left(\prod_{i=1}^n (b_i - a_i)\right)}{(\alpha+1)^n} \cdot \left(\prod_{i=1}^n ((x_i^0 - a_i)^{\alpha+1} + (b_i - x_i^0)^{\alpha+1})\right). \qquad (5.32)$$

And next we observe that

$$\text{R.H.S.}(5.25) = \frac{\alpha^n \cdot \left(\prod_{i=1}^n (b_i - a_i)\right) \cdot \left(\prod_{i=1}^n (\max(b_i - x_i^0, x_i^0 - a_i))^{\alpha-1}\right)}{2^n}$$

$$\cdot \left\{\prod_{i=1}^n ((x_i^0 - a_i)^2 + (b_i - x_i^0)^2)\right\}. \qquad (5.33)$$

Now let $\alpha \to 1$. We find that

$$\lim_{\alpha \to 1} \text{L.H.S.}(5.25) = \frac{\prod_{i=1}^n (b_i - a_i)}{2^n} \cdot \left(\prod_{i=1}^n ((x_i^0 - a_i)^2 + (b_i - x_i^0)^2)\right),$$

and

$$\lim_{\alpha \to 1} \text{R.H.S.}(5.25) = \frac{\prod_{i=1}^n (b_i - a_i)}{2^n} \cdot \left(\prod_{i=1}^n ((x_i^0 - a_i)^2 + (b_i - x_i^0)^2)\right).$$

That is,

$$\lim_{\alpha \to 1} \text{L.H.S.}(5.25) = \lim_{\alpha \to 1} \text{R.H.S.}(5.25), \qquad (5.34)$$

hence proving the sharpness of (5.18). $\qquad\square$

Remark 5.12. Another interesting case for (5.4) and (5.16) is to suppose that for specific (x, y, z) $((x_1, \ldots, x_n)$, respectively) all the marginal integrals of f are equal to zero. Then we find

$$\theta_{2,3} = (A - a)(B - b)(C - c) \cdot f(x, y, z) - \int_a^A \int_a^B \int_c^C f(s, t, r)\,ds\,dt\,dr, \qquad (5.35)$$

and

$$\theta_{2,n} = \left(\prod_{i=1}^n (b_i - a_i) \right) \cdot f(x_1, \ldots, x_n) + (-1)^n \int_{\times_{i=1}^n [a_i, b_i]} f(s_1, \ldots, s_n)\,ds_1 \cdots ds_n.$$

$$(5.36)$$

Hence inequalities (5.17), (5.18), (5.19), and (5.20) become again much simpler.

Chapter 6

Ostrowski Inequalities on Euclidean Domains

The classical Ostrowski inequality for functions on intervals is extended to functions on general domains in Euclidean space. For radial functions on balls the inequality is sharp. This treatment relies on [56].

6.1 Introduction

The classical Ostrowski inequality (of 1938) [196] is

$$\left| \frac{1}{b-a} \int_a^b f(y)dy - f(x) \right| \le \left(\frac{1}{4} + \frac{\left(x - \frac{a+b}{2}\right)^2}{(b-a)^2} \right) (b-a)\|f'\|_\infty,$$

$$\text{for } f \in C^1([a,b]), \ x \in [a,b],$$

and it is a sharp inequality. This was extended from intervals to rectangles in $\mathbb{R}^N$, $N \ge 1$, see [21], p. 507. For other recent results related to Ostrowski's inequality, see [110], [126] and [199], [214], [231].

The extension to general domains in $\mathbb{R}^N$ is presented here. We deduce Ostrowski type inequalities on general bounded domains in $\mathbb{R}^N$, and the inequalities are shown to be sharp on balls.

6.2 Main Results

Let $N > 1$, $B(0,R) := \{x \in \mathbb{R}^N : |x| < R\}$ be the ball in $\mathbb{R}^N$ centered at the origin and of radius $R > 0$. Let $S^{N-1} := \{x \in \mathbb{R}^N : |x| = 1\}$ be the unit sphere in $\mathbb{R}^N$.

Let $d\omega$ be the element of surface measure on S^{N-1} and let $\omega_N = \int_{S^{N-1}} d\omega = \frac{2\pi^{N/2}}{\Gamma\left(\frac{N}{2}\right)}$. For $x \in \mathbb{R}^N - \{0\}$ we can write $x = r\omega$, where $r = |x| > 0$ and $\omega = \frac{x}{r} \in S^{N-1}$. Note that $\int_{B(0,R)} dy = \frac{\omega_N R^N}{N}$ is the Lebesgue measure of the ball.

For $f \in C\left(\overline{B(0,R)}\right)$ let

$$\fint_{B(0,R)} f(y)dy := \frac{1}{\mathrm{Vol}(B(0,R))} \int_{B(0,R)} f(y)dy,$$

93

and

$$\fint_{S^{N-1}} f(r\omega)d\omega = \frac{1}{\omega_N} \int_{S^{N-1}} f(r\omega)d\omega$$

be the averages of f over the ball and the sphere, respectively. Here f can be real or complex valued.

Let

$$\tilde{f}(r) := \fint_{S^{N-1}} f(r\omega)d\omega$$

be the average of $f(x)$ as x ranges over $\{y \in \mathbb{R}^N : |y| = r\}$. Then

$$\mathcal{N}(f) := \sup_{x \in \overline{B(0,R)}} |f(x) - \tilde{f}(r)| = \|f - \tilde{f}\|_\infty \tag{6.1}$$

measures how far f is from being a radial function. More precisely, $\mathcal{N}$ is a seminorm on $C\left(\overline{B(0,R)}\right)$, and $\mathcal{N}(f) = 0$ if and only if f is a radial function, i.e. $f(x) = g(r)$ for some function $g \in C([0,R])$.

We view how close f is to being radial by measuring $\mathcal{N}(f)$; the closer f is to being radial, the smaller $\mathcal{N}(f)$ is and conversely.

Let Ω be a domain in $\mathbb{R}^N$ and let

$$\mathrm{Lip}(\Omega) = \big\{ f \in C(\overline{\Omega}) \colon |f(x) - f(y)| \le K|x - y|$$
$$\text{for some } K > 0 \text{ and all } x, y \in \Omega \big\}. \tag{6.2}$$

The Lipschitz constant of $f \in \mathrm{Lip}(\Omega)$ is

$$\|f\|_{\mathrm{Lip}} = \inf\{K \colon K \text{ as in } (6.2)\}.$$

Then $X := \mathrm{Lip}(\Omega)$ is a Banach space with norm

$$f \to \|f\|_\infty + \|f\|_{\mathrm{Lip}} =: \|f\|_X.$$

Equivalently, X is the Sobolev space $W^{1,\infty}(\Omega)$ (cf. [139]).

We present

Theorem 6.1. *Let* $f \in \mathrm{Lip}(B(0,R)) = W^{1,\infty}(B(0,R))$. *Then for* $x = r\omega$ *as above*

$$\left| f(x) - \fint_{B(0,R)} f(y)dy \right| \le \mathcal{N}(f) + \frac{N}{R^N}\|\nabla f\|_\infty \left[\frac{2|x|^{N+1}}{N(N+1)} + R^N\left(\frac{R}{N+1} - \frac{|x|}{N}\right) \right].$$
$$\tag{6.3}$$

The constants in (6.3) are best possible, equality can be attained for nontrivial radial functions at any $r \in [0, R]$.

Proof. Let $f \in \mathrm{Lip}(B(0,R))$. Then

$$\left| f(x) - \fint_{B(0,R)} f(y)dy \right|$$

$$\leq |f(x) - \tilde{f}(r)| + \left| \fint_{S^{N-1}} f(r\omega')d\omega' - \frac{N}{\omega_N R^N} \int_{S^{N-1}} \int_0^R f(s\omega')s^{N-1}dsd\omega' \right|$$

$$\leq \mathcal{N}(f) + \frac{N}{R^N} \fint_{S^{N-1}} \left[\int_0^R |f(r\omega') - f(s\omega')|S^{N-1}ds \right] d\omega'$$

$$\leq \mathcal{N}(f) + \frac{N}{R^N} \fint_{S^{N-1}} \int_0^R \left\| \frac{\partial f}{\partial r}(\omega') \right\|_{L^\infty((0,R))} |s - r|s^{N-1}dsd\omega'$$

$$\leq \mathcal{N}(f) + \left\| \frac{\partial f}{\partial r} \right\|_{L^\infty(B(0,R))} \frac{N}{R^N} \left(\int_0^R |s - r|s^{N-1}ds \right)$$

$$= \mathcal{N}(f) + \left\| \frac{\partial f}{\partial r} \right\|_{L^\infty(B(0,R))} \frac{N}{R^N} \left[\frac{2r^{N+1}}{N(N+1)} + R^N \left(\frac{R}{N+1} - \frac{r}{N} \right) \right]. \qquad (6.4)$$

Then (6.3) follows since

$$\left\| \frac{\partial f}{\partial r} \right\|_\infty \leq \|\nabla f\|_\infty.$$

In particular, a stronger form of (6.3) actually holds in all cases, replacing $\|\nabla f\|_\infty$ by $\left\| \frac{\partial f}{\partial r} \right\|_\infty$. Let $r \in [0, R]$ and $g^*(z) = |z - r|$. We can view g^* as a radial function on $\overline{B(0,R)}$.

Then

$$g^{*\prime}(z) = \begin{cases} \mathrm{sign}(z - r), & z \neq r, & 0 < r < R \\ 1, & & r = 0 \\ -1 & & r = R. \end{cases}$$

Thus $\|g^{*\prime}\|_\infty = 1$. Therefore

$$\mathrm{L.H.S.}(6.3) = \left| g^*(z) - \frac{N}{R^N} \int_0^R g^*(s)s^{N-1}ds \right| = \left| |z - r| - \frac{N}{R^N} \int_0^R |s - r|s^{N-1}ds \right|$$

$$= \left| |z - r| - \frac{N}{R^N} \left[\frac{2r^{N+1}}{N(N+1)} + R^N \left(\frac{R}{N+1} - \frac{r}{N} \right) \right] \right|.$$

Also

$$\mathrm{R.H.S.}(6.3) = \frac{N}{R^N} \left[\frac{2r^{N+1}}{N(N+1)} + R^N \left(\frac{R}{N+1} - \frac{r}{N} \right) \right].$$

Hence equality holds in (6.3) at $z = r$.

Note that the function $g^*(z) = |z - r|$, is in $C^1([0, R])$ only for $r = 0$ and $r = R$; for $0 < r < R$, $g^* \in \mathrm{Lip}([0, R]) - C^1([0, R])$. Of course for $0 < r < R$, g^* can be approximated by C^1 functions, namely $g_n(z) = |z - r|^{1+\frac{1}{n}}$. ⊔

Remark 6.2. A key step in the proof is the fact that we can evaluate exactly

$$Q(r) = \int_0^R |s - r| p(s) ds$$

for $0 \le r \le R$, where p is a nonnegative continuous function satisfying $\int_0^R p(s) ds = 1$. In the Ostrowski case ($N = 1$), $p(s) = \frac{1}{R}$; in our N-dimensional case, $p(s) = \frac{N s^{N-1}}{R^N}$.

This works for many other cases including linear combinations of $p(s) = \sum_{j=1}^m a_j s^{q_j}$, where $a_j > 0$, $q_j \ge 0$ (not necessarily an integer) and

$$\sum_{j=1}^m a_j \frac{R^{q_j+1}}{q_j + 1} = 1,$$

$p(s) = \sum_{j=1}^{m_1} a_j e^{\lambda_j s}$, where $a_j > 0$, $\lambda_j \in \mathbb{R} - \{0\}$,

$$\sum_{j=1}^{m_1} a_j \left(\frac{e^{\lambda_j R} - 1}{\lambda_j} \right) = 1,$$

$p(s) =$ sums of the form

$$a_j \sin(b_j s + c_j) + d_j \cos(e_j s + f_j),$$

where the coefficients are such that $p(s) \ge 0$ and $\int_0^R p(s) ds = 1$.

The space $\mathrm{Lip}(\Omega) \cap C_0(\Omega)$ consists of all Lipschitz continuous functions on $\overline{\Omega}$ vanishing on the boundary $\partial \Omega$ of Ω. Note that

$$\mathrm{Lip}(\Omega) \cap C_0(\Omega) = \left\{ f \in W^{1,\infty}(\Omega) \cap C(\overline{\Omega}) \colon f = 0 \text{ on } \partial \Omega \right\}$$

(cf. [139]).

Next comes a more general result when we consider functions over general domains.

Theorem 6.3. *Let $f \in \mathrm{Lip}(\Omega) \cap C_0(\Omega)$, where Ω is a bounded domain in $\mathbb{R}^N$. Extend f by zero to F on $B(0, R)$, the smallest ball centered at the origin and containing Ω. Then for all $x \in \Omega$,*

$$\left| f(x) - \fint_\Omega f(y) dy \right| \le \mathcal{N}(F) + \left(1 - \frac{\mathrm{Vol}(\Omega)}{\mathrm{Vol}(B(0,R))} \right) \left| \fint_\Omega f(y) dy \right|$$

$$+ \frac{N}{R^N} \|\nabla f\|_{L^\infty(\Omega)} \left[\frac{2|x|^{N+1}}{N(N+1)} + R^N \left(\frac{R}{N+1} - \frac{|x|}{N} \right) \right]. \tag{6.5}$$

Proof. Let $R := \inf\{R_0 > 0 \colon \Omega \subset B(0, R_0)\}$. Then

$$F(x) := \begin{cases} f(x), & x \in \overline{\Omega}, \\ 0, & x \in \overline{B(0, R)} - \Omega, \end{cases}$$

satisfies

$$F \in \mathrm{Lip}(B(0,R)) \cap C_0(B(0,R)).$$

Then for $x \in \Omega$

$$\left| f(x) - \fint_\Omega f(y)dy \right| \le \left| F(x) - \fint_{B(0,R)} F(y)dy \right| + \left| \fint_{B(0,R)} F(y)dy - \fint_\Omega f(y)dy \right|$$
$$= \mathcal{J}_1 + \mathcal{J}_2,$$

where

$$\mathcal{J}_1 := \left| F(x) - \fint_{B(0,R)} F(y)dy \right|$$

and

$$\mathcal{J}_2 := \left| \left(\frac{1}{\mathrm{Vol}(B(0,R))} - \frac{1}{\mathrm{Vol}(\Omega)} \right) \int_\Omega f(y)dy \right|.$$

By Theorem 6.1,

$$\mathcal{J}_1 \le \mathcal{N}(F) + \frac{N}{R^N} \|\nabla f\|_\infty \left[\frac{2|x|^{N+1}}{N(N+1)} + R^N \left(\frac{R}{N+1} - \frac{|x|}{N} \right) \right].$$

and

$$\mathcal{J}_2 = \left[1 - \frac{\mathrm{Vol}(\Omega)}{\mathrm{Vol}(B(0,R))} \right] \left| \fint_\Omega f(y)dy \right|.$$

This completes the proof of Theorem 6.3. $\qquad\qquad\qquad\qquad\qquad\square$

Remark 6.4. Note that $\mathcal{N}(F)$ appears in (6.5). In this context, $\mathcal{N}(f)$ does not make sense. Also, $\mathcal{N}(F)$ need not be small (of course, it is small if f is approximately spherically symmetric).

Here is a simple example to illustrate that $\mathcal{N}(F)$ can be large. Let $x_0 \in \Omega$ and choose $\varepsilon > 0$, small enough so that $B(x_0,\varepsilon) \subset \Omega$. Let $f \in C^\infty(\Omega)$ with f have support in $B(x_0,\varepsilon)$ and satisfy $f(y) = g(\rho)$ where $\rho = |y - x_0|$, for $0 \le \rho \le \varepsilon$. Assume further that g is nonincreasing and $g(0) = f(x_0) = M > 0$. Fix M, then

$$0 < f(x_0) - \fint_\Omega f(y)dy \to M = \|f\|_\infty \ \ \text{as} \ \varepsilon \to 0+.$$

Chapter 7

High Order Ostrowski Inequalities on Euclidean Domains

The original Ostrowski inequality for functions on intervals estimates the value of the function minus its average in terms of the maximum of its first derivative. This result is extended to functions on general domains using the L^∞ norm of its nth partial derivatives. For radial functions on balls the inequality is sharp. This chapter relies on [57].

7.1 Introduction

The classical Ostrowski inequality (of 1938) [196] is

$$\left| \frac{1}{b-a} \int_a^b f(y)dy - f(x) \right| \le \left(\frac{1}{4} + \frac{\left(x - \frac{a+b}{2}\right)^2}{(b-a)^2} \right) (b-a)\|f'\|_\infty, \qquad (7.1)$$

for $f \in C^1([a,b])$, $x \in [a,b]$, and it is a sharp inequality. Further related work was done in [16]. Inequality (7.1) was extended from intervals to rectangles in $\mathbb{R}^N$, $N \ge 1$, see [21], p. 507. The extension to general domains in $\mathbb{R}^N$ was done [56].

The resulting inequality was sharp, and equality was shown to hold for certain radial functions on balls. The right hand side included the factor $\|\nabla f\|_\infty$. The extension of (7.1) to high order derivative bounds on an interval was obtained in [21], p. 502.

The purpose here is to extend the results of [56] and [21], p. 502, to the higher order case when $\|\nabla f\|_\infty$ is replaced by $\max_{|\alpha|=n} \|D^\alpha f\|_\infty$ for $W^{n,\infty}$ functions on balls and more general domains. The obtained inequalities are sharp on balls.

7.2 Main Results

Let $N > 1$, $B(0,R) := \{x \in \mathbb{R}^N : |x| < R\}$ be the ball in $\mathbb{R}^N$ centered at the origin and of radius $R > 0$. Let $S^{N-1} := \{x \in \mathbb{R}^N : |x| - 1\}$ be the unit sphere in $\mathbb{R}^N$.

99

Let $d\omega$ be the element of surface measure on S^{N-1} and let

$$\omega_N = \int_{S^{N-1}} d\omega = \frac{2\pi^{N/2}}{\Gamma(N/2)}.$$

For $x \in \mathbb{R}^N - \{0\}$ we can write $x = r\omega$, where $r = |x| > 0$ and $\omega = \frac{x}{r} \in S^{N-1}$. Note that $\int_{B(0,R)} dy = \frac{\omega_N R^N}{N}$ is the Lebesgue measure of the ball.

For $f \in C\big(\overline{B(0,R)}\big)$ let

$$\fint_{B(0,R)} f(y)dy := \frac{1}{\mathrm{Vol}(B(0,R))} \int_{B(0,R)} f(y)dy,$$

and

$$\fint_{S^{N-1}} f(r\omega)d\omega = \frac{1}{\omega_N} \int_{S^{N-1}} f(r\omega)d\omega$$

be the averages of f over the ball and the sphere, respectively. Here f can be real or complex valued.

Let

$$\tilde{f}(r) := \fint_{S^{N-1}} f(r\omega)d\omega$$

be the average of $f(x)$ as x ranges over $\{y \in \mathbb{R}^N : |y| = r\}$. Therefore

$$\mathcal{N}(f) := \sup_{x \in \overline{B(0,R)}} |f(x) - \tilde{f}(r)| = \|f - \tilde{f}\|_\infty \tag{7.2}$$

measures how far f is from being a radial function. More precisely, $\mathcal{N}$ is a seminorm on $C\big(\overline{B(0,R)}\big)$, and $\mathcal{N}(f) = 0$ if and only if f is a radial function, i.e. $f(x) = g(r)$ for some function $g \in C([0,R])$. We view how close f is to being radial by measuring $\mathcal{N}(f)$; the closer f is to being radial, the smaller $\mathcal{N}(f)$ is and conversely.

Let Ω be a domain in $\mathbb{R}^N$ and let

$$\mathrm{Lip}(\Omega) = \big\{ f \in C(\overline{\Omega}) \colon |f(x) - f(y)| \le K|x - y|$$
$$\text{for some } K > 0 \text{ and all } x, y \in \Omega \big\}. \tag{7.3}$$

The Lipschitz constant of $f \in \mathrm{Lip}(\Omega)$ is

$$\|f\|_{\mathrm{Lip}} = \inf\{K \colon K \text{ as in } (7.3)\}. \tag{7.4}$$

Let $n \in \mathbb{Z}_+ = \{0, 1, 2, \ldots\}$. Define the Sobolev space $W^{n,\infty}(\Omega)$ in the usual way:

$$W^{n,\infty}(\Omega) := \big\{ u \colon \Omega \to \mathbb{C} \colon \text{the distributional derivative } D^\alpha u \text{ exists}$$
$$\text{and is in } L^\infty(\Omega) \text{ for } 0 \le |\alpha| \le n \big\}$$
$$= \big\{ u \in C^{n-1}(\overline{\Omega}) \colon D^\alpha u \in \mathrm{Lip}(\Omega) \text{ for } |\alpha| = n \big\}. \tag{7.5}$$

This is a Banach space with norm

$$\|u\| := \max_{0 \le |\alpha| \le n} \|D^\alpha u\|_{L^\infty(\Omega)}, \tag{7.6}$$

(see [139]). Clearly $W^{n,\infty}(\Omega) \supset C^n(\overline{\Omega})$.

The first main result is the following.

Theorem 7.1. *Let $f \in W^{n,\infty}(B(0,R))$. Then for $x = r\omega$ as above*

$$\left| f(x) - \fint_{B(0,R)} f(y)dy \right|$$

$$\leq \mathcal{N}(f) + \frac{N}{R^N} \left\{ \left| \sum_{k=1}^{n-1} \frac{1}{k!} \left(\fint_{S^{N-1}} \frac{\partial^k f(r\omega')}{\partial r^k} d\omega' \right) \right. \right.$$

$$\left. \times \left(\sum_{m=0}^{k} \binom{k}{m} (-1)^m r^m \left(\frac{R^{N+k-m}}{N+k-m} \right) \right) \right|$$

$$+ \frac{\left\| \frac{\partial^n f}{\partial r^n} \right\|_{L^\infty(B(0,R))}}{n!} \left(\sum_{m=0}^{n} \binom{n}{m} (-1)^m r^m \frac{R^{N+n-m}}{N+n-m} \right.$$

$$\left. \left. - \beta_n \sum_{m=0}^{n} \binom{n}{m} (-1)^m \frac{r^{N+n}}{N+n-m} \right) \right\}, \tag{7.7}$$

where $n \in \mathbb{N}$, $\beta_n = 0$ or 2, according as n is even or odd.

The constants in (7.7) are best possible, equality can be attained for nontrivial radial functions at any $r \in [0,R]$. We remark that

$$\left\| \frac{\partial^n f}{\partial r^n} \right\|_\infty \leq \|D^n f\|_\infty := \max_{|\alpha|=n} \|D^\alpha f\|_\infty,$$

and the larger bound $\|D^n f\|_\infty$ replacing $\left\| \frac{\partial^n f}{\partial r^n} \right\|_\infty$ gives perhaps a simpler version of the inequality (7.7).

Proof. Let $f \in W^{n,\infty}(B(0,R))$. Hence

$$\left| f(x) - \fint_{B(0,R)} f(y)dy \right| \leq |f(x) - \tilde{f}(r)| + \left| \fint_{S^{N-1}} f(r\omega')d\omega' \right.$$

$$\left. - \frac{N}{\omega_N R^N} \int_{S^{N-1}} \int_0^R f(s\omega')s^{N-1}dsd\omega' \right| \tag{7.8}$$

$$\leq \mathcal{N}(f) + \frac{N}{R^N} \left| \fint_{S^{N-1}} \left[\int_0^R [f(r\omega') - f(s\omega')]s^{N-1}ds \right] d\omega' \right| = \mathcal{N}(f) + (*).$$

$$\tag{7.9}$$

We see by Taylor's formula that

$$f(s\omega') - f(\rho\omega') = \sum_{k=1}^{n-1} \frac{\partial^k f(\rho\omega')}{k!\partial r^k} (s-\rho)^k + R_{n-1}(\rho, s), \tag{7.10}$$

where

$$R_{n-1}(\rho, s) := \int_{\rho}^{s} \left(\frac{\partial^{n-1} f(t\omega')}{\partial \rho^{n-1}} - \frac{\partial^{n-1} f(\rho\omega')}{\partial \rho^{n-1}} \right) \frac{(s-t)^{n-2}}{(n-2)!} dt, \qquad (7.11)$$

for fixed $\omega' \in S^{N-1}$ and all $s, \rho \in (0, R]$. As in p. 500 of [21] we derive

$$|R_{n-1}(\rho, s)| \leq \frac{\left\| \frac{\partial^n f}{\partial r^n} \right\|_{\infty}}{n!} |s - \rho|^n, \quad \forall s, \rho \in [0, R]. \qquad (7.12)$$

Consequently it holds

$$(*) = \frac{N}{R^N} \left| \fint_{S^{N-1}} \left[\int_0^R \left[\sum_{k=1}^{n-1} \frac{\partial^k f(r\omega')}{k! \partial r^k} (s-r)^k + R_{n-1}(r, s) \right] s^{N-1} ds \right] d\omega' \right| \quad (7.13)$$

$$\leq \frac{N}{R^N} \left\{ \left| \sum_{k=1}^{n-1} \fint_{S^{N-1}} \left[\int_0^R \frac{\partial^k f(r\omega')}{k! \partial r^k} (s-r)^k s^{N-1} ds \right] d\omega' \right| \right.$$

$$\left. + \fint_{S^{N-1}} \int_0^R |R_{n-1}(r, s)| s^{N-1} ds d\omega' \right\} \qquad (7.14)$$

$$\leq \frac{N}{R^N} \left\{ \left| \sum_{k=1}^{n-1} \frac{1}{k!} \left(\fint_{S^{N-1}} \frac{\partial^k f(r\omega')}{\partial r^k} d\omega' \right) \left(\int_0^R (s-r)^k s^{N-1} ds \right) \right| \right.$$

$$\left. + \frac{\left\| \frac{\partial^n f}{\partial r^n} \right\|_{L^{\infty}(B(0,R))}}{n!} \int_0^R |s-r|^n s^{N-1} ds \right\} =: A. \qquad (7.15)$$

$$\square$$

To evaluate the integrals in (7.15) we use an elementary calculation as follows.

Lemma 7.2. (i)

$$\int_0^R s^{N-1}(s-r)^k ds = \sum_{m=0}^{k} \binom{k}{m} (-1)^m r^m \frac{R^{N+k-m}}{N+k-m}, \quad k \in \mathbb{N}. \qquad (7.16)$$

(ii)

$$\int_0^R s^{N-1}|s-r|^n ds = \sum_{m=0}^{n} \binom{n}{m} (-1)^m r^m \frac{R^{N+n-m}}{N+n-m}$$

$$- \beta_n \sum_{m=0}^{n} \binom{n}{m} (-1)^m \frac{r^{N+n}}{N+n-m}, \qquad (7.17)$$

where $n \in \mathbb{N}$, $\beta_n = 0$ or 2, according as n is even or odd.

Proof.　(i) We see that

$$\int_0^R s^{N-1}(s-r)^k ds = \int_0^R s^{N-1} \left(\sum_{m=0}^{k} \binom{k}{m} (-1)^m s^{k-m} r^m \right) ds$$

$$= \sum_{m=0}^{k} \binom{k}{m} (-1)^m r^m \int_0^R s^{N+k-m-1} ds$$

$$= \sum_{m=0}^{k} \binom{k}{m} (-1)^m r^m \frac{R^{N+k-m}}{N+k-m}. \qquad (7.18)$$

(ii) Using (i) twice,

$$\int_0^R s^{N-1}|s-r|^n ds = \int_0^r s^{N-1}(r-s)^n ds + \int_r^R s^{N-1}(s-r)^n ds$$

$$= (-1)^n \sum_{m=0}^n \binom{n}{m}(-1)^m \frac{r^{N+n}}{N+n-m}$$

$$+ \sum_{m=0}^n \binom{n}{m}(-1)^m r^m \left(\frac{R^{N+n-m} - r^{N+n-m}}{N+n-m}\right)$$

$$= \sum_{m=0}^n \binom{n}{m}(-1)^m r^m \frac{R^{N+n-m}}{N+n-m}$$

$$+ \sum_{m=0}^n \binom{n}{m}(-1)^m [(-1)^n - 1] \frac{r^{N+n}}{N+n-m}. \tag{7.19}$$

Part (ii) now follows.

Continuing the calculation in (7.15)

$$A = \frac{N}{R^N}\left\{ \left| \sum_{k=1}^{n-1} \frac{1}{k!}\left(\oint_{S^{N-1}} \frac{\partial^k f(r\omega')}{\partial r^k} d\omega' \right) \right. \right.$$

$$\times \left(\sum_{m=0}^k \binom{k}{m}(-1)^m r^m \left(\frac{R^{N+k-m}}{N+k-m} \right) \right) \Bigg|$$

$$+ \frac{\left\| \frac{\partial^n f}{\partial r^n} \right\|_{L^\infty(B(0,R))}}{n!} \left(\sum_{m=0}^n \binom{n}{m}(-1)^m r^m \frac{R^{N+n-m}}{N+n-m} \right.$$

$$\left. \left. - \beta_n \sum_{m=0}^n \binom{n}{m}(-1)^m \frac{r^{N+n}}{N+n-m} \right) \right\}, \tag{7.20}$$

by Lemma 7.2. Hence proving (7.7).

Let n be even. Let f be a radial function so that $f(x) = g(r)$. Then equality in (7.7) is attained by $g(r) = (r-z)^n$ for any fixed z satisfying $0 \le z \le R$. We observe that $g^{(j)}(z) = 0$, $j = 0, 1, \ldots, n-1$ and $g^{(n)}(z) = n!$, $\|g^{(n)}\|_\infty = n!$.

We look at inequality (7.7) evaluating the function at z:

$$\text{L.H.S.}(7.7) = \frac{N}{R^N}\int_0^R (s-z)^n s^{N-1} ds$$

$$\overset{(7.16)}{=} \frac{N}{R^N} \sum_{m=0}^n \binom{n}{m}(-1)^m z^m \frac{R^{N+n-m}}{N+n-m} \tag{7.21}$$

$$= N\left(\sum_{m=0}^n \binom{n}{m}(-1)^m z^m \frac{R^{n-m}}{N+n-m} \right). \tag{7.22}$$

Because $\mathcal{N}(f) = 0$ and $\beta_n = 0$ the right hand side of (7.7) collapses to

$$\text{R.H.S.}(7.7) = \frac{N}{R^N}\left(\sum_{m=0}^{n}\binom{n}{m}(-1)^m z^m \frac{R^{N+n-m}}{N+n-m}\right)$$

$$= N\left(\sum_{m=0}^{n}\binom{n}{m}(-1)^m z^m \frac{R^{n-m}}{N+n-m}\right). \qquad (7.23)$$

That is equality holds in (7.7). Note that $g \in C^n\left(\overline{B(0,R)}\right)$ if $z > 0$, but $g \in W^{n,\infty}(B(0,R)) - C^n\left(\overline{B(0,R)}\right)$ if $z = 0$.

The optimal function in the case of n odd is

$$g(r) := |z - r|^n, \quad 0 \le r \le R, \quad z \text{ is fixed in } [0,R]. \qquad (7.24)$$

Direct calculations show, for n odd,

$$g^{(k)}(r) = n(n-1)\cdots(n-k+1)|r-z|^{n-k}\alpha_k(r), \qquad (7.25)$$

where

$$\begin{aligned}\alpha_k(r) &= 1 && \text{if } k \text{ is even,}\\ \alpha_k(r) &= \text{sign}(r-z) && \text{if } k \text{ is odd and } r \neq z.\end{aligned} \qquad (7.26)$$

It follows that

$$g \in C^\infty([0,R] - \{z\}) \cap C^{n-1}([0,R])$$

and $g^{(n)} \in L^\infty((0,R))$ with a jump discontinuity at $r = z$, whence $g \in W^{n,\infty}((0,R))$. It follows that $f \in W^{n,\infty}(B(0,R))$. Moreover, $g^{(k)}(z) = 0$, for $k = 0, 1, \ldots, n-1$ and $|g^{(k)}(z+\varepsilon)| \to n!$ as $\varepsilon \to 0$ (with $\varepsilon \neq 0$). Thus $\|g^{(n)}\|_\infty = n!$.

The left-hand side of (7.7) is

$$\text{L.H.S.}(7.7) = \frac{N}{R^N}\int_0^R |z-s|^n s^{N-1}ds$$

$$\overset{(7.17)}{=} \frac{N}{R^N}\left(\sum_{m=0}^{n}\binom{n}{m}(-1)^m z^m \frac{R^{N+n-m}}{N+n-m}\right.$$

$$\left. - 2\sum_{m=0}^{n}\binom{n}{m}(-1)^m \frac{z^{N+n}}{N+n-m}\right). \qquad (7.27)$$

Since $\mathcal{N}(f) = 0$ and n is odd, using the calculation on g given above leads to the conclusion that

$$\text{R.H.S.}(7.7) = \frac{N}{R^N}\left\{\frac{n!}{n!}\left(\sum_{m=0}^{n}\binom{n}{m}(-1)^m z^m \frac{R^{N+n-m}}{N+n-m}\right.\right.$$

$$\left.\left. - 2\sum_{m=0}^{n}\binom{n}{m}(-1)^m \frac{z^{N+n}}{N+n-m}\right)\right\}. \qquad (7.28)$$

Hence equality holds in (7.7). This completes the proof of the sharpness of inequality (7.7). The proof of the theorem is now complete. $\qquad\square$

Note that the extremal functions which give equality in (7.7) in the case of n odd are not in $C^n\big(\overline{B(0,R)}\big)$. To find a sequence $\{f_\nu\}_{\nu\in\mathbb{N}}$ of approximate extremal functions in $C^n\big(\overline{B(0,R)}\big)$, simply take $f_\nu(x) = g_\nu(r)$, where

$$g_\nu(r) = |z - r|^{n+\frac{1}{\nu}}, \quad \nu \in \mathbb{N}. \tag{7.29}$$

Next we treat the case of $r = R$.

Theorem 7.3. *Let $f \in W^{n,\infty}(B(0,R))$ such that $\frac{\partial^k f}{\partial r^k}$, $k = 1,\ldots,n-1$ vanish on $\partial B(0,R)$. Then for all $w \in S^{N-1}$,*

$$\left| f(Rw) - \fint_{B(0,R)} f(y)dy \right| \le \mathcal{N}(f) + \frac{R^n \left\| \frac{\partial^n f}{\partial r^n} \right\|_{L^\infty(B(0,R))}}{(N+1)\cdots(N+n)}. \tag{7.30}$$

The constants in (7.30) are best possible, equality can be attained for nontrivial radial functions.

Proof. Here from (7.7) we obtain

$$\left| f(Rw) - \fint_{B(0,R)} f(y)dy \right| \le \mathcal{N}(f)$$

$$+ \frac{NR^n}{n!} \left\| \frac{\partial^n f}{\partial r^n} \right\|_{L^\infty(B(0,R))} (-1)^n \sum_{m=0}^{n} \frac{\binom{n}{m}(-1)^m}{N+n-m} = (*), \tag{7.31}$$

since $1 - \beta_n = (-1)^n$.

Next,

$$\sum_{m=0}^{n} \binom{n}{m}(-1)^m t^{N+n-m-1} = t^{N-1}(t-1)^n, \tag{7.32}$$

by the Binomial Theorem.

Integrate (7.32) over $t \in [0,1]$, the result is

$$\sum_{m=0}^{n} \binom{n}{m} \frac{(-1)^m}{N+n-m} = (-1)^n \int_0^1 (1-t)^n t^{N-1} dt \tag{7.33}$$

$$= (-1)^n B(n+1, N)$$

$$= (-1)^n \frac{n!(N-1)!}{(N+n)!}. \tag{7.34}$$

By (7.34) we derive

$$(*) = \mathcal{N}R^n \left\| \frac{\partial^n f}{\partial r^n} \right\|_{L^\infty(B(0,R))} \frac{(N-1)!}{(N+n)!} = \frac{R^n \left\| \frac{\partial^n f}{\partial r^n} \right\|_{L^\infty(B(0,R))}}{(N+1)\cdots(N+n)}, \tag{7.35}$$

that is proving (7.30).

The function $f(x) = g(r) = (R-r)^n$ gives equality in (7.30), whether n is even or odd. $\qquad\square$

7.3 Functions on General Domains

The next result gives high order Ostrowski type inequalities functions on general bounded domains in $\mathbb{R}^N$.

Theorem 7.4. *Let $f \in W^{n,\infty}(\Omega) \cap C_0^{n-1}(\Omega)$, where Ω is a bounded domain in $\mathbb{R}^N$ and $C_0^{n-1}(\Omega) = \{u \in C^{n-1}(\overline{\Omega}): \ D^\alpha u = 0 \text{ on } \partial\Omega \text{ for } |\alpha| \le n-1\}$. Extend f by zero to F on $B(0,R)$, the smallest ball centered at the origin and containing Ω. Then $F \in W^{n,\infty}(B(0,R))$, and for all $x = r\omega \in \Omega$,*

$$\left| f(x) - \fint_\Omega f(y)dy \right| \le \mathcal{N}(F) + \left(1 - \frac{\mathrm{Vol}(\Omega)}{\mathrm{Vol}(B(0,R))} \right) \left| \fint_\Omega f(y)dy \right|$$

$$+ \frac{N}{R^N} \left\{ \left| \sum_{k=1}^{n-1} \frac{1}{k!} \left(\int_{S^{n-1}} \frac{\partial^k F(r\omega')}{\partial r^k} d\omega' \right) \left(\sum_{m=0}^{k} \binom{k}{m} (-1)^m r^m \left(\frac{R^{N+k-m}}{N+k-m} \right) \right) \right|$$

$$+ \frac{\|D^n f\|_{L^\infty(\Omega)}}{n!} \left(\sum_{m=0}^{n} \binom{n}{m} (-1)^m r^m \frac{R^{N+n-m}}{N+n-m} \right.$$

$$\left. \left. - \beta_n \sum_{m=0}^{n} \binom{n}{m} (-1)^m \frac{r^{N+n}}{N+n-m} \right) \right\}, \tag{7.36}$$

where $n \in \mathbb{N}$, $\beta_n = 0$ or 2, according as n is even or odd. Here

$$\|D^n f\|_{L^\infty(\Omega)} = \max_{|\alpha|=n} \|D^\alpha f\|_{L^\infty(\Omega)}. \tag{7.37}$$

Proof. Let $R := \inf\{R_0 > 0: \Omega \subset B(0,R_0)\}$. Then

$$F(x) := \begin{cases} f(x), & x \in \overline{\Omega} \\ 0, & x \in \overline{B(0,R)} - \Omega, \end{cases} \tag{7.38}$$

satisfies

$$F \in W^{n,\infty}(B(0,R)) \cap C_0^{n-1}(B(0,R)).$$

Then for $x \in \Omega$

$$\left| f(x) - \fint_\Omega f(y)dy \right| \le \left| F(x) - \fint_{B(0,R)} F(y)dy \right|$$

$$+ \left| \fint_{B(0,R)} F(y)dy - \fint_\Omega f(y)dy \right| = \mathcal{J}_1 + \mathcal{J}_2, \tag{7.39}$$

where

$$\mathcal{J}_1 := \left| F(x) - \fint_{B(0,R)} F(y)dy \right| \tag{7.40}$$

and

$$\mathcal{J}_2 := \left| \left(\frac{1}{\mathrm{Vol}(B(0,R))} - \frac{1}{\mathrm{Vol}(\Omega)} \right) \int_{\Omega} f(y) dy \right|$$

$$= \left[1 - \frac{\mathrm{Vol}(\Omega)}{\mathrm{Vol}(B(0,R))} \right] \left| \fint_{\Omega} f(y) dy \right|. \tag{7.41}$$

Theorem 7.1 applies and we may replace $\left\| \frac{\partial^n F}{\partial r^n} \right\|_{L^{\infty}(B(0,R))}$ by its majorant $\|D^n f\|_{L^{\infty}(\Omega)}$.

The theorem then follows. $\qquad\square$

Chapter 8

Ostrowski Inequalities on Spherical Shells

Here are presented Ostrowski type inequalities over spherical shells. These regard sharp or close to sharp estimates to the difference of the average of a multivariate function from its value at a point. This chapter relies on [47].

8.1 Introduction

The famous Ostrowski's inequality (1938), see [196], is

$$\left| \frac{1}{b-a} \int_a^b f(y)dy - f(x) \right| \leq \left(\frac{1}{4} + \frac{(x - \frac{a+b}{2})^2}{(b-a)^2} \right) (b-a) \|f'\|_\infty,$$

for $f \in C^1([a,b])$, $x \in [a,b]$, and it is a sharp inequality.

This was generalized from intervals to boxes in $\mathbb{R}^N, N \geq 1$, see [17], [21], p. 507. Here we establish Ostrowski type inequalities over spherical shells.

We present first the sharp results for the radial functions, then we move to the non-radial case. We use the polar method.

The estimates in both cases involve radial derivatives of arbitrary order of the engaged function.

At the end we give the connection of radial derivative to the ordinary partial derivative of the function.

8.2 Main Results

We make

Remark 8.1. Let A be a *spherical shell* $\subseteq \mathbb{R}^N, N \geq 1, i.e. A := B(0, R_2) - \overline{B(0, R_1)}, 0 < R_1 < R_2$.

Here the ball $B(0, R) := \{x \in \mathbb{R}^N : |x| < R\}, R > 0$, where $|\cdot|$ is the Euclidean norm , also $S^{N-1} := \{x \in \mathbb{R}^N : |x| = 1\}$ is the unit sphere in $\mathbb{R}^N$ with surface area $\omega_N := \frac{2\pi^{N/2}}{\Gamma(N/2)}$. For $x \in \mathbb{R}^N - \{0\}$ one can write uniquely $x = r\omega$, where $r > 0, \omega \in S^{N-1}$.

Let $f \in C^1(\bar{A})$. We suppose first that f is radial i.e $f(x) = g(r)$, where $r = |x|$, $R_1 \le r \le R_2$. Clearly here $g \in C^1([R_1, R_2])$.

In general it holds $\|\frac{\partial f}{\partial r}\|_\infty \le \| \nabla f \|_\infty$, with equality in the radial case.

For $F \in C(\bar{A})$ we have

$$\int_A F(x)dx = \int_{S^{N-1}} \left(\int_{R_1}^{R_2} F(r\omega)r^{N-1}dr \right) d\omega.$$

We notice that

$$\frac{N}{R_2^N - R_1^N} \int_{R_1}^{R_2} s^{N-1}ds = 1, \tag{8.1}$$

and

$$Vol(A) = \frac{\omega_N(R_2^N - R_1^N)}{N}. \tag{8.2}$$

Let $x \in A$. Then by using the polar method we derive

$$\left| f(x) - \frac{\int_A f(y)dy}{Vol(A)} \right| = \left| f(x) - \frac{N \int_{S^{N-1}} \left(\int_{R_1}^{R_2} f(s\omega)s^{N-1}ds \right) d\omega}{\omega_N(R_2^N - R_1^N)} \right| \tag{8.3}$$

$$= \left| g(r) - \frac{N \int_{S^{N-1}} \left(\int_{R_1}^{R_2} g(s)s^{N-1}ds \right) d\omega}{\omega_N(R_2^N - R_1^N)} \right| \tag{8.4}$$

$$= \left| g(r) - \left(\frac{N}{R_2^N - R_1^N} \right) \int_{R_1}^{R_2} g(s)s^{N-1}ds \right| \tag{8.5}$$

$$= \left| \left(\frac{N}{R_2^N - R_1^N} \right) \left(\int_{R_1}^{R_2} g(r)s^{N-1}ds - \int_{R_1}^{R_2} g(s)s^{N-1}ds \right) \right|$$

$$= \left(\frac{N}{R_2^N - R_1^N} \right) \left| \int_{R_1}^{R_2} (g(r) - g(s)) \, s^{N-1}ds \right| \tag{8.6}$$

$$\le \left(\frac{N}{R_2^N - R_1^N} \right) \int_{R_1}^{R_2} |g(r) - g(s)| \, s^{N-1}ds \le$$

$$\left(\frac{N}{R_2^N - R_1^N} \right) \|g'\|_\infty \int_{R_1}^{R_2} |r - s| \, s^{N-1}ds \tag{8.7}$$

$$= \left(\frac{N}{R_2^N - R_1^N} \right) \|g'\|_\infty \left[\int_{R_1}^{r} (r - s)s^{N-1}ds + \int_{r}^{R_2} (s - r)s^{N-1}ds \right] \tag{8.8}$$

$$= \left(\frac{N}{R_2^N - R_1^N} \right) \|g'\|_\infty \left[r \left(\frac{2r^N - (R_1^N + R_2^N)}{N} \right) \right.$$

$$+ \left(\frac{R_1^{N+1} + R_2^{N+1} - 2r^{N+1}}{N+1} \right) \Bigg]. \tag{8.9}$$

So we have established the first main result.

Theorem 8.2. *Let $f \in C^1(\bar{A})$ be radial, i.e. $f(x) = g(r)$, $R_1 \leq r \leq R_2$, $x \in \bar{A}$.*
Then

$$\left| f(x) - \frac{\int_A f(y)dy}{Vol(A)} \right| = \left| g(r) - \left(\frac{N}{R_2^N - R_1^N} \right) \int_{R_1}^{R_2} g(s) s^{N-1} ds \right| \leq$$

$$\left(\frac{N}{R_2^N - R_1^N} \right) \|g'\|_\infty \left[r \left(\frac{2r^N - (R_1^N + R_2^N)}{N} \right) \right.$$

$$\left. + \left(\frac{R_1^{N+1} + R_2^{N+1} - 2r^{N+1}}{N+1} \right) \right] \tag{8.10}$$

$$= \left(\frac{N}{R_2^N - R_1^N} \right) \|\nabla f\|_\infty \left[|x| \left(\frac{2|x|^N - (R_1^N + R_2^N)}{N} \right) \right.$$

$$\left. + \left(\frac{R_1^{N+1} + R_2^{N+1} - 2|x|^{N+1}}{N+1} \right) \right]. \tag{8.11}$$

Optimality comes next

Theorem 8.3. *Inequality (8.10) is sharp. More precisely*
(i) it is asymptotically attained by $g^*(z) := |z-r|^\alpha$, $1 < \alpha \leq k$, when $0 < r < R$.
(ii) It is attained by $g^*(z) = (z - R_1)$, when $r = R_1$.
(iii) It is attained by $g^*(z) = (z - R_2)$, when $r = R_2$.

Proof. (i) We observe that

$$g^{*'}(z) = \alpha |z - r|^{\alpha - 1} sign(z - r)$$

and

$$\|g^{*'}\|_\infty = \alpha (max\{R_2 - r, \, r - R_1\})^{\alpha - 1},$$

along with $g^*(r) = 0$.
We see that

$$L.H.S(8.10) = \left(\frac{N}{R_2^N - R_1^N} \right) \int_{R_1}^{R_2} |s - r|^\alpha \, s^{N-1} ds$$

$$\xrightarrow{\alpha \to 1} \left(\frac{N}{R_2^N - R_1^N} \right) \int_{R_1}^{R_2} |s - r| s^{N-1} ds$$

$$= \left(\frac{N}{R_2^N - R_1^N}\right)\left[r\left(\frac{2r^N - (R_1^N + R_2^N)}{N}\right) + \left(\frac{R_1^{N+1} + R_2^{N+1} - 2r^{N+1}}{N+1}\right)\right]. \quad (8.12)$$

We also have

$$R.H.S(8.10) = \left(\frac{N}{R_2^N - R_1^N}\right)\alpha(max\{R_2 - r,\, r - R_1\})^{\alpha-1}$$

$$\left[r\left(\frac{2r^N - (R_1^N + R_2^N)}{N}\right) + \left(\frac{R_1^{N+1} + R_2^{N+1} - 2r^{N+1}}{N+1}\right)\right] \xrightarrow{\alpha \to 1}$$

$$\left(\frac{N}{R_2^N - R_1^N}\right)\left[r\left(\frac{2r^N - (R_1^N + R_2^N)}{N}\right) + \left(\frac{R_1^{N+1} + R_2^{N+1} - 2r^{N+1}}{N+1}\right)\right]. \quad (8.13)$$

That is

$$\lim_{\alpha \to 1} L.H.S(8.10) = \lim_{\alpha \to 1} R.H.S(8.10),$$

proving sharpness for the case.

(ii) We have $g^*(R_1) = 0$ and $\|g^{*'}\|_\infty = 1$.

Thus

$$L.H.S(8.10) = \left(\frac{N}{R_2^N - R_1^N}\right)\int_{R_1}^{R_2}(s - R_1)s^{N-1}ds$$

$$= \left(\frac{N}{R_2^N - R_1^N}\right)\left[\left(\frac{R_2^{N+1} - R_1^{N+1}}{N+1}\right) - R_1\left(\frac{R_2^N - R_1^N}{N}\right)\right] = R.H.S\,(8.10), \quad (8.14)$$

proving the attainability for the case.

(iii) We have $g^*(R_2) = 0$, and $\|g^{*'}\|_\infty = 1$.

Hence

$$L.H.S(8.10) = \left(\frac{N}{R_2^N - R_1^N}\right)\int_{R_1}^{R_2}(R_2 - s)s^{N-1}ds =$$

$$\left(\frac{N}{R_2^N - R_1^N}\right)\left[R_2\left(\frac{R_2^N - R_1^N}{N}\right) - \left(\frac{R_2^{N+1} - R_1^{N+1}}{N+1}\right)\right] = R.H.S\,(8.10), \quad (8.15)$$

proving attainability of last case. $\qquad\qquad\qquad\qquad\qquad\qquad\qquad\square$

We would line to rewrite Theorem 8.3 for the equivalent inequality (8.10) in terms of f. We have

Theorem 8.4. *Let $x \in \bar{A}$. Inequality (8.10) is sharp as follows:*

(i) Let $0 < |x| < R$, then it is asymptotically attained by

$$f^*(w) := |\,|w| - |x|\,|^\alpha,\, 1 < \alpha \le k.$$

(ii) Let $|x| = R_1$, then it is asymptotically attained by

$$f^*(w) = |w| - R_1.$$

(iii) Let $|x| = R_2$, then it is asymptotically attained by

$$f^*(w) = |w| - R_2.$$

We continue from Remark 8.1 into

Remark 8.5. Now $f \in C^n(\bar{A})$, $n \in \mathbb{N}$, again radial such that $f(x) = g(r)$, where $r = |x|$, $x \in \bar{A}$, $R_1 \leq r \leq R_2$. Hence $g \in C^n([R_1, R_2])$.

Using the polar method we get again

$$E := \left| f(x) - \frac{\int_A f(y)dy}{Vol(A)} \right| = \left(\frac{N}{R_2^N - R_1^N} \right) \left| \int_{R_1}^{R_2} (g(r) - g(s))s^{N-1}ds \right|. \tag{8.16}$$

Let $s, r \in [R_1, R_2]$, then by Taylor's formula we obtain

$$g(s) - g(r) = \sum_{k=1}^{n-1} \frac{g^{(k)}(r)}{k!}(s - r)^k + R_{n-1}(r, s), \tag{8.17}$$

where

$$R_{n-1}(r, s) := \int_r^s \left(g^{(n-1)}(t) - g^{(n-1)}(r) \right) \frac{(s - t)^{n-2}}{(n - 2)!} dt. \tag{8.18}$$

As in [21], p. 500, we find

$$|R_{n-1}(r, s)| \leq \frac{\|g^{(n)}\|_{\infty, [R_1, R_2]}}{n!}|s - r|^n, \tag{8.19}$$

$\forall \, s, r \in [R_1, R_2]$.

Therefore

$$E = \left(\frac{N}{R_2^N - R_1^N} \right) \left| \int_{R_1}^{R_2} \left(\sum_{k=1}^{n-1} \frac{g^{(k)}(r)}{k!}(s - r)^k + R_{n-1}(r, s) \right) s^{N-1}ds \right| \tag{8.20}$$

$$\leq \left(\frac{N}{R_2^N - R_1^N} \right) \left[\sum_{k=1}^{n-1} \frac{|g^{(k)}(r)|}{k!} \left| \int_{R_1}^{R_2} s^{N-1}(s - r)^k ds \right| + \int_{R_1}^{R_2} s^{N-1}|R_{n-1}(r, s)|ds \right]$$

$$\leq \left(\frac{N}{R_2^N - R_1^N} \right) \left[\sum_{k=1}^{n-1} \frac{|g^k(r)|}{k!} \left| \sum_{m=0}^{k} \binom{k}{m}(-1)^m r^m \right. \right.$$

$$\left. \left(\frac{R_2^{N+k-m} - R_1^{N+k-m}}{N + k - m} \right) \right| + \frac{\|g^{(n)}\|_\infty}{n!} \int_{R_1}^{R_2} s^{N-1}|s - r|^n ds \right]. \tag{8.21}$$

But one finds that

$$\int_{R_1}^{R_2} s^{n-1}|s - r|^n ds$$

$$= \sum_{m=0}^{n} \binom{n}{m}(-1)^m \left[r^{n-m}\left(\frac{r^{m+N} - R_1^{m+N}}{m+N}\right) + r^m\left(\frac{R_2^{N+n-m} - r^{N+n-m}}{N+n-m}\right)\right].$$

$$(8.22)$$

Putting all the above together we have derived

Theorem 8.6. *Let $f \in C^n(\bar{A}), n \in \mathbb{N}$, be radial, i.e. $f(x) = g(r), R_1 \le r \le R_2, x \in \bar{A}$.*
    Then

$$\left| f(x) - \frac{\int_A f(y)dy}{Vol(A)} \right| = \left| g(r) - \left(\frac{N}{R_2^N - R_1^N}\right) \int_{R_1}^{R_2} g(s)s^{N-1}ds \right|$$

$$\le \left(\frac{N}{R_2^N - R_1^N}\right)\left[\sum_{k=1}^{n-1} |g^{(k)}(r)| \left| \sum_{m=0}^{k} \frac{(-1)^m r^m}{m!(k-m)!}\left(\frac{R_2^{N+k-m} - R_1^{N+k-m}}{N+k-m}\right)\right| \right.$$

$$+ \left\| g^{(n)} \right\|_\infty \left[\sum_{m=0}^{n} \frac{(-1)^m}{m!(n-m)!}\left[r^{n-m}\left(\frac{r^{m+N} - R_1^{m+N}}{m+N}\right)\right.\right.$$

$$\left.\left.\left. +r^m\left(\frac{R_2^{N+n-m} - r^{N+n-m}}{N+n-m}\right)\right]\right]\right]$$

$$(8.23)$$

$$= \left(\frac{N}{R_2^N - R_1^N}\right)\left\{\sum_{k=1}^{n-1} \left|\frac{\partial^k f(x)}{\partial r^k}\right| \left|\sum_{m=0}^{k} \frac{(-1)^m |x|^m}{m!(k-m)!}\left(\frac{R_2^{N+k-m} - R_1^{N+k-m}}{N+k-m}\right)\right| \right.$$

$$+ \left\|\frac{\partial^n f}{\partial r^n}\right\|_{\infty,\bar{A}} \cdot \left[\sum_{m=0}^{n} \frac{(-1)^m}{m!(n-m)!}\left[|x|^{n-m}\left(\frac{|x|^{m+N} - R_1^{m+N}}{m+N}\right)\right.\right.$$

$$\left.\left.\left. + |x|^m\left(\frac{R_2^{N+n-m} - |x|^{N+n-m}}{N+n-m}\right)\right]\right]\right\}.$$

$$(8.24)$$

We give

Corollary 8.7. *Let $f \in C^n(\bar{A}), n \in \mathbb{N}$, be radial i.e. $f(x) = f(r), R_1 \le r \le R_2, x \in \bar{A}$. Suppose $\frac{\partial^i f(x_0)}{\partial^i r^i} = g^{(i)}(r_0) = 0, i = 1, \ldots, n-1$, for $r_0 \in [R_1, R_2]$, $x_0 = r_0\omega \in \bar{A}$, $\omega \in S^{N-1}$.*
    Then

$$\left| f(x_0) - \frac{\int_A f(y)dy}{Vol(A)} \right| = \left| g(r_0) - \left(\frac{N}{R_2^N - R_1^N}\right) \int_{R_1}^{R_2} g(s)s^{N-1}ds \right|$$

$$\le \frac{N\|g^{(n)}\|_\infty}{R_2^N - R_1^N}\left[\sum_{m=0}^{n} \frac{(-1)^m}{m!(n-m)!}\right.$$

$$\left[r_0^{n-m}\left(\frac{r_0^{m+N} - R_1^{m+N}}{m+N} \right) + r_0^m \left(\frac{R_2^{N+n-m} - r_0^{N+n-m}}{N+n-m} \right) \right] \right] \tag{8.25}$$

$$= \frac{N \left\| \frac{\partial^n f}{\partial r^n} \right\|_{\infty,\overline{A}}}{R_2^N - R_1^N} \left[\sum_{m=0}^{n} \frac{(-1)^m}{m!(n-m)!} \right.$$

$$\left[|x_0|^{n-m}\left(\frac{|x_0|^{m+N} - R_1^{m+N}}{m+N} \right) + |x_0|^m \left(\frac{R_2^{N+n-m} - |x_0|^{N+n-m}}{N+n-m} \right) \right] \right] . \tag{8.26}$$

We also have the extreme cases.

Corollary 8.8. *Let $f \in C^n(\overline{A})$, $n \in \mathbb{N}$, be radial ; $f(x) = g(r)$, $r \in [R_1, R_2]$, $x \in \overline{A}, x = r\omega$. Suppose that $\frac{\partial^i f(x)}{\partial r^i}$, $i = 1,\ldots,n-1$, are zero on $\partial B(0, R_1)$, i.e. $g^{(i)}(R_1) = 0$, $i = 1,\ldots,n-1$. Then for $x \in \partial B(0, R_1)$ we have*

$$\left| f(x) - \frac{\int_A f(y)dy}{Vol(A)} \right| = \left| g(R_1) - \frac{N}{R_2^N - R_1^N} \int_{R_1}^{R_2} g(s)s^{N-1}ds \right|$$

$$\leq \left(\frac{N}{R_2^N - R_1^N} \right) \|g^{(n)}\|_\infty \left[\sum_{m=0}^{n} \frac{(-1)^m}{m!(n-m)!} R_1^m \left(\frac{R_2^{N+n-m} - R_1^{N+n-m}}{N+n-m} \right) \right] \tag{8.27}$$

$$= \left(\frac{N}{R_2^N - R_1^N} \right) \left\| \frac{\partial^n f}{\partial r^n} \right\|_{\infty,\overline{A}} \left[\sum_{m=0}^{n} \frac{(-1)^m}{m!(n-m)!} R_1^m \left(\frac{R_2^{N+n-m} - R_1^{N+n-m}}{N+n-m} \right) \right] . \tag{8.28}$$

Another extreme case follows.

Corollary 8.9. *Let $f \in C^n(\overline{A})$, $n \in \mathbb{N}$, be radial; $f(x) = g(r)$, $r \in [R_1, R_2]$, $x \in bar A; x = r\omega$. Assume that $\frac{\partial^i f(x)}{\partial r^i}, i = 1,\ldots,n-1$, are zero on $\partial B(0, R_2)$, i.e., $g^{(i)}(R_2) = 0, i = 1,\ldots,n-1$. Then for $x \in \partial B(0, R_2)$ we have*

$$\left| f(x) - \frac{\int_A f(y)dy}{Vol(A)} \right| = \left| g(R_2) - \frac{N}{R_2^N - R_1^N} \int_{R_1}^{R_2} g(s)s^{N-1}ds \right|$$

$$\leq \left(\frac{N}{R_2^N - R_1^N} \right) \|g^{(n)}\|_\infty \left[\sum_{m=0}^{n} \frac{(-1)^m}{m!(n-m)!} R_2^{n-m} \left(\frac{R_2^{m+N} - R_1^{m+N}}{m+N} \right) \right] \tag{8.29}$$

$$= \left(\frac{N}{R_2^N - R_1^N} \right) \left\| \frac{\partial^n f}{\partial r^n} \right\|_{\infty,\overline{A}} \left[\sum_{m=0}^{n} \frac{(-1)^m}{m!(n-m)!} R_2^{n-m} \left(\frac{R_2^{m+N} - R_1^{m+N}}{m+N} \right) \right] . \tag{8.30}$$

Optimality follows.

Proposition 8.10. *Inequality (8.25) is sharp, namely it is attained by*

$$g^*(y) := (y - r_0)^n, \quad y, \quad r_0 \in [R_1, R_2], \quad \text{when } n \text{ is even.}$$

Proof.   Notice

$$g^{*^{(j)}}(r_0) = 0, \quad j = 0, 1, \ldots, \quad n-1 \ \text{ and } \ \|g^{*^{(n)}}\|_\infty = n!.$$

We observe

$$L.H.S(8.25) = \left(\frac{N}{R_2^N - R_1^N}\right) \int_{R_1}^{R_2} (s - r_0)^n s^{N-1} ds$$

$$= \left(\frac{N}{R_2^N - R_1^N}\right) \left[\sum_{m=0}^{n} \binom{n}{m}(-1)^m r_0^m \left(\frac{R_2^{N+n-m} - R_1^{N+n-m}}{N+n-m}\right)\right]. \tag{8.31}$$

Next we see that

$$R.H.S\,(8.25) = \left(\frac{N}{R_2^N - R_1^N}\right) \left[\sum_{m=0}^{n} \binom{n}{m}(-1)^m\right.$$

$$\left.\left[\left(\frac{r_0^{n+N} - r_0^{n-m} R_1^{m+N}}{m+N}\right) + \left(\frac{r_0^m R_2^{N+n-m} - r_0^{N+n}}{N+n-m}\right)\right]\right] \tag{8.32}$$

$$= \left(\frac{N}{R_2^N - R_1^N}\right) \left\{\sum_{m=0}^{n} \binom{n}{m}(-1)^m \left[\frac{r_0^m R_2^{N+n-m}}{N+n-m} - \frac{r_0^{n-m} R_1^{m+N}}{m+N}\right]\right.$$

$$\left. + r_0^{n+N} \sum_{m=0}^{n} \binom{n}{m}(-1)^m \left[\frac{1}{N+m} - \frac{1}{N+n-m}\right]\right\} \tag{8.33}$$

$$= \left(\frac{N}{R_2^N - R_1^N}\right) \left\{\sum_{m=0}^{n} \binom{n}{m}(-1)^m \frac{r_0^m R_2^{N+n-m}}{N+n-m}\right.$$

$$\left. - \sum_{m=0}^{n} \binom{n}{m}(-1)^m \frac{r_0^{n-m} R_1^{m+N}}{m+N}\right\} \tag{8.34}$$

$$= \left(\frac{N}{R_2^N - R_1^N}\right) \left\{\sum_{m=0}^{n} \binom{n}{m}(-1)^m \frac{r_0^m R_2^{N+n-m}}{N+n-m}\right.$$

$$\left. - \sum_{m=0}^{n} \binom{n}{m}(-1)^m r_0^m \frac{R_1^{N+n-m}}{N+n-m}\right\} \tag{8.35}$$

$$= \left(\frac{N}{R_2^N - R_1^N}\right) \left\{\sum_{m=0}^{n} \binom{n}{m}(-1)^m r_0^m \frac{(R_2^{N+n-m} - R_1^{N+n-m})}{N+n-m}\right\}$$

$$= L.H.S \quad (8.25). \tag{8.36}$$

That is proving the claim.   □

The other optimal case follows.

Proposition 8.11. *Inequality (8.25) is sharp, namely it is asymptotically attained by* $g^*(y) := |y - r_0|^{n-1+\alpha}$, y, $r_0 \in [R_1, R_2]$, $1 < \alpha \le T$, *in the case of n is odd.*

Proof. It holds $g^{*(k)}(r_0) = 0$, $k = 0, 1, \ldots, n - 1$.

Also we have
$$g^{*(n)}(y) = (n - 1 + \alpha)(n - 2 + \alpha)\cdots(\alpha + 1)\alpha|y - r_0|^{\alpha-1}sign(y - r_0).$$

That is
$$\left|g^{*(n)}(y)\right| = \left(\prod_{j=1}^{n}(n - j + \alpha)\right)|y - r_0|^{\alpha-1},$$

and
$$\left\|g^{*(n)}\right\|_\infty = \left(\prod_{j=1}^{n}(n - j + \alpha)\right)(max\{R_2 - r_0, r_0 - R_1\})^{\alpha-1}. \tag{8.37}$$

Consequently,
$$L.H.S(8.25) = \left(\frac{N}{R_2^N - R_1^N}\right)\int_{R_1}^{R_2}|s - r_0|^{n-1+\alpha}s^{N-1}ds \xrightarrow{\alpha \to 1}$$

$$\left(\frac{N}{R_2^N - R_1^N}\right)\int_{R_1}^{R_2}|s - r_0|^n s^{N-1}ds \tag{8.38}$$

$$= \left(\frac{N}{R_2^N - R_1^N}\right)\left\{\sum_{m=0}^{n}\binom{n}{m}(-1)^m\left[r_0^{n-m}\left(\frac{r_0^{m+N} - R_1^{m+N}}{m + N}\right)\right.\right. \tag{8.39}$$

$$\left.\left.+r_0^m\left(\frac{R_2^{N+n-m} - r_0^{N+n-m}}{N + n - m}\right)\right]\right\} =: \mu. \tag{8.40}$$

Next we derive
$$R.H.S(8.25) = \left(\frac{N}{R_2^N - R_1^N}\right)\frac{\left(\prod_{j=1}^{n}(n - j + \alpha)\right)(max\{R_2 - r_0, r_0 - R_1\})^{\alpha-1}}{n!}$$

$$\left\{\sum_{m=0}^{n}\binom{n}{m}(-1)^m\left\{r_0^{n-m}\left(\frac{r_0^{m+N} - R_1^{m+N}}{m + N}\right) + r_0^m\left(\frac{R_2^{N+n-m} - r_0^{N+n-m}}{N + n - m}\right)\right\}\right\}$$
$$\tag{8.41}$$

$$\xrightarrow{\alpha \to 1} \left(\frac{N}{R_2^N - R_1^N}\right)\left\{\sum_{m=0}^{n}\binom{n}{m}(-1)^m\right.$$

$$\left.\left\{r_0^{n-m}\left(\frac{r_0^{m+N} - R_1^{m+N}}{m + N}\right) + r_0^m\left(\frac{R_2^{N+n-m} - r_0^{N+n-m}}{N + n - m}\right)\right\}\right\} = \mu. \tag{8.42}$$

That is, $L.H.S$ $(8.25), R.H.S(8.25) \to \mu$, proving asymptotic attainability and sharpness of (8.25). $\square$

Optimality of external inequality (8.25) follows.

Proposition 8.12. *External inequality (8.25) is asymptotically attained, that is sharp as follows:*

(i) when n is even, then optimal function is $f^*(w) := (|w| - |x_0|)^n, w \in \bar{A}$.

(ii) when n is odd, then optional function is $f^*(w) :=| |w| - |x_0| |^{n-1+\alpha}, 1 < \alpha \leq T, \quad w \in \bar{A}$.

Similarly we obtain

Proposition 8.13. *Inequalities (8.23), (8.27) and (8.29) are asymptotically attained, therefore sharp, as inequality (8.25).*

A simple but general result follows.

Theorem 8.14. *Let $\emptyset \neq \mathcal{R}$ be a convex bounded region of $\mathbb{R}^N, N \geq 1$. Let $f \in C^1(\overline{\mathcal{R}})$. Then*

$$\left| f(x) - \frac{\int_{\mathcal{R}} f(y)dy}{Vol(\mathcal{R})} \right| \leq \frac{\|\nabla f\|_\infty}{Vol(\mathcal{R})} \int_{\mathcal{R}} |x - y|dy, x \in \overline{\mathcal{R}}. \tag{8.43}$$

Proof.   We observe that

$$\left| f(x) - \frac{\int_{\mathcal{R}} f(y)dy}{Vol(\mathcal{R})} \right| = \frac{1}{Vol(\mathcal{R})} \left| f(x)Vol(\mathcal{R}) - \int_{\mathcal{R}} f(y)dy \right|$$

$$= \frac{1}{Vol(\mathcal{R})} \left| \int_{\mathcal{R}} (f(x) - f(y))dy \right| \tag{8.44}$$

$$\leq \frac{1}{Vol(\mathcal{R})} \int_{\mathcal{R}} |f(x) - f(y)|dy$$

$$= (z \text{ belongs to the line segment from } x \, to \, y)$$

$$\frac{1}{Vol(\mathcal{R})} \int_{\mathcal{R}} |\nabla f(z) \cdot (x - y)|dy \tag{8.45}$$

$$\leq \frac{1}{Vol(\mathcal{R})} \int_{\mathcal{R}} |\nabla f(z)| \cdot |x - y|dy \leq \frac{\|\nabla f\|_\infty}{Vol(\mathcal{R})} \int_{\mathcal{R}} |x - y|dy. \tag{8.46}$$

Specializing on the shell and sphere we have $\qquad\qquad\square$

Proposition 8.15. *Let $f \in C^1(\bar{A})$, or $f \in C^1(\overline{B(0, R)}), R > 0$. Then*

(i)

$$\left| f(x) - \frac{N\Gamma(\frac{N}{2})}{2\pi^{N/2}(R_2^N - R_1^N)} \int_A f(y)dy \right|$$

$$\leq \frac{N\Gamma(\frac{N}{2})}{2\pi^{N/2}(R_2^N - R_1^N)} \|\nabla f\|_\infty \int_A |x - y|\,dy, \quad x \in \bar{A}; \tag{8.47}$$

also it holds

(ii)

$$\left| f(x) - \frac{N\Gamma(\frac{N}{2})}{2\pi^{N/2}R^N} \int_{B(0,R)} f(y)dy \right| \le \frac{N\Gamma(\frac{N}{2})}{2\pi^{N/2}R^N} \|\nabla f\|_\infty \int_{B(0,R)} |x-y|\,dy,$$

$x \in \overline{B(0,R)}$.

More precise Ostrowski type inequalities for general, not necessarily radial functions, follow.

Theorem 8.16. *Let $f \in C^1(\bar{A})$, $x \in \bar{A}$, $x = r\omega$, $r > 0$. Then*

$$\left| f(x) - \frac{\int_A f(y)dy}{Vol(A)} \right| \le \left| f(x) - \frac{\int_{S^{N-1}} f(r\omega)d\omega}{\omega_N} \right|$$

$$+ \left(\frac{N}{R_2^N - R_1^N} \right) \left\| \frac{\partial f}{\partial r} \right\|_\infty \left[|x| \left(\frac{2|x|^N - (R_1^N + R_2^N)}{N} \right) \right.$$

$$\left. + \left(\frac{R_1^{N+1} + R_2^{N+1} - 2|x|^{N+1}}{N+1} \right) \right] \tag{8.48}$$

$$\le \left| f(x) - \frac{\Gamma(\frac{N}{2}) \int_{S^{N-1}} f(r\omega)d\omega}{2\pi^{N/2}} \right| + \left(\frac{N}{R_2^N - R_1^N} \right) \|\nabla f\|_\infty$$

$$\left[|x| \left(\frac{2|x|^N - (R_1^N + R_2^N)}{N} \right) + \left(\frac{R_1^{N+1} + R_2^{N+1} - 2|x|^{N+1}}{N+1} \right) \right]. \tag{8.49}$$

Proof. Applying internal (8.10) to $f(r\omega)$ we get

$$\left| f(r\omega) - \left(\frac{N}{R_2^N - R_1^N} \right) \int_{R_1}^{R_2} f(s\omega)s^{N-1}ds \right|$$

$$\le \left(\frac{N}{R_2^N - R_1^N} \right) \left\| \frac{\partial f(r\omega)}{\partial r} \right\|_{\infty, r \in [R_1, R_2]} \left[r \left(\frac{2r^N - (R_1^N + R_2^N)}{N} \right) \right.$$

$$\left. + \left(\frac{R_1^{N+1} + R_2^{N+1} - 2r^{N+1}}{N+1} \right) \right] \tag{8.50}$$

$$\le \left(\frac{N}{R_2^N - R_1^N} \right) \left\| \frac{\partial f}{\partial r} \right\|_{\infty, \bar{A}} \left[|x| \left(\frac{2|x|^N - (R_1^N + R_2^N)}{N} \right) \right.$$

$$\left. + \left(\frac{R_1^{N+1} + R_2^{N+1} - 2|x|^{N+1}}{N+1} \right) \right]. \tag{8.51}$$

Hence it holds

$$\left| \frac{\int_{S^{N-1}} f(r\omega)d\omega}{\omega_N} - \frac{N}{\omega_N(R_2^N - R_1^N)} \int_{S^{n-1}} \left(\int_{R_1}^{R_2} f(s\omega)s^{N-1}ds \right) d\omega \right|$$

$$\leq \left(\frac{N}{R_2^n - R_1^N} \right) \left\| \frac{\partial f}{\partial r} \right\|_\infty \left[|x| \left(\frac{2|x|^N - (R_1^N + R_2^N)}{N} \right) \right.$$

$$\left. + \left(\frac{R_1^{N+1} + R_2^{N+1} - 2|x|^{N+1}}{N+1} \right) \right], \tag{8.52}$$

proving the claim. $\qquad\qquad\square$

We continue with

Theorem 8.17. *Let $f \in C^n(\bar{A})$, $\quad n \in \mathbb{N}$, $\quad x \in \bar{A}$, $\quad x = r\omega$, $\quad r > 0$. Then*

$$\left| f(x) - \frac{\int_A f(y)dy}{Vol(A)} \right| \leq \left| f(x) - \frac{\Gamma(\frac{N}{2}) \int_{S^{N-1}} f(r\omega)d\omega}{2\pi^{N/2}} \right|$$

$$+ \left(\frac{N}{R_2^N - R_1^N} \right) \left\{ \frac{\Gamma(\frac{N}{2})}{2\pi^{N/2}} \left\{ \sum_{k=1}^{n-1} \left(\int_{S^{N-1}} \left| \frac{\partial^k f(r\omega)}{\partial r^k} \right| d\omega \right) \right.\right.$$

$$\left| \sum_{m=0}^{k} \frac{(-1)^m |x|^m}{m!(k-m)!} \left(\frac{R_2^{N+k-m} - R_1^{N+k-m}}{N+k-m} \right) \right| \right\}$$

$$+ \left\| \frac{\partial^n f}{\partial r^n} \right\|_\infty \left[\sum_{m=0}^{n} \frac{(-1)^m}{m!(n-m)!} \left[|x|^{n-m} \left(\frac{|x|^{m+N} - R_1^{m+N}}{m+N} \right) \right.\right.$$

$$\left.\left.\left.+ |x|^m \left(\frac{R_2^{N+n-m} - |x|^{N+n-m}}{N+n-m} \right) \right] \right] \right\}. \tag{8.53}$$

Proof. Applying internal (8.23) to $f(r\omega)$ we obtain

$$\left| f(r\omega) - \left(\frac{N}{R_2^N - R_1^N} \right) \int_{R_1}^{R_2} f(s\omega)s^{N-1}ds \right|$$

$$\leq \left(\frac{N}{R_2^N - R_1^N} \right) \left\{ \sum_{k=1}^{n-1} \left| \frac{\partial^k f(r\omega)}{\partial r^k} \right| \left| \sum_{m=0}^{k} \frac{(-1)^m |x|^m}{m!\,(k-m)!} \right.\right.$$

$$\left(\frac{R_2^{N+k-m} - R_1^{N+k-m}}{N+k-m} \right) \right| + \left\| \frac{\partial^n f}{\partial r^n} \right\|_{\infty,\bar{A}} \left[\sum_{m=0}^{n} \frac{(-1)^m}{m!(n-m)!} \right.$$

$$\left[|x|^{n-m}\left(\frac{|x|^{m+N}-R_1^{m+N}}{m+N}\right)+|x|^m\left(\frac{R_2^{N+n-m}-|x|^{N+n-m}}{N+n-m}\right)\right]\right]\right\}. \qquad (8.54)$$

Therefore it holds

$$\left|\frac{\Gamma(\frac{N}{2})}{2\pi^{N/2}}\int_{S^{N-1}}f(r\omega)d\omega-\frac{1}{Vol(A)}\int_A f(y)dy\right|$$

$$\leq\left(\frac{N}{R_2-R_1^N}\right)\left\{\frac{\Gamma(N/2)}{2\pi^{N/2}}\left\{\sum_{k=1}^{n-1}\left(\int_{S^{N-1}}\left|\frac{\partial^k f(r\omega)}{\partial r^k}\right|d\omega\right)\right.\right.$$

$$\left.\left|\sum_{m=0}^{k}\frac{(-1)^m|x|^m}{m!(k-m)!}\left(\frac{R_2^{N+k-m}-R_1^{N+k-m}}{N+k-m}\right)\right|\right\}$$

$$+\left\|\frac{\partial^n f}{\partial r^n}\right\|_{\infty,\bar{A}}\left[\sum_{m=0}^{n}\frac{(-1)^m}{m!(n-m)!}\left[|x|^{n-m}\left(\frac{|x|^{m+N}-R_1^{m+N}}{m+N}\right)\right.\right.$$

$$\left.\left.+|x|^m\left(\frac{R_2^{N+n-m}-|x|^{N+n-m}}{N+n-m}\right)\right]\right]\right\}, \qquad (8.55)$$

proving the claim. $\qquad\qquad\Box$

We also give

Proposition 8.18. *Let* $f\in C^n(\bar{A})$, $n\in\mathbb{N}$, *such that* $\frac{\partial^i f}{\partial r^i}$, $i=1,\ldots,n-1$, *are zero on* $\partial B(0,r_0)$, $r_0\in(R_1,R_2)$. *Then for* $x_0\in\partial B(0,r_0)$ *we have*

$$\left|f(x_0)-\frac{\int_A f(y)dy}{Vol(A)}\right|\leq\left|f(x_0)-\frac{\Gamma(N/2)}{2\pi^{N/2}}\int_{S^{N-1}}f(r_0\omega)d\omega\right|$$

$$+\left(\frac{N}{R_2^N-R_1^N}\right)\left\|\frac{\partial^n f}{\partial r^n}\right\|_{\infty}\left[\sum_{m=0}^{n}\frac{(-1)^m}{m!(n-m)!}\left[|x_0|^{n-m}\left(\frac{|x_0|^{m+N}-R_1^{m+N}}{m+N}\right)\right.\right.$$

$$\left.\left.+|x_0|^m\left(\frac{R_2^{N+n-m}-|x_0|^{N+n-m}}{N+n-m}\right)\right]\right]. \qquad (8.56)$$

Proof. Applying internal (8.25) to $f(r\omega)$ we find

$$\left|f(r_0\omega)-\left(\frac{N}{R_2^N-R_1^N}\right)\int_{R_1}^{R_2}f(s\omega)s^{N-1}ds\right|$$

$$\leq\left(\frac{N}{R_2^N-R_1^N}\right)\left\|\frac{\partial^n f}{\partial r^n}\right\|_{\infty,A}\left[\sum_{m=0}^{n}\frac{(-1)^m}{m!(n-m)!}\right.$$

$$\left[|x_0|^{n-m}\left(\frac{|x_0|^{m+N}-R_1^{m+N}}{m+N}\right)+|x_0|^m\left(\frac{R_2^{N+n-m}-|x_0|^{N+n-m}}{N+n-m}\right)\right]\right]. \tag{8.57}$$

Therefore

$$\left|\frac{\Gamma(\frac{N}{2})}{2\pi^{N/2}}\int_{S^{N-1}}f(r_0\omega)d\omega-\frac{1}{Vol(A)}\int_A f(y)dy\right|$$

$$\leq\left(\frac{N}{R_2^N-R_1^N}\right)\left\|\frac{\partial^n f}{\partial r^n}\right\|_\infty\left[\sum_{m=0}^n\frac{(-1)^m}{m!(n-m)!}\left[|x_0|^{n-m}\left(\frac{|x_0|^{m+N}-R_1^{m+N}}{m+N}\right)\right.\right.$$

$$\left.\left.+|x_0|^m\left(\frac{R_2^{N+n-m}-|x_0|^{N+n-m}}{N+n-m}\right)\right]\right]. \tag{8.58}$$

Claim is clear. $\qquad\square$

We present the extreme cases.

Proposition 8.19. *Let* $f\in C^n(\bar{A})$, $n\in\mathbb{N}$ *such that* $\frac{\partial^i f}{\partial r^i}$, $i=1,\ldots,n-1$, *are zero on* $\partial B(0,R_1)$. *Then for* $x_0\in\partial B(0,R_1)$ *it holds*

$$\left|f(x_0)-\frac{\int_A f(y)dy}{Vol(A)}\right|\leq\left|f(x_0)-\frac{\Gamma(\frac{N}{2})}{2\pi^{N/2}}\int_{S^{N-1}}f(R_1\omega)d\omega\right|$$

$$+\left(\frac{N}{R_2^N-R_1^N}\right)\left\|\frac{\partial^n f}{\partial r^n}\right\|_\infty\left[\sum_{m=0}^n\frac{(-1)^m}{m!(n-m)!}R_1^m\left(\frac{R_2^{N+n-m}-R_1^{N+n-m}}{N+n-m}\right)\right] \tag{8.59}$$

Proof. By internal (8.27). $\qquad\square$

We finish the main results with

Proposition 8.20. *Let* $f\in C^n(\bar{A})$, $n\in\mathbb{N}$, *such that* $\frac{\partial^i f}{\partial r^i}$, $i=1,\ldots,n-1$, *are zero on* $\partial B(0,R_2)$. *Then for* $x_0\in\partial B(0,R_2)$ *we find*

$$\left|f(x_0)-\frac{\int_A f(y)dy}{Vol(A)}\right|\leq\left|f(x_0)-\frac{\Gamma\frac{(N)}{2}}{2\pi^{N/2}}\int_{S^{N-1}}f(R_2\omega)d\omega\right|$$

$$+\left(\frac{N}{R_2^N-R_1^N}\right)\left\|\frac{\partial^n f}{\partial r^n}\right\|_\infty\left[\sum_{m=0}^n\frac{(-1)^m}{m!(n-m)!}R_2^{n-m}\left(\frac{R_2^{m+N}-R_1^{m+N}}{m+N}\right)\right]. \tag{8.60}$$

Proof. By internal (8.29). $\qquad\square$

The radial derivatives appearing in the right hand sides of the inequalities can be expressed and estimated by regular partial derivatives in terms of $x_1,\ldots,x_N$. See Addendum next.

8.3 Addendum

Let $u \in C^n(\overline{B(0,R)})$, the open ball $B(0,R) \subseteq \mathbb{R}^N$, $n, N \in \mathbb{N}$. Here $x = (x_1, \cdots, x_N) \in \overline{B(0,R)}$ and the *radial derivative* of u is given also by

$$\frac{\partial u(x)}{\partial r} = \nabla u(x) \cdot \frac{x}{|x|}, \quad x \neq 0.$$

That is,

$$\frac{\partial u(x)}{\partial r} = \frac{1}{|x|} \left(\sum_{i=1}^{N} \frac{\partial u(x)}{\partial x_i} x_i \right), \quad x \neq 0. \tag{8.61}$$

In general for $1 \leq l \leq n$ it holds (by induction) that

$$\frac{\partial^l u(x)}{\partial r^l} = \frac{1}{|x|^l} \left[\sum_{k_1;\ldots;k_N, \sum_{j=1}^{N} k_j = l, k_j \in \mathbb{Z}_+} \frac{l!}{\Pi_{j=1}^{N} k_j!} \cdot \frac{\partial^l u(x)}{\Pi_{j=1}^{N} \partial^{k_j} x_j} \Pi_{j=1}^{N} x_j^{k_j} \right], x \neq 0. \tag{8.62}$$

For example, when $n = N = 2$ we obtain

$$\frac{\partial^2 u(x)}{\partial r^2} = \frac{1}{|x|^2} \left[\frac{\partial^2 u(x)}{\partial x_1^2} x_1^2 + 2 \frac{2\partial^2 u(x)}{\partial x_1 \partial x_2} x_1 x_2 + \frac{\partial^2 u(x)}{\partial x_2^2} x_2^2 \right], x \neq 0. \tag{8.63}$$

Thus we find

$$\left| \frac{\partial^l u(x)}{\partial r^l} \right| \leq \sum_{k_1;\ldots;k_N, \sum_{j=1}^{N} k_j = l, k_j \in \mathbb{Z}_+} \frac{l!}{\Pi_{j=1}^{N} k_j!} \cdot \left| \frac{\partial^l u(x)}{\Pi_{j=1}^{N} \partial^{k_j} x_j} \right|, x \neq 0, \tag{8.64}$$

or better in brief,

$$\left| \frac{\partial^l u(x)}{\partial r^l} \right| \leq \left(\sum_{i=1}^{N} \left| \frac{\partial}{\partial x_i} \right| \right)^l (u(x)), \tag{8.65}$$

all $x \in \overline{B(0,R)} - \{0\}$, all $1 \leq l \leq n$.

So if all the u partial derivatives vanish then the corresponding radial derivative is zero. Consequently, from (8.65) it holds for the essential supreme $\|\cdot\|_\infty$ that

$$\left\| \frac{\partial^l u}{\partial r^l} \right\|_{\infty, \overline{B(0,R)}} \leq \left(\sum_{i=1}^{N} \left\| \frac{\partial}{\partial x_i} \right\|_{\infty, \overline{B(0,R)}} \right)^l (u) < +\infty. \tag{8.66}$$

II. Continuing and specializing in the study of higher order radial derivatives of radial functions.

Let now u be radial i.e. $u(x) = g(|x|) = g(r)$; $x = r\omega$, $r \in \mathbb{R}_+$, $\omega \in S^{N-1}$. Here we will suppose $x \neq 0$ and $N = 2$. Let further $x_1, x_2 \neq 0$, then by chain rule one has

$$g'(r) = \frac{\partial u}{\partial x_1} \frac{r}{x_1} = \frac{\partial u}{\partial x_2} \frac{r}{x_2}, \quad \text{where } r = \sqrt{x_1^2 + x_2^2}. \tag{8.67}$$

That is,

$$g'(r) = \frac{1}{2} \left[\frac{\partial u}{\partial x_1} \frac{|x|}{x_1} + \frac{\partial u}{\partial x_2} \frac{|x|}{x_2} \right]. \tag{8.68}$$

So one has

$$|\nabla u(x)| = |g'(r)| = \left| \frac{\partial u(x)}{\partial r} \right|, \tag{8.69}$$

for any $x \in \overline{B(0, R)}$.

But if u radial, not necessarily $\frac{\partial u}{\partial x_j}$ is radial. Here we put $x_1 = r \cos \theta$, $x_2 = r \sin \theta$. Again for $x_1, x_2 \neq 0$ and via chain rule we derive

$$g''(r) = \frac{\partial^2 u(x)}{\partial x_1^2} + \frac{\partial^2 u(x)}{\partial x_2 \partial x_1} tan\,\theta = \frac{\partial^2 u(x)}{\partial x_2^2} + \frac{\partial^2 u(x)}{\partial x_1 \partial x_2} cot\,\theta. \tag{8.70}$$

That is

$$g''(r) = \frac{1}{2} \left[\frac{\partial^2 u(x)}{\partial x_1^2} + \frac{\partial^2 u(x)}{\partial x_2^2} + \frac{\partial^2 u(x)}{\partial x_1 \partial x_2} (tan\,\theta + cot\,\theta) \right]. \tag{8.71}$$

Or better, by using the Laplacian Δ we find

$$g''(r) = \frac{1}{2} \Delta u(x) + \frac{\partial^2 u(x)}{\partial x_1 \partial x_2} csc(2\,\theta), \tag{8.72}$$

or

$$g''(r) = \frac{1}{2} \left(\Delta u(x) + \frac{\partial^2 u(x)}{\partial x_1 \partial x_2} \frac{|x|^2}{x_1 x_2} \right), \tag{8.73}$$

$x_1, x_2 \neq 0$.

Similarly, one has that

$$g'''(r) = \frac{\partial^3 u(x)}{\partial x_1^3} cos\,\theta + 2 \frac{\partial^3 u(x)}{\partial x_2 \partial x_1^2} sin\,\theta + \frac{\partial^3 u(x)}{\partial x_2^2 \partial x_1} \frac{sin^2 \theta}{cos\,\theta}, \quad x_1, x_2 \neq 0. \tag{8.74}$$

Also we get

$$g'''(r) = 2 \frac{\partial^3 u(x)}{\partial x_1 \partial x_2^2} cos\,\theta + \frac{\partial^3 u(x)}{\partial x_2^3} sin\,\theta + \frac{\partial^3 u(x)}{\partial x_1^2 \partial x_2} \frac{cos^2 \theta}{sin\,\theta}, \quad x_1, x_2 \neq 0. \tag{8.75}$$

That is, by (8.74) and (8.75) we have

$$g'''(r) = \frac{1}{2} \left[\frac{\partial^3 u(x)}{\partial x_1^3} cos\,\theta + \frac{\partial^3 u(x)}{\partial x_2^3} sin\,\theta + \right.$$

$$2 \left(\frac{\partial^3 u(x)}{\partial x_2 \partial x_1^2} sin\,\theta + \frac{\partial^3 u(x)}{\partial x_1 \partial x_2^2} cos\,\theta \right) + \frac{\partial^3 u(x)}{\partial x_2^2 \partial x_1} \frac{sin^2 \theta}{cos\,\theta} + \left. \frac{\partial^3 u(x)}{\partial x_1^2 \partial x_2} \frac{cos^2 \theta}{sin\,\theta} \right], x_1, x_2 \neq 0. \tag{8.76}$$

Clearly, it holds

$$g'''(r) = \frac{1}{2} \left[\frac{\partial^3 u(x)}{\partial x_1^3} \frac{x_1}{|x|} + \frac{\partial^3 u(x)}{\partial x_2^3} \frac{x_2}{|x|} + \right.$$

$$\frac{\partial^3 u(x)}{\partial x_1^2 \partial x_2} \left(\frac{x_1^2}{x_2 |x|} + 2 \frac{x_2}{|x|} \right) + \left. \frac{\partial^3 u(x)}{\partial x_2^2 \partial x_1} \left(\frac{x_2^2}{x_1 |x|} + 2 \frac{x_1}{|x|} \right) \right], x_1, x_2 \neq 0. \tag{8.77}$$

Not general formula, as in (8.62), can be derived for $g^{(l)}(r)$, $l \in \mathbb{N}$, in the radial case. Of course (8.62) is valid for both non-radial and radial cases. Notice in (8.67), (8.70), (8.74) and (8.75) are used less number of terms than in the corresponding non-radial cases.

Chapter 9

Ostrowski Inequalities on Balls and Shells Via Taylor–Widder Formula

The classical Ostrowski inequality for functions on intervals estimates the value of the function minus its average in terms of the maximum of its first derivative. This result is extended to higher order over shells and balls of $\mathbb{R}^N$, $N \geq 1$, with respect to an extended complete Tschebyshev system and the generalized radial derivatives of Widder type. We treat radial and non-radial functions. This chapter relies on [42].

9.1 Introduction

The classical Ostrowski inequality (of 1938, see [196]) is

$$\left| \frac{1}{b-a} \int_a^b f(y)dy - f(x) \right| \leq \left(\frac{1}{4} + \frac{\left(x - \dfrac{a+b}{2} \right)^2}{(b-a)^2} \right) (b-a)\|f'\|_\infty,$$

for $f \in C^1([a,b])$, $x \in [a,b]$, and it is a sharp inequality. This was extended to $\mathbb{R}^N$, $N \geq 1$, over balls and shells in [56], [57], [47]. Also this extension was done over boxes and rectangles, see [21], pp. 507-520, and [17], and [16]. The produced Ostrowski inequalities, in the above mentioned references, were mostly sharp and they involved the first and higher order derivatives of the engaged function f.

Here we derive a set of very general higher order Ostrowski type inequalities over shells and balls with respect to an extended complete Tschebyshev system (see [163])and generalized derivatives if Widder type (see [243]). The proofs are based on the polar method and the general Taylor-Widder formula (see [243], 1928). These results generalize the higher order Ostrowski type inequalities established in the above mentioned references.

125

9.2 Background

The following are taken from [243]. Let $f, u_0, u_1, \ldots, u_n \in C^{n+1}([a,b]), n \geq 0$, and the Wronskians

$$W_i(x) := W[u_0(x), u_1(x), \ldots, u_i(x)] := \begin{vmatrix} u_0(x) & u_1(x) & \ldots & u_i(x) \\ u_0'(x) & u_1'(x) & \ldots & u_i'(x) \\ \vdots & & & \\ u_0^{(i)}(x) & u_1^{(i)}(x) & \ldots & u_i^{(i)}(x) \end{vmatrix}, \; i = 0, 1, \ldots, n.$$

Assume $W_i(x) > 0$ over $[a,b]$. Clearly then

$$\phi_0(x) := W_0(x) = u_0(x), \; \phi_1(x) := \frac{W_1(x)}{(W_0(x))^2}, \ldots, \phi_i(x) := \frac{W_i(x)W_{i-2}(x)}{(W_{i-1}(x))^2},$$

$i = 2, 3, \ldots, n$, are positive on $[a,b]$.

For $i \geq 0$, the linear differentiable operator of order i:

$$L_i f(x) := \frac{W[u_0(x), u_1(x), \ldots, u_{i-1}(x), f(x)]}{W_{i-1}(x)}, \quad i = 1, \ldots, n+1;$$

$$L_0 f(x) := f(x), \; \forall \; x \in [a,b].$$

Then for $i = 1, \ldots, n+1$ we have

$$L_i f(x) = \phi_0(x)\phi_1(x) \ldots \phi_{i-1}(x) \frac{d}{dx} \frac{1}{\phi_{i-1}(x)} \frac{d}{dx} \frac{1}{\phi_{i-2}(x)} \frac{d}{dx} \cdots \frac{d}{dx} \frac{1}{\phi_1(x)} \frac{d}{dx} \frac{f(x)}{\phi_0(x)}.$$

Consider also

$$g_i(x,t) := \frac{1}{W_i(t)} \begin{vmatrix} u_0(t) & u_1(t) & \ldots & u_i(t) \\ u_0'(t) & u_1'(t) & \ldots & u_i'(t) \\ \ldots & \ldots & \ldots & \ldots \\ u_0^{(i-1)}(t) & u_1^{(i-1)}(t) & \ldots & u_i^{(i-1)}(t) \\ u_0(x) & u_1(x) & \ldots & u_i(x) \end{vmatrix},$$

$i = 1, 2, \ldots, n$; $g_0(x,t) := \dfrac{u_0(x)}{u_0(t)}, \; \forall \; x, t \in [a,b]$.

Note that $g_i(x,t)$ as a function of x is a linear combination of $u_0(x), u_1(x), \ldots, u_i(x)$ and it holds

$$g_i(x,t)$$

$$= \frac{\phi_0(x)}{\phi_0(t) \ldots \phi_i(t)} \int_t^x \phi_1(x) \int_t^{x_1} \cdots \int_t^{x_{i-2}} \phi_{i-1}(x_{i-1}) \int_t^{x_{i-1}} \phi_i(x_i) dx_i dx_{i-1} \ldots dx_1$$

$$= \frac{1}{\phi_0(t) \ldots \phi_i(t)} \int_t^x \phi_0(s) \ldots \phi_i(s) g_{i-1}(x,s) ds, \; i = 1, 2, \ldots, n.$$

Example 9.1. ([243]). The sets

$$\{1, x, x^2, \ldots, x^n\}, \{1, \sin x, -\cos x, -\sin 2x, \cos 2x, \ldots, (-1)^{n-1} \sin nx, (-1)^n \cos nx\}$$

fulfill the above theory.

We mention

Theorem 9.2. (Karlin and Studded (1966), see p. 376, [163]). *Let $u_0, u_1, \ldots, u_n \in C^n([a, b])$, $n \geq 0$. Then $\{u_i\}_{i=0}^n$ is an extended complete Tschebyshev system on $[a, b]$ iff $W_i(x) > 0$ on $[a, b]$, $i = 0, 1, \ldots, n$.*

We also mention

Theorem 9.3. (D. Widder, p. 138, [243]). *Let the functions*

$$f(x), u_0(x), u_1(x), \ldots, u_n(x) \in C^{n+1}([a, b]),$$

and the Wronskians $W_0(x), W_1(x), \ldots, W_n(x) > 0$ on $[a, b]$, $x \in [a, b]$. Then for $t \in [a, b]$ we have

$$f(x) = f(t)\frac{u_0(x)}{u_0(t)} + L_1 f(t)g_1(x, t) + \cdots + L_n f(t)g_n(x, t) + R_n(x),$$

where

$$R_n(x) := \int_t^x g_n(x, s)L_{n+1}f(s)ds.$$

For example one could take $u_0(x) = c > 0$. If $u_i(x) = x^i$, $i = 0, 1, \ldots, n$, defined on $[a, b]$, then

$$L_i f(t) = f^{(i)}(t) \quad \text{and} \quad g_i(x, t) = \frac{(x - t)^i}{i!}, \quad t \in [a, b].$$

So under the assumptions of Theorem 9.3 we get

$$f(x) = f(y)\frac{u_0(x)}{u_0(y)} + \sum_{i=1}^n L_i f(y)g_i(x, y) + \int_y^x g_n(x, t)L_{n+1}f(t)dt, \ \forall \ x, y \in [a, b].$$

$$(9.1)$$

If $u_0(x) = c > 0$, then

$$f(x) = f(y) + \sum_{i=1}^n L_i f(y)g_i(x, y) + \int_y^x g_n(x, t)L_{n+1}f(t)dt, \ \forall \ x, y \in [a, b]. \quad (9.2)$$

We call L_i the generalized Widder-type derivative.

We need

Notation 9.4. Let A be a spherical shell $\subseteq \mathbb{R}^N$, $N \geq 1$, i.e. $A := B(0, R_2) - \overline{B(0, R_1)}$, $0 < R_1 < R_2$.

Here the ball $B(0, R) := \{x \in \mathbb{R}^N : |x| < R\}$, $R > 0$, where $|\cdot|$ is the Euclidean norm, also $S^{N-1} := \{x \in \mathbb{R}^N : |x| = 1\}$ is the unit sphere in $\mathbb{R}^N$ with surface area $\omega_N := \frac{2\pi^{N/2}}{\Gamma(N/2)}$. For $x \in \mathbb{R}^N - \{0\}$ one can write uniquely $x = r\omega$, where $r > 0$, $\omega \in S^{N-1}$.

Let $f \in C^{n+1}(\overline{A})$, $n \geq 0$. If f is radial i.e. $f(x) = g(r)$, where $r = |x|$, $R_1 \leq r \leq R_2$, then $g \in C^{n+1}([R_1, R_2])$.

For radial f define

$$\theta_i f(x) := L_i g(r), \quad \text{all} \quad i = 1, \ldots, n+1, \ \forall \ x \in \overline{A}. \tag{9.3}$$

Here

$$g^{(i)}(r) = \frac{\partial^i f(x)}{\partial r^i}, \quad i = 1, \ldots, n+1.$$

For $F \in C(\overline{A})$ we have

$$\int_A F(x)dx = \int_{S^{N-1}} \left(\int_{R_1}^{R_2} F(r\omega)r^{N-1}dr \right) d\omega. \tag{9.4}$$

We notice that

$$\frac{N}{R_2^N - R_1^N} \int_{R_1}^{R_2} s^{N-1}ds = 1, \tag{9.5}$$

and

$$Vol(A) = \frac{\omega_N(R_2^N - R_1^N)}{N}.$$

9.3 Results on the Shell

We make

Remark 9.5. Let here $u_0, u_1, \ldots, u_n \in C^{n+1}([R_1, R_2])$, and $W_0, W_1, \ldots, W_n > 0$ on $[R_1, R_2]$, $0 < R_1 < R_2$, $n \geq 0$ integer, with $u_0(r) = c > 0$.

Let also $f \in C^{n+1}(\overline{A})$. We suppose first that f is radial, i.e. there exists g such that $f(x) = g(r)$, $r = |x|$, $R_1 \leq r \leq R_2$. Clearly $g \in C^{n+1}([R_1, R_2])$.

Let $x \in \overline{A}$. Then by using the polar method (9.4) we obtain

$$E(x) := \left| f(x) - \frac{\int_A f(y)dy}{Vol(A)} \right| = \left| f(x) - \frac{N \int_{S^{N-1}} \left(\int_{R_1}^{R_2} f(s\omega)s^{N-1}ds \right) d\omega}{\omega_N(R_2^N - R_1^N)} \right| \tag{9.6}$$

$$= \left| g(r) - \frac{N \int_{S^{N-1}} \left(\int_{R_1}^{R_2} g(s)s^{N-1}ds \right) d\omega}{\omega_N(R_2^N - R_1^N)} \right| \tag{9.7}$$

$$= \left| g(r) - \left(\frac{N}{R_2^N - R_1^N} \right) \int_{R_1}^{R_2} g(s)s^{N-1}ds \right| \tag{9.8}$$

$$= \left| \left(\frac{N}{R_2^N - R_1^N} \right) \left(\int_{R_1}^{R_2} g(r)s^{N-1}ds - \int_{R_1}^{R_2} g(s)s^{N-1}ds \right) \right| \tag{9.9}$$

$$= \left(\frac{N}{R_2^N - R_1^N}\right)\left|\int_{R_1}^{R_2}(g(r) - g(s))s^{N-1}ds\right| =: (*).\tag{9.10}$$

Let $s, r \in [R_1, R_2]$, then by generalized Taylor's formula (9.2) we have

$$g(s) - g(r) = \sum_{i=1}^{n} L_i g(r)g_i(s,r) + R_n(r,s),\tag{9.11}$$

where

$$R_n(r,s) := \int_r^s g_n(s,t)L_{n+1}g(t)dt.\tag{9.12}$$

But it holds

$$|R_n(r,s)| \le \left|\int_r^s |g_n(s,t)|dt\right| \|L_{n+1}g\|_{\infty,[R_1,R_2]}.\tag{9.13}$$

By calling

$$N_n(r,s) := \left|\int_r^s |g_n(s,t)|dt\right|, \ \forall \ s,r \in [R_1,R_2],\tag{9.14}$$

we find

$$|R_n(r,s)| \le N_n(r,s)\|L_{n+1}g\|_{\infty,[R_1,R_2]}, \ \forall \ s,r \in [R_1,R_2].\tag{9.15}$$

Therefore by (9.10), (9.11), we have

$$(*) = \left(\frac{N}{R_2^N - R_1^N}\right)\left|\int_{R_1}^{R_2}\left[\sum_{i=1}^{n}L_i g(r)g_i(s,r) + R_n(r,s)\right]s^{N-1}ds\right|\tag{9.16}$$

$$\le \left(\frac{N}{R_2^N - R_1^N}\right)\left[\sum_{i=1}^{n}\left|\int_{R_1}^{R_2}L_i g(r)g_i(s,r)s^{N-1}ds\right| + \int_{R_1}^{R_2}|R_n(r,s)|s^{N-1}ds\right]$$

$$\overset{\text{by (9.15)}}{\le} \left(\frac{N}{R_2^N - R_1^N}\right)\left[\sum_{i=1}^{n}|L_i g(r)|\left|\int_{R_1}^{R_2}g_i(s,r)s^{N-1}ds\right|\right.$$

$$\left. + (\|L_{n+1}g\|_{\infty,[R_1,R_2]})\int_{R_1}^{R_2}N_n(r,s)s^{N-1}ds\right].\tag{9.17}$$

We have established the following result.

Theorem 9.6. *Let $u_0, u_1, \ldots, u_n \in C^{n+1}([R_1, R_2])$, $W_0, W_1, \ldots, W_n > 0$ on $[R_1, R_2]$, $0 < R_1 < R_2$, $n \ge 0$ integer, with $u_0(r) = c > 0$. Let $f \in C^{n+1}(\overline{A})$ be radial, i.e. there exists g such that $f(x) = g(r)$, $r = |x|$, $R_1 \le r \le R_2$, $x \in \overline{A}$.*
Then

$$E(x) := \left|f(x) - \frac{\int_A f(y)dy}{Vol(A)}\right| = \left|g(r) - \left(\frac{N}{R_2^N - R_1^N}\right)\int_{R_1}^{R_2}g(s)s^{N-1}ds\right|$$

$$\leq \left(\frac{N}{R_2^N - R_1^N}\right)\left[\sum_{i=1}^{n}|L_i g(r)|\left|\int_{R_1}^{R_2} g_i(s,r)s^{N-1}ds\right|\right.$$

$$\left. +(\|L_{n+1}g\|_{\infty,[R_1,R_2]})\int_{R_1}^{R_2} N_n(r,s)s^{N-1}ds\right] \tag{9.18}$$

$$= \left(\frac{N}{R_2^N - R_1^N}\right)\left[\sum_{i=1}^{n}|\theta_i f(x)|\left|\int_{R_1}^{R_2} g_i(s,|x|)s^{N-1}ds\right|\right.$$

$$\left. +(\|\theta_{n+1}f\|_{\infty,\overline{A}})\int_{R_1}^{R_2} N_n(|x|,s)s^{N-1}ds\right].$$

We give

Corollary 9.7. *Same terms and assumptions as in Theorem 9.6. Suppose further that $L_i g(r_0) = 0$, $i = 1,\ldots,n$, for a fixed $r_0 \in [R_1,R_2]$; considered all $x_0 = r_0\omega \in \overline{A}$, for any $\omega \in S^{N-1}$. Then*

$$E(x_0) = \left|f(x_0) - \frac{\displaystyle\int_A f(y)dy}{Vol(A)}\right| = \left|g(r_0) - \left(\frac{N}{R_2^N - R_1^N}\right)\int_{R_1}^{R_2} g(s)s^{N-1}ds\right|$$

$$\leq \left(\frac{N}{R_2^N - R_1^N}\right)(\|L_{n+1}g\|_{\infty,[R_1,R_2]})\left(\int_{R_1}^{R_2} N_n(r_0,s)s^{N-1}ds\right)$$

$$= \left(\frac{N}{R_2^N - R_1^N}\right)(\|\theta_{n+1}f\|_{\infty,\overline{A}})\left(\int_{R_1}^{R_2} N_n(|x_0|,s)s^{N-1}ds\right). \tag{9.19}$$

Interesting cases also arise when $r_0 = R_1$ or R_2.

We continue Remark 9.5 with

Remark 9.8. Let now $f \in C^{n+1}(\overline{A})$, $n \geq 0$, $x \in \overline{A}$, $x = r\omega$, $r > 0$. Clearly for fixed $\omega \in S^{N-1}$, the function $f(r\omega)$, $r \in [R_1,R_2]$ is radial, it also belongs to $C^{n+1}([R_1,R_2])$.

By applying internal inequality (9.18) we obtain

$$\left|f(r\omega) - \left(\frac{N}{R_2^N - R_1^N}\right)\int_{R_1}^{R_2} f(s\omega)s^{N-1}ds\right|$$

$$\leq \left(\frac{N}{R_2^N - R_1^N}\right)\left[\sum_{i=1}^{n}|(L_i f(\cdot\omega))(r)|\left|\int_{R_1}^{R_2} g_i(s,r)s^{N-1}ds\right|\right.$$

$$+(\|L_{n+1}f(\cdot\omega)\|_{\infty,[R_1,R_2]})\int_{R_1}^{R_2}N_n(r,s)s^{N-1}ds\Bigg].\tag{9.20}$$

For non-radial f we define again

$$\theta_i f(x)=\theta_i f(r\omega):=(L_i f(\cdot\omega))(r),\quad\text{all}\quad i=1,\dots,n+1,\ \forall\ x\in\overline{A}.\tag{9.21}$$

Here the involved

$$\frac{\partial^i f(r\omega)}{\partial r^i}=\frac{\partial^i f(x)}{\partial r^i},\quad i=1,\dots,n+1,$$

are the radial derivatives. In a sense θ_i is a generalized radial derivative of Widder-type.

Hence

$$R.H.S.(9.20)\le\left(\frac{N}{R_2^N-R_1^N}\right)\left[\sum_{i=1}^{n}|\theta_i f(r\omega)|\right.$$

$$\left|\int_{R_1}^{R_2}g_i(s,r)s^{N-1}ds\right|+\|\theta_{n+1}f\|_{\infty,\overline{A}}\int_{R_1}^{R_2}N_n(|x|,s)s^{N-1}ds\Bigg].\tag{9.22}$$

Therefore, by (9.20) and (9.22) we have

$$\left|\frac{\Gamma\left(\dfrac{N}{2}\right)}{2\pi^{N/2}}\int_{S^{N-1}}f(r\omega)d\omega-\frac{1}{Vol(A)}\int_A f(y)dy\right|$$

$$\le\left(\frac{N}{R_2^N-R_1^N}\right)\left[\frac{\Gamma(N/2)}{2\pi^{N/2}}\left\{\sum_{i=1}^{n}\left(\int_{S^{N-1}}|\theta_i f(r\omega)|d\omega\right)\left|\int_{R_1}^{R_2}g_i(s,r)s^{N-1}ds\right|\right\}\right.$$

$$+\|\theta_{n+1}f\|_{\infty,\overline{A}}\int_{R_1}^{R_2}N_n(|x|,s)s^{N-1}ds\Bigg].\tag{9.23}$$

We have established

Theorem 9.9. *Let* $u_0,u_1,\dots,u_n\in C^{n+1}([R_1,R_2])$; $W_0,W_1,\dots,W_n>0$ *on* $[R_1,R_2]$, $0<R_1<R_2$, $n\ge 0$ *integer, with* $u_0(r)=c>0$. *Let* $f\in C^{n+1}(\overline{A})$, $x\in\overline{A}$; $x=r\omega$, $r\in[R_1,R_2]$, $\omega\in S^{N-1}$.
Then

$$E(x)=\left|f(x)-\frac{\displaystyle\int_A f(y)dy}{Vol(A)}\right|\le\left|f(x)-\frac{\Gamma\left(\dfrac{N}{2}\right)\int_{S^{N-1}}f(r\omega)d\omega}{2\pi^{N/2}}\right|$$

$$+\left(\frac{N}{R_2^N-R_1^N}\right)\left[\frac{\Gamma\left(\dfrac{N}{2}\right)}{2\pi^{N/2}}\left\{\sum_{i=1}^{n}\left(\int_{S^{N-1}}|\theta_i f(r\omega)|d\omega\right)\left|\int_{R_1}^{R_2}g_i(s,r)s^{N-1}ds\right|\right\}\right.$$

$$+\|\theta_{n+1}f\|_{\infty,\overline{A}}\int_{R_1}^{R_2} N_n(|x|,s)s^{N-1}ds\Bigg]\,,\ \forall\ x\in\overline{A}. \qquad (9.24)$$

We give

Corollary 9.10. *Same terms and assumptions as in Theorem 9.9. Suppose that*

$$\theta_i f(r_0\omega)=0,\ i=1,\ldots,n,\ \text{for a fixed } r_0\in[R_1,R_2],\ \forall\ \omega\in S^{N-1};$$

also consider all $x_0=r_0\omega\in\overline{A}$ *for any* $\omega\in S^{N-1}$. *Then*

$$E(x_0)=\left| f(x_0)-\frac{\displaystyle\int_A f(y)dy}{Vol(A)}\right|\leq\left| f(x_0)-\frac{\Gamma\left(\dfrac{N}{2}\right)\displaystyle\int_{S^{N-1}}f(r_0\omega)d\omega\Big)}{2\pi^{N/2}}\right|$$

$$+\left(\frac{N}{R_2^N-R_1^N}\right)\|\theta_{n+1}f\|_{\infty,\overline{A}}\left(\int_{R_1}^{R_2} N_n(|x_0|,s)s^{N-1}ds\right). \qquad (9.25)$$

When $r_0=R_1$ or R_2 is of special interest.

9.4   Results on the Sphere

Notation 9.11. Let $N\geq 1$, $B(0,R):=\{x\in\mathbb{R}^N:|x|<R\}$ be the ball in $\mathbb{R}^N$ centered at the origin and of radius $R>0$. Note that $Vol(B(0,R))=\dfrac{\omega_N R^N}{N}$.

Let f from $\overline{B(0,R)}$ into $\mathbb{R}$ and consider f be radial, i.e. $f(x)=g(r)$, where $r=|x|$, $0\leq r\leq R$, we assume $g\in C^{n+1}([0,R])$, $n\geq 0$. Clearly then $f\in C(\overline{B(0,R)})$.

For $F\in C(\overline{B(0,R)})$ we have

$$\int_{B(0,R)}F(x)dx=\int_{S^{N-1}}\left(\int_0^R F(r\omega)r^{N-1}dr\right)d\omega. \qquad (9.26)$$

We notice that

$$\frac{N}{R^N}\int_0^R s^{N-1}ds=1. \qquad (9.27)$$

The operator θ_i in the radial case, is defined as in (9.3), now $\forall\ x\in\overline{B(0,R)}$. In the non-radial case, for $f\in C^{n+1}(\overline{B(0,R)})$, θ_i is defined as in (9.21), $\forall\ x\in\overline{B(0,R)}-\{0\}$.

We make

Remark 9.12. Let here $u_0,u_1,\ldots,u_n\in C^{n+1}([0,R])$ and $W_0,W_1,\ldots,W_n>0$ on $[0,R]$, $R>0$, $n\geq 0$ integer, with $u_0(r)=c>0$.

We again first suppose that f is radial on $\overline{B(0,R)}$, i.e. there exists g such that $f(x) = g(r)$, $r = |x|$, $0 \le r \le R$. Assume further that $g \in C^{n+1}([0,R])$. Let $x \in \overline{B(0,R)}$. Then by using the polar method (9.26) we obtain

$$E(x) := \left| f(x) - \frac{\displaystyle\int_{B(0,R)} f(y)dy}{Vol(B(0,R))} \right| = \left| g(r) - \frac{N \displaystyle\int_{S^{N-1}} \left(\int_0^R g(s)s^{n-1}ds \right) d\omega}{\omega_N R^N} \right|$$

$$(9.28)$$

$$= \left| g(r) - \frac{N}{R^N} \int_0^R g(s)s^{n-1}ds \right| \overset{\text{(by 9.27)}}{=} \frac{N}{R^N} \left| \int_0^R (g(r) - g(s))s^{N-1}ds \right| =: (*).$$

$$(9.29)$$

Let $s, r \in [0, R]$, then by generalized Taylor's formula (9.2) we have

$$g(s) - g(r) = \sum_{i=1}^n L_i g(r)g_i(s,r) + R_n(r,s),$$

$$(9.30)$$

where

$$R_n(r,s) := \int_r^s g_n(s,t)L_{n+1}g(t)dt.$$

$$(9.31)$$

But it holds

$$|R_n(r,s)| \le \left| \int_r^s |g_n(s,t)|dt \right| \|L_{n+1}g\|_{\infty,[0,R]}.$$

$$(9.32)$$

By calling

$$N_n(r,s) := \left| \int_r^s |g_n(s,t)|dt \right|, \ \forall\, s, r \in [0,R],$$

$$(9.33)$$

we find

$$|R_n(r,s)| \le N_n(r,s)\|L_{n+1}g\|_{\infty,[0,R]}, \ \forall\, s, r \in [0,R].$$

$$(9.34)$$

Therefore by (9.29) and (9.30), we have

$$(*) = \frac{N}{R^N} \left| \int_0^R \left[\sum_{i=1}^n L_i g(r)g_i(s,r) + R_n(r,s) \right] s^{N-1}ds \right|$$

$$\le \frac{N}{R^N} \left[\sum_{i=1}^n \left| \int_0^R L_i g(r)g_i(s,r)s^{N-1}ds \right| + \int_0^R |R_n(r,s)|s^{N-1}ds \right]$$

$$(9.35)$$

$$\overset{\text{by (9.34)}}{\le} \frac{N}{R^N} \left[\sum_{i=1}^n |L_i g(r)| \left| \int_0^R g_i(s,r)s^{N-1}ds \right| \right.$$

$$\left. + (\|L_{n+1}g\|_{\infty,[0,R]}) \int_0^R N_n(r,s)s^{N-1}ds \right].$$

$$(9.36)$$

We have established the next result.

Theorem 9.13. *Let $u_0, u_1, \ldots, u_n \in C^{n+1}([0, R])$, $W_0, W_1, \ldots, W_n > 0$ on $[0, R]$, $R > 0$, $n \geq 0$ integer, with $u_0(r) = c > 0$. Let f from $\overline{B(0, R)}$ into $\mathbb{R}$ be radial, i.e. there exists g such that $f(x) = g(r)$, $r = |x|$, $0 \leq r \leq R$, $\forall\, x \in \overline{B(0, R)}$; further suppose that $g \in C^{n+1}([0, R])$.*

Then

$$E(x) := \left| f(x) - \frac{\displaystyle\fint_{B(0,R)} f(y)dy}{Vol(B(0, R))} \right| = \left| g(r) - \frac{N}{R^N} \int_0^R g(s)s^{n-1}ds \right|$$

$$\leq \frac{N}{R^N} \left[\sum_{i=1}^n |L_i g(r)| \left| \int_0^R g_i(s, r)s^{N-1}ds \right| + (\|L_{n+1}g\|_{\infty,[0,R]}) \int_0^R N_n(r, s)s^{N-1}ds \right]$$

$$= \frac{N}{R^N} \left[\sum_{i=1}^n |\theta_i f(x)| \left| \int_0^R g_i(s, |x|)s^{N-1}ds \right| \right.$$

$$\left. + (\|\theta_{n+1}f\|_{\infty,\overline{B(0,R)}}) \int_0^R N_n(|x|, s)s^{N-1}ds \right]. \tag{9.37}$$

We give

Corollary 9.14. *Same terms and assumptions as in Theorem 9.13. Assume further that $L_i g(r_0) = 0$, $i = 1, \ldots, n$, for a fixed $r_0 \in [0, R]$; consider all $x_0 = r_0\omega \in \overline{B(0, R)}$, any $\omega \in S^{N-1}$.*

Then

$$E(x_0) := \left| f(x_0) - \frac{\displaystyle\fint_{B(0,R)} f(y)dy}{Vol(B(0, R))} \right| = \left| g(r_0) - \frac{N}{R^N} \int_0^R g(s)s^{n-1}ds \right|$$

$$\leq \frac{N}{R^N}(\|L_{n+1}g\|_{\infty,[0,R]}) \left(\int_0^R N_n(r_0, s)s^{N-1}ds \right)$$

$$= \frac{N}{R^N}(\|\theta_{n+1}f\|_{\infty,\overline{B(0,R)}}) \left(\int_0^R N_n(|x_0|, s)s^{N-1}ds \right). \tag{9.38}$$

Interesting cases especially arise when $r_0 = 0$ or R.

We continue Remark 9.12 with

Remark 9.15. Let f be non-radial.

Here assume that $f \in C^{n+1}(\overline{B(0,R)})$. Consider $x \in \overline{B(0,R)} - \{0\}$, i.e. uniquely $x = r\omega$, $r \in (0,R]$, $\omega \in S^{N-1}$. Then by using again the polar method (9.26) we get

$$\left| \frac{\int_{S^{N-1}} f(r\omega)d\omega}{\omega_N} - \frac{\int_{B(0,R)} f(y)dy}{Vol(B(0,R))} \right| \tag{9.39}$$

$$= \left| \frac{\int_{S^{N-1}} f(r\omega)d\omega}{\omega_N} - \frac{N\int_{S^{N-1}} \left(\int_0^R f(s\omega)s^{N-1}ds \right) d\omega}{\omega_N R^N} \right| \tag{9.40}$$

$$\overset{\text{by } (9.27)}{=} \left| \frac{N}{\omega_N R^N} \left(\int_{S^{N-1}} \left(\int_0^R f(r\omega)s^{N-1}ds \right) d\omega \right) \right.$$

$$\left. - \frac{N}{\omega_N R^N} \left(\int_{S^{N-1}} \left(\int_0^R f(s\omega)s^{N-1}ds \right) d\omega \right) \right|$$

$$= \frac{N}{\omega_N R^N} \left| \int_{S^{N-1}} \left(\int_0^R (f(r\omega) - f(s\omega))s^{N-1}ds \right) d\omega \right| =: (*). \tag{9.41}$$

Clearly here $f(\cdot\omega) \in C^{n+1}((0,R])$.
Let $\rho \in (0,R]$, then by (9.2) we obtain

$$f(\rho\omega) - f(r\omega) = \sum_{i=1}^n ((L_i(f(\cdot\omega)))(r))g_i(\rho,r) + R_n(r,\rho), \tag{9.42}$$

where

$$R_n(r,\rho) := \int_r^\rho g_n(\rho,t)(L_{n+1}(f(\cdot\omega)))(t)dt. \tag{9.43}$$

That is

$$f(\rho\omega) - f(r\omega) = \sum_{i=1}^n (\theta_i f(r\omega))g_i(\rho,r) + R_n(r,\rho), \tag{9.44}$$

with

$$R_n(r,\rho) = \int_r^\rho g_n(\rho,t)\theta_{n+1}f(t\omega)dt. \tag{9.45}$$

We further assume that

$$\theta := \|\theta_{n+1}f\|_{\infty,\overline{B(0,R)}-\{0\}} < +\infty. \tag{9.46}$$

Therefore we derive

$$|R_n(r,\rho)| \le \left| \int_r^\rho |g_n(\rho,t)|dt \right| \theta. \tag{9.47}$$

By calling

$$N_n(r, \rho) := \left| \int_r^\rho |g_n(\rho, t)| dt \right|, \ \forall \ \rho, r \in [0, R], \tag{9.48}$$

we get

$$|R_n(r, \rho)| \le N_n(r, \rho)\theta. \tag{9.49}$$

Hence by (9.44) we obtain

$$|f(\rho\omega) - f(r\omega)| \le \sum_{i=1}^n |\theta_i f(r\omega)||g_i(\rho, r)| + |R_n(r, \rho)|$$

$$\le \sum_{i=1}^n |\theta_i f(r\omega)||g_i(\rho, r)| + N_n(r, \rho)\theta. \tag{9.50}$$

By continuity of f and g_i, $i = 1, \ldots, n$, and by taking the limit as $\rho \longrightarrow 0$ in the external inequality (9.50), we find

$$|f(0) - f(r\omega)| \le \sum_{i=1}^n |\theta_i f(r\omega)||g_i(0, r)| + N_n(r, 0)\theta. \tag{9.51}$$

Notice here $g_n(\rho, t)$ is jointly continuous in $(\rho, t) \in [0, R]^2$, hence $N_n(r, \rho)$ is continuous in $\rho \in [0, R]$.

That is, $\forall \ s \in [0, R]$ we get

$$|f(s\omega) - f(r\omega)| \le \sum_{i=1}^n |\theta_i f(r\omega)||g_i(s, r)| + N_n(r, s)\theta. \tag{9.52}$$

Consequently by (9.41) and (9.52) we find

$$(*) \le \frac{N}{\omega_N R^N} \int_{S^{N-1}} \left(\int_0^R |f(s\omega) - f(r\omega)| s^{N-1} ds \right) d\omega \tag{9.53}$$

$$\le \frac{N}{\omega_N R^N} \left[\sum_{i=1}^n \left(\int_{S^{N-1}} \left(\int_0^R |\theta_i f(r\omega)||g_i(s, r)| s^{N-1} ds \right) d\omega \right) \right.$$

$$\left. + \theta \int_{S^{N-1}} \left(\int_0^R N_n(r, s) s^{N-1} ds \right) d\omega \right] \tag{9.54}$$

$$= \frac{\sum_{i=1}^n \left(\int_{S^{N-1}} \left(\int_0^R |\theta_i f(r\omega)||g_i(s, r)| s^{N-1} ds \right) d\omega \right)}{Vol(B(0, R))}$$

$$+ \frac{\theta N \left(\int_0^R N_n(r, s) s^{N-1} ds \right)}{R^N} \tag{9.55}$$

$$= \sum_{i=1}^{N} \left(\frac{\left(\int_{S^{N-1}} |\theta_i f(r\omega)| d\omega \right) \left(\int_0^R |g_i(s,r)| s^{N-1} ds \right)}{Vol(B(0,R))} \right)$$

$$+ \frac{\theta N \left(\int_0^R N_n(r,s) s^{N-1} ds \right)}{R^N}. \tag{9.56}$$

Consequently,

$$\Delta := \left| \frac{\int_{S^{N-1}} f(r\omega) d\omega}{\omega_N} - \frac{\int_{B(0,R)} f(y) dy}{Vol(B(0,R))} \right|$$

$$\leq \sum_{i=1}^{n} \left[\frac{\left(\int_{S^{N-1}} |\theta_i f(r\omega)| d\omega \right) \left(\int_0^R |g_i(s,r)| s^{N-1} ds \right)}{Vol(V(0,R))} \right]$$

$$+ \frac{\theta N \left(\int_0^R N_n(r,s) s^{N-1} ds \right)}{R^N}. \tag{9.57}$$

We have proved

Theorem 9.16. *Let $u_0, u_1, \ldots, u_n \in C^{n+1}([0,R])$, $W_0, W_1, \ldots, W_n > 0$ on $[0,R]$, $R > 0$, $n \geq 0$ integer, with $u_0(r) = c > 0$. Let $f \in C^{n+1}(\overline{B(0,R)})$. Suppose that*

$$\theta := \|\theta_{n+1} f\|_{\infty, \overline{B(0,R)} - \{0\}} < +\infty. \tag{9.58}$$

Let $x \in \overline{B(0,R)} - \{0\}$, i.e. uniquely $x = r\omega$, $r \in (0,R]$, $\omega \in S^{N-1}$. Then

$$E(x) := \left| f(x) - \frac{\int_{B(0,R)} f(y) dy}{Vol(B(0,R))} \right| \leq \left| f(x) - \frac{\Gamma(N/2)}{2\pi^{N/2}} \int_{S^{N-1}} f(r\omega') d\omega' \right|$$

$$+ \sum_{i=1}^{n} \left[\frac{\left(\int_{S^{N-1}} |\theta_i f(r\omega')| d\omega' \right) \left(\int_0^R |g_i(s,r)| s^{N-1} ds \right)}{Vol(B(0,R))} \right]$$

$$+ \frac{\theta N \left(\int_0^R N_n(r,s) s^{N-1} ds \right)}{R^N}. \tag{9.59}$$

We finally give

Corollary 9.17. *Same terms and assumptions as in Theorem 9.16. Assume further that $\theta_i f(r_0 \omega') = 0$, $\forall\, \omega' \in S^{N-1}$, for some $r_0 \in (0, R]$, for all $i = 1, \ldots, n$. Consider all $x_0 = r_0 \omega \in \overline{B(0,R)} - \{0\}$, any $\omega \in S^{N-1}$.*

Then

$$E(x_0) := \left| f(x_0) - \frac{\int_{B(0,r)} f(y) dy}{Vol(B(0,R))} \right| \leq \left| f(x_0) - \frac{\Gamma(N/2)}{2\pi^{N/2}} \int_{S^{N-1}} f(|x_0|\omega') d\omega' \right|$$

$$+ \frac{\theta N \left(\int_0^R N_n(|x_0|, s) s^{N-1} ds \right)}{R^N}. \tag{9.60}$$

An interesting case is when $r_0 = R$.

9.5   Addendum

We give

Proposition 9.18. *Let $f \in C^1(\overline{B(0,R)})$ such that f is radial, i.e. $f(x) = g(r)$, $r = |x|$, $0 \leq r \leq R$, $\forall\, x \in \overline{B(0,R)}$. Then*

(i) $\exists\, g' \in C((0,R])$,   (9.61)

(ii) $\exists\, g'(0) = 0$.   (9.62)

Proof.   (i) is obvious.

(ii) Let u be a unit vector, $h > 0$, then

$$\frac{g(h) - g(0)}{h} = \frac{f(hu) - f(0)}{h}$$

$$(\text{by first order multivariate Taylor's formula}) = \frac{\nabla f(0) \cdot hu + o(h)}{h}$$

$$= \frac{\nabla f(0) \cdot hu}{h} + \frac{o(h)}{h} = \nabla f(0) \cdot u + \frac{o(h)}{h}$$

$$\overset{h \to 0}{\longrightarrow} \nabla f(0) \cdot u, \quad \text{by } \frac{o(h)}{h} \to 0.$$

The last is true for any unit vector u. Thus $\nabla f(0) = 0$, proving the claim. $\square$

Comment 9.19. By Proposition 9.18 we see that, it may well be that g' is discontinuous at zero, if only $f \in C^1(\overline{B(0,R)})$.

Therefore the assumption that $g \in C^{n+1}([0,R])$ in Theorem 9.13 seems to be the best.

Chapter 10

Multivariate Opial Type Inequalities for Functions Vanishing at an Interior Point

In this chapter we generalize Opial inequalities in the multivariate case on the balls. The inequalities carry weights and are proved to be sharp. The functions we study vanish at the center of the ball. This treatment relies on [58].

10.1 Introduction

Z. Opial [195] and C. Olech [194] in 1960 proved the following famous inequality.

Theorem 10.1. *Let $c > 0$, and $y(x)$ be real, continuously differentiable on $[0, c]$, with $y(0) = y(c) = 0$. Then*

$$\int_0^c |y(x)y'(x)|\, dx \le \frac{c}{4} \int_0^c (y'(x))^2\, dx.$$

Equality holds for the function $y(x) = x$ on $\left[0, \frac{c}{2}\right]$, and $y(x) = c - x$ on $\left[\frac{c}{2}, c\right]$.

In 1962 P. Beesack [75] gave the following improvement.

Theorem 10.2. *Let $b > 0$. If $y(x)$ is real, continuously differentiable on $[0, b]$, and $y(0) = 0$, then*

$$\int_0^b |y(x)y'(x)|\, dx \le \frac{b}{2} \int_0^b (y'(x))^2\, dx.$$

Equality holds only for $y(x) = mx$, where m is a constant.

Since then many people have worked on this type of inequalities in many directions; for an account see the important monograph of 1995 by R. Agarwal and P. Pang [6]. One motivation for this chapter is the interesting article of W. Troy [237] of 2001. His relevant result follows.

Theorem 10.3. *Let $p > -1$. Let $a, b \in \mathbb{R}$ with $0 \le a < b$. If $y(x)$ is continuously differentiable on $[a, b]$, and $y(a) = 0$, then*

$$\int_a^b t^p |y(t)y'(t)|\, dt \le \frac{1}{2\sqrt{p+1}} \int_a^b (b^{p+1} - at^p)(y'(t))^2\, dt.$$

139

Let $|\cdot|$ denote the Euclidean norm in $\mathbb{R}^N$, $N \geq 1$. Let $B(0, R) := \{x \in \mathbb{R}^N : |x| < R\}$ be the open ball of radius R with center 0 in $\mathbb{R}^N$. Let $S^{N-1} := \{x \in \mathbb{R}^N : |x| = 1\}$ be the unit sphere in $\mathbb{R}^N$, centered at zero. Let $\omega_N = \frac{N\pi^{N/2}}{\left(\frac{N}{2}\right)\Gamma\left(\frac{N}{2}\right)}$ (see p. 220, [142]) be its surface area.

In this chapter we estimate the integral

$$I = \int_{B(0,R)} |x|^p |u(x)| \, |\nabla u(x)| \, dx,$$

for $p \in \mathbb{R}$ and $u \in C^1(\overline{B(0, R)})$. Using polar coordinates we obtain

$$I = \int_{S^{N-1}} \left(\int_0^R r^p |u(r\omega)| \, |\nabla u(r\omega)| r^{N-1} dr \right) d\omega,$$

where $0 \neq x = r\omega$ with $r := |x|$ and $\omega := \frac{x}{r}$. For radial (spherically symmetric) functions u, this reduces to

$$I = \omega_N \int_0^R r^{p+N-1} |u(r)| \left| \frac{\partial u}{\partial r}(r) \right| dr,$$

where here $|\nabla u| = \left| \frac{\partial u}{\partial r} \right|$, with $\frac{\partial u(x)}{\partial r} = \nabla u(x) \cdot \frac{x}{|x|}$ the radial derivative of u. Spherically symmetric function means that

$$u(x) = u(r\omega) = u(r).$$

In general one has

$$\left| \frac{\partial u(x)}{\partial r} \right| \leq |\nabla u(x)|, \quad \text{for any } u \in C^1(\overline{B(0, R)}).$$

We will prove that

$$\int_{B(0,R)} |x|^p |u(x)| \, |\nabla u(x)| \, dx \leq C \int_{B(0,R)} |x|^p |\nabla u(x)|^2 \, dx$$

for some constant C and functions vanishing at the origin. The idea is to use

$$\int_{B(0,R)} F(x) \, dx = \int_{S^{N-1}} \left(\int_0^R F(r\omega) r^{N-1} \, dr \right) d\omega,$$

and to do first a one-dimensional analysis on the inner integral with ω fixed. The interior constraint $(u(0) = 0)$ becomes a boundary condition $(F(0) = 0)$.

10.2   Main Results

We present a basic result.

Theorem 10.4. *Let $R > 0$, $N \geq 1$, $B(0, R)$ the ball centered at 0 of radius R in $\mathbb{R}^N$. Let $p \in \mathbb{R}$ such that $p + N \in (0, 2)$. Consider $u \in C^1(\overline{B(0, R)})$ such that $u(0) = 0$. Then*

$$\int_{B(0,R)} |x|^p |u(x)| \, |\nabla u(x)| \, dx \leq \left(\frac{R}{2\sqrt{2 - p - N}} \right) \left(\int_{B(0,R)} |x|^p |\nabla u(x)|^2 \, dx \right),$$

$$\tag{10.1}$$

where ∇ is the gradient operator.

Proposition 10.5. *Inequality (10.1) is sharp, more precisely it is attained by $u(x) = |x|$, for all $x \in \overline{B(0, R)}$, when $p + N = 1$.*

Proof. Call $r := |x|$, $0 \le r \le R$. Clearly $u(0) = 0$. Notice that

$$\frac{\partial u}{\partial r} = 1 \quad \text{and} \quad |\nabla u(x)| = 1.$$

We observe that

$$\text{L.H.S.}(10.1) = \omega_N \frac{R^{p+N+1}}{p+N+1},$$

and

$$\text{R.H.S}(10.1) = \omega_N \frac{R^{p+N+1}}{2(p+N)\sqrt{2-(p+N)}} .$$

Thus

$$\text{L.H.S.}(10.1) = \text{R.H.S.}(10.1) \quad \text{iff}$$

$$(p+N)+1 = 2(p+N)\sqrt{2-(p+N)} \quad \text{iff}$$

(calling $y := p + N$, $y \in (0, 2)$)

$$y + 1 = 2y\sqrt{2-y} \quad \text{iff}$$

$$g(y) := 4y^3 - 7y^2 + 2y + 1 = 0, \quad y \in (0, 2).$$

See that $g(1) = 0$, $g(0) = 1$, $g(2) = 9$, and

$$g'(y) = 12(y-1)\left(y - \frac{1}{6}\right).$$

Thus g has critical numbers 1, $\frac{1}{6}$ with local maximum $g\left(\frac{1}{6}\right) = 1.1574078$, and minimum $g(1) = 0$. So $y = p + N = 1$ is the only optimal value making inequality (10.1) attained. $\qquad\square$

Proof of Theorem 10.4. The integral in the R.H.S(10.1) is finite since $2 > p + N > 0$. We can rewrite inequality (10.1) by the use of polar coordinates as follows, cf. p. 217, [142]:

$$\int_{S^{N-1}} \left(\int_0^R r^p |u(r\omega)|\, |\nabla u(r\omega)| r^{N-1}\, dr \right) d\omega$$

$$\le \frac{R}{2\sqrt{2-p-N}} \left(\int_{S^{N-1}} \left(\int_0^R r^p |\nabla u(r\omega)|^2 r^{N-1}\, dr \right) d\omega \right). \qquad (10.2)$$

Here $0 \neq x \in \overline{B(0,R)}$ is written as $x := r\omega$ with $r := |x|$, $0 < r \leq R$, and $\omega = \frac{x}{|x|} \in S^{N-1}$. So it is enough to prove that

$$\int_0^R r^{p+N-1} |u(r\omega)| \, |\nabla u(r\omega)| \, dr \leq \left(\frac{R}{2\sqrt{2-p-N}} \right) \left(\int_0^R r^{p+N-1} |\nabla u(r\omega)|^2 \, dr \right).$$
$$(10.3)$$

We set

$$z(r) := \int_0^r s^{p+N-1} |\nabla u(s\omega)|^2 \, ds, \quad 0 \leq r \leq R.$$

Here $z(r) \geq 0$ and $z(0) = 0$. Therefore

$$z'(r) = r^{p+N-1} |\nabla u(r\omega)|^2 \geq 0, \quad 0 < r \leq R.$$

Whence

$$r^{\frac{p+N-1}{2}} |\nabla u(r\omega)| = (z'(r))^{1/2}, \quad 0 < r \leq R. \tag{10.4}$$

By the fundamental theorem of calculus we have

$$u(r\omega) = \int_0^r \frac{\partial u(s\omega)}{\partial s} \, ds.$$

Consequently it holds

$$|u(r\omega)| \leq \int_0^r \left| \frac{\partial u}{\partial s}(s\omega) \right| ds = \int_0^r s^{-\left(\frac{p+N-1}{2} \right)} s^{\left(\frac{p+N-1}{2} \right)} \left| \frac{\partial u}{\partial s}(s\omega) \right| ds$$
$$\text{(by the Cauchy–Schwarz inequality)}$$
$$\leq \left(\int_0^r s^{-(p+N-1)} \, ds \right)^{1/2} \left(\int_0^r s^{(p+N-1)} \left| \frac{\partial u}{\partial s}(s\omega) \right|^2 ds \right)^{1/2}$$
$$\leq \left(\frac{r^{2-p-N}}{2-p-N} \right)^{1/2} \left(\int_0^r s^{(p+N-1)} |\nabla u(s\omega)|^2 \, ds \right)^{1/2}$$
$$= \left(\frac{r^{2-p-N}}{2-p-N} \right)^{1/2} (z(r))^{1/2}.$$

So we have proved that

$$|u(r\omega)| \leq \left(\frac{r^{2-p-N}}{2-p-N} \right)^{1/2} (z(r))^{1/2}, \quad \text{all } 0 \leq r \leq R.$$

Furthermore,

$$r^{\left(\frac{p+N-1}{2} \right)} |u(r\omega)| \leq \frac{\sqrt{r}}{\sqrt{2-p-N}} (z(r))^{1/2}, \quad \text{all } 0 \leq r \leq R. \tag{10.5}$$

Consequently by (10.4) and (10.5) we derive

$$r^{(p+N-1)} |u(r\omega)| \, |\nabla u(r\omega)| \leq \frac{\sqrt{r}}{\sqrt{2-p-N}} (z(r))^{1/2} (z'(r))^{1/2}, \text{all } 0 < r \leq R. \tag{10.6}$$

Next we integrate (10.6) and use the Cauchy–Schwarz inequality to get

$$\int_0^R r^{p+N-1}|u(r\omega)|\,|\nabla u(r\omega)|\,dr \leq \frac{1}{\sqrt{2-p-N}}\int_0^R \sqrt{r}(z(r))^{1/2}(z'(r))^{1/2}\,dr$$

$$\leq \frac{R}{\sqrt{2}\sqrt{2-p-N}}\left(\int_0^R z(r)\,dz(r)\right)^{1/2}$$

$$= \frac{R}{2\sqrt{2-p-N}}z(R) = \frac{R}{2\sqrt{2-p-N}}\left(\int_0^R r^{p+N-1}|\nabla u(r\omega)|^2 dr\right),$$

establishing (10.3). $\qquad\qquad\square$

A similar result follows.

Theorem 10.6. *Let $R > 0$, $N \geq 1$, $B(0,R)$ the ball centered at 0 of radius R in $\mathbb{R}^N$. Let $p \in \mathbb{R}$ such that $p + N > 0$. Consider $u \in C^1(\overline{B(0,R)})$ such that $u(0) = 0$. Then*

$$\int_{B(0,R)} |x|^p|u(x)|\,|\nabla u(x)|\,dx \leq \frac{R^{p+N}}{2\sqrt{p+N}}\int_{B(0,R)} |x|^{1-N}|\nabla u(x)|^2\,dx. \qquad (10.7)$$

Proposition 10.7. *Inequality (10.7) is sharp, namely it is attained by $u(x) = |x|$, for all $x \in \overline{B(0,R)}$, when $p + N = 1$.*

Proof.   Call $r := |x|$, $0 \leq r \leq R$. Clearly $u(0) = 0$. Note that $\frac{\partial u}{\partial r} = 1$ and $|\nabla u(x)| = 1$. We have that

$$\text{L.H.S.}(10.7) = \omega_N \frac{R^{p+N+1}}{p+N+1},$$

and

$$\text{R.H.S.}(10.7) = \omega_N \frac{R^{p+N+1}}{2\sqrt{p+N}}.$$

Thus

$$\text{L.H.S.}(10.7) = \text{R.H.S.}(10.7) \quad \text{iff}$$
$$(p+N) + 1 = 2\sqrt{p+N} \quad \text{iff}$$
$$y + 1 = 2\sqrt{y}, \quad \text{where } y := p + N > 0, \quad \text{iff}$$
$$y = 1.$$

$\qquad\qquad\square$

Proof of Theorem 10.6.   Clearly the R.H.S.(10.7) is finite. As in the proof of Theorem 10.4 it is enough to prove

$$\int_0^R r^p|u(r\omega)|\,|\nabla u(r\omega)|r^{N-1}\,dr \leq \frac{R^{p+N}}{2\sqrt{p+N}}\left(\int_0^R |\nabla u(r\omega)|^2\,dr\right). \qquad (10.8)$$

Therefore we begin with

$$\int_0^R r^p |u(r\omega)| \, |\nabla u(r\omega)| r^{N-1} \, dr = \int_0^R r^{p+N-1} |u(r\omega)| \, |\nabla u(r\omega)| \, dr$$

$$= \int_0^R \left(r^{\left(\frac{p+N}{2}\right)} |\nabla u(r\omega)| \right) \left(r^{\left(\frac{p+N-2}{2}\right)} |u(r\omega)| \, dr \right)$$

$$\text{(by the Cauchy–Schwarz inequality)}$$

$$\le \left(\int_0^R r^{p+N} |\nabla u(r\omega)|^2 dr \right)^{1/2} \left(\int_0^R r^{(p+N-1)} r^{-1} (u(r\omega))^2 \, dr \right)^{1/2} =: (*).$$

We have again

$$u(r\omega) = \int_0^r \frac{\partial u}{\partial s}(s\omega) \, ds, \quad 0 \le r \le R.$$

Hence

$$|u(r\omega)| \le \int_0^r 1 \cdot \left| \frac{\partial u}{\partial s}(s\omega) \right| ds$$

$$\text{(by the Cauchy–Schwarz inequality)}$$

$$\le \sqrt{r} \left(\int_0^r \left| \frac{\partial u}{\partial s}(s\omega) \right|^2 ds \right)^{1/2} \le \sqrt{r} \left(\int_0^r |\nabla u(s\omega)|^2 ds \right)^{1/2}.$$

That is

$$(u(r\omega))^2 \le r \int_0^r |\nabla u(s\omega)|^2 \, ds. \tag{10.9}$$

Thus by (10.9) we derive

$$(*) \le \left(\int_0^R r^{p+N} |\nabla u(r\omega)|^2 \, dr \right)^{1/2} \cdot \left(\int_0^R r^{(p+N-1)} \left(\int_0^r |\nabla u(s\omega)|^2 \, ds \right) dr \right)^{1/2}$$

$$= \left(\int_0^R r^{p+N} |\nabla u(r\omega)|^2 \, dr \right)^{1/2} \cdot \left[\frac{R^{p+N}}{p+N} \left(\int_0^R |\nabla u(r\omega)|^2 \, dr \right) \right.$$

$$\left. - \frac{1}{p+N} \int_0^R r^{p+N} |\nabla u(r\omega)|^2 \, dr \right]^{1/2} =: (**).$$

If $A \ge 0$, $B \ge 0$ and $\varepsilon > 0$ we have

$$(AB)^{1/2} \le \frac{\varepsilon}{2} A + \frac{1}{2\varepsilon} B. \tag{10.10}$$

Therefore

$$(**) \leq \frac{\varepsilon}{2} \left(\int_0^R r^{p+N} |\nabla u(r\omega)|^2 \, dr \right) + \frac{1}{2\varepsilon} \left[\frac{R^{p+N}}{p+N} \left(\int_0^R |\nabla u(r\omega)|^2 \, dr \right) \right.$$

$$\left. - \frac{1}{p+N} \left(\int_0^R r^{p+N} |\nabla u(r\omega)|^2 \, dr \right) \right]$$

$$= \frac{1}{2} \left(\varepsilon - \frac{1}{\varepsilon(p+N)} \right) \left(\int_0^R r^{p+N} |\nabla u(r\omega)|^2 \, dr \right)$$

$$+ \frac{R^{p+N}}{2\varepsilon(p+N)} \left(\int_0^R |\nabla u(r\omega)|^2 \, dr \right)$$

$$\left(\text{choosing } \varepsilon = \frac{1}{\sqrt{p+N}} \right) = \frac{R^{p+N}}{2\sqrt{p+N}} \left(\int_0^R |\nabla u(r\omega)|^2 \, dr \right).$$

We have established (10.8). $\qquad\square$

Finally we present a generalization and extension of Theorem 10.4.

Theorem 10.8. *Let $R > 0$, $N \geq 1$, $\rho > 1$, $\alpha, \beta > 0$, $\rho \geq \alpha + \beta$ and $p \in \mathbb{R}$ such that $0 < p + N < \rho$. Consider $u \in C^1(\overline{B(0,R)})$ such that $u(0) = 0$. Then*

$$\int_{B(0,R)} |x|^{\left[(p+N-1)\left(\frac{\alpha+\beta}{\rho}\right) + (1-N) \right]} |u(x)|^\beta \, |\nabla u(x)|^\alpha \, dx$$

$$\leq L \omega_N^{\left(\frac{\rho - \alpha - \beta}{\rho} \right)} \left(\int_{B(0,R)} |x|^p \, |\nabla u(x)|^\rho \, dx \right)^{\left(\frac{\alpha+\beta}{\rho} \right)}, \quad (10.11)$$

where

$$L := \left(\frac{\rho - 1}{\rho - p - N} \right)^{\beta \left(\frac{\rho - 1}{\rho} \right)} \left(\frac{\rho - a}{\beta(\rho - 1) + (\rho - \alpha)} \right)^{\left(\frac{\rho - \alpha}{\rho} \right)} \left(\frac{\alpha}{\alpha + \beta} \right)^{\frac{\alpha}{\rho}} R^{\left(\frac{\beta(\rho-1) + (\rho-\alpha)}{\rho} \right)}.$$

$$(10.12)$$

Proposition 10.9. *Inequality (10.11) is sharp, namely it is attained by $u(x) = |x|$, for all $x \in \overline{B(0,R)}$, when $p + N = 1$, $\rho = \alpha + \beta$ and $\alpha^{\left(\frac{\alpha}{\alpha+\beta} \right)} = \frac{\alpha+\beta}{1+\beta}$.*

Proof. Call $r := |x|$, $0 \leq r \leq R$. Clearly $u(0) = 0$. Note that $\frac{\partial u}{\partial r} = 1$ and $|\nabla u(x)| = 1$. We have

$$\text{L.H.S.}(10.11) = \omega_N \frac{R^{\beta+1}}{\beta+1},$$

and

$$\text{R.H.S.}(10.11) = \omega_N \frac{\alpha^{\alpha/\rho}}{\rho} R^{\beta+1}.$$

It is obvious now that (10.11) holds as equality. $\qquad\square$

Proof of Theorem 10.8. The integral in the R.H.S.(10.11) is finite by $p + N > 0$. Here for $0 \neq x \in \overline{B(0, R)}$ we set $\omega := \frac{x}{|x|} = \frac{x}{r}$, where $r := |x|$, $0 < r \leq R$. That is $x = r\omega$. By the fundamental theorem of calculus we have

$$u(r\omega) = \int_0^r \frac{\partial u}{\partial s}(s\omega)\, ds, \quad 0 \leq r \leq R.$$

We see that

$$|u(r\omega)| \leq \int_0^r \left| \frac{\partial u}{\partial s}(s\omega) \right| ds = \int_0^r s^{-\left(\frac{p+N-1}{\rho} \right)} s^{\left(\frac{p+N-1}{\rho} \right)} \left| \frac{\partial u}{\partial s}(s\omega) \right| ds$$

$$\text{(using Hölder's inequality with indices } \rho \text{ and } \tfrac{\rho}{\rho-1})$$

$$\leq \left(\int_0^r s^{-\frac{(p+N-1)}{\rho-1}} ds \right)^{\left(\frac{\rho-1}{\rho} \right)} \left(\int_0^r s^{(p+N-1)} \left| \frac{\partial u}{\partial s}(s\omega) \right|^\rho ds \right)^{1/\rho}$$

$$\leq \left(\frac{\rho - 1}{\rho - p - N} \right)^{\left(\frac{\rho-1}{\rho} \right)} r^{\frac{(\rho-p-N)}{\rho}} \left(\int_0^r s^{(p+N-1)} |\nabla u(s\omega)|^\rho ds \right)^{1/\rho}.$$

That is,

$$|u(r\omega)| \leq \left(\frac{\rho - 1}{\rho - p - N} \right)^{\left(\frac{\rho-1}{\rho} \right)} r^{\left(\frac{\rho-p-N}{\rho} \right)} z(r)^{1/\rho}, \quad 0 \leq r \leq R.$$

Here we call

$$+\infty > z(r) := \int_0^r s^{(p+N-1)} |\nabla u(s\omega)|^\rho ds \geq 0,$$

with $z(0) = 0$. We have

$$z'(r) = r^{p+N-1} |\nabla u(r\omega)|^\rho \geq 0, \quad 0 < r \leq R.$$

and

$$r^{\left(\frac{p+N-1}{\rho} \right)\alpha} |\nabla u(r\omega)|^\alpha = (z'(r))^{\alpha/\rho}, \quad 0 < r \leq R. \tag{10.13}$$

We observe that

$$|u(r\omega)|^\beta \leq \left(\frac{\rho - 1}{\rho - p - N} \right)^{\beta\left(\frac{\rho-1}{\rho} \right)} r^{\beta \frac{(\rho-p-N)}{\rho}} (z(r))^{\beta/\rho},$$

and

$$r^{\left(\frac{p+N-1}{\rho} \right)\beta} |u(r\omega)|^\beta \leq \left(\frac{\rho - 1}{\rho - p - N} \right)^{\beta\left(\frac{\rho-1}{\rho} \right)} \cdot r^{\beta\left(\frac{\rho-1}{\rho} \right)} z(r)^{\beta/\rho}, \quad 0 \leq r \leq R. \tag{10.14}$$

By multiplying (10.13) and (10.14) we get

$$r^{(p+N-1)\left(\frac{\alpha+\beta}{\rho} \right)} |u(r\omega)|^\beta |\nabla u(r\omega)|^\alpha$$

$$\leq \left(\frac{\rho - 1}{\rho - p - N} \right)^{\beta\left(\frac{\rho-1}{\rho} \right)} r^{\beta\left(\frac{\rho-1}{\rho} \right)} (z(r))^{\beta/\rho} (z'(r))^{\alpha/\rho}, \quad 0 < r \leq R. \tag{10.15}$$

Integrate (10.15) and use Hölder's inequality with indices $\frac{\rho}{\alpha}$, $\frac{\rho}{\rho-\alpha}$ to find

$$\int_0^R r^{(p+N-1)\left(\frac{\alpha+\beta}{\rho}\right)}|u(r\omega)|^\beta|\nabla u(r\omega)|^\alpha\, dr$$

$$\leq \left(\frac{\rho-1}{\rho-p-N}\right)^{\beta\left(\frac{\rho-1}{\rho}\right)}\int_0^R r^{\beta\left(\frac{\rho-1}{\rho}\right)}(z(r))^{\beta/\rho}(z'(r))^{\alpha/\rho}\, dr$$

$$\leq \left(\frac{\rho-1}{\rho-p-N}\right)^{\beta\left(\frac{\rho-1}{\rho}\right)}\left(\int_0^R r^{\beta\left(\frac{\rho-1}{\rho-\alpha}\right)}\, dr\right)^{\left(\frac{\rho-\alpha}{\rho}\right)}\cdot\left(\int_0^R (z(r))^{\beta/\alpha}z'(r)\, dr\right)^{\alpha/\rho}$$

$$\overset{(10.12)}{=} L\cdot z(R)^{\frac{\alpha+\beta}{\rho}} = L\left(\int_0^R r^{p+N-1}|\nabla u(r\omega)|^\rho\, dr\right)^{\left(\frac{\alpha+\beta}{\rho}\right)}.$$

We have established

$$\left(\int_0^R r^{\left[(p+N-1)\left(\frac{\alpha+\beta}{\rho}\right)+(1-N)\right]}|u(r\omega)|^\beta|\nabla u(r\omega)|^\alpha r^{N-1}\, dr\right)$$

$$\leq L\left(\int_0^R r^p|\nabla u(r\omega)|^\rho r^{N-1}dr\right)^{\left(\frac{\alpha+\beta}{\rho}\right)}. \tag{10.16}$$

Integrating (10.16) over S^{N-1} we find

$$\int_{S^{N-1}}\left(\int_0^R r^{\left[(p+N-1)\left(\frac{\alpha+\beta}{\rho}\right)+(1-N)\right]}|u(r\omega)|^\beta|\nabla u(r\omega)|^\alpha r^{N-1}\, dr\right)d\omega$$

$$\leq L\cdot\int_{S^{N-1}}\left(\int_0^R r^p|\nabla u(r\omega)|^\rho r^{N-1}\, dr\right)^{\left(\frac{\alpha+\beta}{\rho}\right)}d\omega =: (***). \tag{10.17}$$

If $\rho > \alpha+\beta$ then apply again Hölder's inequality with indices $\left(\frac{\rho}{\rho-\alpha-\beta}\right)$ abd $\left(\frac{\rho}{\alpha+\beta}\right)$ to obtain

$$(***)$$

$$\leq L\left(\int_{S^{N-1}} 1^{\frac{\rho}{\rho-\alpha-\beta}}\, d\omega\right)^{\left(\frac{\rho-\alpha-\beta}{\rho}\right)}\cdot\left(\int_{S^{N-1}}\left(\int_0^R r^p|\nabla u(r\omega)|^\rho r^{N-1}\, dr\right)d\omega\right)^{\left(\frac{\alpha+\beta}{\rho}\right)}$$

$$= L\omega_N^{\left(\frac{\rho-\alpha-\beta}{\rho}\right)}\cdot\left(\int_{S^{N-1}}\left(\int_0^R r^p|\nabla u(r\omega)|^\rho r^{N-1}dr\right)d\omega\right)^{\left(\frac{\alpha+\beta}{\rho}\right)}. \tag{10.18}$$

From (10.17) and (10.18) we conclude (10.11). $\qquad\square$

The work of Nečeav [193] in 1973 is related to this chapter, see also [6, p. 275].

Theorem 10.10 [193]. *Let $u \in C^1(\overline{B(0,R)})$ be such that $u(0) = 0$, and $N+p < 2$, $N \geq 1$. Then*

$$\int_{B(0,R)} |x|^{1-N}|u(x)|\,|\nabla u(x)|\, dx \leq \frac{R^{2-N-p}}{2(2-N-p)}\int_{B(0,R)} |x|^p|\nabla u(x)|^2\, dx,$$

with equality holding when

$$u(x) = c|x|^{2-N-p}$$

for a real constant c.

This result can be compared to Theorem 10.6 if and only if $p = 1 - N$, in which case the conclusions are the same, with the common constant being $R/2$.

Chapter 11

General Multivariate Weighted Opial Inequalities

In this chapter we expose Opial type weighted multivariate inequalities on balls and arbitrary smooth bounded domains. The inequalities are mostly sharp. The functions we consider vanish on the boundary. This chapter relies on [59].

11.1   Introduction

Z. Opial [195] and C. Olech [194] in 1960 proved the following well-known inequality.

Theorem 11.1. *Let $c > 0$, and $y(x)$ be real, continuously differentiable on $[0, c]$, with $y(0) = y(c) = 0$. Then*

$$\int_0^c |y(x)y'(x)|\, dx \le \frac{c}{4} \int_0^c (y'(x))^2\, dx.$$

Equality holds for the function $y(x) = x$ on $\left[0, \frac{c}{2}\right]$, and $y(x) = c - x$ on $\left[\frac{c}{2}, c\right]$.

In 1962 P. Beesack [75] gave the following variant.

Theorem 11.2. *Let $b > 0$. If $y(x)$ is real, continuously differentiable on $[0, b]$, and $y(0) = 0$, then*

$$\int_0^b |y(x)y'(x)|\, dx \le \frac{b}{2} \int_0^b (y'(x))^2\, dx.$$

Equality holds only for $y = mx$, where m is a constant.

Since then many researchers have been extending this type of inequalities in many directions, for a thorough account see the important monograph of 1995 by R. Agarwal and P. Pang [6]. An inspiration for this chapter is the interesting article by W. Troy [237] of 2001. His relevant result is as follows.

Theorem 11.3. *Let $p > -1$. Let b and c be real with $0 \le b < c$. If $y(x)$ is continuously differentiable on $[b, c]$, and $y(c) = 0$, then*

$$\int_b^c t^p |y(t)y'(t)|\, dt \le \frac{1}{2\sqrt{p+1}} \int_b^c (ct^p - b^{p+1})(y'(t))^2\, dt.$$

R. Agarwal in 1981, see [4] and p. 208 of [6], proved the following two dimensional result.

Theorem 11.4. *If $u(t,s) \in C^{(1,1)}([a,T] \times [c,S])$, $u(a,s) = u(t,c) = 0$, then*

$$\int_a^T \int_c^S |u(t,s)u_{ts}(t,s)|\, dtds \leq \frac{(T-a)(S-c)}{2\sqrt{2}} \int_a^T \int_c^S |u_{ts}(t,s)|^2\, dtds.$$

In 1982, G.S. Yang [247] proved the following Opial-type inequality in two variables:

Theorem 11.5. *If $f(s,t)$, $f_1(s,t)$, and $f_{12}(s,t)$ are continuous functions on $[a,b] \times [c,d]$, and if $f(a,t) = f(b,t) = f_1(s,c) = f_1(s,d) = 0$, for $a \leq s \leq b$, $c \leq t \leq d$, then*

$$\int_a^b \int_c^d |f(s,t)|\,|f_{12}(s,t)|\, dtds \leq \frac{(b-a)(d-c)}{8} \int_a^b \int_c^d (f_{12}(s,t))^2\, dtds.$$

In 1983, C.T. Lin and G.S. Yang [179] generalized Theorem 11.5 in the following form:

Theorem 11.6. *If $f(s,t)$, $f_1(s,t)$, and $f_{12}(s,t)$ are continuous functions on $[a,b] \times [c,d]$, and if $f(a,t) = f(b,t) = f_1(s,c) = f_1(s,d) = 0$, for $a \leq s \leq b$, $c \leq t \leq d$, then*

$$\int_a^b \int_c^d |f(s,t)|^m |f_{12}(s,t)|^n\, dtds$$

$$\leq \left(\frac{n}{m+n}\right) \left[\frac{(b-a)(d-c)}{4}\right]^m \int_a^b \int_c^d |f_{12}(s,t)|^{m+n}\, dtds.$$

The above Theorems 11.4 – 11.6 motivate this chapter.

Here we prove weighted Opial type inequalities over a ball (or other domain) in $\mathbb{R}^N$, $N \geq 1$, where the functions under consideration vanish on the boundary.

Let $|\cdot|$ denote the Euclidean distance for vectors in $\mathbb{R}^N$. Let $B(0,R) := \{x \in \mathbb{R}^N : |x| < R\}$ be the open ball of radius R with center 0 in $\mathbb{R}^N$.

Let $S^{N-1} := \{x \in \mathbb{R}^N : |x| = 1\}$ be the unit sphere in $\mathbb{R}^N$ centered at zero, with $\omega_N = \frac{N\pi^{N/2}}{\left(\frac{N}{2}\right)\Gamma\left(\frac{N}{2}\right)}$ to be its surface area (cf. [142], p. 220).

The basic results here are extended to *Sobolev spaces of order 1*.

We mention

Definition 11.7 (cf. [139]). The *Sobolev space of order* 1 denoted by $H_0^1(\Omega)$ is the completion of $C_c^1(\Omega)$. Here $C_c^1(\Omega)$ is the space of one time continuously differentiable functions with compact support on the bounded domain $\Omega \subset \mathbb{R}^N$, $N \geq 1$. The norm in $H_0^1(\Omega)$ is given by

$$\|u\|_* := \left(\int_\Omega |u(x)|^2\, dx + \int_\Omega |\nabla u(x)|^2\, dx\right)^{1/2}, \quad \text{for } u \in H_0^1(\Omega).$$

In this chapter we estimate the integral

$$I = \int_{B(0,R)} |x|^p |u(x)|\, |\nabla u(x)|\, dx,$$

for $p \in \mathbb{R}$ and $u \in C_c^1(B(0,R))$. Using polar coordinates we have

$$I = \int_{S^{N-1}} \left(\int_0^R r^p |u(r\omega)|\, |\nabla u(r\omega)| r^{N-1} dr \right) d\omega,$$

where $x = r\omega$ with $r := |x|$ and $\omega := \frac{x}{r}$. For radial (spherically symmetric) functions u, this reduces to

$$I = \omega_N \int_0^R r^{p+N-1} |u(r)| \left| \frac{\partial u}{\partial r}(r) \right| dr,$$

where here $|\nabla u| = \left| \frac{\partial u}{\partial r} \right|$, with $\frac{\partial u(x)}{\partial r} = \nabla u(x) \cdot \frac{x}{|x|}$ the radial derivative of u. Spherically symmetric function means that

$$u(x) = u(r\omega) = u(r).$$

In general one has

$$\left| \frac{\partial u(x)}{\partial r} \right| \le |\nabla u(x)|, \quad \text{for any } u \in C_c^1(B(0,R)).$$

We will prove that

$$\int_{B(0,R)} |x|^p |u(x)|\, |\nabla u(x)|\, dx \le C \int_{B(0,R)} |x|^p |\nabla u(x)|^2\, dx$$

for some constant C and functions u as above. The idea is to use

$$\int_{B(0,R)} F(x)\, dx = \int_{S^{N-1}} \left(\int_0^R F(r\omega) r^{N-1}\, dr \right) d\omega,$$

and to do first a one-dimensional analysis on the inner integral with ω fixed. The boundary condition $u = 0$ on $\partial B(0,R)$ becomes the boundary condition $F(R) = 0$.

11.2 Main Results

We present a basic result.

Theorem 11.8. *Let $R > 0$, $N \ge 1$, $B(0,R)$ the ball centered at 0 of radius R in $\mathbb{R}^N$. Let $p \in \mathbb{R}$ such that $p + N \ge 1$. Let $u \in C^1(\overline{B(0,R)})$ be such that $u(\partial B(0,R)) = 0$. Then*

$$\int_{B(0,R)} |x|^p |u(x)|\, |\nabla u(x)|\, dx \le \frac{R}{2} \int_{B(0,R)} |x|^p |\nabla u(x)|^2\, dx, \tag{11.1}$$

where ∇ is the gradient operator.

Note that Theorem 11.8 holds for more general choices of u. By a simple approximation argument, the condition $u \in C^1(\overline{B(0,R)})$ can be weakened to

$$u \in W^{1,2}_{loc}(B(0,R) - \{0\}) \cap C(\overline{B(0,R)})$$

such that $u = 0$ on $\partial B(0,R)$ and the right hand side of (11.1) is finite. The extremal in the following proposition satisfies this weaker hypothesis. Similar remarks apply to the Theorems 11.13 and 11.17 below.

Proposition 11.9. *Inequality (11.1) is sharp, namely it is attained by* $u(x) = R - |x|$, *for all* $x \in B(0,R)$, *when* $p + N = 1$.

Proof. Call $r := |x|$, $0 \le r \le R$. Notice $u(\partial B(0,R)) = 0$. Since $u(x) = R - r$ we find $\frac{\partial u}{\partial r} = -1$ and $|\nabla u(x)| = 1$. Clearly we get

$$\text{L.H.S.}(11.1) = \omega_N \frac{R^{p+N+1}}{(p+N)(p+N+1)},$$

and

$$\text{R.H.S.}(11.1) = \omega_N \frac{R^{p+N+1}}{2(p+N)},$$

where ω_N is the surface area of the unit sphere S^{N-1} in $\mathbb{R}^N$. Thus L.H.S.(11.1) = R.H.S.(11.1) iff $p + N = 1$. $\qquad\square$

Proof of Theorem 11.8. The integral in the R.H.S.(11.1) is finite since $p + N > 0$. We can rewrite inequality (11.1) by the use of polar coordinates as follows

$$\int_{S^{N-1}} \left(\int_0^R r^p |u(r\omega)| \, |\nabla u(r\omega)| r^{N-1} \, dr \right) d\omega$$

$$\le \frac{R}{2} \int_{S^{N-1}} \left(\int_0^R r^p |\nabla u(r\omega)|^2 r^{N-1} \, dr \right) d\omega. \tag{11.2}$$

Here $x \in B(0,R)$ is given in polar coordinates by $x = r\omega$ with $r := |x|$, $0 \le r \le R$. Clearly $\omega = \frac{x}{|x|}$ if $r > 0$. So it is enough to prove that, for each ω,

$$\int_0^R r^p |u(r\omega)| \, |\nabla u(r\omega)| r^{N-1} \, dr \le \frac{R}{2} \int_0^R r^p |\nabla u(r\omega)|^2 r^{N-1} \, dr. \tag{11.3}$$

We set

$$z(r) := \int_r^R s^{p+N-1} |\nabla u(s\omega)|^2 \, ds, \quad 0 \le r \le R.$$

Here $z(r) \ge 0$ and $z(R) = 0$. Then

$$-z'(r) = r^{p+N-1} |\nabla u(r\omega)|^2 \ge 0, \quad 0 \le r \le R.$$

Thus

$$r^{\frac{p+N-1}{2}} |\nabla u(r\omega)| = (-z'(r))^{1/2}. \tag{11.4}$$

Since $u = 0$ on $\partial B(0, R)$ we find

$$u(r\omega) = -\int_r^R \frac{\partial u}{\partial s}(s\omega)\, ds.$$

We observe that

$$r^{\frac{p+N-1}{2}}|u(r\omega)| \le \int_r^R r^{\left(\frac{p+N-1}{2}\right)}\left|\frac{\partial u}{\partial s}(s\omega)\right| ds$$

$$\le \int_r^R 1 \cdot s^{\left(\frac{p+N-1}{2}\right)}\left|\frac{\partial u}{\partial s}(s\omega)\right| ds$$

$$\text{(by the Cauchy–Schwarz inequality)}$$

$$\le \sqrt{R-r}\left(\int_r^R s^{p+N-1}\left|\frac{\partial u}{\partial s}(s\omega)\right|^2 ds\right)^{1/2}$$

$$\le \sqrt{R-r}\left(\int_r^R s^{p+N-1}|\nabla u(s\omega)|^2\, ds\right)^{1/2} = \sqrt{R-r}(z(r))^{1/2}.$$

That is, we have

$$r^{\frac{p+N-1}{2}}|u(r\omega)| \le \sqrt{R-r}(z(r))^{1/2}, \tag{11.5}$$

all $0 \le r \le R$. Multiplying (11.4) by (11.5) we obtain

$$r^{p+N-1}|u(r\omega)|\,|\nabla u(r\omega)| \le \sqrt{R-r}(z(r))^{1/2}(-z'(r))^{1/2}, \tag{11.6}$$

for all $0 \le r \le R$.

Therefore by integrating (11.6) and using the Cauchy–Schwarz inequality we get

$$\int_0^R r^{p+N-1}|u(r\omega)|\,|\nabla u(r\omega)|\, dr$$

$$\le \int_0^R \sqrt{R-r}(z(r))^{1/2}(-z'(r))^{1/2}\, dr$$

$$\le \left(\int_0^R (R-r)\, dr\right)^{1/2}\left(-\int_0^R z(r)z'(r)\, dr\right)^{1/2}$$

$$= \frac{R}{2}z(0) = \frac{R}{2}\int_0^R r^{p+N-1}|\nabla u(r\omega)|^2\, dr.$$

Clearly we have established (11.3), and (11.1) follows. $\qquad\square$

We have

Corollary 11.10. *Let $p \in \mathbb{R}$ be such that $p + N \ge 1$ and let $u \in H_0^1(B(0, R))$. Then*

$$\int_{B(0,R)} |x|^p|u(x)|\,|\nabla u(x)|\, dx \le \frac{R}{2}\int_{B(0,R)} |x|^p|\nabla u(x)|^2\, dx. \tag{11.7}$$

Proof. By (11.1), inequality (11.7) is valid for $u \in C^1(\overline{B(0, R)})$ with $u = 0$ on $\partial B(0, R)$. If $u \in H_0^1(B(0, R))$ then there exists a sequence $\{u_m\}$ from $C^1(\overline{B(0, R)})$ such that $u_m = 0$ on $\partial B(0, R)$ and $u_m \xrightarrow{\|\cdot\|_*} u$. Then the corollary follows by using (11.1) for u_m and letting $m \to \infty$. $\qquad\square$

Next we are ready to give

Theorem 11.11. *Let Ω be a smooth bounded domain in $\mathbb{R}^N$, $N \geq 1$. Let $p \in \mathbb{R}$ such that $p + N \geq 1$. Then there exists a constant $K = K(\Omega)$ such that*

$$\int_\Omega |x|^p |u(x)| \, |\nabla u(x)| \, dx \leq K(\Omega) \int_\Omega |x|^p |\nabla u(x)|^2 \, dx, \qquad (11.8)$$

for all $u \in H_0^1(\Omega)$. Here

$$K(\Omega) := \frac{R}{2},$$

where R is the radius of the smallest ball centered at the origin and containing Ω.

Proof. Since $\partial\Omega$ is smooth any function u in $H_0^1(\Omega)$ can be viewed as a member of $H_0^1(B(0,R))$ by defining it to be zero on $B(0,R) - \Omega$. For the last step, see p. 245 of [139]. Now Corollary 11.10 applies. $\square$

Remark 11.12. (i) When u is spherically symmetric then (11.7) reduces to

$$\int_0^R r^{p+N-1} |u(r)| \, |u'(r)| \, dr \leq \frac{R}{2} \int_0^R r^{p+N-1} (u'(r))^2 \, dr. \qquad (11.9)$$

(ii) When $p = 0$ then (11.8) collapses into

$$\int_\Omega |u(x)| \, |\nabla u(x)| \, dx \leq K(\Omega) \int_\Omega |\nabla u(x)|^2 \, dx, \qquad (11.10)$$

for all $u \in H_0^1(\Omega)$.

The counterpart of Theorem 11.8 follows

Theorem 11.13. *Let $R > 0$, $N \geq 1$, $B(0,R)$ the ball centered at 0 of radius R in $\mathbb{R}^N$. Let $p \in \mathbb{R}$ be such that $p + N > 1$. Let $u \in C^1(\overline{B(0,R)})$ such that $u(\partial B(0,R)) = 0$. Then*

$$\int_{B(0,R)} |x|^p |u(x)| \, |\nabla u(x)| \, dx \leq \frac{R\sqrt{p+N}}{2(p+N-1)} \int_{B(0,R)} |x|^p |\nabla u(x)|^2 \, dx. \qquad (11.11)$$

Proof. The integral in the R.H.S.(11.11) is finite by $p + N > 0$. We estimate

$$\int_0^R r^{p+N-1} |u(r\omega)| \, |\nabla u(r\omega)| \, dr = \int_0^R \left(r^{\frac{p+N}{2}} |\nabla u(r\omega)| \right) \left(r^{\frac{p+N-2}{2}} |u(r\omega)| \right) dr$$

(by the Cauchy–Schwarz inequality)

$$\leq \left(\int_0^R r^{p+N} |\nabla u(r\omega)|^2 \, dr \right)^{1/2} \cdot \left(\int_0^R r^{p+N-2} |u(r\omega)|^2 \, dr \right)^{1/2} =: (\ast\ast).$$

Since $u = 0$ on $\partial B(0,R)$ we find

$$u(r\omega) = - \int_r^R \frac{\partial u}{\partial s}(s\omega) \, ds.$$

Therefore

$$|u(r\omega)| \le \int_r^R 1 \cdot \left| \frac{\partial u}{\partial s}(s\omega) \right| ds \le \left(\int_r^R 1^2 \, ds \right)^{1/2} \left(\int_r^R \left| \frac{\partial u}{\partial s}(s\omega) \right|^2 ds \right)^{1/2}$$

$$\le \sqrt{R-r} \left(\int_r^R |\nabla u(s\omega)|^2 \, ds \right)^{1/2}.$$

That is

$$|u(r\omega)|^2 \le (R-r) \left(\int_r^R |\nabla u(s\omega)|^2 \, ds \right).$$

Consequently we have

$$(**) \le \left(\int_0^R r^{p+N} |\nabla u(r\omega)|^2 \, ds \right)^{1/2}$$

$$\cdot \left(\int_0^R r^{p+N-2}(R-r) \left(\int_r^R |\nabla u(s\omega)|^2 \, ds \right) dr \right)^{1/2}$$

$$= \left(\int_0^R r^{p+N} |\nabla u(r\omega)|^2 \, dr \right)^{1/2} \cdot \left(\int_0^R \left(\int_0^s r^{p+N-2}(R-r) \, dr \right) |\nabla u(s\omega)|^2 \, ds \right)^{1/2}$$

$$= \left(\int_0^R r^{p+N} |\nabla u(r\omega)|^2 \, dr \right)^{1/2} \cdot \left(\int_0^R \left(\frac{Rs^{p+N-1}}{p+N-1} - \frac{s^{p+N}}{p+N} \right) |\nabla u(s\omega)|^2 \, ds \right)^{1/2}$$

$$\left(\text{using the inequality } (AB)^{1/2} \le \frac{\varepsilon}{2} A + \frac{1}{2\varepsilon} B, \text{ for any } A \ge 0, \, B \ge 0 \text{ and } \varepsilon > 0 \right)$$

$$\le \frac{\varepsilon}{2} \left(\int_0^R r^{p+N} |\nabla u(r\omega)|^2 \, dr \right) + \frac{1}{2\varepsilon} \left[\left(\frac{R}{p+N-1} \right) \left(\int_0^R r^{p+N-1} |\nabla u(r\omega)|^2 \, dr \right) \right.$$

$$\left. - \frac{1}{p+N} \left(\int_0^R r^{p+N} |\nabla u(r\omega)|^2 \, dr \right) \right]$$

$$= \frac{1}{2} \left(\varepsilon - \frac{1}{\varepsilon(p+N)} \right) \left(\int_0^R r^{p+N} |\nabla u(r\omega)|^2 \, dr \right)$$

$$+ \frac{R}{2\varepsilon(p+N-1)} \left(\int_0^R r^{p+N-1} |\nabla u(r\omega)|^2 \, dr \right)$$

$$\left(\text{by choosing } \varepsilon = \frac{1}{\sqrt{p+N}} \right)$$

$$= \frac{R\sqrt{p+N}}{2(p+N-1)} \left(\int_0^R r^{p+N-1} |\nabla u(r\omega)|^2 \, dr \right).$$

156   *ADVANCED INEQUALITIES*

We have proved that

$$\int_0^R r^{p+N-1} |u(r\omega)| \, |\nabla u(r\omega)| \, dr \leq \frac{R\sqrt{p+N}}{2(p+N-1)} \left(\int_0^R r^{p+N-1} |\nabla u(r\omega)|^2 \, dr \right).$$

Finally we obtain

$$\int_{S^{N-1}} \left(\int_0^R r^{p+N-1} |u(r\omega)| \, |\nabla u(r\omega)| \, dr \right) dw$$

$$\leq \frac{R\sqrt{p+N}}{2(p+N-1)} \int_{S^{N-1}} \left(\int_0^R r^{p+N-1} |\nabla u(r\omega)|^2 \, dr \right) d\omega,$$

proving (11.11). $\qquad\qquad\qquad\qquad\qquad\qquad\qquad\qquad\qquad\qquad\qquad\qquad\square$

Corollary 11.14. *Let $p \in \mathbb{R}$ be such that $p + N > 1$ and let $u \in H_0^1(B(0, R))$. Then*

$$\int_{B(0,R)} |x|^p |u(x)| \, |\nabla u(x)| \, dx \leq \frac{R\sqrt{p+N}}{2(p+N-1)} \int_{B(0,R)} |x|^p |\nabla u(x)|^2 \, dx. \qquad (11.12)$$

Proof.   Similar to Corollary 11.10. $\qquad\qquad\qquad\qquad\qquad\qquad\qquad\qquad\quad\square$

Next we are ready to give

Theorem 11.15. *Let Ω be a smooth bounded domain in $\mathbb{R}^N$, $N \geq 1$. Let $p \in \mathbb{R}$ be such that $p + N > 1$. Then there exists a constant $K = K(\Omega, p, N)$ such that*

$$\int_\Omega |x|^p |u(x)| \, |\nabla u(x)| \, dx \leq K \int_\Omega |x|^p |\nabla u(x)|^2 \, dx, \qquad (11.13)$$

for all $u \in H_0^1(\Omega)$. Here

$$K := \frac{R\sqrt{p+N}}{2(p+N-1)},$$

where R is the radius of the smallest ball centered at the origin and containing Ω.

Proof.   Similar to Theorem 11.11 by applying Corollary 11.14. $\qquad\qquad\quad\square$

Remark 11.16. (i) When u is spherically symmetric then (11.12) reduces to

$$\int_0^R r^{p+N-1} |u(r)| \, |u'(r)| \, dr \leq \frac{R\sqrt{p+N}}{2(p+N-1)} \int_0^R r^{p+N-1} (u'(r))^2 \, dr. \qquad (11.14)$$

(ii) When $p = 0$ then (11.13) collapses into

$$\int_\Omega |u(x)| \, |\nabla u(x)| \, dx \leq \tilde{K} \int_\Omega |\nabla u(x)|^2 \, dx, \qquad (11.15)$$

for all $u \in H_0^1(\Omega)$, where

$$\tilde{K} := \frac{R\sqrt{N}}{2(N-1)}, \quad N \geq 2.$$

Theorem 11.8 is generalized as follows.

Theorem 11.17 *Let $R > 0$, $N \geq 1$, $\rho > 1$, $\alpha, \beta > 0$, $\rho \geq \alpha + \beta$, $p \geq 1 - N$, and $u \in C^1(\overline{B(0,R)})$ with $u(\partial B(0,R)) = 0$. Then*

$$\int_{B(0,R)} |x|^{\left[(p+N-1)\left(\frac{\alpha+\beta}{\rho}\right)+(1-N)\right]} |u(x)|^\beta |\nabla u(x)|^\alpha \, dx$$

$$\leq M\omega_N^{\left(\frac{\rho-\alpha-\beta}{\rho}\right)} \left(\int_{B(0,R)} |x|^p |\nabla u(x)|^\rho \, dx\right)^{\left(\frac{\alpha+\beta}{\rho}\right)}, \qquad (11.16)$$

where

$$M := \left(\frac{\rho - \alpha}{(\rho-1)\beta + (\rho - \alpha)}\right)^{\left(\frac{\rho-\alpha}{\rho}\right)} \cdot \left(\frac{\alpha}{\alpha+\beta}\right)^{\alpha/\rho} \cdot R^{\left(\frac{(\rho-1)\beta+(\rho-\alpha)}{\rho}\right)}. \qquad (11.17)$$

Proposition 11.18. *Inequality (11.16) is sharp, namely it is attained by $u(x) := R - |x|$, for $x \in B(0,R)$, when $p + N = 1$, $\rho = \alpha + \beta$ and $\alpha = 1$.*

Proof. Notice that

$$\text{L.H.S.}(11.16) = \omega_N \frac{R^{\beta+1}}{\beta+1},$$

and

$$\text{R.H.S.}(11.16) = \omega_N \frac{\alpha^{\alpha/\rho}}{\rho} R^{\beta+1}.$$

Equality holds in (11.16) if

$$\alpha^{\frac{\alpha}{\alpha+\beta}} = \frac{\alpha+\beta}{1+\beta}.$$

But this implies $\alpha = 1$.

Here is a proof that if $\alpha > 0$ and $\beta > 0$, then $\alpha^{\frac{\alpha}{\alpha+\beta}} = \frac{\alpha+\beta}{1+\beta}$ implies that $\alpha = 1$.

Put $\frac{\alpha}{\alpha+\beta} = t$ and $\frac{1}{\alpha} = s$. Then $0 < t < 1$ and $s > 0$. The condition $\alpha^{\frac{\alpha}{\alpha+\beta}} = \frac{\alpha+\beta}{1+\beta}$ yields the equation $s^t - st = 1 - t$. Put $g(x) = x^t - xt$, for $x > 0$. Note that $g'(x) = tx^{t-1} - t = t(x^{t-1} - 1)$. It follows that g has a unique global maximum at $x = 1$. Since $g(s) = g(1)$, we infer that $s = 1$, and hence $\alpha = \frac{1}{s} = 1$. $\qquad \square$

Proof of Theorem 11.17. The integral in the R.H.S.(11.16) is finite by $p + N > 0$. We have

$$u(r\omega) = -\int_r^R \frac{\partial u}{\partial s}(s\omega) \, ds, \quad \text{all } 0 \leq r \leq R.$$

Next we apply Hölder's inequality with indices ρ and $\frac{\rho}{\rho-1}$ to derive

$$r^{\frac{p+N-1}{\rho}} |u(r\omega)| \leq \int_r^R r^{\frac{p+N-1}{\rho}} \left|\frac{\partial u(s\omega)}{\partial s}\right| ds \leq \int_r^R 1 \cdot s^{\frac{p+N-1}{\rho}} \left|\frac{\partial u}{\partial s}(s\omega)\right| ds$$

$$\leq \left(\int_r^R ds\right)^{\frac{\rho-1}{\rho}} \left(\int_r^R s^{p+N-1} \left|\frac{\partial u}{\partial s}(s\omega)\right|^\rho ds\right)^{1/\rho}$$

$$\leq (R - r)^{\frac{\rho-1}{\rho}} \left(\int_r^R s^{p+N-1} |\nabla u(s\omega)|^\rho \, ds\right)^{1/\rho}.$$

So by calling

$$z(r) := \int_r^R s^{p+N-1}|\nabla u(s\omega)|^\rho \, ds \qquad (11.18)$$

we got

$$r^{\left(\frac{p+N-1}{\rho}\right)}|u(r\omega)| \le (R-r)^{\frac{\rho-1}{\rho}}(z(r))^{\frac{1}{\rho}}, \qquad (11.19)$$

and

$$r^{(p+N-1)\left(\frac{\beta}{\rho}\right)}|u(r\omega)|^\beta \le (R-r)^{(\rho-1)\left(\frac{\beta}{\rho}\right)}(z(r))^{\frac{\beta}{\rho}}, \qquad (11.20)$$

all $0 \le r \le R$.

Here notice that $z(r) \ge 0$ and $z(R) = 0$. Also we have

$$-z(r) = \int_R^r s^{p+N-1}|\nabla u(s\omega)|^\rho \, ds$$

and

$$-z'(r) = r^{p+N-1}|\nabla u(r\omega)|^\rho \ge 0.$$

Consequently

$$r^{\left(\frac{p+N-1}{\rho}\right)}|\nabla u(r\omega)| = (-z'(r))^{\frac{1}{\rho}}$$

and

$$r^{(p+N-1)\left(\frac{\alpha}{\rho}\right)}|\nabla u(r\omega)|^\alpha = (-z'(r))^{\frac{\alpha}{\rho}}, \quad \text{for all } 0 \le r \le R. \qquad (11.21)$$

Multiplying (11.20) by (11.21) we get

$$r^{(p+N-1)\left(\frac{\alpha+\beta}{\rho}\right)}|u(r\omega)|^\beta|\nabla u(r\omega)|^\alpha \le (R-r)^{(\rho-1)\left(\frac{\beta}{\rho}\right)}(z(r))^{\frac{\beta}{\rho}}(-z'(r))^{\frac{\alpha}{\rho}}, \qquad (11.22)$$

for all $0 \le r \le R$.

Next we integrate (11.22) and apply Hölder's inequality with indices $\frac{\rho}{\alpha}$ and $\frac{\rho}{(\rho-\alpha)}$ to find

$$\int_0^R r^{(p+N-1)\left(\frac{\alpha+\beta}{\rho}\right)}|u(r\omega)|^\beta|\nabla u(r\omega)|^\alpha \, dr$$

$$\le \int_0^R (R-r)^{(\rho-1)\left(\frac{\beta}{\rho}\right)}(z(r))^{\frac{\beta}{\rho}}(-z'(r))^{\frac{\alpha}{\rho}} \, dr$$

$$\le \left(\int_0^R \left((R-r)^{(\rho-1)\left(\frac{\beta}{\rho}\right)}\right)^{\left(\frac{\rho}{\rho-\alpha}\right)} dr\right)^{\left(\frac{\rho-\alpha}{\rho}\right)} \cdot \left(\int_0^R \left((z(r))^{\frac{\beta}{\rho}}(-z'(r))^{\frac{\alpha}{\rho}}\right)^{\frac{\rho}{\alpha}} dr\right)^{\frac{\alpha}{\rho}}$$

$$= \left(\int_0^R (R-r)^{\frac{(\rho-1)\beta}{(\rho-\alpha)}} dr\right)^{\frac{\rho-\alpha}{\rho}} \cdot \left(-\int_0^R (z(r))^{\frac{\beta}{\alpha}}z'(r) \, dr\right)^{\frac{\alpha}{\rho}}$$

$$= \left(\frac{\rho-\alpha}{(\rho-1)\beta+(\rho-\alpha)}\right)^{\left(\frac{\rho-\alpha}{\rho}\right)} R^{\left(\frac{(\rho-1)\beta+(\rho-\alpha)}{\rho}\right)} \left(\frac{\alpha}{\alpha+\beta}\right)^{\frac{\alpha}{\rho}} (z(0))^{\left(\frac{\alpha+\beta}{\rho}\right)}$$

$$= \left(\frac{\rho-\alpha}{(\rho-1)\beta+(\rho-\alpha)}\right)^{\left(\frac{\rho-\alpha}{\rho}\right)} \left(\frac{\alpha}{\alpha+\beta}\right)^{\frac{\alpha}{\rho}} R^{\left(\frac{(\rho-1)\beta+(\rho-\alpha)}{\rho}\right)}$$

$$\left(\int_0^R r^{p+N-1}|\nabla u(r\omega)|^\rho \, dr\right)^{\left(\frac{\alpha+\beta}{\rho}\right)}.$$

We have proved that

$$\int_0^R r^{(p+N-1)\left(\frac{\alpha+\beta}{\rho}\right)} |u(r\omega)|^\beta |\nabla u(r\omega)|^\alpha \, dr$$

$$\leq M \left(\int_0^R r^{p+N-1} |\nabla u(r\omega)|^\rho \, dr \right)^{\left(\frac{\alpha+\beta}{\rho}\right)}. \tag{11.23}$$

Finally we derive

$$\int_{S^{N-1}} \left(\int_0^R r^{\left[(p+N-1)\left(\frac{\alpha+\beta}{\rho}\right)+(1-N) \right]} |u(r\omega)|^\beta |\nabla u(r\omega)|^\alpha r^{N-1} \, dr \right) d\omega$$

$$\leq M \int_{S^{N-1}} \left(\int_0^R r^p |\nabla u(r\omega)|^\rho r^{N-1} \, dr \right)^{\frac{(\alpha+\beta)}{\rho}} d\omega$$

(if $\rho > \alpha + \beta$ then we apply again Hölder's inequality with indices $\left(\frac{\rho}{\rho-\alpha-\beta}\right)$ and $\left(\frac{\rho}{\alpha+\beta}\right)$ to find)

$$\leq M \left(\int_{S^{N-1}} 1^{\left(\frac{\rho}{\rho-\alpha-\beta}\right)} d\omega \right)^{\left(\frac{\rho-\alpha-\beta}{\rho}\right)}$$

$$\cdot \left(\int_{S^{N-1}} \left(\int_0^R r^p |\nabla u(r\omega)|^\rho r^{N-1} \, dr \right) d\omega \right)^{\left(\frac{\alpha+\beta}{\rho}\right)}$$

$$= M \omega_N^{\left(\frac{\rho-\alpha-\beta}{\rho}\right)} \left(\int_{S^{N-1}} \left(\int_0^R r^p |\nabla u(r\omega)|^\rho r^{N-1} \, dr \right) d\omega \right)^{\left(\frac{\alpha+\beta}{\rho}\right)}.$$

We have established inequality (11.16). $\qquad \square$

Corollary 11.19. *Let all the conditions of Theorem 11.17 hold, except now consider* $u \in H_0^1(B(0,R))$. *Then*

$$\int_{B(0,R)} |x|^{\left[(p+N-1)\left(\frac{\alpha+\beta}{\rho}\right)+(1-N) \right]} |u(x)|^\beta |\nabla u(x)|^\alpha \, dx$$

$$\leq M \omega_N^{\left(\frac{\rho-\alpha-\beta}{\rho}\right)} \left(\int_{B(0,R)} |x|^p |\nabla u(x)|^\rho \, dx \right)^{\left(\frac{\alpha+\beta}{\rho}\right)}. \tag{11.24}$$

Proof. Similar to Corollary 11.10. $\qquad \square$

Next we give the final main result of this chapter.

Theorem 11.20. *Let Ω be a smooth bounded domain in $\mathbb{R}^N$, $N \geq 1$. Let $\rho > 1$, $\alpha, \beta > 0$, $\rho \geq \alpha + \beta$, $p \geq 1 - N$. Then there exists a constant $K = K(\Omega, \alpha, \beta, \rho, N)$ such that*

$$\int_\Omega |x|^{\left[(p+N-1)\left(\frac{\alpha+\beta}{\rho}\right)+(1-N) \right]} |u(x)|^\beta |\nabla u(x)|^\alpha \, dx$$

$$\leq K \left(\int_\Omega |x|^p |\nabla u(x)|^\rho \, dx \right)^{\left(\frac{\alpha+\beta}{\rho}\right)}, \tag{11.25}$$

for all $u \in H_0^1(\Omega)$. Here

$$K = M\omega_N^{\left(\frac{\rho-\alpha-\beta}{\rho}\right)} \tag{11.26}$$

is the constant of inequality (11.16).

The M in (11.26) is given by (11.17), where R is the radius of the smallest ball centered at the origin and containing Ω.

Proof. Similar to Theorem 11.11 by applying Corollary 11.19. $\qquad\square$

Remark 11.21. (i) When u is spherically symmetric then (11.24) reduces to

$$\int_0^R r^{(p+N-1)\left(\frac{\alpha+\beta}{\rho}\right)}|u(r)|^\beta|u'(r)|^\alpha \, dr \le M \left(\int_0^R r^{p+N-1}|u'(r)|^\rho \, dr\right)^{\left(\frac{\alpha+\beta}{\rho}\right)}, \tag{11.27}$$

where M is given by (11.17).

(ii) When $p = 0$ then (11.25) collapses into

$$\int_\Omega |x|^{(N-1)\left(\frac{\alpha+\beta-\rho}{\rho}\right)}|u(x)|^\beta|\nabla u(x)|^\alpha \, dx \le K \left(\int_\Omega |\nabla u(x)|^\rho \, dx\right)^{\left(\frac{\alpha+\beta}{\rho}\right)}, \tag{11.28}$$

for all $u \in H_0^1(\Omega)$.

Chapter 12

Opial Inequalities for Widder Derivatives

Various L_p form Opial type inequalities are given for Widder derivatives. This treatment relies on [36].

12.1 Introduction

This chapter is greatly motivated by the article of Z. Opial [195].

Theorem 12.1. (Opial [195]) *Let $x(t) \in C^1([0,h])$ be such that $x(0) = x(h) = 0$, and $x(t) > 0$ in $(0,h)$. Then,*

$$\int_0^h |x(t)x'(t)|dt \le \frac{h}{4} \int_0^h (x'(t))^2 dt. \qquad (12.1)$$

In the last inequality the constant $h/4$ is the best possible.

Opial type inequalities have applications in establishing uniqueness of solution to initial value problems in differential equations, see [244].

12.2 Background

The following are taken from [243].

Let $f, u_0, u_1, \ldots, u_n \in C^{n+1}([a,b])$, $n \ge 0$, and the Wronskians

$$W_i(x) := W[u_0(x), u_1(x), \ldots, u_i(x)] := \begin{vmatrix} u_0(x) & u_1(x) & \ldots & u_i(x) \\ u_0'(x) & u_1'(x) & \ldots & u_i'(x) \\ & \vdots & & \\ u_0^{(i)}(x) & u_1^{(i)}(x) & \ldots & u_i^{(i)}(x) \end{vmatrix}, \qquad (12.2)$$

$i = 0, 1, \ldots, n$. Here $W_0(x) = u_0(x)$. Assume $W_i(x) > 0$ over $[a,b]$, $i = 0, 1, \ldots, n$. For $i \ge 0$, the differential operator of order i (Widder derivative):

$$L_i f(x) := \frac{W[u_0(x), u_1(x), \ldots, u_{i-1}(x), f(x)]}{W_{i-1}(x)}, \qquad (12.3)$$

161

$i = 1, \ldots, n + 1$; $L_0 f(x) := f(x)$, $\forall\, x \in [a, b]$.

Consider also

$$g_i(x, t) := \frac{1}{W_i(t)} \begin{vmatrix} u_0(t) & u_1(t) & \cdots & u_i(t) \\ u_0'(t) & u_1'(t) & \cdots & u_i'(t) \\ \vdots & & & \\ u_0^{(i-1)}(t) & u_1^{(i-1)}(t) & \cdots & u_i^{(i-1)}(t) \\ u_0(x) & u_1(x) & \cdots & u_i(x) \end{vmatrix}, \tag{12.4}$$

$i = 1, 2, \ldots, n$; $g_0(x, t) := \dfrac{u_0(x)}{u_0(t)}$, $\forall\, x, t \in [a, b]$.

Example ([243]). Sets of the form $\{u_0, u_1, \ldots, u_n\}$ are $\{1, x, x^2, \ldots, x^n\}$,

$$\{1, \sin x, -\cos x, -\sin 2x, \cos 2x, \ldots, (-1)^{n-1} \sin nx, (-1)^n \cos nx\}, \text{ etc.}$$

We also mention the generalized Widder–Taylor's formula, see [243].

Theorem 12.2. *Let the functions* $f, u_0, u_1, \ldots, u_n \in C^{n+1}([a, b])$, *and the Wronskians* $W_0(x), W_1(x), \ldots, W_n(x) > 0$ *on* $[a, b]$, $x \in [a, b]$. *Then for* $t \in [a, b]$ *we have*

$$f(x) = f(t)\frac{u_0(x)}{u_0(t)} + L_1 f(t) g_1(x, t) + \cdots + L_n f(t) g_n(x, t) + R_n(x), \tag{12.5}$$

where

$$R_n(x) := \int_t^x g_n(x, s) L_{n+1} f(s) ds. \tag{12.6}$$

For example ([243]) one could take $u_0(x) = c > 0$. If $u_i(x) = x^i$, $i = 0, 1, \ldots, n$, defined on $[a, b]$, then

$$L_i f(t) = f^{(i)}(t) \quad \text{and} \quad g_i(x, t) = \frac{(x - t)^i}{i!}, \quad t \in [a, b].$$

We need

Corollary 12.3 (on Theorem 12.2). *By additionally assuming for fixed* $x_0 \in [a, b]$ *that* $L_i f(x_0) = 0$, $i = 0, 1, \ldots, n$, *we get that*

$$f(x) = \int_{x_0}^x g_n(x, t) L_{n+1} f(t) dt, \quad \forall\, x \in [a, b]. \tag{12.7}$$

12.3   Results

From now on we are working under the terms and assumptions of Theorem 12.2 and Corollary 12.3.

We first present the following

Theorem 12.4. *Let $x \geq x_0$, $x_0, x \in [a, b]$ and $p, q > 1$: $\dfrac{1}{p} + \dfrac{1}{q} = 1$. Then*

$$\int_{x_0}^{x} |f(w)||L_{n+1}f(w)|dw \leq 2^{-1/q} \left(\int_{x_0}^{x} \left(\int_{x_0}^{w} (g_n(w,t))^p dt \right) dw \right)^{1/p}$$

$$\times \left(\int_{x_0}^{x} |L_{n+1}f(w)|^q dw \right)^{2/q}. \tag{12.8}$$

Proof. By Hölder's inequality we have

$$|f(x)| \leq \left(\int_{x_0}^{x} |g_n(x,t)|^p dt \right)^{1/p} \left(\int_{x_0}^{x} |L_{n+1}f(t)|^q dt \right)^{1/q}. \tag{12.9}$$

Call

$$z(w) := \int_{x_0}^{w} |L_{n+1}f(t)|^q dt, \quad x_0 \leq w \leq x \quad (z(x_0) = 0).$$

Thus

$$z'(w) = |L_{n+1}f(w)|^q$$

and

$$|L_{n+1}f(w)| = (z'(w))^{1/q}.$$

From (12.9) we obtain

$$|f(w)||L_{n+1}f(w)| \leq \left(\int_{x_0}^{w} |g_n(w,t)|^p dt \right)^{1/p} (z(w)z'(w))^{1/q}.$$

Integrating the last inequality over $[x_0, x]$ we obtain

$$\int_{x_0}^{x} |f(w)||L_{n+1}f(w)|dw \leq \int_{x_0}^{x} \left(\int_{x_0}^{w} |g_n(w,t)|^p dt \right)^{1/p} (z(w)z'(w))^{1/q} dw$$

$$\leq \left(\int_{x_0}^{x} \left(\int_{x_0}^{w} |g_n(w,t)|^p dt \right) dw \right)^{1/p} \left(\int_{x_0}^{x} z(w)z'(w)dw \right)^{1/q}$$

$$= \left(\int_{x_0}^{x} \left(\int_{x_0}^{w} |g_n(w,t)|^p dt \right) dw \right)^{1/p} \frac{(z(x))^{2/q}}{2^{1/q}},$$

proving the claim of Theorem 12.4. See also Remark 22.11 next. $\square$

The counter part of the previous result follows.

Theorem 12.5. *Let $x \leq x_0$, $x_0 \in [a, b]$ and $p, q > 1$: $\dfrac{1}{p} + \dfrac{1}{q} = 1$. Then is*

$$\int_{x}^{x_0} |f(w)||L_{n+1}f(w)|dw < 2^{-1/q} \left(\int_{x_0}^{x} \left(\int_{x_0}^{w} |g_n(w,t)|^p dt \right) dw \right)^{1/p}$$

$$\times \left(\int_x^{x_0} |L_{n+1} f(w)|^q dw \right)^{2/q}. \tag{12.10}$$

Proof. Here take $x \le x_0$ and $p, q > 1$ such that $\dfrac{1}{p} + \dfrac{1}{q} = 1$. From Hölder's inequality we have

$$|f(x)| = \left| \int_x^{x_0} g_n(x,t) L_{n+1} f(t) dt \right|$$

$$\le \int_x^{x_0} |g_n(x,t)||L_{n+1} f(t)| dt \le \left(\int_x^{x_0} |g_n(x,t)|^p dt \right)^{1/p} \left(\int_x^{x_0} |L_{n+1} f(t)|^q dt \right)^{1/q}. \tag{12.11}$$

Call

$$z(x) := \int_x^{x_0} |L_{n+1} f(t)|^q dt \ge 0, \quad z(x_0) = 0.$$

That is

$$-z(x) = \int_{x_0}^{x} |L_{n+1} f(t)|^q dt \le 0,$$

and

$$-z'(x) = |L_{n+1} f(x)|^q \ge 0,$$

and

$$|L_{n+1} f(x)| = (-z'(x))^{1/q}, \quad x \in [a, x_0].$$

Hence by (12.11) $(x \le w \le x_0)$

$$|f(w)||L_{n+1} f(w)| \le \left(\int_w^{x_0} |g_n(w,t)|^p dt \right)^{1/p} (z(w)(-z'(w))^{1/q}.$$

Integrating the last inequality over $[x, x_0]$ we obtain

$$\int_x^{x_0} |f(w)||L_{n+1} f(w)| dw \le \int_x^{x_0} \left(\int_w^{x_0} |g_n(w,t)|^p dt \right)^{1/p} (z(w)(-z'(w)))^{1/q} dw$$

$$\le \left(\int_x^{x_0} \left(\int_w^{x_0} |g_n(w,t)|^p dt \right) dw \right)^{1/p} \left(\int_x^{x_0} z(w)(-z'(w)) dw \right)^{1/q}$$

$$= 2^{-1/q} \left(\int_{x_0}^{x} \left(\int_{x_0}^{w} |g_n(w,t)|^p dt \right) dw \right)^{1/p} (z(x))^{2/q},$$

proving the claim. $\qquad\qquad\square$

Next we study the extreme cases.

Theorem 12.6. *Here $p = 1$, $q = \infty$ and $x \geq x_0$. Then*

$$\int_{x_0}^{x} |f(w)||L_{n+1}f(w)|dw \leq \left(\int_{x_0}^{x} \left(\int_{x_0}^{w} g_n(w,t)dt \right) dw \right) \|L_{n+1}f\|_\infty^2. \qquad (12.12)$$

Proof. By

$$f(w) = \int_{x_0}^{w} g_n(w,t)L_{n+1}f(t)dt$$

we get

$$|f(w)| \leq \left(\int_{x_0}^{w} |g_n(w,t)|dt \right) \|L_{n+1}f\|_\infty,$$

and

$$|f(w)||L_{n+1}f(w)| \leq \left(\int_{x_0}^{w} |g_n(w,t)|dt \right) \|L_{n+1}f\|_\infty^2.$$

Integrating the last inequality we find (12.12). See also Remark 12.11 next. $\square$

Theorem 12.7. *Again $p = 1$, $q = \infty$, but $x \leq x_0$. Then*

$$\int_{x}^{x_0} |f(w)||L_{n+1}f(w)|dw \leq \left(\int_{x_0}^{x} \left(\int_{x_0}^{w} g_n(w,t)|dt \right) dw \right) (\|L_{n+1}f\|_\infty^2). \qquad (12.13)$$

Proof. Here $x \leq w \leq x_0$, and

$$|f(w)| = \left| \int_{x_0}^{w} g_n(w,t)L_{n+1}f(t)dt \right|$$

$$= \left| \int_{w}^{x_0} g_n(w,t)L_{n+1}f(t)dt \right| \leq \left(\int_{w}^{x_0} |g_n(w,t)|dt \right) \|L_{n+1}f\|_\infty.$$

Thus

$$|f(w)||L_{n+1}f(w)| \leq \left(\int_{w}^{x_0} |g_n(w,t)|dt \right) \|L_{n+1}f\|_\infty^2,$$

and

$$\int_{x}^{x_0} |f(w)||L_{n+1}f(w)|dw \leq \left(\int_{x}^{x_0} \left(\int_{w}^{x_0} |g_n(w,t)|dt \right) dw \right) \|L_{n+1}f\|_\infty^2,$$

proving the claim. $\square$

Corollary 12.8. *Let $p, q > 1$: $\dfrac{1}{p} + \dfrac{1}{q} = 1$. Then*

$$\left| \int_{x_0}^{x} |f(w)||L_{n+1}f(w)|dw \right| \leq 2^{-1/q} \left(\int_{x_0}^{x} \left(\int_{x_0}^{w} |g_n(w,t)|^p dt \right) dw \right)^{1/p}$$

$$\times \left(\left| \int_{x_0}^{x} |L_{n+1}f(w)|^q dw \right| \right)^{2/q}. \qquad (12.14)$$

Proof. By Theorems 12.4, 12.5. $\square$

In particular when $p = q = 2$ we have

Corollary 12.9. *It holds*

$$\left| \int_{x_0}^{x} |f(w)| |L_{n+1}f(w)| dw \right| \leq 2^{-1/2} \left(\int_{x_0}^{x} \left(\int_{x_0}^{w} (g_n(w,t))^2 dt \right) dw \right)^{1/2}$$

$$\times \left| \int_{x_0}^{x} (L_{n+1}f(w))^2 dw \right|. \tag{12.15}$$

Furthermore we get

Corollary 12.10. *Let $p = 1$, $q = \infty$. Then*

$$\left| \int_{x_0}^{x} |f(w)| |L_{n+1}f(w)| dw \right| \leq \left(\int_{x_0}^{x} \left(\int_{x_0}^{w} |g_n(w,t)| dt \right) dw \right) \|L_{n+1}f\|_{\infty}^2. \tag{12.16}$$

Proof.   By Theorems 12.6, 12.7. $\square$

We need to make

Remark 12.11. We define (see [243])

$$\phi_0(x) := W_0(x), \;\; \phi_1(x) := \frac{W_1(x)}{(W_0(x))^2}, \ldots,$$

in general

$$\phi_k(x) := \frac{W_k(x)W_{k-2}(x)}{(W_{k-1}(x))^2}, \quad k = 2, 3, \ldots, n.$$

The functions $\phi_i(x)$ are positive on $[a, b]$. According to [243] we get, for x, x_0 not fixed, that

$$g_n(x, x_0) = \frac{\phi_0(x)}{\phi_0(x_0) \ldots \phi_n(x_0)} \int_{x_0}^{x} \phi_1(x_1) \int_{x_0}^{x_1} \ldots \int_{x_0}^{x_{n-2}} \phi_{n-1}(x_{n-1})$$

$$\times \int_{x_0}^{x_{n-1}} \phi_n(x_n) dx_1 dx_2 \ldots dx_n$$

$$= \frac{1}{\phi_0(x_0) \ldots \phi_n(x_0)} \int_{x_0}^{x} \phi_0(s) \ldots \phi_n(s) g_{n-1}(x, s) ds. \tag{12.17}$$

We get that $g_n(x, x) = 0$, all $x \in [a, b]$, and $g_n(x, x_0) > 0$, $x > x_0$, $x, x_0 \in [a, b]$, $\forall \, n \geq 1$. Also $g_0(x, x_0) > 0$ for any $x, x_0 \in [a, b]$.

To complete the chapter we present

Theorem 12.12. *Let $0 < p < 1$ and $q < 0$ be such that $\dfrac{1}{p} + \dfrac{1}{q} = 1$, and $x > x_0$, $x, x_0 \in [a, b]$. Assume that $L_{n+1}f$ is of fixed sign and nowhere zero. Then*

$$\int_{x_0}^{x} |f(w)| |L_{n+1}f(w)| dw \geq 2^{-1/q} \left(\int_{x_0}^{x} \left(\int_{x_0}^{w} (g_n(w,t))^p dt \right) dw \right)^{1/p}$$

$$\times \left(\int_{x_0}^{x} |L_{n+1}f(w)|^q dw \right)^{2/q}. \tag{12.18}$$

Proof. Here for $x_0 \leq w \leq x$ we have

$$|f(w)| = \int_{x_0}^{w} g_n(w,t)|L_{n+1}f(t)|dt, \tag{12.19}$$

by Remark 22.11.

From (12.19) by Hölder's inequality we have

$$|f(w)| \geq \left(\int_{x_0}^{x} (g_n(w,t))^p dt \right)^{1/p} \left(\int_{x_0}^{w} |L_{n+1}f(t)|^q dt \right)^{1/q}, \tag{12.20}$$

for $w > x_0$. Consider

$$z(w) := \int_{x_0}^{w} |L_{n+1}f(t)|^q dt > 0, \quad z(x_0) = 0.$$

So that $z'(w) = |L_{n+1}f(w)|^q > 0$ and $|L_{n+1}f(w)| = (z'(w))^{1/q}$, all $x_0 \leq w \leq x$. Thus by (12.20) we get

$$|f(w)||L_{n+1}f(w)| \geq \left(\int_{x_0}^{w} (g_n(w,t))^p dt \right)^{1/p} (z(w)z'(w))^{1/q},$$

all $x_0 < w \leq x$. Let $x_0 < \theta \leq w \leq x$ and $\theta \downarrow x_0$, then by integration of the last inequality we obtain

$$\int_{x_0}^{x} |f(w)||L_{n+1}f(w)|dw = \lim_{\theta \downarrow x_0} \int_{\theta}^{x} |f(w)||L_{n+1}f(w)|dw$$

$$\geq \lim_{\theta \downarrow x_0} \left(\int_{\theta}^{x} \left(\int_{x_0}^{w} (g_n(w,t))^p dt \right)^{1/p} (z(w)z'(w))^{1/q} dw \right)$$

$$\geq \lim_{\theta \downarrow x_0} \left(\int_{\theta}^{x} \left(\int_{x_0}^{w} (g_n(w,t))^p dt \right) dw \right)^{1/p} \lim_{\theta \downarrow x_0} \left(\left(\int_{\theta}^{x} z(w)z'(w)dw \right)^{1/q} \right)$$

$$= 2^{-1/q} \left(\int_{x_0}^{x} \left(\int_{x_0}^{w} (g_n(w,t))^p dt \right) dw \right)^{1/p} \lim_{\theta \downarrow x_0} (z^2(x) - z^2(\theta))^{1/q}$$

$$= 2^{-1/q} \left(\int_{x_0}^{x} \left(\int_{x_0}^{w} (g_n(w,t))^p dt \right) dw \right)^{1/p} (z(x))^{2/q}.$$

That is establishing (12.18). $\qquad \sqcup$

We make

Remark 12.13. We notice by (12.17) that

$$g_n(x,t) < 0, \quad x < t, \quad n \text{ odd},$$

$$g_n(x,t) > 0, \quad x < t, \quad n \text{ even},$$

where $x, t \in [a,b]$.

We give

Theorem 12.14. *Let $0 < p < 1$ and $q < 0$ such that $\dfrac{1}{p} + \dfrac{1}{q} = 1$, and $x < x_0$, $x, x_0 \in [a,b]$. Let n be odd. Also assume that $L_{n+1}f$ is of fixed sign and nowhere zero. Then*

$$\int_x^{x_0} |f(w)||L_{n+1}f(w)|dw \geq 2^{-1/q} \left(\int_{x_0}^{x} \left(\int_{x_0}^{w} (-g_n(w,t))^p dt \right) dw \right)^{1/p}$$

$$\times \left(\int_x^{x_0} |(L_{n+1}f)(w)|^q dw \right)^{2/q}. \tag{12.21}$$

Proof.   Here for $x \leq w \leq x_0$ we have

$$|f(w)| = \left| \int_{x_0}^{w} g_n(w,t) L_{n+1}f(t)dt \right|$$

$$= \left| \int_{w}^{x_0} g_n(w,t) L_{n+1}f(t)dt \right| = \left| \int_{w}^{x_0} (-g_n(w,t)) L_{n+1}f(t)dt \right|$$

$$= \int_{w}^{x_0} (-g_n(w,t)) |L_{n+1}f(t)|dt. \tag{12.22}$$

From (12.22) by Hölder's inequality we get

$$|f(w)| \geq \left(\int_{w}^{x_0} (-g_n(w,t))^p dt \right)^{1/p} \left(\int_{w}^{x_0} |L_{n+1}f(t)|^q dt \right)^{1/q},$$

for $w < x_0$. That is

$$|f(w)| \geq \left(\int_{w}^{x_0} |g_n(w,t)|^p dt \right)^{1/p} \left(\int_{w}^{x_0} |L_{n+1}f(t)|^q dt \right)^{1/q} \tag{12.23}$$

for $w < x_0$.

Consider

$$z(w) := \int_{w}^{x_0} |L_{n+1}f(t)|^q dt > 0, \quad z(x_0) = 0.$$

So that

$$-z(w) = \int_{x_0}^{w} |L_{n+1}f(t)|^q dt$$

and

$$-z'(w) = |L_{n+1}f(w)|^q > 0,$$

i.e. $z'(w) < 0$.

Hence

$$|L_{n+1}f(w)| = (-z'(w))^{1/q} > 0, \text{ all } x \le w \le x_0.$$

Therefore by (12.23) we derive

$$|f(w)||L_{n+1}f(w)| \ge \left(\int_w^{x_0} |g_n(w,t)|^p dt\right)^{1/p} (z(w)(-z'(w))^{1/q},$$

all $x \le w < x_0$. Let $x \le w \le \theta < x_0$ and $\theta \uparrow x_0$, then by integration of the last inequality we obtain

$$\int_x^{x_0} |f(w)||L_{n+1}f(w)|dw = \lim_{\theta \uparrow x_0} \int_x^{\theta} |f(w)||L_{n+1}f(w)|dw$$

$$\ge \lim_{\theta \uparrow x_0} \left(\int_x^{\theta} \left(\int_w^{x_0} |g_n(w,t)|^p dt\right)^{1/p} (z(w)(-z'(w)))^{1/q} dw\right)$$

$$\ge \lim_{\theta \uparrow x_0} \left(\int_x^{\theta} \left(\int_w^{x_0} |g_n(w,t)|^p dt\right) dw\right)^{1/p} \lim_{\theta \uparrow x_0} \left(\int_x^{\theta} z(w)(-z'(w))dw\right)^{1/q}$$

$$= \left(\int_x^{x_0} \left(\int_w^{x_0} |g_n(w,t)|^p dt\right) dw\right)^{1/p} \lim_{\theta \uparrow x_0} \left(-\int_x^{\theta} z(w)dz(w)\right)^{1/q}$$

$$= 2^{-1/q} \left(\int_x^{x_0} \left(\int_w^{x_0} |g_n(w,t)|^p dt\right) dw\right)^{1/p} \lim_{\theta \uparrow x_0} (z^2(x) - z^2(\theta))^{1/q}$$

$$= 2^{-1/q} \left(\int_x^{x_0} \left(\int_w^{x_0} |g_n(w,t)|^p dt\right) dw\right)^{1/p} (z(x))^{2/q}.$$

Hence

$$\int_x^{x_0} |f(w)||L_{n+1}f(w)|dw \ge 2^{-1/q} \left(\int_{x_0}^x \left(\int_{x_0}^w |g_n(w,t)|^p dt\right) dw\right)^{1/p}$$

$$\times \left(\int_x^{x_0} |L_{n+1}f(w)|^q dw\right)^{2/q},$$

that is proving (12.21). $\square$

Putting things together we have

Corollary 12.15. *Let* $0 < p < 1$ *and* $q < 0$ *be such that* $\dfrac{1}{p} + \dfrac{1}{q} = 1$, *and* $x_0, x \in [a, b]$ *such that* $x \neq x_0$. *Let* n *be odd. Assume that* $L_{n+1}f$ *is of fixed sign and nowhere zero. Then*

$$\left| \int_{x_0}^{x} |f(w)| |L_{n+1}f(w)| dw \right| \geq 2^{-1/q} \left(\int_{x_0}^{x} \left(\int_{x_0}^{w} |g_n(w,t)|^p dt \right) dw \right)^{1/p}$$

$$\times \left(\left| \int_{x_0}^{x} |L_{n+1}f(w)|^q dw \right| \right)^{2/q}. \tag{12.24}$$

Proof. By Theorems 12.12, 12.14. $\qquad\qquad\qquad\qquad\qquad\qquad\qquad\qquad\square$

We give without proof the similar Theorem 12.14, see Remark 12.13.

Theorem 12.16. *Let* $0 < p < 1$ *and* $q < 0$ *such that* $\dfrac{1}{p} + \dfrac{1}{q} = 1$, *and* $x < x_0$, $x, x_0 \in [a, b]$. *Let* n *be even. Also assume that* $L_{n+1}f$ *is of fixed sign and nowhere zero. Then*

$$\int_{x}^{x_0} |f(w)| |L_{n+1}f(w)| dw \geq 2^{-1/q} \left(\int_{x_0}^{x} \left(\int_{x_0}^{w} (g_n(w,t))^p dt \right) dw \right)^{1/p}$$

$$\times \left(\int_{x}^{x_0} |L_{n+1}f(w)|^q dw \right)^{2/q}. \tag{12.25}$$

We finish here with

Corollary 12.17. *Let* $0 < p < 1$ *and* $q < 0$ *be such that* $\dfrac{1}{p} + \dfrac{1}{q} = 1$, *and* $x_0, x \in [a, b]$ *such that* $x \neq x_0$. *Let* n *be even. Assume that* $L_{n+1}f$ *is of fixed sign and nowhere zero. Then*

$$\left| \int_{x_0}^{x} |f(w)| |L_{n+1}f(w)| dw \right| \geq 2^{-1/q} \left(\int_{x_0}^{x} \left(\int_{x_0}^{w} (g_n(w,t))^p dt \right) dw \right)^{1/p}$$

$$\times \left(\left| \int_{x_0}^{x} |L_{n+1}f(w)|^q dw \right| \right)^{2/q}. \tag{12.26}$$

Proof. By Theorems 12.12, 12.16. $\qquad\qquad\qquad\qquad\qquad\qquad\qquad\qquad\square$

Chapter 13

Opial Inequalities for Linear Differential Operators

Various L_p form Opial inequalities [195] are given for a Linear Differential Operator L, involving its related initial value problem solution y, Ly, the associated Green's function H and initial conditions point $x_0 \in \mathbf{R}$. This chapter follows [18].

13.1 Background

Here we follow [169], pp. 145–154.

Let I be a closed interval of $\mathbf{R}$. Let $a_i(x)$, $i = 0, 1, \ldots, n - 1$ $(n \in \mathbf{N})$, $h(x)$ be continuous functions on I and let $L = D^n + a_{n-1}(x)D^{n-1} + \cdots + a_0(x)$ be a fixed linear differential operator on $C^n(I)$. Let $y_1(x), \ldots, y_n(x)$ be a set of linear independent solutions to $Ly = 0$. Here the associated Green's function for L is

$$
H(x, t) := \left. \begin{vmatrix} y_1(t) \cdots y_n(t) \\ y_1'(t) \ldots y_n'(t) \\ \vdots \\ y_1^{(n-2)}(t) \cdots y_n^{(n-2)}(t) \\ y_1(x) \cdots y_n(x) \end{vmatrix} \middle/ \begin{vmatrix} y_1(t) \cdots y_n(t) \\ y_1'(t) \cdots y_n'(t) \\ \vdots \\ y_1^{(n-2)}(t) \cdots y_n^{(n-2)}(t) \\ y_1^{(n-1)}(t) \cdots y_n^{(n-1)}(t) \end{vmatrix} \right. ,
$$

which is a continuous function on I^2. Consider fixed $x_0 \in I$, then

$$
y(x) = \int_{x_0}^x H(x, t)h(t)dt, \quad \text{all } x \in I
$$

is the unique solution to the initial value problem

$$
Ly = h; \quad y^{(i)}(x_0) = 0, \quad i = 0, 1, \ldots, n - 1.
$$

13.2 Results

We first give

171

Proposition 13.1. *Let $x \geq x_0$; $x_0, x \in I$ and $p, q > 1$: $\frac{1}{p} + \frac{1}{q} = 1$. Then*

$$\int_{x_0}^{x} |y(w)||(Ly)(w)|dw \leq 2^{-1/q} \cdot \left(\int_{x_0}^{x} \left(\int_{x_0}^{w} |H(w,t)|^p dt \right) dw \right)^{1/p}$$

$$\cdot \left(\int_{x_0}^{x} |(Ly)(w)|^q dw \right)^{2/q} . \tag{13.1}$$

Proof.   Here take $x \geq x_0$ and $p, q > 1$ such that $\frac{1}{p} + \frac{1}{q} = 1$. From Hölder's inequality we have

$$|y(x)| \leq \left(\int_{x_0}^{x} |H(x,t)|^p dt \right)^{1/p} \left(\int_{x_0}^{x} |h(t)|^q dt \right)^{1/q} . \tag{13.2}$$

Set

$$z(w) := \int_{x_0}^{w} |h(t)|^q dt, \quad x_0 \leq w \leq x \quad (z(x_0) = 0).$$

Thus

$$z'(w) = |h(w)|^q$$

and

$$|h(w)| = (z'(w))^{1/q}.$$

From (13.2) we obtain

$$|y(w)|\,|h(w)| \leq \left(\int_{x_0}^{w} |H(w,t)|^p dt \right)^{1/p} \cdot (z(w) \cdot z'(w))^{1/q}.$$

Integrating the last inequality over $[x_0, x]$ we derive

$$\int_{x_0}^{x} |y(w)|\,|h(w)|dw \leq \int_{x_0}^{x} \left(\int_{x_0}^{w} |H(w,t)|^p dt \right)^{1/p} \cdot (z(w) \cdot z'(w))^{1/q} dw$$

$$\leq \left(\int_{x_0}^{x} \left(\int_{x_0}^{w} |H(w,t)|^p dt \right) dw \right)^{1/p} \cdot \left(\int_{x_0}^{x} z(w) \cdot z'(w)dw \right)^{1/q}$$

$$= \left(\int_{x_0}^{x} \left(\int_{x_0}^{w} |H(w,t)|^p dt \right) dw \right)^{1/p} \cdot \frac{(z(x))^{2/q}}{2^{1/q}},$$

proving the claim of Proposition 13.1. $\square$

The counterpart of the previous result follows.

Proposition 13.2. *Let $x \leq x_0$; $x_0, x \in I$ and $p, q > 1$: $\frac{1}{p} + \frac{1}{q} = 1$. Then*

$$\int_{x}^{x_0} |y(w)|\,|(Ly)(w)|dw \leq 2^{-1/q} \cdot \left(\int_{x_0}^{x} \left(\int_{x_0}^{w} |H(w,t)|^p dt \right) dw \right)^{1/p}$$

$$\cdot \left(\int_{x}^{x_0} |(Ly)(w)|^q dw \right)^{2/q} . \tag{13.3}$$

Proof. Here take $x \leq x_0$ and $p, q > 1$ such that $\frac{1}{p} + \frac{1}{q} = 1$. From Hölder's inequality we have

$$|y(x)| = \left| \int_x^{x_0} H(x,t)h(t)dt \right| \leq \int_x^{x_0} |H(x,t)|\,|h(t)|dt$$

$$\leq \left(\int_x^{x_0} |H(x,t)|^p dt \right)^{1/p} \left(\int_x^{x_0} |h(t)|^q dt \right)^{1/q}. \tag{13.4}$$

Set

$$z(x) := \int_x^{x_0} |h(t)|^q dt \geq 0, \quad z(x_0) = 0.$$

That is

$$-z(x) = \int_{x_0}^x |h(t)|^q dt \leq 0,$$

and

$$-z'(x) = |h(x)|^q \geq 0,$$

and

$$|h(x)| = (-z'(x))^{1/q}, \quad x \in I.$$

Hence by (13.4) $(x \leq w \leq x_0)$

$$|y(w)| \cdot |h(w)| \leq \left(\int_w^{x_0} |H(w,t)|^p dt \right)^{1/p} \cdot (z(w) \cdot (-z'(w)))^{1/q}.$$

Integrating the last inequality over $[x, x_0]$ we find

$$\int_x^{x_0} |y(w)|\,|h(w)|dw \leq \int_x^{x_0} \left(\int_w^{x_0} |H(w,t)|^p dt \right)^{1/p}$$
$$\cdot (z(w) \cdot (-z'(w)))^{1/q} \cdot dw$$
$$\leq \left(\int_x^{x_0} \left(\int_w^{x_0} |H(w,t)|^p dt \right) \cdot dw \right)^{1/p} \left(\int_x^{x_0} (-z'(w) \cdot z(w)) \cdot dw \right)^{1/q}$$
$$= 2^{-1/q} \cdot \left(\int_{x_0}^x \left(\int_{x_0}^w |H(w,t)|^p dt \right) \cdot dw \right)^{1/p} \cdot (z(x))^{2/q},$$

proving the claim of Proposition 13.2. $\qquad\square$

Extreme cases come next.

Proposition 13.3. *Here* $p = 1$, $q = \infty$ *and* $x \geq x_0$. *Then*

$$\int_{x_0}^x |y(w)|\,|(Ly)(w)|dw$$

$$\leq \left(\int_{x_0}^x \left(\int_{x_0}^w |H(w,t)|dt \right) dw \right) \cdot \|Ly\|_\infty^2. \tag{13.5}$$

Proof. From

$$y(w) = \int_{x_0}^{w} H(w,t)h(t)dt$$

we obtain

$$|y(w)| \leq \left(\int_{x_0}^{w} |H(w,t)|dt \right) \|h\|_{\infty},$$

and

$$|y(w)|\,|(Ly)(w)| \leq \left(\int_{x_0}^{w} |H(w,t)|dt \right) \cdot (\|Ly\|_{\infty})^2.$$

Integrating the last inequality we obtain (13.5). $\qquad\square$

Proposition 13.4. *Again $p=1$, $q=\infty$, but $x \leq x_0$. Then*

$$\int_{x}^{x_0} |y(w)|\,|(Ly)(w)|dw$$

$$\leq \left(\int_{x_0}^{x} \left(\int_{x_0}^{w} |H(w,t)|dt \right) dw \right) \cdot (\|(Ly)\|_{\infty}^2). \tag{13.6}$$

Proof. Here $x \leq w \leq x_0$, and

$$|y(w)| = \left| \int_{x_0}^{w} H(w,t)h(t)dt \right| = \left| \int_{w}^{x_0} H(w,t)h(t)dt \right|$$

$$\leq \left(\int_{w}^{x_0} |H(w,t)|dt \right) \|h\|_{\infty}.$$

Therefore

$$|y(w)|\,|(Ly)(w)| \leq \left(\int_{w}^{x_0} |H(w,t)|dt \right) (\|Ly\|_{\infty})^2,$$

and

$$\int_{x}^{x_0} |y(w)|\,|(Ly)(w)|dw \leq \left(\int_{x}^{x_0} \left(\int_{w}^{x_0} |H(w,t)|dt \right) dw \right) \cdot (\|Ly\|_{\infty})^2.$$

$\qquad\square$

Corollary 13.5. *Let $p,q > 1$: $\frac{1}{p} + \frac{1}{q} = 1$. Then*

$$\left| \int_{x_0}^{x} |y(w)|\,|(Ly)(w)|dw \right| \leq 2^{-1/q} \cdot \left(\int_{x_0}^{x} \left(\int_{x_0}^{w} |H(w,t)|^p dt \right) dw \right)^{1/p}$$

$$\cdot \left(\left| \int_{x_0}^{x} |(Ly)(w)|^q dw \right| \right)^{2/q}. \tag{13.7}$$

Proof. By Propositions 13.1, 13.2. $\qquad\square$

In particular when $p = q = 2$ we obtain

Corollary 13.6. *It holds*

$$\left| \int_{x_0}^{x} |y(w)| \, |(Ly)(w)| dw \right| \leq 2^{-1/2} \cdot \left(\int_{x_0}^{x} \left(\int_{x_0}^{w} (H(w,t))^2 dt \right) dw \right)^{1/2}$$

$$\cdot \left| \int_{x_0}^{x} ((Ly)(w))^2 dw \right|. \tag{13.8}$$

Furthermore we have

Corollary 13.7 *Let $p = 1$, $q = \infty$. Then*

$$\left| \int_{x_0}^{x} |y(w)| \, |(Ly)(w)| dw \right|$$

$$\leq \left(\int_{x_0}^{x} \left(\int_{x_0}^{w} |H(w,t)| dt \right) dw \right) \cdot \|Ly\|_{\infty}^2. \tag{13.9}$$

Proof. By Propositions 13.3 and 13.4. $\square$

To complete the chapter we present

Proposition 13.8. *Let $0 < p < 1$ and q be such that $\frac{1}{p} + \frac{1}{q} = 1$, and $x > x_0$, $x_0, x \in I$. Suppose that*

$$H(w,t) \geq 0 \quad for \ x_0 \leq t \leq w, \ w \in I.$$

Also assume that $Ly = h$ is of fixed sign and nowhere zero. Then

$$\int_{x_0}^{x} |y(w)| \, |(Ly)(w)| dw \geq 2^{-1/q} \left(\int_{x_0}^{x} \left(\int_{x_0}^{w} (H(w,t))^p dt \right) dw \right)^{1/p}$$

$$\cdot \left(\int_{x_0}^{x} |(Ly)(w)|^q dw \right)^{2/q}. \tag{13.10}$$

Proof. Here for $x_0 \leq w \leq x$ we have

$$|y(w)| = \int_{x_0}^{w} H(w,t)|h(t)| dt. \tag{13.11}$$

From (13.11) by Hölder's inequality we derive

$$|y(w)| \geq \left(\int_{x_0}^{w} (H(w,t))^p dt \right)^{1/p} \cdot \left(\int_{x_0}^{w} |h(t)|^q dt \right)^{1/q}, \tag{13.12}$$

for $w > x_0$. Consider

$$z(w) := \int_{x_0}^{w} |h(t)|^q dt, \quad z(x_0) = 0.$$

So that $z'(w) = |h(w)|^q$ and $|h(w)| = (z'(w))^{1/q}$, all $x_0 \le w \le x$. Hence by (13.12) we get

$$|y(w)|\,|h(w)| \ge \left(\int_{x_0}^{w} (H(w,t))^p dt \right)^{1/p} \cdot (z(w) \cdot z'(w))^{1/q},$$

all $x_0 < w \le x$. Let $x_0 < \theta \le w \le x$ and $\theta \downarrow x_0$, then by integration of the last inequality we obtain

$$\int_{x_0}^{x} |y(w)|\,|h(w)|dw = \lim_{\theta \downarrow x_0} \int_{\theta}^{x} |y(w)|\,|h(w)|dw$$

$$\ge \lim_{\theta \downarrow x_0} \left(\int_{\theta}^{x} \left(\int_{x_0}^{w} (H(w,t))^p dt \right)^{1/p} \cdot (z(w) \cdot z'(w))^{1/q} \cdot dw \right)$$

$$\ge \lim_{\theta \downarrow x_0} \left(\int_{\theta}^{x} \left(\int_{x_0}^{w} (H(w,t))^p dt \right) dw \right)^{1/p} \cdot \lim_{\theta \downarrow x_0} \left(\left(\int_{\theta}^{x} z(w)z'(w)dw \right)^{1/q} \right)$$

$$= 2^{-1/q} \cdot \left(\int_{x_0}^{x} \left(\int_{x_0}^{w} (H(w,t))^p dt \right) dw \right)^{1/p} \cdot \lim_{\theta \downarrow x_0} (z^2(x) - z^2(\theta))^{1/q}$$

$$= 2^{-1/q} \left(\int_{x_0}^{x} \left(\int_{x_0}^{w} (H(w,t))^p dt \right) dw \right)^{1/p} \cdot (z(x))^{2/q}.$$

That is establishing (13.10). $\qquad\square$

Proposition 13.9. *Let $0 < p < 1$ and q be such that $\frac{1}{p} + \frac{1}{q} = 1$, and $x < x_0$, $x_0, x \in I$. Suppose that*

$$H(w,t) \le 0 \quad for\ w \le t \le x_0,\ w \in I.$$

Also assume that $Ly = h$ is of fixed sign and nowhere zero. Then

$$\int_{x}^{x_0} |y(w)|\,|(Ly)(w)|dw \ge 2^{-1/q} \cdot \left(\int_{x_0}^{x} \left(\int_{x_0}^{w} |H(w,t)|^p dt \right) dw \right)^{1/p}$$

$$\cdot \left(\int_{x}^{x_0} |(Ly)(w)|^q dw \right)^{2/q}. \tag{13.13}$$

Proof. Here for $x \le w \le x_0$ we have

$$|y(w)| = \left| \int_{x_0}^{w} H(w,t)h(t)dt \right| = \left| \int_{w}^{x_0} H(w,t)h(t)dt \right|$$

$$= \left| \int_{w}^{x_0} (-H(w,t))h(t)dt \right| = \int_{w}^{x_0} (-H(w,t))|h(t)|dt. \tag{13.14}$$

From (13.14) by Hölder's inequality we find

$$|y(w)| \ge \left(\int_{w}^{x_0} (-H(w,t))^p dt \right)^{1/p} \left(\int_{w}^{x_0} |h(t)|^q dt \right)^{1/q},$$

for $w < x_0$. That is

$$|y(w)| \geq \left(\int_w^{x_0} |H(w,t)|^p dt \right)^{1/p} \left(\int_w^{x_0} |h(t)|^q dt \right)^{1/q}, \qquad (13.15)$$

for $w < x_0$. Consider

$$z(w) := \int_w^{x_0} |h(t)|^q dt, \quad z(x_0) = 0.$$

So that

$$-z(w) = \int_{x_0}^w |h(t)|^q dt$$

and

$$-z'(w) = |h(w)|^q,$$

with

$$|h(w)| = (-z'(w))^{1/q}, \quad \text{all } x \leq w \leq x_0.$$

Therefore by (13.15) we find

$$|y(w)|\,|h(w)| \geq \left(\int_w^{x_0} |H(w,t)|^p dt \right)^{1/p} \cdot (z(w) \cdot (-z'(w)))^{1/q},$$

all $x \leq w < x_0$. Let $x \leq w \leq \theta < x_0$ and $\theta \uparrow x_0$, then by integration of the last inequality we obtain

$$\int_x^{x_0} |y(w)|\,|h(w)|dw = \lim_{\theta \uparrow x_0} \int_x^{\theta} |y(w)|\,|h(w)|dw$$

$$\geq \lim_{\theta \uparrow x_0} \left(\int_x^{\theta} \left(\int_w^{x_0} |H(w,t)|^p dt \right)^{1/p} \cdot (z(w) \cdot (-z'(w)))^{1/q} \cdot dw \right)$$

$$\geq \lim_{\theta \uparrow x_0} \left(\int_x^{\theta} \left(\int_w^{x_0} |H(w,t)|^p dt \right) dw \right)^{1/p} \cdot \lim_{\theta \uparrow x_0} \left(\int_x^{\theta} (-z(w))(z'(w)) \cdot dw \right)^{1/q}$$

$$= \left(\int_x^{x_0} \left(\int_w^{x_0} |H(w,t)|^p dt \right) dw \right)^{1/p} \cdot \lim_{\theta \uparrow x_0} \left(- \int_x^{\theta} z(w)dz(w) \right)^{1/q}$$

$$= 2^{-1/q} \cdot \left(\int_x^{x_0} \left(\int_w^{x_0} |H(w,t)|^p dt \right) dw \right)^{1/p} \cdot \lim_{\theta \uparrow x_0} (z^2(x) - z^2(\theta))^{1/q}$$

$$= 2^{-1/q} \cdot \left(\int_x^{x_0} \left(\int_w^{x_0} |H(w,t)|^p dt \right) dw \right)^{1/p} \cdot (z(x))^{2/q}.$$

Therefore

$$\int_x^{x_0} |y(w)|\,|h(w)|dw \geq 2^{-1/q} \cdot \left(\int_{x_0}^{x} \left(\int_{x_0}^{w} |H(w,t)|^p dt \right) dw \right)^{1/p}$$

$$\cdot \left(\int_x^{x_0} |h(w)|^q dw \right)^{2/q},$$

that is proving (13.13). $\qquad \square$

Putting things together we have

Corollary 13.10. *Let $0 < p < 1$ and q be such that $\frac{1}{p} + \frac{1}{q} = 1$, and $x_0, x \in I$ that $x \neq x_0$. Suppose that*

$$(w - x_0) \cdot H(w, t) \geq 0,$$

for all t between $x_0, w \in I$. Also assume that $Ly = h$ is of fixed sign and nowhere zero. Then

$$\left| \int_{x_0}^{x} |y(w)| \, |(Ly)(w)| dw \right| \geq 2^{-1/q} \cdot \left(\int_{x_0}^{x} \left(\int_{x_0}^{w} |H(w,t)|^p dt \right) dw \right)^{1/p}$$

$$\cdot \left(\left| \int_{x_0}^{x} |(Ly)(w)|^q dw \right| \right)^{2/q}. \tag{13.16}$$

Proof. By Propositions 13.8 and 13.9. $\square$

Chapter 14

Opial Inequalities for Vector Valued Functions

Various L_p form Opial type inequalities are given for functions valued in a Banach vector space with applications in $\mathbb{C}$. This chapter relies on [52].

14.1 Introduction

This chapter is greatly motivated by the article of Z. Opial [195].

Theorem 14.1. (Opial [195]) *Let $x(t) \in C^1([0, h])$ be such that $x(0) = x(h) = 0$, and $x(t) > 0$ in $(0, h)$. Then*

$$\int_0^h |x(t)x'(t)| dt \leq \frac{h}{4} \int_0^h (x'(t))^2 dt. \tag{14.1}$$

In the last inequality the constant $h/4$ is the best possible.

Equality holds for the function

$$x(t) = t \quad \text{on } [0, h/2]$$

and

$$x(t) = h - t \quad \text{on } [h/2, h].$$

Opial type inequalities have applications in establishing uniqueness of solution to initial value problems in differential equations, see [244]. We are also motivated by [16], [21].

We need

14.2 Background

(see [230], pp. 83–94)

Let $f(t)$ be a function defined on $[a, b] \subseteq \mathbb{R}$ taking values in a real or complex normed linear space $(X, \|\cdot\|)$. Then $f(t)$ is said to be *differentiable at a point* $t_0 \in [a, b]$ if the limit

$$f'(t_0) := \lim_{h \to 0} \frac{f(t_0 + h) - f(t_0)}{h} \tag{14.2}$$

179

exists in X, the convergence is in $\|\cdot\|$. This is called the *derivative* of $f(t)$ at $t = t_0$.

We call $f(t)$ *differentiable* on [a,b], iff there exists $f'(t) \in X$ for all $t \in [a, b]$.

Similarly and inductively are defined higher order derivatives of f, denoted $f'', f^{(3)}, ..., f^{(k)}, k \in \mathbb{N}$, just as for numerical functions.

For all the properties of derivatives see [230], pp. 83–86.

Let now $(X, \|\cdot\|)$ be a Banach space, and $f : [a, b] \to X$.

We define the *vector valued Riemann integral* $\int_a^b f(t)dt \in X$ as the limit of the vector valued Riemann sums in X, convergence is in $\|\cdot\|$. The definition is as for the numerical valued functions.

If $\int_a^b f(t)dt \in X$ we call f *integrable* on [a,b].

For the properties of vector valued Riemann integrals see [230], pp. 86–91.

We define the space $C^n([a, b], X), n \in \mathbb{N}$, of n-times continuously differentiable functions from [a,b] into X; here continuity is with respect to $\|\cdot\|$ and defined in the usual way as for numerical functions.

Let $(X, \|\cdot\|)$ be a Banach space and $f \in C^n([a, b], X)$, then we have the *vector valued Taylor's formula*, see [230], pp. 93–94, and also [227], (IV, 9; 47):

It holds

$$E_n(x, y) := f(y) - f(x) - f'(x)(y - x)$$

$$-\frac{1}{2}f''(x)(y - x)^2 - \cdots - \frac{1}{(n-1)!}f^{(n-1)}(x)(y - x)^{n-1}$$

$$= \frac{1}{(n-1)!} \int_x^y (y - t)^{n-1} f^{(n)}(t)dt, \quad \forall \, x, y \in [a, b]. \tag{14.3}$$

In particular (14.3) is true when $X = \mathbb{R}^m, \mathbb{C}^m, m \in \mathbb{N}$, etc.

In case of some $x_0 \in [a, b]$ such that $f^{(k)}(x_0) = 0, k = 0, 1, ..., n - 1$, then

$$f(y) = \frac{1}{(n-1)!} \int_{x_0}^y (y - t)^{n-1} f^{(n)}(t)dt, \, \forall \, y \in [a, b]. \tag{14.4}$$

In that case $E_n(x_0, y) = f(y)$.

14.3   Results

Here we consider always X to be a Banach space, $n \in \mathbb{N}$, and $f \in C^n([a, b], X)$, $[a, b] \subseteq \mathbb{R}$. We fix $x_0 \in [a, b]$.

We present the first result of the chapter.

Theorem 14.2. *Let $p, q > 1 : \frac{1}{p} + \frac{1}{q} = 1, \, y \geq x_0; \, y, x_0 \in [a, b]$.*

Then

$$I_1 := \int_{x_0}^y \|E_n(x_0, w)\| \, \|f^{(n)}(w)\|dw$$

$$\le \frac{(y-x_0)^{n-1+\frac{2}{p}}}{2^{1/q}(n-1)![(p(n-1)+1)(p(n-1)+2)]^{1/p}} \left(\int_{x_0}^{y} \|f^{(n)}(w)\|^q dw\right)^{2/q}. \quad (14.5)$$

We give

Corollary 14.3. *For $y \in [x_0, b]$ we have*

$$I_1 \le \frac{(y-x_0)^n}{2(n-1)!\sqrt{n(2n-1)}} \left(\int_{x_0}^{y} \|f^{(n)}(w)\|^2 dw\right). \quad (14.6)$$

Proof of Theorem 14.2. We have by (14.3) that

$$E_n(x_0, y) = \frac{1}{(n-1)!} \int_{x_0}^{y} (y-t)^{n-1} f^{(n)}(t) dt, \ \forall \ y \ge x_0; x_0, y \in [a,b]. \quad (14.7)$$

By Hölder's inequality we derive

$$\|E_n(x_0, y)\| \le \frac{1}{(n-1)!} \int_{x_0}^{y} (y-t)^{n-1} \|f^{(n)}(t)\| dt$$

$$\le \frac{1}{(n-1)!} \left(\int_{x_0}^{y} (y-t)^{p(n-1)} dt\right)^{1/p} \left(\int_{x_0}^{y} \|f^{(n)}(t)\|^q dt\right)^{1/q}$$

$$= \frac{1}{(n-1)!} \frac{(y-x_0)^{n-1+\frac{1}{p}}}{(p(n-1)+1)^{1/p}} (z(y))^{1/q}, \quad (14.8)$$

where

$$z(y) := \int_{x_0}^{y} \|f^{(n)}(t)\|^q dt, \ x_0 \le y \le b \ (z(x_0) = 0). \quad (14.9)$$

Thus

$$z'(y) = \|f^{(n)}(y)\|^q$$

and

$$\|f^{(n)}(y)\| = (z'(y))^{1/q}. \quad (14.10)$$

Therefore we obtain

$$\|E_n(x_0, y)\| \, \|f^{(n)}(y)\| \le \frac{1}{(n-1)!} \frac{(y-x_0)^{n-1+\frac{1}{p}}}{(p(n-1)+1)^{1/p}} (z(y)z'(y))^{1/q}. \quad (14.11)$$

Integrating the last inequality we have

$$\int_{x_0}^{y} \|E_n(x_0, w)\| \, \|f^{(n)}(w)\| dw$$

$$\le \frac{1}{(n-1)!(p(n-1)+1)^{1/p}} \int_{x_0}^{y} (w-x_0)^{n-1+\frac{1}{p}} (z(w)z'(w))^{1/q} dw$$

$$\leq \frac{1}{(n-1)!(p(n-1)+1)^{1/p}} \left(\int_{x_0}^{y} (w-x_0)^{p(n-1)+1} dw\right)^{1/p} \left(\int_{x_0}^{y} z(w)z'(w)dw\right)^{1/q}$$

$$= \frac{(y-x_0)^{n-1+\frac{2}{p}}}{2^{1/q}(n-1)![(p(n-1)+1)(p(n-1)+2)]^{1/p}} (z(y))^{2/q}, \tag{14.12}$$

proving the claim of the theorem. $\qquad\qquad\square$

We continue with

Theorem 14.4. *Let* $p, q > 1 : \frac{1}{p} + \frac{1}{q} = 1,\ y \in [a, x_0]$.
Then

$$I_2 := \int_{y}^{x_0} \|E_n(x_0, w)\|\ \|f^{(n)}(w)\| dw$$

$$\leq \frac{(x_0-y)^{n-1+\frac{2}{p}}}{2^{1/q}(n-1)![(p(n-1)+1)(p(n-1)+2)]^{1/p}} \left(\int_{y}^{x_0} \|f^{(n)}(w)\|^q dw\right)^{2/q}. \tag{14.13}$$

We give

Corollary 14.5. *For* $y \in [a, x_0]$ *we have*

$$I_2 \leq \frac{(x_0-y)^n}{2(n-1)!\sqrt{n(2n-1)}} \left(\int_{y}^{x_0} \|f^{(n)}(w)\|^2 dw\right). \tag{14.14}$$

Proof of Theorem 14.4. We have by (14.3) that

$$\|E_n(x_0, y)\| = \frac{1}{(n-1)!} \left\|\int_{y}^{x_0} (y-t)^{n-1} f^{(n)}(t)dt\right\|$$

$$\leq \frac{1}{(n-1)!} \int_{y}^{x_0} (t-y)^{n-1} \|f^{(n)}(t)\| dt$$

$$\leq \frac{1}{(n-1)!} \left(\int_{y}^{x_0} (t-y)^{p(n-1)} dt\right)^{1/p} \left(\int_{y}^{x_0} \|f^{(n)}(t)\|^q dt\right)^{1/q}$$

$$= \frac{(x_0-y)^{n-1+\frac{1}{p}}}{(n-1)!(p(n-1)+1)^{1/p}} (z(y))^{1/q}. \tag{14.15}$$

Here

$$z(y) := \int_{y}^{x_0} \|f^{(n)}(t)\|^q dt, \quad (z(x_0) = 0), \tag{14.16}$$

and

$$-z(y) = \int_{x_0}^{y} \|f^{(n)}(t)\|^q\ dt \leq 0, \tag{14.17}$$

$$-z'(y) = \|f^{(n)}(y)\|^q \geq 0, \tag{14.18}$$

and

$$\|f^{(n)}(y)\| = (-z'(y))^{1/q}, \quad a \le y \le x_0. \tag{14.19}$$

Therefore

$$\|E_n(x_0, y)\| \, \|f^{(n)}(y)\|$$

$$\le \frac{(x_0 - y)^{n-1+\frac{1}{p}}}{(n-1)!(p(n-1)+1)^{1/p}} \, (z(y)(-z'(y)))^{1/q}, \quad a \le y \le x_0. \tag{14.20}$$

Hence we find

$$\int_y^{x_0} \|E_n(x_0, w)\| \, \|f^{(n)}(w)\| dw$$

$$\le \frac{1}{(n-1)!(p(n-1)+1)^{1/p}} \int_y^{x_0} (x_0 - w)^{n-1+\frac{1}{p}} (z(w)(-z'(w)))^{1/q} dw \tag{14.21}$$

$$\le \frac{1}{(n-1)!(p(n-1)+1)^{1/p}} \left(\int_y^{x_0} (x_0 - w)^{p(n-1)+1} dw \right)^{1/p}$$

$$\left(\int_y^{x_0} z(w)(-z'(w)) dw \right)^{1/q} \tag{14.22}$$

$$= \frac{(x_0 - y)^{n-1+\frac{2}{p}}}{2^{1/q}(n-1)![(p(n-1)+1)(p(n-1)+2)]^{1/p}} (z(y))^{2/q}, \forall \, y \in [a, x_0], \tag{14.23}$$

proving the claim of the theorem. $\qquad\square$

Combining Theorems 14.2 and 14.4 we obtain

Theorem 14.6. *Let* $p, q > 1 : \frac{1}{p} + \frac{1}{q} = 1$ *and* $y, x_0 \in [a, b]$.
Then

$$I := \left| \int_{x_0}^y \|E_n(x_0, w)\| \, \|f^{(n)}(w)\| dw \right|$$

$$\le \frac{|y - x_0|^{n-1+\frac{2}{p}}}{2^{1/q}(n-1)![(p(n-1)+1)(p(n-1)+2)]^{1/p}} \left| \int_{x_0}^y \|f^{(n)}(w)\|^q dw \right|^{2/q}. \tag{14.24}$$

Combining Corollaries 14.3 and 14.5 we derive

Corollary 14.7. *For* $y, x_0 \in [a, b]$ *it holds*

$$I \le \frac{|y - x_0|^n}{2(n-1)!\sqrt{n(2n-1)}} \left| \int_{x_0}^y \|f^{(n)}(w)\|^2 dw \right|. \tag{14.25}$$

A special but important case follows

Theorem 14.8. *Let $p, q > 1 : \frac{1}{p} + \frac{1}{q} = 1$ and $y, x_0 \in [a, b]$. Suppose further that*
$f^{(k)}(x_0) = 0, \ k = 0, 1, ..., n - 1.$
Then

$$\left| \int_{x_0}^{y} \|f(w)\| \, \|f^{(n)}(w)\| \, dw \right|$$

$$\leq \min \left(\frac{|y - x_0|^{n-1+\frac{2}{p}}}{2^{1/q}(n-1)![(p(n-1)+1)(p(n-1)+2)]^{1/p}} \left| \int_{x_0}^{y} \|f^{(n)}(w)\|^q dw \right|^{2/q}, \right.$$

$$\left. \frac{|y - x_0|^n}{2(n-1)!\sqrt{n(2n-1)}} \left| \int_{x_0}^{y} \|f^{(n)}(w)\|^2 dw \right| \right). \tag{14.26}$$

Proof. By Theorem 14.6 and Corollary 14.7. $\square$

We continue with

Theorem 14.9. *Let $p = 1$, $q = \infty$ and $y \in [x_0, b]$.*
Then

$$\int_{x_0}^{y} \|E_n(x_0, w)\| \, \|f^{(n)}(w)\| \, dw \leq \frac{(y - x_0)^{n+1}}{(n+1)!} \|f^{(n)}\|_{\infty,[x_0,b]}^2. \tag{14.27}$$

Proof. We have by (14.3) that

$$\|E_n(x_0, y)\| \leq \frac{1}{(n-1)!} \int_{x_0}^{y} (y - t)^{n-1} \|f^{(n)}(t)\| \, dt \leq \|f^{(n)}\|_{\infty,[x_0,b]} \frac{(y - x_0)^n}{n!}. \tag{14.28}$$

Consequently it holds

$$\|E_n(x_0, y)\| \, \|f^{(n)}(y)\| \leq \|f^{(n)}\|_{\infty,[x_0,b]}^2 \frac{(y - x_0)^n}{n!}, \tag{14.29}$$

all $x_0 \leq y \leq b$.
Hence we derive

$$\int_{x_0}^{y} \|E_n(x_0, w)\| \, \|f^{(n)}(w)\| \, dw \leq \frac{\|f^{(n)}\|_{\infty,[x_0,b]}^2}{n!} \int_{x_0}^{y} (w - x_0)^n dw$$

$$= \frac{(y - x_0)^{n+1}}{(n+1)!} \|f^{(n)}\|_{\infty,[x_0,b]}^2. \tag{14.30}$$
$\square$

The counterpart of Theorem 14.9 follows.

Theorem 14.10. *Let $p = 1$, $q = \infty$ and $y \in [a, x_0]$.*
Then

$$\int_y^{x_0} \|E_n(x_0, w)\| \, \|f^{(n)}(w)\| dw \leq \frac{(x_0 - y)^{n+1}}{(n+1)!} \|f^{(n)}\|_{\infty,[a,x_0]}^2.$$ (14.31)

Proof. We have by (14.3) that

$$\|E_n(x_0, y)\| = \frac{1}{(n-1)!} \left\| \int_{x_0}^y (y - t)^{n-1} f^{(n)}(t) dt \right\|$$

$$= \frac{1}{(n-1)!} \left\| \int_y^{x_0} (y - t)^{n-1} f^{(n)}(t) dt \right\| \leq \frac{1}{(n-1)!} \int_y^{x_0} (t - y)^{n-1} \|f^{(n)}(t)\| dt$$ (14.32)

$$\leq \frac{\|f^{(n)}\|_{\infty,[a,x_0]}}{(n-1)!} \int_y^{x_0} (t - y)^{n-1} dt = \|f^{(n)}\|_{\infty,[a,x_0]} \frac{(x_0 - y)^n}{n!}.$$ (14.33)

Hence it holds

$$\|E_n(x_0, y)\| \, \|f^{(n)}(y)\| \leq \|f^{(n)}\|_{\infty,[a,x_0]}^2 \frac{(x_0 - y)^n}{n!},$$ (14.34)

all $a \leq y \leq x_0$.

Consequently

$$\int_y^{x_0} \|E_n(x_0, w)\| \|f^{(n)}(w)\| dw \leq \frac{\|f^{(n)}\|_{\infty,[a,x_0]}^2}{n!} \int_y^{x_0} (x_0 - w)^n \, dw$$

$$= \|f^{(n)}\|_{\infty,[a,x_0]}^2 \frac{(x_0 - y)^{n+1}}{(n+1)!}.$$ (14.35)

$\square$

Combining Theorems 14.9 and 14.10 we obtain

Proposition 14.11. *Let* $p = 1$, $q = \infty$, $y, x_0 \in [a, b]$.
Then

$$\left| \int_{x_0}^y \|E_n(x_0, w)\| \, \|f^{(n)}(w)\| dw \right| \leq \frac{|y - x_0|^{n+1}}{(n+1)!} \|f^{(n)}\|_\infty^2.$$ (14.36)

In particular we get

Proposition 14.12. *Let* $p = 1$, $q = \infty$, $y, x_0 \in [a, b]$. *Further suppose that* $f^{(k)}(x_0) = 0$, $k = 0, 1, ..., n - 1$.
Then

$$\left| \int_{x_0}^y \|f(w)\| \, \|f^{(n)}(w)\| dw \right| \leq \frac{|y - x_0|^{n+1}}{(n+1)!} \|f^{(n)}\|_\infty^2.$$ (14.37)

14.4 Applications

Here $X = \mathbb{C}$, $f \in C^n([a,b], \mathbb{C})$, $n \in \mathbb{N}$, $x_0, y \in [a,b]$. Furthermore suppose that $f^{(k)}(x_0) = 0$, $k = 0, 1, ..., n-1$.

Applying Theorem 14.8 we obtain

Theorem 14.13. *Let $p, q > 1 : \frac{1}{p} + \frac{1}{q} = 1$.*
Then

$$\left| \int_{x_0}^{y} |f(w)| \, |f^{(n)}(w)| dw \right|$$

$$\leq \min \left(\frac{|y - x_0|^{n-1+\frac{2}{p}}}{2^{1/q}(n-1)![(p(n-1)+1)(p(n-1)+2)]^{1/p}} \left| \int_{x_0}^{y} |f^{(n)}(w)|^q dw \right|^{2/q}, \right.$$

$$\left. \frac{|y - x_0|^n}{2(n-1)!\sqrt{n(2n-1)}} \left| \int_{x_0}^{y} |f^{(n)}(w)|^2 dw \right| \right). \tag{14.38}$$

By Proposition 14.12 we derive

Proposition 14.14. *Let $p = 1$, $q = \infty$.*
Then

$$\left| \int_{x_0}^{y} |f(w)| \, |f^{(n)}(w)| dw \right| \leq \frac{|y - x_0|^{n+1}}{(n+1)!} \|f^{(n)}\|_{\infty}^2. \tag{14.39}$$

Let now $f \in C^1([a,b], \mathbb{C})$, $x_0, y \in [a,b]$ with $f(x_0) = 0$. Applying Theorem 14.13 we get

Theorem 14.15. *Let $p, q > 1 : \frac{1}{p} + \frac{1}{q} = 1$.*
Then

$$\left| \int_{x_0}^{y} |f(w)| \, |f'(w)| dw \right|$$

$$\leq \frac{1}{2} \min \left(|y - x_0|^{2/p} \left| \int_{x_0}^{y} |f'(w)|^q dw \right|^{2/q}, |y - x_0| \left| \int_{x_0}^{y} |f'(w)|^2 dw \right| \right). \tag{14.40}$$

Finally by applying Proposition 14.14 we find

Proposition 14.16. *Let $p = 1$, $q = \infty$.*
Then

$$\left| \int_{x_0}^{y} |f(w)| \, |f'(w)| dw \right| \leq \frac{(y - x_0)^2}{2} \|f'\|_{\infty}^2. \tag{14.41}$$

Chapter 15

Opial Inequalities for Semigroups

Various L_p form Opial type inequalities are given for semigroups with applications. This chapter relies on [31].

15.1 Introduction

The chapter is also greatly motivated by the article of Z. Opial ([195]):

Theorem 15.1 (Opial [195]) *Let* $x(t) \in C^1([0, h])$ *be such that* $x(0) = x(h) = 0$, *and* $x(t) > 0$ *in* $(0, h)$. *The following inequality holds*

$$\int_0^h |x(t)x'(t)|dt \leq (h/4) \int_0^h (x'(t))^2 dt. \tag{15.1}$$

In the last inequality the constant $h/4$ is the best possible.

Opial type inequalities have applications in establishing uniqueness of solution to initial value problems in differential equations, see [244].

We are also motivated by [16], [21].

The generalized Opial type inequalities we prove here are in Theorems 15.8, 15.10 and 15.15.

15.2 Background

All this background comes from [79] (in general see also [147]).

Let X a real or complex Banach space with elements $f, g, \cdots$ having norm $\|f\|, \|g\|, \cdots$ and let $\mathcal{E}(X)$ be the Banach algebra of endomorphisms of X.

If $T \in \mathcal{E}(X)$, $\|T\|$ denotes the norm of T.

Definition 15.2. If $T(t)$ is an operator function on the non- negative real axis $0 \leq t < \infty$ to the Banach algebra $\mathcal{E}(X)$ satisfying the following conditions:

187

$$\begin{cases} (i)\ T(t_1 + t_2) = T(t_1)T(t_2), & (t_1, t_2 \geq 0), \\ (ii)\ T(0) = I & (I = identity\ operator), \end{cases} \tag{15.2}$$

then $\{T(t);\ 0 \leq t < \infty\}$ is called a one-parameter semi-group of operators in $\mathcal{E}(X)$.

The semi-group $\{T(t);\ 0 \leq t < \infty\}$ is said to be of class $\mathcal{C}_0$ if it satisfies the further property

$$(iii)s - \lim_{t \to 0+} T(t)f = f, \quad (f \in X) \tag{15.3}$$

referred to as the strong continuity of $T(t)$ at the origin.

In this chapter we shall assume that the family of bounded linear operators $\{T(t);\ 0 \leq t < \infty\}$ mapping X to itself is a semi-group of class $\mathcal{C}_0$, thus that all three conditions of the above definition are satisfied.

Proposition 15.3. *(a) $\|T(t)\|$ is bounded on every finite subinterval of $[0, \infty)$.*

(b) For each $f \in X$, the vector-valued function $T(t)f$ on $[0, \infty)$ is strongly continuous.

Definition 15.4. The infinitesimal generator A of the semi-group $\{T(t);\ 0 \leq t < \infty\}$ is defined by

$$Af = s - \lim_{\tau \to 0+} A_\tau f, \quad A_\tau f = \frac{1}{\tau}[T(\tau) - I] \tag{15.4}$$

whenever the limit exists; the domain of A, in symbols $D(A)$, is the set of elements f for which the limit exists.

Proposition 15.5. *(a) $D(A)$ is a linear manifold in X and A is a linear operator.*

(b) If $f \in D(A)$, then $T(t)f \in D(A)$ for each $t \geq 0$ and

$$\frac{d}{dt}T(t)f = AT(t)f = T(t)Af \quad (t \geq 0); \tag{15.5}$$

furthermore,

$$T(t)f - f = \int_0^t T(u)Af\,du \quad (t > 0). \tag{15.6}$$

(c) $D(A)$ is dense in X, i.e. $\overline{D(A)} = X$, and A is a closed operator.

Definition 15.6. For $r = 0, 1, 2, \ldots$ the operator A^r is defined inductively by the relations $A^0 = I$, $A^1 = A$, and

$$D(A^r) = \{f;\ f \in D(A^{r-1})\ and\ A^{r-1}f \in D(A)\}$$

$$A^r f = A(A^{r-1}f) = s - \lim_{\tau \to 0+} A_\tau(A^{r-1}f) \quad (f \in D(A^r)). \tag{15.7}$$

For the operator A^r and its domain $D(A^r)$ we have the following

Proposition 15.7. *(a) $D(A^r)$ is a linear subspace in X and A^r is a linear operator.*
(b) If $f \in D(A^r)$, so does $T(t)f$ for each $t \geq 0$ and

$$\frac{d^r}{dt^r} T(t)f = A^r T(t)f = T(t)A^r f. \tag{15.8}$$

Furthermore

$$T(t)f - \sum_{k=0}^{r-1} \frac{t^k}{k!} A^k f = \frac{1}{(r-1)!} \int_0^t (t-u)^{r-1} T(u) A^r f du, \tag{15.9}$$

the Taylor's formula for semigroups.
Additionally it holds

$$[T(t) - I]^r = \int_0^t \int_0^t \cdots \int_0^t T(u_1 + u_2 + \cdot + u_r) A^r f du_1 du_2 \ldots du_r. \tag{15.10}$$

(c) $D(A^r)$ is dense in X for $r = 1, 2 \ldots$; moreover, $\cap_{r=1}^{\infty} D(A^r)$ is dense in
X. *A^r is a closed operator.*

Integrals in (15.9) and (15.10) are vector valued Riemann integrals, see [79], [164].

15.3 Results

Here always we consider $T(t), A$ as in the Background of this chapter and $f \in D(A^r)$, $r \in \mathbf{N}$.

We present the first main result.

Theorem 15.8. *Let $p, q > 1 : \frac{1}{p} + \frac{1}{q} = 1$.*
One has

$$\int_0^t \|\Delta(w)f\| \, \|T(w)A^r f\| dw \leq$$

$$\frac{t^{r-1+\frac{2}{p}}}{2^{1/q}(r-1)! \, [(p(r-1)+1)(p(r-1)+2)]^{1/p}} \left(\int_0^t \|T(w)A^r f\|^q dw \right)^{2/q}, \tag{15.11}$$

for all $t \in \mathbf{R}_+ = [0, \infty)$, where

$$\Delta(t)f := T(t)f - \sum_{k=0}^{r-1} \frac{t^k}{k!} A^k f. \tag{15.12}$$

We give

Corollary 15.9. *Let $p = q = 2$. Then*

$$\int_0^t \|\Delta(w)f\| \, \|T(w)A^r f\| dw \leq$$

$$\frac{t^r}{2(r-1)! \sqrt{r(2r-1)}} \left(\int_0^t \|T(w)A^r f\|^2 dw \right), \textit{for all } t \in \mathbf{R}_+. \tag{15.13}$$

Proof of Theorem 15.8. By Taylor's formula (15.9) we have

$$\Delta(t)f = \frac{1}{(r-1)!} \int_0^t (t-u)^{r-1} T(u) A^r f du. \tag{15.14}$$

Therefore

$$\|\Delta(t)f\| = \frac{1}{(r-1)!} \left\| \int_0^t (t-u)^{r-1} T(u) A^r f du \right\|$$

$$\leq \frac{1}{(r-1)!} \int_0^t (t-u)^{r-1} \|T(u) A^r f\| du$$

$$\leq \frac{1}{(r-1)!} \left(\int_0^t (t-u)^{p(r-1)} du \right)^{1/p} \left(\int_0^t \|T(u) A^r f\|^q du \right)^{1/q}$$

$$= \frac{t^{r-1+\frac{1}{p}}}{(r-1)! \, (p(r-1)+1)^{1/p}} \left(\int_0^t \|T(u) A^r f\|^q du \right)^{1/q}.$$

So far we have found that

$$\|\Delta(t)f\| \leq \frac{t^{r-1+\frac{1}{p}}}{(r-1)! \, (p(r-1)+1)^{1/p}} \, (z(t))^{1/q}, \tag{15.15}$$

where

$$z(t) := \int_0^t \|T(u) A^r f\|^q du, \text{ for all } t \in \mathbf{R}_+. \tag{15.16}$$

Clearly

$$z'(w) := \|T(w) A^r f\|^q, \; 0 \leq w \leq t, \; z(0) = 0. \tag{15.17}$$

In particular

$$\|T(w) A^r f\| = (z'(w))^{1/q}.$$

Hence we have that

$$\|\Delta(w)f\| \, \|T(w) A^r f\| \leq \frac{w^{r-1+\frac{1}{p}}}{(r-1)! \, (p(r-1)+1)^{1/p}} \, (z(w) z'(w))^{1/q}. \tag{15.18}$$

Consequently it follows

$$\int_0^t \|\Delta(w)f\| \, \|T(w) A^r f\| dw \leq \frac{1}{(r-1)! \, (p(r-1)+1)^{1/p}}$$

$$\int_0^t w^{r-1+\frac{1}{p}} \, (z(w) z'(w))^{1/q} \, dw \leq \frac{1}{(r-1)! \, (p(r-1)+1)^{1/p}}$$

$$\cdot \left(\int_0^t \left(w^{r-1+\frac{1}{p}} \right)^p dw \right)^{1/p} \left(\int_0^t z(w) z'(w) dw \right)^{1/q}$$

$$= \frac{1}{(r-1)! \, (p(r-1)+1)^{1/p}} \cdot \frac{t^{r-1+\frac{2}{p}}}{(p(r-1)+2)^{1/p}} \left(\frac{z^2(w)}{2} \Big|_0^t \right)^{1/q}$$

$$= \frac{1}{(r-1)! \, (p(r-1)+1)^{1/p}} \cdot \frac{t^{r-1+\frac{2}{p}}}{2^{1/q}(p(r-1)+2)^{1/p}} \left(z(t) \right)^{2/q}$$

$$= \frac{1}{(r-1)! \, (p(r-1)+1)^{1/p}} \cdot \frac{t^{r-1+\frac{2}{p}}}{2^{1/q}(p(r-1)+2)^{1/p}} \left(\int_0^t \|T(u)A^r f\|^q du \right)^{2/q},$$

$$(15.19)$$

proving the claim. $\square$

We give

Theorem 15.10. *Here* $p = 1$, $q = \infty$ *and suppose*

$$\Big\| \, \|T(u)A^r f\| \, \Big\|_\infty := \sup_{u \in \mathbf{R}_+} \|T(u)A^r f\| < \infty. \tag{15.20}$$

Then

$$\int_0^t \|\Delta(w)f\| \, \|T(w)A^r f\| dw \le \frac{t^{r+1}}{(r+1)!} \Big\| \, \|T(u)A^r f\| \, \Big\|_\infty^2, \text{ for all } t \in \mathbf{R}_+,$$

where

$$\Delta(t)f := T(t)f - \sum_{k=0}^{r-1} \frac{t^k}{k!} A^k f. \tag{15.21}$$

Proof. We have

$$\|\Delta(t)f\| \le \frac{1}{(r-1)!} \int_0^t (t-u)^{r-1} \|T(u)A^r f\| du$$

$$\le \frac{1}{(r-1)!} \left(\int_0^t (t-u)^{r-1} du \right) \Big\| \, \|T(u)A^r f\| \, \Big\|_\infty$$

$$= \frac{t^r}{r!} \Big\| \, \|T(u)A^r f\| \, \Big\|_\infty, \text{ for all } t \in \mathbf{R}_+.$$

That is

$$\|\Delta(t)f\| \le \frac{t^r}{r!} \Big\| \, \|T(u)A^r f\| \, \Big\|_\infty, \text{ for all } t \in \mathbf{R}_+.$$

Therefore

$$\|\Delta(w)f\| \, \|T(w)A^r f\| \le \frac{w^r}{r!} \Big\| \, \|T(u)A^r f\| \, \Big\|_\infty^2, \text{ all } 0 \le w \le t. \tag{15.22}$$

Hence

$$\int_0^t \|\Delta(w)f\| \, \|T(w)A^r f\| dw \le \left(\int_0^t w^r dw \right) \frac{\Big\| \, \|T(u)A^r f\| \, \Big\|_\infty^2}{r!}$$

$$= \frac{t^{r+1}}{(r+1)!} \Big\| \, \|T(u)A^r f\| \, \Big\|_\infty^2, \text{ for all } t \in \mathbf{R}_+, \tag{15.23}$$

proving the claim. $\square$

We give

Application 15.11 (see also [80]). It is known that the classical diffusion equation

$$\frac{\partial W}{\partial t} = \frac{\partial^2 W}{\partial x^2}, \quad -\infty < x < \infty, \ t > 0 \tag{15.24}$$

with initial condition

$$\lim_{t \to 0+} W(x,t) = f(x), \tag{15.25}$$

has under general conditions its solution given by

$$W(x,t,f) = [T(t)f](x) = \frac{1}{2\sqrt{\pi t}} \int_{-\infty}^{\infty} f(x+u)e^{-u^2/4t}du, \tag{15.26}$$

the so called Gauss–Weierstrass singular integral.

The infinitesimal generator of the semigroup $\{T(t); \ 0 \le t < \infty\}$ is $A = \partial^2/\partial x^2$ ([159], p. 578).

Here we suppose that $f, f^{(2k)}, k = 1, \ldots, r$, all belong to the Banach space $UCB(\mathbf{R})$, the space of bounded and uniformly continuous functions from $\mathbf{R}$ into itself, with norm

$$\|f\|_C := \sup_{x \in \mathbf{R}} |f(x)|. \tag{15.27}$$

Here we define

$$\overline{\Delta}(t)f(x) := W(x,t,f) - \sum_{k=0}^{r-1} \frac{t^k}{k!} f^{(2k)}(x), \text{ for all } x \in \mathbf{R}. \tag{15.28}$$

Therefore by Theorem 15.8 we derive

Proposition 15.12. *Let* $p, q > 1 : \frac{1}{p} + \frac{1}{q} = 1.$
  Then

$$I_t := \int_0^t \|\overline{\Delta}(t)f\|_C \ \|W(\cdot, w, f^{(2r)})\|_C dw \le$$

$$\frac{t^{r-1+\frac{2}{p}}}{2^{1/q}(r-1)! \ [(p(r-1)+1)(p(r-1)+2)]^{1/p}}$$

$$\cdot \left(\int_0^t \|W(\cdot, w, f^{(2r)})\|_C^q dw \right)^{2/q}, \textit{for all } t \in \mathbf{R}_+. \tag{15.29}$$

We make

Remark 15.13. We notice here that

$$|W(x,t,f^{(2r)})| = \frac{1}{2\sqrt{\pi t}} \left| \int_{-\infty}^{\infty} f^{(2r)}(x+u)e^{-u^2/4t}du \right|$$

$$\leq \frac{1}{2\sqrt{\pi t}} \int_{-\infty}^{\infty} |f^{(2r)}(x+u)|e^{-u^2/4t}du$$

$$\leq \|f^{(2r)}\|_C \frac{1}{2\sqrt{\pi t}} \int_{-\infty}^{\infty} e^{-u^2/4t}du$$

$$= \|f^{(2r)}\|_C \cdot 1 = \|f^{(2r)}\|_C < \infty, \text{for all } t \in \mathbf{R}. \tag{15.30}$$

That is

$$\|W(\cdot, t, f^{(2r)})\|_C \leq \|f^{(2r)}\|_C. \tag{15.31}$$

And thus we obtain

$$\Phi := \left\| \|W(\cdot, t, f^{(2r)})\|_C \right\|_\infty.$$

$$:= \sup_{t \in \mathbf{R}_+} \|W(\cdot, t, f^{(2r)})\|_C \leq \|f^{(2r)}\|_C < \infty. \tag{15.32}$$

We give

Proposition 15.14. *Here $p = 1$, $q = \infty$.*
Then

$$I_t \leq \frac{t^{r+1}}{(r+1)!}\Phi^2, \text{for all } t \in \mathbf{R}_+. \tag{15.33}$$

Proof. By Theorem 15.10. $\qquad\qquad\qquad\qquad\qquad\qquad\square$

We finally get

Theorem 15.15. *Let $p, q > 1 : \frac{1}{p} + \frac{1}{q} = 1$, $t \geq 0$.*
Choose $x_t^* \in [0, rt]$ such that

$$\rho_t := \|T(x_t^*)A^r f\| = \sup_{x \in [0, rt]} \|T(x)A^r f\|. \tag{15.34}$$

Then

$$\int_0^t \|(T(w) - I)^r f\|\,\|T(rw)A^r f\|dw$$

$$\leq \frac{\rho_t t^{r+(1/p)}}{(rp+1)^{1/p}} \left(\int_0^t \|T(rw)A^r f\|^q dw \right)^{1/q}. \tag{15.35}$$

Proof. Call

$$\Gamma(t) := T(t) - I, \ t \geq 0.$$

Then by (15.10) we have

$$\Gamma^r(t)f = \int_0^t \int_0^t \cdots \int_0^t T(u_1 + u_2 + \cdot + u_r)A^r f\, du_1 du_2 \ldots du_r, \ t \geq 0. \qquad (15.36)$$

Therefore it follows that

$$\|\Gamma^r(t)f\| \leq \int_0^t \int_0^t \cdots \int_0^t \|T(u_1 + u_2 + \cdot + u_r)A^r f\|\, du_1 du_2 \ldots du_r$$

$$\leq \left(\int_0^t \int_0^t \cdots \int_0^t 1^p du_1 du_2 \ldots du_r \right)^{1/p}$$

$$\cdot \left(\int_0^t \int_0^t \cdots \int_0^t \|T(u_1 + u_2 + \cdot + u_r)A^r f\|^q du_1 du_2 \ldots du_r \right)^{1/q}$$

$$= t^{r/p} \left(\int_0^t \cdots \int_0^t \|T(u_1 + u_2 + \cdot + u_r)A^r f\|^q du_1 du_2 \ldots du_r \right)^{1/q}.$$

That is we derive

$$\|\Gamma^r(t)f\| \leq t^{r/p} \left(\int_0^t \int_0^t \cdots \int_0^t \|T(u_1 + u_2 + \cdot + u_r)A^r f\|^q du_1 du_2 \ldots du_r \right)^{1/q},$$

$$t \geq 0. \qquad (15.37)$$

Set

$$G(t_1, t_2, \ldots, t_r) := \int_0^{t_1} \int_0^{t_2} \cdots \int_0^{t_r} \|T(u_1 + u_2 + \cdot + u_r)A^r f\|^q du_1 du_2 \ldots du_r,$$

$$where \quad t_1, t_2, \ldots, t_r \geq 0. \qquad (15.38)$$

Thus

$$\|\Gamma^r(t)f\| \leq t^{r/p} \left(G(t,t,\ldots,t) \right)^{1/q}, \ t \geq 0. \qquad (15.39)$$

Clearly we find

$$\frac{\partial^r G(t_1, t_2, \ldots, t_r)}{\partial t_1 \partial t_2 \ldots \partial t_r} = \|T(t_1 + t_2 + \ldots + t_r)A^r f\|^q, \ all \ t_i \geq 0, \ i = \overline{1,r}, \qquad (15.40)$$

and in particular we get

$$\frac{\partial^r G(t,t,\ldots,t)}{\partial t_1 \partial t_2 \ldots \partial t_r} = \|T(rt)A^r f\|^q, \qquad (15.41)$$

and

$$\|T(rt)A^r f\| = \left(\frac{\partial^r G(t,t,\ldots,t)}{\partial t_1 \partial t_2 \ldots \partial t_r} \right)^{1/q}, \ \text{for all } t \geq 0. \qquad (15.42)$$

Consequently we have

$$\|\Gamma^r(t)f\|\,\|T(rt)A^r f\| \le$$

$$t^{r/p}\left(G(t,t,\ldots,t)\frac{\partial^r G(t,t,\ldots,t)}{\partial t_1 \partial t_2 \ldots \partial t_r}\right)^{1/q}, \quad \text{for all } t \ge 0. \tag{15.43}$$

Thus

$$\int_0^t \|\Gamma^r(w)f\|\,\|T(rw)A^r f\|dw \le$$

$$\int_0^t w^{r/p}\left(G(w,w,\ldots,w)\frac{\partial^r G(w,w,\ldots,w)}{\partial t_1 \partial t_2 \ldots \partial t_r}\right)^{1/q} dw =: \mathcal{J} \tag{15.44}$$

Notice here that

$$G(w,w,\ldots,w) = \int_0^w \int_0^w \cdots \int_0^w \|T(u_1 + u_2 + \cdots + u_r)A^r f\|^q du_1 du_2 \ldots du_r$$

$$\le w^r \sup_{x \in [0,rt]} \|T(x)A^r f\|^q = w^r \|T(x_t^*)A^r f\|^q, \tag{15.45}$$

for some $x_t^* \in [0, rt]$, $0 \le w \le t$.

That is

$$G(w,w,\ldots,w) \le w^r \|T(x_t^*)A^r f\|^q. \tag{15.46}$$

Hence we derive

$$\mathcal{J} \le \int_0^t w^r \|T(x_t^*)A^r f\| \left(\frac{\partial^r G(w,w,\ldots,w)}{\partial t_1 \partial t_2 \ldots \partial t_r}\right)^{1/q} dw$$

$$= \|T(x_t^*)A^r f\| \int_0^t w^r \left(\frac{\partial^r G(w,w,\ldots,w)}{\partial t_1 \partial t_2 \ldots \partial t_r}\right)^{1/q} dw$$

$$\le \|T(x_t^*)A^r f\| \left(\int_0^t w^{rp} dw\right)^{1/p} \left(\int_0^t \left(\frac{\partial^r G(w,w,\ldots,w)}{\partial t_1 \partial t_2 \ldots \partial t_r}\right) dw\right)^{1/q}$$

$$= \|T(x_t^*)A^r f\| \frac{t^{r+(1/p)}}{(rp+1)^{1/p}} \left(\int_0^t \|T(rw)A^r f\|^q dw\right)^{1/q}. \tag{15.47}$$

We have established that

$$\int_0^t \|\Gamma^r(w)f\|\,\|T(rw)A^r f\|dw$$

$$\le \frac{t^{r+(1/p)}}{(rp+1)^{1/p}} \|T(x_t^*)A^r f\| \left(\int_0^t \|T(rw)A^r f\|^q dw\right)^{1/q}, \tag{15.48}$$

that is proving the claim. $\qquad\square$

Chapter 16

Opial Inequalities for Cosine and Sine Operator Functions

Various L_p form Opial type inequalities are given for Cosine and Sine Operator functions with applications. This chapter relies on [45].

16.1 Introduction

This chapter is also greatly motivated by the article of Z. Opial [195].

Theorem 16.1. (Opial [195]) *Let $x(t) \in C^1([0, h])$ be such that $x(0) = x(h) = 0$, and $x(t) > 0$ in $(0, h)$. Then,*

$$\int_0^h |x(t)x'(t)|dt \le \frac{h}{4} \int_0^h (x'(t))^2 dt. \tag{16.1}$$

In the last inequality the constant $h/4$ is the best possible.

Opial type inequalities have applications in establishing uniqueness of solution to initial value problems in differential equations, see [244]. We are also motivated by [16], [21].

We need

16.2 Background

(see [93], [147], [191], [192])

Let $(X, \|\cdot\|)$ be a real or complex Banach space. By definition, a *Cosine operator function* is a family $\{C(t); t \in \mathbb{R}\}$ of bounded linear operators from X into itself, satisfying

(i) $C(0) = I, I$ the identity operator;

(ii)

$$C(t + s) + C(t - s) = 2C(t)C(s), \ \forall\, t, s \in \mathbb{R}; \tag{16.2}$$

(the last product is composition)

(iii) $C(\cdot)f$ is continuous an $\mathbb{R}, \ \forall\, f \in X$.

197

Notice that $C(t) = C(-t)\ \forall\, t \in \mathbb{R}$.

The associated *Sine operator function* $S(\cdot)$ is defined by

$$S(t)f := \int_0^t C(s)f\,ds, \quad \forall\, t \in \mathbb{R}, \quad \forall\, f \in X. \tag{16.3}$$

The Cosine operator function $C(\cdot)$ is such that $\|C(t)\| \le M e^{\omega t}$, for some $M \ge 1,\ \omega \ge 0,\ \forall\, t \in \mathbb{R}_+$, here $\|\cdot\|$ is the norm of the operator.

The *infinitesimal generator* A of $C(\cdot)$ is the operator from X into itself defined as

$$Af := \lim_{t \to 0+} \frac{2}{t^2}\left(C(t) - I\right)f \tag{16.4}$$

with domain $D(A)$. Operator A is closed and $D(A)$ is dense in X, i.e. $\overline{D(A)} = X$, and satisfies:

$$\int_0^t S(s)f\,ds \in D(A) \ \text{ and } \ A\int_0^t S(s)f\,ds = C(t)f - f, \forall f \in X. \tag{16.5}$$

Also it holds $A = C''(0)$, and $D(A)$ is the set of $f \in X : C(t)f$ is twice differentiable at $t = 0$; equivalently,

$$D(A) = \{f \in X : C(\cdot)f \in C^2(\mathbb{R}, X)\}. \tag{16.6}$$

If $f \in D(A)$, then $C(t)f \in D(A)$, and $C''(t)f = C(t)Af = AC(t)f,\ \forall\, t \in \mathbb{R}$; $C'(0)f = 0$, see [140], [232].

We define $A^0 = I, A^2 = A \circ A, ..., A^n = A \circ A^{n-1}, n \in \mathbb{N}$. Let $f \in D(A^n)$, then $C(t)f \in C^{2n}(\mathbb{R}, X)$, and $C^{(2n)}(t)f = C(t)A^n f = A^n C(t)f, \forall\, t \in \mathbb{R}$, and $C^{(2k-1)}(0)f = 0, 1 \le k \le n$, see [191].

For $f \in D(A^n), t \in \mathbb{R}$, we have the Cosine operator function's Taylor formula ([191], [192]) saying that

$$T_n(t)f := C(t)f - \sum_{k=0}^{n-1} \frac{t^{2k}}{(2k)!}A^k f = \int_0^t \frac{(t-s)^{2n-1}}{(2n-1)!}C(s)A^n f\,ds. \tag{16.7}$$

By integrating (16.7) we obtain the Sine operator function's Taylor formula

$$M_n(t)f := S(t)f - ft - \frac{t^3}{3!}Af - \cdots - \frac{t^{2n-1}}{(2n-1)!}A^{n-1}f =$$

$$\int_0^t \frac{(t-s)^{2n}}{(2n)!}\,C(s)A^n f\,ds, \forall\, t \in \mathbb{R}, \tag{16.8}$$

all $f \in D(A^n)$.

Integrals in (16.7) and (16.8) are vector valued Riemann integrals, see [79], [164].

16.3   Results

Here, always, we consider $C(t), S(t), A$ as in the Background and $f \in D(A^n), n \in \mathbb{N}$.

Theorem 16.2. *Let $p, q > 1 : \frac{1}{p} + \frac{1}{q} = 1$, $t \in \mathbb{R}_+$. Then*
   i)

$$I := \int_0^t \|T_n(w)f\| \, \|C(w)A^n f\| dw \leq$$

$$\frac{t^{2n-1+\frac{2}{p}}}{2^{1/q}(2n-1)![(p(2n-1)+1)(p(2n-1)+2)]^{1/p}} \left(\int_0^t \|C(w)A^n f\|^q dw \right)^{2/q},$$
$$(16.9)$$

and
   ii)

$$J := \int_0^t \|M_n(w)f\| \, \|C(w)A^n f\| dw \leq$$

$$\frac{t^{2(n+\frac{1}{p})}}{2(2n)![(2pn+1)(pn+1)]^{1/p}} \left(\int_0^t \|C(w)A^n f\|^q dw \right)^{2/q}. \qquad (16.10)$$

We give

Corollary 16.3. *It holds*
   i)

$$I \leq \frac{t^{2n}}{2(2n-1)!\sqrt{2n(4n-1)}} \int_0^t \|C(w)A^n f\|^2 dw, \, \forall t \in \mathbb{R}_+, \qquad (16.11)$$

and
   ii)

$$J \leq \frac{t^{2n+1}}{2(2n)!\sqrt{(2n+1)(4n+1)}} \int_0^t \|C(w)A^n f\|^2 dw, \, \forall t \in \mathbb{R}_+. \qquad (16.12)$$

Proof of Theorem 16.2. For convenience we call $m := 2n - 1$. By (16.7) we have

$$T_n(t)f = \int_0^t \frac{(t-s)^m}{m!} C(s)A^n f \, ds, \, t \in \mathbb{R}_+. \qquad (16.13)$$

Thus

$$\|T_n(t)f\| \leq \frac{1}{m!} \int_0^t (t-s)^m \|C(s)A^n f\| ds \leq$$

$$\frac{1}{m!} \left(\int_0^t (t-s)^{pm} ds \right)^{1/p} \left(\int_0^t \|C(s)A^n f\|^q ds \right)^{1/q}$$

$$= \frac{t^{m+\frac{1}{p}}}{m!(pm+1)^{1/p}} \left(\int_0^t \|C(s)A^n f\|^q ds \right)^{1/q}. \tag{16.14}$$

So far we have found that

$$\|T_n(t)f\| \leq \frac{t^{m+\frac{1}{p}}}{m!(pm+1)^{1/p}} (z(t))^{1/q}, \tag{16.15}$$

where

$$z(t) := \int_0^t \|C(s)A^n f\|^q ds, \ \forall t \in \mathbb{R}_+. \tag{16.16}$$

Clearly

$$z'(w) = \|C(w)A^n f\|^q, \ 0 \leq w \leq t, \ z(0) = 0. \tag{16.17}$$

Also

$$\|C(w)A^n f\| = (z'(w))^{1/q}. \tag{16.18}$$

Therefore we have that

$$\|T_n(w)f\| \, \|C(w)A^n f\| \leq \frac{w^{m+\frac{1}{p}}}{m!(pm+1)^{1/p}} (z(w)z'(w))^{1/q}. \tag{16.19}$$

Consequently it holds

$$\int_0^t \|T_n(w)f\| \, \|C(w)A^n f\| dw \leq \frac{1}{m!(pm+1)^{1/p}} \int_0^t w^{m+\frac{1}{p}} (z(w)z'(w))^{1/q} dw \tag{16.20}$$

$$\leq \frac{1}{m!(pm+1)^{1/p}} \left(\int_0^t w^{pm+1} dw \right)^{1/p} \left(\int_0^t z(w)z'(w) dw \right)^{1/q}$$

$$= \frac{1}{m!(pm+1)^{1/p}} \frac{t^{m+\frac{2}{p}}}{(pm+2)^{1/p}} \left(\frac{z^2(w)}{2} \Big|_0^t \right)^{1/q} \tag{16.21}$$

$$= \frac{1}{m!(pm+1)^{1/p}} \frac{t^{m+\frac{2}{p}}}{2^{1/q}(pm+2)^{1/p}} \left(\int_0^t \|C(w)A^n f\|^q dw \right)^{2/q}, \tag{16.22}$$

proving (16.9).

Setting now $m := 2n$ and working similarly with (16.8) we get (16.10).

The proof of the theorem is now complete. $\qquad \square$

We present

Theorem 16.4. *Here $p = 1, q = \infty$, and*

$$\| \, \|C(s)A^n f\| \, \|_{\infty,+} := \sup_{s \in \mathbb{R}_+} \|C(s)A^n f\| < \infty. \tag{16.23}$$

Then

i)

$$I \leq \frac{t^{2n+1}}{(2n+1)!} \,\| \,\|C(s)A^n f\| \,\|^2_{\infty,+} \,, \forall t \in \mathbb{R}_+, \tag{16.24}$$

and

ii)

$$J \leq \frac{t^{2(n+1)}}{(2(n+1))!} \,\| \,\|C(s)A^n f\| \,\|^2_{\infty,+} \,, \; \forall t \in \mathbb{R}_+. \tag{16.25}$$

Proof. We have from (16.7) that

$$\|T_n(t)f\| \leq \int_0^t \frac{(t-s)^{2n-1}}{(2n-1)!} \|C(s)A^n f\| ds \leq \frac{t^{2n}}{(2n)!} \| \,\|C(s)A^n f\| \,\|_{\infty,+}, \; \forall t \in \mathbb{R}_+. \tag{16.26}$$

Thus

$$\|T_n(w)f\| \,\|C(w)A^n f\| \leq \frac{w^{2n}}{(2n)!} \| \,\|C(s)A^n f\| \,\|^2_{\infty,+} \tag{16.27}$$

all $0 \leq w \leq t$.

So that

$$I = \int_0^t \|T_n(w)f\| \,\|C(w)A^n f\| dw \leq$$

$$\left(\int_0^t w^{2n} dw \right) \frac{\| \,\|C(s)A^n f\| \,\|^2_{\infty,+}}{(2n)!} = \frac{t^{2n+1}}{(2n+1)!} \| \,\|C(s)A^n f\| \,\|^2_{\infty,+}, \tag{16.28}$$

proving (16.24).

Similarly from (16.8) we obtain

$$\|M_n(t)f\| \leq \frac{t^{2n+1}}{(2n+1)!} \| \,\|C(s)A^n f\| \,\|_{\infty,+}, \; \forall t \in \mathbb{R}_+. \tag{16.29}$$

Furthermore it holds

$$\|M_n(w)f\| \,\|C(w)A^n f\| \leq \frac{w^{2n+1}}{(2n+1)!} \| \,\|C(s)A^n f\| \,\|^2_{\infty,+}, \tag{16.30}$$

all $0 \leq w \leq t$.

Therefore

$$J = \int_0^t \|M_n(w)f\| \,\|C(w)A^n f\| dw \leq \frac{t^{2(n+1)}}{(2(n+1))!} \| \,\|C(s)A^n f\| \,\|^2_{\infty,+}, \forall t \in \mathbb{R}_+. \tag{16.31}$$

That is establishing (16.25). The theorem is proved. $\qquad\square$

We continue with

Theorem 16.5. *Let $p, q > 1 : \frac{1}{p} + \frac{1}{q} = 1$, $t \in \mathbb{R}_-$. Then*
  i)

$$I^* := \int_t^0 \|T_n(w)f\| \, \|C(w)A^n f\| dw$$

$$\leq \frac{(-t)^{2n-1+\frac{2}{p}}}{2^{1/q}(2n-1)![(p(2n-1)+1)(p(2n-1)+2)]^{1/p}} \left(\int_t^0 \|C(w)A^n f\|^q dw \right)^{2/q},$$
$$\tag{16.32}$$

and
  ii)

$$J^* := \int_t^0 \|M_n(w)f\| \, \|C(w)A^n f\| dw$$

$$\leq \frac{(-t)^{2(n+\frac{1}{p})}}{2(2n)![(pn+1)(2pn+1)]^{1/p}} \left(\int_t^0 \|C(w)A^n f\|^q dw \right)^{2/q}. \tag{16.33}$$

We give

Corollary 16.6. *It holds*
  i)

$$I^* \leq \frac{t^{2n}}{2(2n-1)!\sqrt{2n(4n-1)}} \int_t^0 \|C(w)A^n f\|^2 dw, \forall t \in \mathbb{R}_-, \tag{16.34}$$

and
  ii)

$$J^* \leq \frac{(-t)^{2n+1}}{2(2n)!\sqrt{(2n+1)(4n+1)}} \int_t^0 \|C(w)A^n f\|^2 dw, \forall t \in \mathbb{R}_-. \tag{16.35}$$

Proof of Theorem 16.5. For convenience call $m := 2n - 1$.
  By (16.7) we have

$$T_n(t)f = \int_0^t \frac{(t-s)^m}{m!} C(s)A^n f \, ds, \ t \in \mathbb{R}_-. \tag{16.36}$$

Thus

$$\|T_n(t)f\| = \frac{1}{m!} \left\| \int_0^t (t-s)^m C(s)A^n f \, ds \right\|$$

$$= \frac{1}{m!} \left\| \int_t^0 (t-s)^m C(s)A^n f \, ds \right\| \leq \frac{1}{m!} \int_t^0 (s-t)^m \|C(s)A^n f\| ds \tag{16.37}$$

$$\leq \frac{1}{m!} \left(\int_t^0 (s-t)^{pm} ds \right)^{1/p} \left(\int_t^0 \|C(s)A^n f\|^q ds \right)^{1/q}$$

$$= \frac{1}{m!} \frac{(-t)^{m+\frac{1}{p}}}{(pm+1)^{1/p}} \left(\int_t^0 \|C(s)A^n f\|^q ds \right)^{1/q}. \tag{16.38}$$

So far we have found that

$$\|T_n(t)f\| \le \frac{(-t)^{m+\frac{1}{p}}}{m!(pm+1)^{1/p}} \, (z(t))^{1/q}, \tag{16.39}$$

where

$$z(t) := \int_t^0 \|C(s)A^n f\|^q ds, \ z(0) = 0. \tag{16.40}$$

That is

$$-z(t) = \int_0^t \|C(s)A^n f\|^q ds \ge 0, \tag{16.41}$$

and

$$-z'(t) = \|C(t)A^n f\|^q \ge 0, \tag{16.42}$$

and

$$\|C(t)A^n f\| = (-z'(t))^{1/q}, \ t \in \mathbb{R}_- \tag{16.43}$$

Hence we have

$$\|T_n(w)f\| \, \|C(w)A^n f\| \le \frac{(-w)^{m+\frac{1}{p}}}{m!(pm+1)^{1/p}} (z(w)(-z'(w)))^{1/q}. \tag{16.44}$$

Integrating the last inequality we derive

$$\int_t^0 \|T_n(w)f\| \, \|C(w)A^n f\| dw$$

$$\le \frac{1}{m!(pm+1)^{1/p}} \int_t^0 (-w)^{m+\frac{1}{p}} (z(w)(-z'(w)))^{1/q} dw \tag{16.45}$$

$$\le \frac{1}{m!(pm+1)^{1/p}} \left(\int_t^0 (-w)^{pm+1} dw \right)^{1/p} \left(\int_t^0 z(w)(-z'(w)) dw \right)^{1/q} \tag{16.46}$$

$$= \frac{1}{m!(pm+1)^{1/p}} \frac{(-t)^{m+\frac{2}{p}}}{(pm+2)^{1/p}} \frac{(z(t))^{2/q}}{2^{1/q}} \tag{16.47}$$

$$= \frac{(-t)^{m+\frac{2}{p}}}{2^{1/q}m![(pm+1)(pm+2)]^{1/p}} \left(\int_t^0 \|C(w)A^n f\|^q dw \right)^{2/q}, \tag{16.48}$$

proving (16.32).

Setting now $m := 2n$ and working similarly with (16.8) we get (16.33).

The proof of the theorem is now complete. $\qquad\square$

Combining Theorems 16.2 and 16.5 we derive

Proposition 16.7. *Let $p, q > 1 : \frac{1}{p} + \frac{1}{q} = 1, t \in \mathbb{R}$. Then*

i)

$$\bar{I} := \left| \int_0^t \|T_n(w)f\| \, \|C(w)A^n f\| dw \right|$$

$$\leq \frac{|t|^{2n-1+\frac{2}{p}}}{2^{1/q}(2n-1)![(p(2n-1)+1)(p(2n-1)+2)]^{1/p}} \left| \int_0^t \|C(w)A^n f\|^q dw \right|^{2/q},$$

(16.49)

and

ii)

$$\bar{J} := \left| \int_0^t \|M_n(w)f\| \, \|C(w)A^n f\| dw \right|$$

$$\leq \frac{|t|^{2(n+\frac{1}{p})}}{2(2n)![(pn+1)(2pn+1)]^{1/p}} \left| \int_0^t \|C(w)A^n f\|^q dw \right|^{2/q}.$$

(16.50)

Combining Corollaries 16.3 and 16.6 we obtain

Corollary 16.8. *It holds*

i)

$$\bar{I} \leq \frac{t^{2n}}{2(2n-1)!\sqrt{2n(4n-1)}} \left| \int_0^t \|C(w)A^n f\|^2 dw \right|, \forall t \in \mathbb{R},$$

(16.51)

and

ii)

$$\bar{J} \leq \frac{|t|^{2n+1}}{2(2n)!\sqrt{(2n+1)(4n+1)}} \left| \int_0^t \|C(w)A^n f\|^2 dw \right|, \forall t \in \mathbb{R}.$$

(16.52)

Next we present

Theorem 16.9. *Here $p = 1, q = \infty$, and*

$$\| \, \|C(s)A^n f\| \, \|_{\infty,-} := \sup_{s \in \mathbb{R}_-} \|C(s)A^n f\| < \infty.$$

Then

i)

$$I^* \leq \frac{(-t)^{2n+1}}{(2n+1)!} \| \, \|C(s)A^n f\| \, \|_{\infty,-}^2, \forall t \in \mathbb{R}_-,$$

(16.53)

and

ii)

$$J^* \leq \frac{t^{2(n+1)}}{(2(n+1))!} \| \, \|C(s)A^n f\| \, \|_{\infty,-}^2, \forall t \in \mathbb{R}_-.$$

(16.54)

Proof. For convenience call $m := 2n - 1$. By (16.7) we have

$$T_n(t)f = \int_0^t \frac{(t-s)^m}{m!} \, C(s)A^n f \, ds, \ t \in \mathbb{R}_-. \tag{16.55}$$

Thus

$$\|T_n(t)f\| = \frac{1}{m!} \left\| \int_t^0 (t-s)^m C(s)A^n f \, ds \right\|$$

$$\leq \frac{1}{m!} \int_t^0 (s-t)^m \|C(s)A^n f\| ds \tag{16.56}$$

$$\leq \frac{\| \, \|C(s)A^n f\| \, \|_{\infty,-}}{m!} \left(\int_t^0 (s-t)^m ds \right) = \| \, \|C(s)A^n f\| \, \|_{\infty,-} \frac{(-t)^{m+1}}{(m+1)!}. \tag{16.57}$$

Hence

$$\|T_n(t)f\| \leq \| \, \|C(s)A^n f\| \, \|_{\infty,-} \frac{(-t)^{m+1}}{(m+1)!}, \forall t \in \mathbb{R}_-. \tag{16.58}$$

So that it holds

$$\|T_n(w)f\| \, \|C(w)A^n f\| \leq \| \, \|C(s)A^n f\| \, \|_{\infty,-}^2 \frac{(-t)^{m+1}}{(m+1)!}, \ t \leq w \leq 0 \tag{16.59}$$

Consequently we get

$$I^* = \int_t^0 \|T_n(w)f\| \, \|C(w)A^n f\| dw \leq \frac{\| \, \|C(s)A^n f\| \, \|_{\infty,-}^2}{(m+1)!} \int_t^0 (-w)^{m+1} dw \tag{16.60}$$

$$= \frac{(-t)^{m+2}}{(m+2)!} \| \, \|C(s)A^n f\| \, \|_{\infty,-}^2 \tag{16.61}$$

proving (16.53).

Similarly, from (16.8), for $m = 2n$ we obtain (16.54). $\qquad\square$

Finally we give

Proposition 16.10. *Here $p = 1, q = \infty$, and*

$$\| \, \|C(s)A^n f\| \, \|_\infty := \sup_{s \in \mathbb{R}} \|C(s)A^n f\| < \infty. \tag{16.62}$$

Then

i)

$$\left| \int_0^t \|T_n(w)f\| \, \|C(w)A^n f\| dw \right| \leq \frac{|t|^{2n+1}}{(2n+1)!} \| \, \|C(s)A^n f\| \, \|_\infty^2, \forall t \in \mathbb{R}, \tag{16.63}$$

and

ii)

$$\left| \int_0^t \|M_n(w)f\| \, \|C(w)A^n f\| dw \right| \leq \frac{t^{2(n+1)}}{(2(n+1))!} \| \, \|C(s)A^n f\| \, \|_\infty^2, \forall t \in \mathbb{R}. \tag{16.64}$$

Proof. By Theorems 16.4, 16.9. $\qquad\square$

16.4 Applications

(see [147], p. 121)

Let $\mathcal{X}$ be the Banach space of odd, 2π-periodic real functions in the space of bounded uniformly continuous functions from $\mathbb{R}$ into itself : $BUC(\mathbb{R})$. Let $A := \frac{d^2}{dx^2}$ with $D(A^n) = \{f \in \mathcal{X} : f^{(2k)} \in \mathcal{X}, k = 1, ..., n\}, n \in \mathbb{N}$. A generates a Cosine functions C^* given by

$$C^*(t)f(x) = \frac{1}{2}\left[f(x+t) + f(x-t)\right], \quad \forall x, t \in \mathbb{R}. \tag{16.65}$$

The corresponding Sine function S^* is given by

$$S^*(t)f(x) = \frac{1}{2}\left[\int_0^t f(x+s)ds + \int_0^t f(x-s)ds\right], \quad \forall x, t \in \mathbb{R}. \tag{16.66}$$

Here we consider $f \in D(A^n), n \in \mathbb{N}$, as above. By (16.7) we obtain

$$T_n^*(t)f := \frac{1}{2}\left[f(\cdot + t) + f(\cdot - t)\right] - \sum_{k=0}^{n-1} \frac{t^{2k}}{(2k)!} f^{(2k)}$$

$$= \int_0^t \frac{(t-s)^{2n-1}}{2(2n-1)!} \left[f^{(2n)}(\cdot + s) + f^{(2n)}(\cdot - s)\right] ds, \forall t \in \mathbb{R}. \tag{16.67}$$

By (16.8) we get

$$M_n^*(t)f := \frac{1}{2}\left[\int_0^t f(\cdot + s)ds + \int_0^t f(\cdot - s)ds\right] - \sum_{k=1}^{n} \frac{t^{2k-1}}{(2k-1)!} f^{(2(k-1))}$$

$$= \int_0^t \frac{(t-s)^{2n}}{2(2n)!} \left[f^{(2n)}(\cdot + s) + f^{(2n)}(\cdot - s)\right] ds, \forall t \in \mathbb{R}. \tag{16.68}$$

Let $g \in BUC(\mathbb{R})$, we define $\|g\| = \|g\|_\infty := \sup_{x \in \mathbb{R}} |g(x)| < \infty$.

Notice also that

$$\| \|C^*(s)A^n f\|_\infty \|_\infty = \| \|C^*(s)f^{(2n)}\|_\infty \|_\infty$$

$$= \frac{1}{2}\| \|(f^{(2n)}(\cdot + s) + f^{(2n)}(\cdot - s))\|_\infty \|_\infty$$

$$\leq \frac{1}{2}\left[\| \|f^{(2n)}(\cdot + s)\|_\infty \|_\infty + \| \|f^{(2n)}(\cdot - s)\|_\infty \|_\infty\right] \leq \Theta < \infty, \tag{16.69}$$

where

$$\Theta := \|f^{(2n)}\|_\infty. \tag{16.70}$$

We have the following applications. From Proposition 16.7 we derive

Proposition 16.11. *Let* $p, q > 1 : \frac{1}{p} + \frac{1}{q} = 1, t \in \mathbb{R}$. *Then*

i)

$$\varepsilon_1 := \left| \int_0^t \|T_n^*(w)f\|_\infty \left\|C^*(w)f^{(2n)}\right\|_\infty dw \right|$$

$$\leq \frac{|t|^{2n-1+\frac{2}{p}}}{2^{1/q}(2n-1)![(p(2n-1)+1)(p(2n-1)+2)]^{1/p}} \left| \int_0^t \left\|C^*(w)f^{(2n)}\right\|_\infty^q dw \right|^{2/q},$$

$$\forall t \in \mathbb{R}, \tag{16.71}$$

and

ii)

$$\varepsilon_2 := \left| \int_0^t \|M_n^*(w)f\|_\infty \left\|C^*(w)f^{(2n)}\right\|_\infty dw \right|$$

$$\leq \frac{|t|^{2(n+\frac{1}{p})}}{2(2n)![(pn+1)(2pn+1)]^{1/p}} \left| \int_0^t \left\|C^*(w)f^{(2n)}\right\|_\infty^q dw \right|^{2/q}, \forall t \in \mathbb{R}. \tag{16.72}$$

From Corollary 16.8 we get

Corollary 16.12. *It holds*

i)

$$\varepsilon_1 \leq \frac{t^{2n}}{2(2n-1)!\sqrt{2n(4n-1)}} \left| \int_0^t \left\|C^*(w)f^{(2n)}\right\|_\infty^2 dw \right|, \forall t \in \mathbb{R}, \tag{16.73}$$

and

ii)

$$\varepsilon_2 \leq \frac{|t|^{2n+1}}{2(2n)!\sqrt{(2n+1)(4n+1)}} \left| \int_0^t \left\|C^*(w)f^{(2n)}\right\|_\infty^2 dw \right|, \forall t \in \mathbb{R}. \tag{16.74}$$

Finally from Proposition 16.10 we find

Proposition 16.13. *Here $p=1, q=\infty$. Then*

i)

$$\varepsilon_1 \leq \frac{|t|^{2n+1}}{(2n+1)!} \left\|f^{(2n)}\right\|_\infty^2, \forall t \in \mathbb{R}, \tag{16.75}$$

and

ii)

$$\varepsilon_2 \leq \frac{t^{2(n+1)}}{(2(n+1))!} \left\|f^{(2n)}\right\|_\infty^2, \forall t \in \mathbb{R}. \tag{16.76}$$

Chapter 17

Poincaré Like Inequalities for Linear Differential Operators

Various L_p form Poincaré like inequalities [2], forward and reverse, are given for a linear differential operator L, involving its related initial value problem solution y, Ly, the associated Green's function H and initial conditions at point $x_0 \in \mathbb{R}$. This chapter follows [48].

17.1 Background

Here we use [169], pp. 145–154.

Let $[a, b] \subset \mathbb{R}$, $a_i(x)$, $i = 0, 1, \ldots, n - 1$ $(n \in \mathbb{N})$, $h(x)$ be continuous functions on $[a, b]$ and let $L = D^n + a_{n-1}(x) D^{n-1} + \ldots + a_0(x)$ be a fixed linear differential operator on $C^n([a, b])$. Let $y_1(x), \ldots, y_n(x)$ be a set of linear independent solutions to $Ly = 0$. Here the associated Green's functions for L is

$$
H(x, t) := \frac{\begin{vmatrix} y_1(t) & \cdots & y_n(t) \\ y_1'(t) & \cdots & y_n'(t) \\ \cdots & \cdots & \cdots \\ y_1^{(n-2)}(t) & \cdots & y_n^{(n-2)}(t) \\ y_1(x) & \cdots & y_n(x) \end{vmatrix}}{\begin{vmatrix} y_1(t) & \cdots & y_n(t) \\ y_1'(t) & \cdots & y_n'(t) \\ \cdots & \cdots & \cdots \\ y_1^{(n-2)}(t) & \cdots & y_n^{(n-2)}(t) \\ y_1^{(n-1)}(t) & \cdots & y_n^{(n-1)}(t) \end{vmatrix}}, \tag{17.1}
$$

which is a continuous function on $[a, b]^2$.

Consider a fixed $x_0 \in [a, b]$, then

$$
y(x) = \int_{x_0}^{x} H(x, t) h(t) \, dt, \quad \forall x \in [a, b] \tag{17.2}
$$

is the unique solution to the initial value problem

$$
Ly = h, \quad y^{(i)}(x_0) = 0, \quad i = 0, 1, \ldots, n - 1. \tag{17.3}
$$

209

Next we suppose all of the above.

17.2   Results

We present the following Poincaré like inequalities.

Theorem 17.1. *Let $x_0 < b$ and $x \in [x_0, b]$, and $p, q > 1 : \frac{1}{p} + \frac{1}{q} = 1; \nu > 0$.*
Then

$$1) \ \|y\|_{L_\nu(x_0,b)} \le \left(\int_{x_0}^{b} \left(\int_{x_0}^{x} |H(x,t)|^p \, dt \right)^{\nu/p} dx \right)^{1/\nu} \|Ly\|_{L_q(x_0,b)}. \tag{17.4}$$

When $\nu = q$ we have

$$2) \ \|y\|_{L_q(x_0,b)} \le \left(\int_{x_0}^{b} \left(\int_{x_0}^{x} |H(x,t)|^p \, dt \right)^{q/p} dx \right)^{1/q} \|Ly\|_{L_q(x_0,b)}. \tag{17.5}$$

When $\nu = p = q = 2$ we obtain

$$3) \ \|y\|_{L_2(x_0,b)} \le \left(\int_{x_0}^{b} \left(\int_{x_0}^{x} H^2(x,t) \, dt \right) dx \right)^{1/2} \|Ly\|_{L_2(x_0,b)}. \tag{17.6}$$

Proof. From (17.2) we have

$$|y(x)| \le \int_{x_0}^{x} |H(x,t)| \, |h(t)| \, dt$$

$$\le \left(\int_{x_0}^{x} |H(x,t)|^p \, dt \right)^{1/p} \left(\int_{x_0}^{x} |h(t)|^q \, dt \right)^{1/q}$$

$$\le \left(\int_{x_0}^{x} |H(x,t)|^p \, dt \right)^{1/p} \left(\int_{x_0}^{b} |h(t)|^q \, dt \right)^{1/q}. \tag{17.7}$$

That is

$$|y(x)|^\nu \le \left(\int_{x_0}^{x} |H(x,t)|^p \, dt \right)^{\nu/p} \|Ly\|_{L_q(x_0,b)}^\nu, \tag{17.8}$$

Hence

$$\int_{x_0}^{b} |y(x)|^\nu \, dx \le \left(\int_{x_0}^{b} \left(\int_{x_0}^{x} |H(x,t)|^p \, dt \right)^{\nu/p} dx \right) \|Ly\|_{L_q(x_0,b)}^\nu, \tag{17.9}$$

proving the claim. $\qquad\qquad\square$

We continue with

Theorem 17.2. *Let $x_0 > a$ and $x \in [a, x_0]$, and $p, q > 1 : \frac{1}{p} + \frac{1}{q} = 1; \nu > 0$.*
Then

$$1) \ \|y\|_{L_\nu(a,x_0)} \leq \left(\int_a^{x_0} \left(\int_x^{x_0} |H(x,t)|^p \, dt \right)^{\nu/p} dx \right)^{1/\nu} \|Ly\|_{L_q(a,x_0)}. \qquad (17.10)$$

When $\nu = q$ we obtain

$$2) \ \|y\|_{L_q(a,x_0)} \leq \left(\int_a^{x_0} \left(\int_x^{x_0} |H(x,t)|^p \, dt \right)^{q/p} dx \right)^{1/q} \|Ly\|_{L_q(a,x_0)}. \qquad (17.11)$$

When $\nu = p = q = 2$ we obtain

$$3) \ \|y\|_{L_2(a,x_0)} \leq \left(\int_a^{x_0} \left(\int_x^{x_0} H^2(x,t) \, dt \right) dx \right)^{1/2} \|Ly\|_{L_2(a,x_0)}. \qquad (17.12)$$

Proof. From (17.2) we have

$$|y(x)| \leq \int_x^{x_0} |H(x,t)| \, |h(t)| \, dt$$

$$\leq \left(\int_x^{x_0} |H(x,t)|^p \, dt \right)^{1/p} \left(\int_x^{x_0} |h(t)|^q \, dt \right)^{1/q}$$

$$\leq \left(\int_x^{x_0} |H(x,t)|^p \, dt \right)^{1/p} \left(\int_a^{x_0} |h(t)|^q \, dt \right)^{1/q}. \qquad (17.13)$$

That is

$$|y(x)|^\nu \leq \left(\int_x^{x_0} |H(x,t)|^p \, dt \right)^{\nu/p} \|Ly\|_{L_q(a,x_0)}^\nu. \qquad (17.14)$$

Therefore

$$\int_a^{x_0} |y(x)|^\nu \, dx \leq \left(\int_a^{x_0} \left(\int_x^{x_0} |H(x,t)|^p \, dt \right)^{\nu/p} dx \right) \|Ly\|_{L_q(a,x_0)}^\nu, \qquad (17.15)$$

proving the claim. $\qquad \square$

Extreme cases follow

Proposition 17.3. *Here $x_0 < b$, $x \in [x_0, b]$, and $p = 1$, $q = \infty$.*
Then

$$1) \ \|y\|_{L_\nu(x_0,b)} \leq \left(\int_{x_0}^b \left(\int_{x_0}^x |H(x,t)| \, dt \right)^\nu dx \right)^{1/\nu} \|Ly\|_{L_\infty(x_0,b)}. \qquad (17.16)$$

When $\nu = 1$ we have

$$2) \ \|y\|_{L_1(x_0,b)} \leq \left(\int_{x_0}^b \left(\int_{x_0}^x |H(x,t)| \, dt \right) dx \right) \|Ly\|_{L_\infty(x_0,b)}. \qquad (17.17)$$

Proof.   From (17.2) we have

$$|y(x)| \le \int_{x_0}^{x} |H(x,t)|\,|h(t)|\,dt$$

$$\le \left(\int_{x_0}^{x} |H(x,t)|\,dt \right) \|h\|_{L_\infty(x_0,b)} . \tag{17.18}$$

That is

$$|y(x)|^\nu \le \left(\int_{x_0}^{x} |H(x,t)|\,dt \right)^\nu \|Ly\|_{L_\infty(x_0,b)}^\nu , \tag{17.19}$$

and

$$\int_{x_0}^{b} |y(x)|^\nu \,dx \le \left(\int_{x_0}^{b} \left(\int_{x_0}^{x} |H(x,t)|\,dt \right)^\nu dx \right) \|Ly\|_{L_\infty(x_0,b)}^\nu , \tag{17.20}$$

proving the claim. $\square$

We continue with

Proposition 17.4. *Here $x_0 > a$, $x \in [a, x_0]$, and $p = 1$, $q = \infty$.*
Then

$$1)\ \ \|y\|_{L_\nu(a,x_0)} \le \left(\int_{a}^{x_0} \left(\int_{x}^{x_0} |H(x,t)|\,dt \right)^\nu dx \right)^{1/\nu} \|Ly\|_{L_\infty(a,x_0)} . \tag{17.21}$$

When $\nu = 1$ we derive

$$2)\ \ \|y\|_{L_1(a,x_0)} \le \left(\int_{a}^{x_0} \left(\int_{x}^{x_0} |H(x,t)|\,dt \right) dx \right) \|Ly\|_{L_\infty(a,x_0)} . \tag{17.22}$$

Proof.   From (17.2) we have

$$|y(x)| \le \int_{x}^{x_0} |H(x,t)|\,|h(t)|\,dt$$

$$\le \left(\int_{x}^{x_0} |H(x,t)|\,dt \right) \|h\|_{L_\infty(a,x_0)} . \tag{17.23}$$

That is

$$|y(x)|^\nu \le \left(\int_{x}^{x_0} |H(x,t)|\,dt \right)^\nu \|Ly\|_{L_\infty(a,x_0)}^\nu , \tag{17.24}$$

and

$$\int_{a}^{x_0} |y(x)|^\nu \,dx \le \left(\int_{a}^{x_0} \left(\int_{x}^{x_0} |H(x,t)|\,dt \right)^\nu dx \right) \|Ly\|_{L_\infty(a,x_0)}^\nu , \tag{17.25}$$

proving the claim. $\square$

Next we give reverse Poincaré like inequalities.

Theorem 17.5. *Let $x_0 < b$, $x \in [x_0, b]$, and $0 < p < 1$, $q < 0 : \frac{1}{p} + \frac{1}{q} = 1$, $\nu > 0$.*
Suppose $H(x, t) \geq 0$ for $x_0 \leq t \leq x$, and $Ly = h$ is of fixed sign and nowhere zero.

Then

$$1) \ \|y\|_{L_\nu(x_0,b)} \geq \left(\int_{x_0}^{b} \left(\int_{x_0}^{x} (H(x,t))^p \, dt \right)^{\nu/p} dx \right)^{1/\nu} \|Ly\|_{L_q(x_0,b)} . \qquad (17.26)$$

When $\nu = p$ we obtain

$$2) \ \|y\|_{L_p(x_0,b)} \geq \left(\int_{x_0}^{b} \left(\int_{x_0}^{x} (H(x,t))^p \, dt \right) dx \right)^{1/p} \|Ly\|_{L_q(x_0,b)} . \qquad (17.27)$$

When $\nu = 1$ we get

$$3) \ \|y\|_{L_1(x_0,b)} \geq \left(\int_{x_0}^{b} \left(\int_{x_0}^{x} (H(x,t))^p \, dt \right)^{1/p} dx \right) \|Ly\|_{L_q(x_0,b)} . \qquad (17.28)$$

Proof. By (17.2) we have

$$|y(x)| = \int_{x_0}^{x} H(x,t) \, |h(t)| \, dt, \ \text{ for all } x_0 \leq x \leq b. \qquad (17.29)$$

From (17.29) by reverse Hölder's inequality we find

$$|y(x)| \geq \left(\int_{x_0}^{x} (H(x,t))^p \, dt \right)^{1/p} \left(\int_{x_0}^{x} |h(t)|^q \, dt \right)^{1/q}$$

$$\geq \left(\int_{x_0}^{x} (H(x,t))^p \, dt \right)^{1/p} \left(\int_{x_0}^{b} |h(t)|^q \, dt \right)^{1/q}, \qquad (17.30)$$

for all $x_0 < x \leq b$.

That is it holds

$$|y(x)|^\nu \geq \left(\int_{x_0}^{x} (H(x,t))^p \, dt \right)^{\nu/p} \|h\|_{L_q(x_0,b)}^\nu , \qquad (17.31)$$

for all $x_0 \leq x \leq b$,

and

$$\int_{x_0}^{b} |y(x)|^\nu \, dx \geq \left(\int_{x_0}^{b} \left(\int_{x_0}^{x} (H(x,t))^p \, dt \right)^{\nu/p} dx \right) \|h\|_{L_q(x_0,b)}^\nu , \qquad (17.32)$$

proving the claim. $\qquad\qquad\square$

We continue with

Theorem 17.6. *Let $x_0 > a$, $x \in [a, x_0]$, and $0 < p < 1$, $q < 0 : \frac{1}{p} + \frac{1}{q} = 1$, $\nu > 0$.*
Suppose $H(x, t) \leq 0$ for $x \leq t \leq x_0$, and $Ly = h$ is of fixed sign and nowhere zero.

Then

$$1) \ \|y\|_{L_\nu(a, x_0)} \geq \left(\int_a^{x_0} \left(\int_x^{x_0} (-H(x, t))^p \, dt \right)^{\nu/p} dx \right)^{1/\nu} \|Ly\|_{L_q(a, x_0)}. \tag{17.33}$$

When $\nu = p$ we obtain

$$2) \ \|y\|_{L_p(a, x_0)} \geq \left(\int_a^{x_0} \left(\int_x^{x_0} (-H(x, t))^p \, dt \right) dx \right)^{1/p} \|Ly\|_{L_q(a, x_0)}. \tag{17.34}$$

When $\nu = 1$ we have

$$3) \ \|y\|_{L_1(a, x_0)} \geq \left(\int_a^{x_0} \left(\int_x^{x_0} (-H(x, t))^p \, dt \right)^{1/p} dx \right) \|Ly\|_{L_q(a, x_0)}. \tag{17.35}$$

Proof. From (17.2) we have

$$|y(x)| = \left| \int_{x_0}^x H(x, t) h(t) \, dt \right|$$

$$= \left| \int_x^{x_0} H(x, t) h(t) \, dt \right|$$

$$= \left| \int_x^{x_0} (-H(x, t)) h(t) \, dt \right|$$

$$= \int_x^{x_0} (-H(x, t)) |h(t)| \, dt. \tag{17.36}$$

From (17.36) by reverse Hölder's inequality we find

$$|y(x)| \geq \left(\int_x^{x_0} (-H(x, t))^p \, dt \right)^{1/p} \left(\int_x^{x_0} |h(t)|^q \, dt \right)^{1/q}$$

$$\geq \left(\int_x^{x_0} (-H(x, t))^p \, dt \right)^{1/p} \left(\int_a^{x_0} |h(t)|^q \, dt \right)^{1/q} \tag{17.37}$$

for all $a \leq x < x_0$.

That is it holds

$$|y(x)|^\nu \geq \left(\int_x^{x_0} (-H(x, t))^p \, dt \right)^{\nu/p} \|Ly\|_{L_q(a, x_0)}^\nu, \tag{17.38}$$

for all $a \leq x \leq x_0$,

and

$$\int_a^{x_0} |y(x)|^\nu \, dx \geq \left(\int_a^{x_0} \left(\int_x^{x_0} (-H(x, t))^p \, dt \right)^{\nu/p} dx \right) \|Ly\|_{L_q(a, x_0)}^\nu, \tag{17.39}$$

proving the claim. $\qquad \square$

Chapter 18

Poincaré and Sobolev Like Inequalities for Widder Derivatives

Various L_p form Poincaré and Sobolev like inequalities, forward and reverse, are given involving Widder derivative ([243]). This chapter follows [54].

18.1 Background

The following come from [243].

Let $f, u_0, u_1, \ldots, u_n \in C^{n+1}([a,b])$, $n \geq 0$, and the Wronskians

$$W_i(x) := W[u_0(x), u_1(x), \ldots, u_i(x)] :=$$

$$\begin{vmatrix} u_0(x) & u_1(x) & \ldots & u_i(x) \\ u_0'(x) & u_1'(x) & \ldots & u_i'(x) \\ & \vdots & & \\ u_0^{(i)}(x) & u_1^{(i)}(x) & \ldots & u_i^{(i)}(x) \end{vmatrix}, \tag{18.1}$$

$i = 0, 1, \ldots, n$. Here $W_0(x) = u_0(x)$. Assume $W_i(x) > 0$ over $[a,b]$, $i = 0, 1, \ldots, n$. For $i \geq 0$, the differential operator of order i (Widder derivative):

$$L_i f(x) := \frac{W[u_0(x), u_1(x), \ldots, u_{i-1}(x), f(x)]}{W_{i-1}(x)}, \tag{18.2}$$

$i = 1, \ldots, n+1$; $L_0 f(x) := f(x)$, $\forall x \in [a,b]$.

Consider also

$$g_i(x,t) := \frac{1}{W_i(t)} \begin{vmatrix} u_0(t) & u_1(t) & \ldots & u_i(t) \\ u_0'(t) & u_1'(t) & \ldots & u_i'(t) \\ & \vdots & & \\ u_0^{(i-1)}(t) & u_1^{(i-1)}(t) & \ldots & u_i^{(i-1)}(t) \\ u_0(x) & u_1(x) & \ldots & u_i(x) \end{vmatrix}, \tag{18.3}$$

$i = 1, 2, \ldots, n$; $g_0(x,t) := \frac{u_0(x)}{u_0(t)}$, $\forall x, t \in [a,b]$.

Example ([243]). Sets of the form $\{u_0, u_1, \ldots, u_n\}$ are

215

$$\left\{ 1, x, x^2, \ldots, x^n \right\},$$

$$\left\{ 1, \sin x, -\cos x, -\sin 2x, \cos 2x, \ldots, (-1)^{n-1} \sin nx, (-1)^n \cos nx \right\}, \text{ etc.}$$

We also mention the generalized Widder–Taylor's formula, see [243].

Theorem 18.1. *Let the functions* $f, u_0, u_1, \ldots, u_n \in C^{n+1}([a,b])$, *and the Wronskians* $W_0(x), W_1(x), \ldots, W_n(x) > 0$ *on* $[a,b]$, $x \in [a,b]$. *Then for* $t \in [a,b]$ *we have*

$$f(x) = f(t)\frac{u_0(x)}{u_0(t)} + L_1 f(t) g_1(x,t) + \ldots + L_n f(t) g_n(x,t) + R_n(x), \quad (18.4)$$

where

$$R_n(x) := \int_t^x g_n(x,s) L_{n+1} f(s)\, ds. \quad (18.5)$$

For example ([243]) one could take $u_0(x) = c > 0$. If $u_i(x) = x^i$, $i = 0, 1, \ldots, n$, defined on $[a,b]$, then

$$L_i f(t) = f^{(i)}(t) \text{ and } g_i(x,t) = \frac{(x-t)^i}{i!}, \quad t \in [a,b]. \quad (18.6)$$

We need

Corollary 18.2. (to Theorem 18.1) *Additionally suppose that for fixed* $x_0 \in [a,b]$ *we have* $L_i f(x_0) = 0$, $i = 0, 1, \ldots, n$. *Then*

$$f(x) = \int_{x_0}^x g_n(x,t) L_{n+1} f(t)\, dt, \quad \forall x \in [a,b]. \quad (18.7)$$

Corollary 18.3. (to Theorem 18.1) *Let* $f, u_0 \in C^1([a,b])$, $u_0(x) > 0$, *for all* $x \in [a,b]$. *Then*

$$f(x) = f(t)\frac{u_0(x)}{u_0(t)} + u_0(x)\int_t^x \frac{L_1 f(s)}{u_0(s)}\, ds, \quad \forall x, t \in [a,b], \quad (18.8)$$

where

$$L_1 f(s) = \frac{W[u_0(s), f(s)]}{u_0(s)} = \frac{u_0(s) f'(s) - u_0'(s) f(s)}{u_0(s)}. \quad (18.9)$$

We need to make

Remark 18.4. We define (see [243])

$$\phi_0(x) := W_0(x), \quad \phi_1(x) := \frac{W_1(x)}{(W_0(x))^2}, \ldots,$$

in general

$$\phi_k(x) := \frac{W_k(x) W_{k-2}(x)}{(W_{k-1}(x))^2}, \quad k = 2, 3, \ldots, n. \quad (18.10)$$

The functions $\phi_i(x)$ are positive on $[a,b]$. According to [243] we get, for any $x, x_0 \in [a,b]$ that

$$g_n(x, x_0) = \frac{\phi_0(x)}{\phi_0(x_0)\dots\phi_n(x_0)} \int_{x_0}^{x} \phi_1(x_1) \int_{x_0}^{x_1} \dots$$

$$\int_{x_0}^{x_{n-2}} \phi_{n-1}(x_{n-1}) \int_{x_0}^{x_{n-1}} \phi_n(x_n)\, dx_1 dx_2 \dots dx_n$$

$$= \frac{1}{\phi_0(x_0)\dots\phi_n(x_0)} \int_{x_0}^{x} \phi_0(s)\dots\phi_n(s)\, g_{n-1}(x,s)\, ds. \tag{18.11}$$

We get that $g_n(x,x) = 0$, all $x \in [a,b]$, and $g_n(x,x_0) > 0$, $x > x_0$, $x, x_0 \in [a,b]$, any $n \in \mathbb{N}$. Also $g_0(x,x_0) > 0$ for any $x, x_0 \in [a,b]$.

By (18.11) we notice that

$$g_n(x,t) < 0, \quad x < t, \quad n \text{ odd},$$

$$g_n(x,t) > 0, \quad x < t, \quad n \text{ even}, \tag{18.12}$$

where $x, t \in [a,b]$.

In the next we work under the terms and assumptions of Theorem 18.1 and Corollary 18.2, and the rest of the above conclusions.

18.2   Results

We give the following weighted Neumann–Poincaré like inequality.

Theorem 18.5. *Let $p, q > 1 : \frac{1}{p} + \frac{1}{q} = 1$, $\nu > 0$. Consider $f, u_0 \in C^1\left([a,b]\right)$, $u_0 > 0$ on $[a,b]$.*

Then

$$1) \quad \left\| \frac{f}{u_0} - \frac{1}{b-a} \int_a^b \frac{f(t)}{u_0(t)} dt \right\|_{L_\nu(a,b)} \leq (b-a)^{\left(\frac{1}{p}+\frac{1}{\nu}\right)} \left\| \frac{L_1 f}{u_0} \right\|_{L_q(a,b)}. \tag{18.13}$$

When $\nu = q$ we obtain

$$2) \quad \left\| \frac{f}{u_0} - \frac{1}{b-a} \int_a^b \frac{f(t)}{u_0(t)} dt \right\|_{L_q(a,b)} \leq (b-a) \left\| \frac{L_1 f}{u_0} \right\|_{L_q(a,b)}. \tag{18.14}$$

When $\nu = p = q = 2$ we have

$$3) \quad \left\| \frac{f}{u_0} - \frac{1}{b-a} \int_a^b \frac{f(t)}{u_0(t)} dt \right\|_{L_2(a,b)} \leq (b-a) \left\| \frac{L_1 f}{u_0} \right\|_{L_2(a,b)}. \tag{18.15}$$

Equivalently,

$$\int_a^b \left(\frac{f(x)}{u_0(x)} - \frac{1}{b-a} \int_a^b \frac{f(t)}{u_0(t)} dt \right)^2 dx$$

$$\leq (b-a)^2 \int_a^b \left(\frac{L_1 f(x)}{u_0(x)}\right)^2 dx. \qquad (18.15^*)$$

Proof.   Let $A = \frac{1}{b-a}\int_a^b \frac{f(t)}{u_0(t)} dt$. By the intermediate value theorem there exists $c \in [a,b]$ with $\frac{f(c)}{u_0(c)} = A$.

Thus by (18.8) we get

$$\frac{f(x)}{u_0(x)} - \frac{f(c)}{u_0(c)} = \int_c^x \frac{L_1 f(s)}{u_0(s)} ds, \qquad (18.16)$$

i.e.

$$\frac{f(x)}{u_0(x)} - A = \int_c^x \frac{L_1 f(s)}{u_0(s)} ds, \quad \forall x \in [a,b]. \qquad (18.17)$$

Hence

$$\left|\frac{f(x)}{u_0(x)} - A\right| \leq \left|\int_c^x \frac{|L_1 f(s)|}{u_0(s)} ds\right| \leq \int_a^b \frac{|L_1 f(s)|}{u_0(s)} ds$$

$$\leq (b-a)^{1/p} \left(\int_a^b \left(\frac{|L_1 f(s)|}{u_0(s)}\right)^q ds\right)^{1/q}, \quad \forall x \in [a,b]. \qquad (18.18)$$

Therefore

$$\left|\frac{f(x)}{u_0(x)} - A\right|^\nu \leq (b-a)^{\frac{\nu}{p}} \left\|\frac{L_1 f}{u_0}\right\|_{L_q(a,b)}^\nu, \quad \forall x \in [a,b]. \qquad (18.19)$$

Consequently we obtain

$$\int_a^b \left|\frac{f(x)}{u_0(x)} - A\right|^\nu dx \leq (b-a)^{\left(\frac{\nu}{p}+1\right)} \left\|\frac{L_1 f}{u_0}\right\|_{L_q(a,b)}^\nu, \qquad (18.20)$$

proving the claim. $\square$

Corollary 18.6. (to Theorem 18.5) *Let* $\int_a^b \frac{f(t)}{u_0(t)} dt = 0$. *Then*

$$1) \quad \left\|\frac{f}{u_0}\right\|_{L_\nu(a,b)} \leq (b-a)^{\left(\frac{1}{p}+\frac{1}{\nu}\right)} \left\|\frac{L_1 f}{u_0}\right\|_{L_q(a,b)}, \qquad (18.21)$$

$$2) \quad \left\|\frac{f}{u_0}\right\|_{L_q(a,b)} \leq (b-a) \left\|\frac{L_1 f}{u_0}\right\|_{L_q(a,b)}, \qquad (18.22)$$

$$3) \quad \left\|\frac{f}{u_0}\right\|_{L_2(a,b)} \leq (b-a) \left\|\frac{L_1 f}{u_0}\right\|_{L_2(a,b)}, \qquad (18.23)$$

equivalently,

$$\int_a^b \left(\frac{f(x)}{u_0(x)}\right)^2 dx \leq (b-a)^2 \int_a^b \left(\frac{L_1 f(x)}{u_0(x)}\right)^2 dx. \qquad (18.23^*)$$

It follows the related result.

Proposition 18.7. *Let $\nu > 0$ and consider $f, u_0 \in C^1\left([a,b]\right)$, $u_0 > 0$ on $[a,b]$. Then*

1) $$\left\| \frac{f}{u_0} - \frac{1}{b-a} \int_a^b \frac{f(t)}{u_0(t)} dt \right\|_{L_\nu(a,b)} \leq (b-a)^{1/\nu} \left\| \frac{L_1 f}{u_0} \right\|_{L_1(a,b)}. \qquad (18.24)$$

When $\nu = 1$ we have

2) $$\left\| \frac{f}{u_0} - \frac{1}{b-a} \int_a^b \frac{f(t)}{u_0(t)} dt \right\|_{L_1(a,b)} \leq (b-a) \left\| \frac{L_1 f}{u_0} \right\|_{L_1(a,b)}. \qquad (18.25)$$

Proof. By (18.18) we have

$$\left| \frac{f(x)}{u_0(x)} - \frac{1}{b-a} \int_a^b \frac{f(t)}{u_0(t)} dt \right|^\nu$$

$$\leq \left\| \frac{L_1 f}{u_0} \right\|_{L_1(a,b)}^\nu, \quad \forall x \in [a,b]. \qquad (18.26)$$

So that

$$\int_a^b \left| \frac{f(x)}{u_0(x)} - \frac{1}{b-a} \int_a^b \frac{f(t)}{u_0(t)} dt \right|^\nu dx \leq (b-a) \left\| \frac{L_1 f}{u_0} \right\|_{L_1(a,b)}^\nu, \qquad (18.27)$$

proving the claim. $\qquad \square$

We continue with generalized Dirichlet–Poincaré like inequalities.

Theorem 18.8. *Let $f, u_0, u_1, \ldots, u_n \in C^{n+1}\left([a,b]\right)$, $n \in \mathbb{Z}_+$; $W_0, W_1, \ldots, W_n > 0$ on $[a,b]$, and for fixed $x_0 \in [a,b]$ suppose that $L_i f(x_0) = 0$, $i = 0, 1, \ldots, n$. Let p, $q > 1 : \frac{1}{p} + \frac{1}{q} = 1$, $\nu > 0$. Then*

1) $$\|f\|_{L_\nu(x_0,b)} \leq \left(\int_{x_0}^b \left(\int_{x_0}^x (g_n(x,t))^p dt \right)^{\nu/p} dx \right)^{1/\nu} \|L_{n+1} f\|_{L_q(x_0,b)}. \qquad (18.28)$$

2) $$\|f\|_{L_q(x_0,b)} \leq \left(\int_{x_0}^b \left(\int_{x_0}^x (g_n(x,t))^p dt \right)^{q/p} dx \right)^{1/q} \|L_{n+1} f\|_{L_q(x_0,b)}. \qquad (18.29)$$

Also,

3) $$\|f\|_{L_2(x_0,b)} \leq \left(\int_{x_0}^b \left(\int_{x_0}^x (g_n(x,t))^2 dt \right) dx \right)^{1/2} \|L_{n+1} f\|_{L_2(x_0,b)}. \qquad (18.30)$$

Proof. By (18.7) we have

$$|f(x)| = \left| \int_{x_0}^x g_n(x,t) L_{n+1} f(t) dt \right|$$

$$\leq \int_{x_0}^{x} g_n\left(x,t\right) \left|L_{n+1} f\left(t\right)\right| dt$$

$$\leq \left(\int_{x_0}^{x} \left(g_n\left(x,t\right)\right)^p dt \right)^{1/p} \left(\int_{x_0}^{x} \left|L_{n+1} f\left(t\right)\right|^q dt \right)^{1/q}$$

$$\leq \left(\int_{x_0}^{x} \left(g_n\left(x,t\right)\right)^p dt \right)^{1/p} \left\|L_{n+1} f\right\|_{L_q\left(x_0,b\right)}, \quad \forall x \in \left[x_0,b\right]. \tag{18.31}$$

That is

$$\left|f\left(x\right)\right|^{\nu} \leq \left(\int_{x_0}^{x} \left(g_n\left(x,t\right)\right)^p dt \right)^{\nu/p} \left\|L_{n+1} f\right\|_{L_q\left(x_0,b\right)}^{\nu}, \tag{18.32}$$

and

$$\int_{x_0}^{b} \left|f\left(x\right)\right|^{\nu} dx \leq \left(\int_{x_0}^{b} \left(\int_{x_0}^{x} \left(g_n\left(x,t\right)\right)^p dt \right)^{\nu/p} dx \right) \left\|L_{n+1} f\right\|_{L_q\left(x_0,b\right)}^{\nu}, \tag{18.33}$$

proving the claim. $\qquad\square$

We give

Theorem 18.9. *Same assumptions as in Theorem 18.8.*
Then

$$1)\ \left\|f\right\|_{L_{\nu}\left(a,x_0\right)} \leq \left(\int_{a}^{x_0} \left(\int_{x}^{x_0} \left|g_n\left(x,t\right)\right|^p dt \right)^{\nu/p} dx \right)^{1/\nu} \left\|L_{n+1} f\right\|_{L_q\left(a,x_0\right)}. \tag{18.34}$$

$$2)\ \left\|f\right\|_{L_q\left(a,x_0\right)} \leq \left(\int_{a}^{x_0} \left(\int_{x}^{x_0} \left|g_n\left(x,t\right)\right|^p dt \right)^{q/p} dx \right)^{1/q} \left\|L_{n+1} f\right\|_{L_q\left(a,x_0\right)}. \tag{18.35}$$

Also,

$$3)\ \left\|f\right\|_{L_2\left(a,x_0\right)} \leq \left(\int_{a}^{x_0} \left(\int_{x}^{x_0} \left(g_n\left(x,t\right)\right)^2 dt \right) dx \right)^{1/2} \left\|L_{n+1} f\right\|_{L_2\left(a,x_0\right)}. \tag{18.36}$$

Proof. By (18.7) we have

$$\left|f\left(x\right)\right| = \left| \int_{x_0}^{x} g_n\left(x,t\right) L_{n+1} f\left(t\right) dt \right|$$

$$= \left| \int_{x}^{x_0} g_n\left(x,t\right) L_{n+1} f\left(t\right) dt \right|$$

$$\leq \int_{x}^{x_0} \left|g_n\left(x,t\right)\right| \left|L_{n+1} f\left(t\right)\right| dt$$

$$\leq \left(\int_{x}^{x_0} \left|g_n\left(x,t\right)\right|^p dt \right)^{1/p} \left(\int_{x}^{x_0} \left|L_{n+1} f\left(t\right)\right|^q dt \right)^{1/q}$$

$$\leq \left(\int_{x}^{x_0} |g_n(x,t)|^p \, dt \right)^{1/p} \|L_{n+1}f\|_{L_q(a,x_0)} . \tag{18.37}$$

That is

$$|f(x)|^{\nu} \leq \left(\int_{x}^{x_0} |g_n(x,t)|^p \, dt \right)^{\nu/p} \|L_{n+1}f\|_{L_q(a,x_0)}^{\nu} , \tag{18.38}$$

and

$$\int_{a}^{x_0} |f(x)|^{\nu} \, dx \leq \left(\int_{a}^{x_0} \left(\int_{x}^{x_0} |g_n(x,t)|^p \, dt \right)^{\nu/p} dx \right) \|L_{n+1}f\|_{L_q(a,x_0)}^{\nu} , \tag{18.39}$$

proving the claim. $\qquad\square$

We continue with

Proposition 18.10. Let $f, u_0, u_1, \ldots, u_n \in C^{n+1}([a,b])$, $n \in \mathbb{Z}_+$; $W_0, W_1, \ldots, W_n > 0$ on $[a,b]$, and for fixed $x_0 \in [a,b]$ suppose that $L_i f(x_0) = 0$, $i = 0, 1, \ldots, n$; $\nu > 0$.
Then

1) $\|f\|_{L_\nu(x_0,b)} \leq \left(\int_{x_0}^{b} \left(\int_{x_0}^{x} g_n(x,t) \, dt \right)^{\nu} dx \right)^{1/\nu} \|L_{n+1}f\|_{L_\infty(x_0,b)} . \tag{18.40}$

2) $\|f\|_{L_1(x_0,b)} \leq \left(\int_{x_0}^{b} \left(\int_{x_0}^{x} g_n(x,t) \, dt \right) dx \right) \|L_{n+1}f\|_{L_\infty(x_0,b)} . \tag{18.41}$

Proof. By (18.7) we have

$$|f(x)| \leq \int_{x_0}^{x} g_n(x,t) \, |L_{n+1}f(t)| \, dt$$

$$\leq \left(\int_{x_0}^{x} g_n(x,t) \, dt \right) \|L_{n+1}f\|_{\infty,[x_0,b]} . \tag{18.42}$$

That is

$$|f(x)|^{\nu} \leq \left(\int_{x_0}^{x} g_n(x,t) \, dt \right)^{\nu} \|L_{n+1}f\|_{\infty,[x_0,b]}^{\nu} , \tag{18.43}$$

and

$$\int_{x_0}^{b} |f(x)|^{\nu} \, dx \leq \left(\int_{x_0}^{b} \left(\int_{x_0}^{x} g_n(x,t) \, dt \right)^{\nu} dx \right) \|L_{n+1}f\|_{\infty,[x_0,b]}^{\nu} , \tag{18.44}$$

proving the claim. $\qquad\square$

We give

Proposition 18.11. *All as in Proposition 18.10.*
Then

1) $\|f\|_{L_\nu(a,x_0)} \leq \left(\int_a^{x_0} \left(\int_x^{x_0} |g_n(x,t)| \, dt \right)^\nu dx \right)^{1/\nu} \|L_{n+1}f\|_{L_\infty(a,x_0)} .$　　　(18.45)

2) $\|f\|_{L_1(a,x_0)} \leq \left(\int_a^{x_0} \left(\int_x^{x_0} |g_n(x,t)| \, dt \right) dx \right) \|L_{n+1}f\|_{L_\infty(a,x_0)} .$　　　(18.46)

Proof.　By (18.7) we have

$$|f(x)| \leq \int_x^{x_0} |g_n(x,t)| \, |L_{n+1}f(t)| \, dt$$

$$\leq \left(\int_x^{x_0} |g_n(x,t)| \, dt \right) \|L_{n+1}f\|_{\infty,[a,x_0]} .$$　　　(18.47)

That is

$$|f(x)|^\nu \leq \left(\int_x^{x_0} |g_n(x,t)| \, dt \right)^\nu \|L_{n+1}f\|_{\infty,[a,x_0]}^\nu ,$$　　　(18.48)

and

$$\int_a^{x_0} |f(x)|^\nu \, dx \leq \left(\int_a^{x_0} \left(\int_x^{x_0} |g_n(x,t)| \, dt \right)^\nu dx \right) \|L_{n+1}f\|_{\infty,[a,x_0]}^\nu ,$$　　　(18.49)

proving the claim.　　　　　　　　　　　　　　　　　　　　　　　　　$\square$

We continue with reverse Dirichlet–Poincaré like inequalities.

Theorem 18.12. *Let* $f, u_0, u_1, \ldots, u_n \in C^{n+1}([a,b])$, $n \in \mathbb{Z}_+$; $W_0, W_1, \ldots, W_n >$
0 *on* $[a,b]$, *and for a fixed* $a \leq x_0 < b$ *assume that* $L_i f(x_0) = 0$, $i = 0, 1, \ldots, n$. *Let*
$0 < p < 1$, $q < 0 : \frac{1}{p} + \frac{1}{q} = 1$, $\nu > 0$. *Further suppose that* $L_{n+1}f$ *is of fixed sign*
and nowhere zero on $[x_0, b]$.
Then

1) $\|f\|_{L_\nu(x_0,b)} \geq \left(\int_{x_0}^b \left(\int_{x_0}^x (g_n(x,t))^p \, dt \right)^{\nu/p} dx \right)^{1/\nu} \|L_{n+1}f\|_{L_q(x_0,b)} .$　　　(18.50)

2) $\|f\|_{L_p(x_0,b)} \geq \left(\int_{x_0}^b \left(\int_{x_0}^x (g_n(x,t))^p \, dt \right) dx \right)^{1/p} \|L_{n+1}f\|_{L_q(x_0,b)} .$　　　(18.51)

3) $\|f\|_{L_1(x_0,b)} \geq \left(\int_{x_0}^b \left(\int_{x_0}^x (g_n(x,t))^p \, dt \right)^{1/p} dx \right) \|L_{n+1}f\|_{L_q(x_0,b)} .$　　　(18.52)

Proof.　Here we have by (18.7) and assumption that

$$|f(x)| = \int_{x_0}^x g_n(x,t) \, |L_{n+1}f(t)| \, dt,$$　　　(18.53)

$\forall x \in [x_0, b]$.

Hence by reverse Hölder's inequality we derive

$$|f(x)| \geq \left(\int_{x_0}^{x} (g_n(x,t))^p \, dt \right)^{1/p} \left(\int_{x_0}^{x} |L_{n+1}f(t)|^q \, dt \right)^{1/q}$$

$$\geq \left(\int_{x_0}^{x} (g_n(x,t))^p \, dt \right)^{1/p} \left(\int_{x_0}^{b} |L_{n+1}f(t)|^q \, dt \right)^{1/q}, \tag{18.54}$$

true for all $x \in (x_0, b]$.

That is

$$|f(x)|^\nu \geq \left(\int_{x_0}^{x} (g_n(x,t))^p \, dt \right)^{\nu/p} \|L_{n+1}f\|_{L_q(x_0,b)}^\nu, \tag{18.55}$$

true for all $x \in [x_0, b]$,

and

$$\int_{x_0}^{b} |f(x)|^\nu \, dx \geq \left(\int_{x_0}^{b} \left(\int_{x_0}^{x} (g_n(x,t))^p \, dt \right)^{\nu/p} dx \right) \|L_{n+1}f\|_{L_q(x_0,b)}^\nu, \tag{18.56}$$

proving the claim. $\square$

We give

Theorem 18.13. *Let $f, u_0, u_1, \ldots, u_n \in C^{n+1}([a,b])$, n is odd; $W_0, W_1, \ldots, W_n > 0$ on $[a,b]$, and for a fixed $a < x_0 \leq b$ assume that $L_i f(x_0) = 0$, $i = 0, 1, \ldots, n$. Let $0 < p < 1$, $q < 0 : \frac{1}{p} + \frac{1}{q} = 1$, $\nu > 0$. Further suppose that $L_{n+1}f$ is of fixed sign and nowhere zero on $[a, x_0]$.*

Then

$$1) \ \|f\|_{L_\nu(a,x_0)} \geq \left(\int_{a}^{x_0} \left(\int_{x}^{x_0} (-g_n(x,t))^p \, dt \right)^{\nu/p} dx \right)^{1/\nu} \|L_{n+1}f\|_{L_q(a,x_0)}.$$

$$\tag{18.57}$$

$$2) \ \|f\|_{L_p(a,x_0)} \geq \left(\int_{a}^{x_0} \left(\int_{x}^{x_0} (-g_n(x,t))^p \, dt \right) dx \right)^{1/p} \|L_{n+1}f\|_{L_q(a,x_0)}. \tag{18.58}$$

$$3) \ \|f\|_{L_1(a,x_0)} \geq \left(\int_{a}^{x_0} \left(\int_{x}^{x_0} (-g_n(x,t))^p \, dt \right)^{1/p} dx \right) \|L_{n+1}f\|_{L_q(a,x_0)}. \tag{18.59}$$

Proof. Here by (18.7) and assumption we have

$$|f(x)| = \left| \int_{x_0}^{x} g_n(x,t) \, L_{n+1}f(t) \, dt \right|$$

$$- \left| \int_{x}^{x_0} g_n(x,t) \, L_{n+1}f(t) \, dt \right|$$

$$\overset{(18.12)}{=} \left| \int_x^{x_0} \left(-g_n\left(x,t\right) \right) L_{n+1} f\left(t\right) dt \right|$$

$$= \int_x^{x_0} \left(-g_n\left(x,t\right) \right) \left| L_{n+1} f\left(t\right) \right| dt. \tag{18.60}$$

So by reverse Hölder's inequality we find

$$\left| f\left(x\right) \right| \geq \left(\int_x^{x_0} \left(-g_n\left(x,t\right) \right)^p dt \right)^{1/p} \left(\int_x^{x_0} \left| L_{n+1} f\left(t\right) \right|^q dt \right)^{1/q}$$

$$\geq \left(\int_x^{x_0} \left(-g_n\left(x,t\right) \right)^p dt \right)^{1/p} \left\| L_{n+1} f \right\|_{L_q(a,x_0)}, \tag{18.61}$$

true for all $x \in [a, x_0)$,

and

$$\left| f\left(x\right) \right|^\nu \geq \left(\int_x^{x_0} \left(-g_n\left(x,t\right) \right)^p dt \right)^{\nu/p} \left\| L_{n+1} f \right\|_{L_q(a,x_0)}^\nu, \tag{18.62}$$

true for all $x \in [a, x_0]$.

Thus

$$\int_a^{x_0} \left| f\left(x\right) \right|^\nu dx \geq \left(\int_a^{x_0} \left(\int_x^{x_0} \left(-g_n\left(x,t\right) \right)^p dt \right)^{\nu/p} dx \right) \left\| L_{n+1} f \right\|_{L_q(a,x_0)}^\nu, \tag{18.63}$$

proving the claim. $\qquad\square$

We add

Theorem 18.14. *Now n is even, the rest as in Theorem 18.13.*
Then

1) $$\left\| f \right\|_{L_\nu(a,x_0)} \geq \left(\int_a^{x_0} \left(\int_x^{x_0} \left(g_n\left(x,t\right) \right)^p dt \right)^{\nu/p} dx \right)^{1/\nu} \left\| L_{n+1} f \right\|_{L_q(a,x_0)}. \tag{18.64}$$

2) $$\left\| f \right\|_{L_p(a,x_0)} \geq \left(\int_a^{x_0} \left(\int_x^{x_0} \left(g_n\left(x,t\right) \right)^p dt \right) dx \right)^{1/p} \left\| L_{n+1} f \right\|_{L_q(a,x_0)}. \tag{18.65}$$

3) $$\left\| f \right\|_{L_1(a,x_0)} \geq \left(\int_a^{x_0} \left(\int_x^{x_0} \left(g_n\left(x,t\right) \right)^p dt \right)^{1/p} dx \right) \left\| L_{n+1} f \right\|_{L_q(a,x_0)}. \tag{18.66}$$

Proof. Similar to Theorem 18.13. $\qquad\square$

We continue with Sobolev like inequalities.

Theorem 18.15. *Same assumptions as in Theorem 18.8. Call*

$$M_{\nu,1} := \max_{0 \leq j \leq n} \left\{ \left(\int_{x_0}^b \left(\int_{x_0}^x \left(g_j\left(x,t\right) \right)^p dt \right)^{\nu/p} dx \right)^{1/\nu} \right\}, \tag{18.67}$$

$\nu > 0$.

Then

$$1) \ \|f\|_{L_\nu(x_0,b)} \le \left(\frac{M_{\nu,1}}{n+1}\right) \left(\sum_{j=0}^{n} \|L_{j+1}f\|_{L_q(x_0,b)}\right), \tag{18.68}$$

$$2) \ \|f\|_{L_q(x_0,b)} \le \left(\frac{M_{q,1}}{n+1}\right) \left(\sum_{j=0}^{n} \|L_{j+1}f\|_{L_q(x_0,b)}\right), \tag{18.69}$$

and when $\nu = p = q = 2$ we obtain

$$3) \ \|f\|_{L_2(x_0,b)} \le \left(\frac{M_{2,1}}{n+1}\right) \left(\sum_{j=0}^{n} \|L_{j+1}f\|_{L_2(x_0,b)}\right), \tag{18.70}$$

where

$$M_{2,1} := \max_{0 \le j \le n} \left\{ \left(\int_{x_0}^{b} \left(\int_{x_0}^{x} (g_j(x,t))^2 \, dt \right) dx \right)^{1/2} \right\}. \tag{18.71}$$

Proof. The assumptions of Theorem 18.8 are fulfilled for all $j = 0, 1, \ldots, n$. Thus by (18.28) we derive

$$\|f\|_{L_\nu(x_0,b)} \le \left(\int_{x_0}^{b} \left(\int_{x_0}^{x} (g_j(x,t))^p \, dt \right)^{\nu/p} dx \right)^{1/\nu} \|L_{j+1}f\|_{L_q(x_0,b)}$$

$$\le M_{\nu,1} \|L_{j+1}f\|_{L_q(x_0,b)}, \tag{18.72}$$

for all $j = 0, 1, \ldots, n$.

From (18.72) by addition we get

$$(n+1) \|f\|_{L_\nu(x_0,b)} \le M_{\nu,1} \left(\sum_{j=0}^{n} \|L_{j+1}f\|_{L_q(x_0,b)} \right), \tag{18.73}$$

proving the claim. $\qquad\square$

We continue with

Theorem 18.16. *Same assumptions as in Theorem 18.8. Call*

$$M_{\nu,2} := \max_{0 \le j \le n} \left\{ \left(\int_{a}^{x_0} \left(\int_{x}^{x_0} |g_j(x,t)|^p \, dt \right)^{\nu/p} dx \right)^{1/\nu} \right\}, \tag{18.74}$$

$\nu > 0$.

Then

$$1) \ \|f\|_{L_\nu(a,x_0)} \le \left(\frac{M_{\nu,2}}{n+1}\right) \left(\sum_{j=0}^{n} \|L_{j+1}f\|_{L_q(a,x_0)}\right), \tag{18.75}$$

$$2) \quad \|f\|_{L_q(a,x_0)} \leq \left(\frac{M_{q,2}}{n+1}\right)\left(\sum_{j=0}^{n}\|L_{j+1}f\|_{L_q(a,x_0)}\right), \tag{18.76}$$

and when $\nu = p = q = 2$ we obtain

$$3) \quad \|f\|_{L_2(a,x_0)} \leq \left(\frac{M_{2,2}}{n+1}\right)\left(\sum_{j=0}^{n}\|L_{j+1}f\|_{L_2(a,x_0)}\right), \tag{18.77}$$

where

$$M_{2,2} := \max_{0 \leq j \leq n}\left\{\left(\int_a^{x_0}\left(\int_x^{x_0}(g_j(x,t))^2\,dt\right)dx\right)^{1/2}\right\}. \tag{18.78}$$

Proof.   Similar to Theorem 18.15, based on Theorem 18.9. $\square$

We continue with L_∞ results.

Proposition 18.17. *All as in Proposition 18.10. Call*

$$K_{\nu,1} := \max_{0 \leq j \leq n}\left\{\left(\int_{x_0}^{b}\left(\int_{x_0}^{x}g_j(x,t)\,dt\right)^{\nu}dx\right)^{1/\nu}\right\},$$

$\nu > 0.$
   Then

$$1) \quad \|f\|_{L_\nu(x_0,b)} \leq \left(\frac{K_{\nu,1}}{n+1}\right)\left(\sum_{j=0}^{n}\|L_{j+1}f\|_{L_\infty(x_0,b)}\right), \tag{18.79}$$

and

$$2) \quad \|f\|_{L_1(x_0,b)} \leq \left(\frac{K_{1,1}}{n+1}\right)\left(\sum_{j=0}^{n}\|L_{j+1}f\|_{L_\infty(x_0,b)}\right). \tag{18.80}$$

Proof.   Similar to Theorem 18.15, based on Proposition 18.10. $\square$

Proposition 18.18. *All as in Proposition 18.10. Call*

$$K_{\nu,2} := \max_{0 \leq j \leq n}\left\{\left(\int_a^{x_0}\left(\int_x^{x_0}|g_n(x,t)|\,dt\right)^{\nu}dx\right)^{1/\nu}\right\}, \tag{18.81}$$

$\nu > 0.$
   Then

$$1) \quad \|f\|_{L_\nu(a,x_0)} \leq \left(\frac{K_{\nu,2}}{n+1}\right)\left(\sum_{j=0}^{n}\|L_{j+1}f\|_{L_\infty(a,x_0)}\right), \tag{18.82}$$

and

$$2) \quad \|f\|_{L_1(a,x_0)} \leq \left(\frac{K_{1,2}}{n+1}\right)\left(\sum_{j=0}^{n}\|L_{j+1}f\|_{L_\infty(a,x_0)}\right). \tag{18.83}$$

Proof.   Based on Proposition 18.11. $\square$

We continue with reverse Sobolev like inequalities.

Theorem 18.19. *Assume here that $L_{j+1}f$ is of fixed sign and nowhere zero on $[x_0, b]$, for $j = 0, 1, \ldots, n$. The rest are supposed as in Theorem 18.12. Call*

$$S_{\nu,1} := \min_{0 \le j \le n} \left\{ \left(\int_{x_0}^{b} \left(\int_{x_0}^{x} (g_j(x,t))^p \, dt \right)^{\nu/p} dx \right)^{1/\nu} \right\}, \tag{18.84}$$

$\nu > 0$.

   Then

$$1) \quad \|f\|_{L_\nu(x_0,b)} \ge \left(\frac{S_{\nu,1}}{n+1} \right) \left(\sum_{j=0}^{n} \|L_{j+1}f\|_{L_q(x_0,b)} \right), \tag{18.85}$$

$$2) \quad \|f\|_{L_p(x_0,b)} \ge \left(\frac{S_{p,1}}{n+1} \right) \left(\sum_{j=0}^{n} \|L_{j+1}f\|_{L_q(x_0,b)} \right), \tag{18.86}$$

and

$$3) \quad \|f\|_{L_1(x_0,b)} \ge \left(\frac{S_{1,1}}{n+1} \right) \left(\sum_{j=0}^{n} \|L_{j+1}f\|_{L_q(x_0,b)} \right). \tag{18.87}$$

Proof. The assumptions of Theorem 18.12 are fulfilled for all $j = 0, 1, \ldots, n$. Thus by (18.50) we obtain

$$\|f\|_{L_\nu(x_0,b)} \ge \left(\int_{x_0}^{b} \left(\int_{x_0}^{x} (g_j(x,t))^p \, dt \right)^{\nu/p} dx \right)^{1/\nu} \|L_{j+1}f\|_{L_q(x_0,b)}$$

$$\ge S_{\nu,1} \|L_{j+1}f\|_{L_q(x_0,b)}, \tag{18.88}$$

for all $j = 0, 1, \ldots, n$.

   From (18.88) by addition we obtain

$$(n+1) \|f\|_{L_\nu(x_0,b)} \ge S_{\nu,1} \left(\sum_{j=0}^{n} \|L_{j+1}f\|_{L_q(x_0,b)} \right), \tag{18.89}$$

proving the claim. $\square$

   We continue with

Theorem 18.20. *Assume all as in Theorem 18.13. Here $n = 2k + 1$, $k \in \mathbb{Z}_+$; $\nu > 0$. Further suppose that $L_{j+1}f$ is of fixed sign and nowhere zero on $[a, x_0]$ for all odds $j \in [1, n]$. Call*

$$S_{\nu,2} := \min_{odd \; j \in [1,n]} \left\{ \left(\int_{a}^{x_0} \left(\int_{x}^{x_0} (-g_j(x,t))^p \, dt \right)^{\nu/p} dx \right)^{1/\nu} \right\}. \tag{18.90}$$

Then

$$1) \quad \|f\|_{L_\nu(a,x_0)} \geq \left(\frac{S_{\nu,2}}{k+1}\right) \left(\sum_{\substack{j\in[1,n]\\ odd}} \|L_{j+1}f\|_{L_q(a,x_0)}\right), \tag{18.91}$$

$$2) \quad \|f\|_{L_p(a,x_0)} \geq \left(\frac{S_{p,2}}{k+1}\right) \left(\sum_{\substack{j\in[1,n]\\ odd}} \|L_{j+1}f\|_{L_q(a,x_0)}\right), \tag{18.92}$$

and

$$3) \quad \|f\|_{L_1(a,x_0)} \geq \left(\frac{S_{1,2}}{k+1}\right) \left(\sum_{\substack{j\in[1,n]\\ odd}} \|L_{j+1}f\|_{L_q(a,x_0)}\right). \tag{18.93}$$

Proof. As in Theorem 18.19, based on Theorem 18.13. Since $n = 2k+1$, $k \in \mathbb{Z}_+$, there are $(k+1)$ odd numbers in $[1,n]$, so we apply (18.57) $(k+1)$ times. $\qquad\square$

We finish with

Theorem 18.21. *Assume all as in Theorem 18.14. Here $n = 2k$, $k \in \mathbb{Z}_+$; $\nu > 0$. Further suppose that $L_{j+1}f$ is of fixed sign and nowhere zero on $[a, x_0]$ for all evens $j \in [0, n]$. Call*

$$S_{\nu,3} := \min_{even\ j\in[0,n]} \left\{ \left(\int_a^{x_0} \left(\int_x^{x_0} (g_j(x,t))^p \, dt\right)^{\nu/p} dx\right)^{1/\nu} \right\}. \tag{18.94}$$

Then

$$1) \quad \|f\|_{L_\nu(a,x_0)} \geq \left(\frac{S_{\nu,3}}{k+1}\right) \left(\sum_{\substack{j\in[0,n]\\ even}} \|L_{j+1}f\|_{L_q(a,x_0)}\right), \tag{18.95}$$

$$2) \quad \|f\|_{L_p(a,x_0)} \geq \left(\frac{S_{p,3}}{k+1}\right) \left(\sum_{\substack{j\in[0,n]\\ even}} \|L_{j+1}f\|_{L_q(a,x_0)}\right), \tag{18.96}$$

and

$$3) \quad \|f\|_{L_1(a,x_0)} \geq \left(\frac{S_{1,3}}{k+1}\right) \left(\sum_{\substack{j\in[0,n]\\ even}} \|L_{j+1}f\|_{L_q(a,x_0)}\right). \tag{18.97}$$

Proof. As in Theorem 18.19, based on Theorem 18.14. Since $n = 2k$, $k \in \mathbb{Z}_+$, there are $(k+1)$ even numbers in $[0,n]$, so we apply (18.64) $(k+1)$ times. $\qquad\square$

Chapter 19

Poincaré and Sobolev Like Inequalities for Vector Valued Functions

Various L_p form Poincaré and Sobolev like inequalities are given for functions valued in a Banach vector space with applications on $\mathbb{C}$. This chapter relies on [44].

19.1 Introduction

This chapter is motivated by the famous Poincaré inequality, see [2]: Given a bounded domain $\Omega \subset \mathbb{R}^n$, $n \in \mathbb{N}$, it holds

$$\|u\|_{L^p(\Omega)} \leq C \|\nabla u\|_{L^p(\Omega)}$$

for real valued functions u with vanishing mean value over Ω, $1 \leq p \leq \infty$, under very general assumptions on Ω, where $\|\nabla u\|_{L^p(\Omega)}$ is defined as the L^p-norm of the euclidian norm of ∇u.

Especially in [2] is proved for a convex domain $\Omega \subset \mathbb{R}^n$ with diameter d that

$$\|u\|_{L^1(\Omega)} \leq \frac{d}{2} \|\nabla u\|_{L^1(\Omega)}$$

for any u with zero mean value on Ω, also the constant $1/2$ is optimal.

We are also motivated by [92], where the authors prove the following: Let $1 < p < \infty$, $-\infty < a < b < \infty$. The best constant C (independent of a, b) for which the $1-$dimensional Poincaré inequality

$$\left\| f - \frac{\int_a^b f\left(t\right) dt}{b - a} \right\|_{L^1([a,b])} \leq C\left(b - a\right)^{2 - \frac{1}{p}} \|f'\|_{L^p([a,b])}$$

holds for all real valued Lipschitz continuous functions f, is $C = \frac{1}{2}\left(1 + p'\right)^{-1/p'}$, where $p' > 1 : \frac{1}{p} + \frac{1}{p'} = 1$.

This chapter is also motivated by the famous Sobolev inequality of the following form, see [139], p. 263: (the Gagliardo–Nirenberg–Sobolev inequality). *Assume* $1 \leq p \leq n$. *Then exists a constant* C, *depending only on* p *and* n, *such that*

$$\|u\|_{L^{p^*}(\mathbb{R}^n)} \leq C \|Du\|_{L^p(\mathbb{R}^n)},$$

229

for all real valued $u \in C_c^1 (\mathbb{R}^n)$.

Here $p^* := \frac{np}{n-p}$, $p^* > p$, and Du is the gradient of u.

Also we are motivated by the following result, p. 265, [139]: estimates for Sobolev space $W^{1,p}$, $1 \leq p < n$. *Let U be a bounded, open subset of $\mathbb{R}^n$, and suppose the boundary ∂U is C^1. Assume $1 \leq p < n$, and real valued $u \in W^{1,p}(U)$. Then $u \in L^{p^*}(U)$, with the estimate*

$$\|u\|_{L^{p^*}(U)} \leq C \|u\|_{W^{1,p}(U)},$$

the constant C depending only on p, n and U.

So here we derive Poincaré and Sobolev like inequalities of L_p form for Banach vector space valued functions. Most of the Poincaré and Sobolev like inequalities here are of Dirichlet type. We assume in this case initial conditions prescribed equal to zero.

But we give also a Neumann–Poincaré like inequality involving the average of the engaged function, see Theorem 19.3.

At the end we apply the results to $\mathbb{C}$-valued functions.

We need

19.2   Background

(see [230], pp. 83–94)

Let $f(t)$ be a function defined on $[a, b] \subseteq \mathbb{R}$ taking values in a real or complex normed linear space $(X, \|\cdot\|)$. Then $f(t)$ is said to be differentiable at a point $t_0 \in [a, b]$ if the limit

$$f'(t_0) := \lim_{h \to 0} \frac{f(t_0 + h) - f(t_0)}{h} \tag{19.1}$$

exists in X, the convergence is in $\|\cdot\|$. This is called the derivative of $f(t)$ at $t = t_0$.

We call $f(t)$ differentiable on $[a, b]$, iff there exists $f'(t) \in X$ for all $t \in [a, b]$.

Similarly and inductively are defined higher order derivatives of f, denoted f'', $f^{(3)}, \ldots, f^{(k)}$, $k \in \mathbb{N}$, just as for numerical functions.

For all the properties of derivatives see [230], pp. 83–86.

Let now $(X, \|\cdot\|)$ be a Banach space, and $f : [a, b] \to X$.

We define the vector valued Riemann integral $\int_a^b f(t)\, dt \in X$ as the limit of the vector valued Riemann sums in X, convergence is in $\|\cdot\|$. The definition is as for the numerical valued functions.

If $\int_a^b f(t)\, dt \in X$ we call f integrable on $[a, b]$. If $f \in C([a, b], X)$, then f is integrable, [230], p. 87.

For all the properties of vector valued Riemann integrals see [230], pp. 86–91.

We define the space $C^n([a, b], X)$, $n \in \mathbb{N}$, of n-times continuously differentiable functions from $[a, b]$ into X; here continuity is with respect to $\|\cdot\|$ and defined in the usual way as for numerical functions.

Let $(X, \|\cdot\|)$ be a Banach space and $f \in C^n\left([a,b], X\right)$, then we have the vector valued Taylor's formula, see [230], pp. 93–94, and also [227], (IV, 9; 47).

It holds

$$E_n\left(x, y\right) := f\left(y\right) - f\left(x\right) - f'\left(x\right)\left(y - x\right) -$$

$$\frac{1}{2} f''\left(x\right)\left(y - x\right)^2 - \ldots - \frac{1}{(n-1)!} f^{(n-1)}\left(x\right)\left(y - x\right)^{n-1}$$

$$= \frac{1}{(n-1)!} \int_x^y \left(y - t\right)^{n-1} f^{(n)}\left(t\right) dt, \quad \forall x, y \in [a, b]. \tag{19.2}$$

In particular (19.2) is true when $X = \mathbb{R}^m$, $\mathbb{C}^m$, $m \in \mathbb{N}$, etc.

In case of some $x_0 \in [a, b]$ such that $f^{(k)}\left(x_0\right) = 0$, $k = 0, 1, \ldots, n - 1$, then

$$f\left(y\right) = \frac{1}{(n-1)!} \int_{x_0}^y \left(y - t\right)^{n-1} f^{(n)}\left(t\right) dt, \quad \forall y \in [a, b], \tag{19.3}$$

see also [52].

In that case $E_n\left(x_0, y\right) = f\left(y\right).$

19.3   Results

Here we consider always X to be a Banach space, $n \in \mathbb{N}$, and $f \in C^n\left([a,b], X\right)$, $[a, b] \subseteq \mathbb{R}$. We fix $x_0 \in [a, b]$; $p, q > 1 : \frac{1}{p} + \frac{1}{q} = 1, \nu > 0$.

We present the first results which are of Dirichlet–Poincaré like inequalities.

Theorem 19.1. *Let* $y \in [x_0, b]$, $x_0 < b$. *Assume* $f^{(k)}\left(x_0\right) = 0$, $k = 0, 1, \ldots, n - 1$.
Then

1) $\|\|f\|\|_{L_\nu(x_0,b)} \leq \dfrac{(b - x_0)^{\left(n - 1 + \frac{1}{p} + \frac{1}{\nu}\right)} \left\|\left\|f^{(n)}\right\|\right\|_{L_q(x_0,b)}}{((n-1)!)\left(p\left(n-1\right) + 1\right)^{1/p}\left(\nu\left(n - 1 + \frac{1}{p}\right) + 1\right)^{1/\nu}}. \tag{19.4}$

When $\nu = q$ *we obtain*

2) $\|\|f\|\|_{L_q(x_0,b)} \leq \dfrac{(b - x_0)^n \left\|\left\|f^{(n)}\right\|\right\|_{L_q(x_0,b)}}{((n-1)!)\left(p\left(n-1\right) + 1\right)^{1/p}\left(qn\right)^{1/q}}. \tag{19.5}$

And for $\nu = q = p = 2$ *we have*

3) $\|\|f\|\|_{L_2(x_0,b)} \leq \dfrac{(b - x_0)^n \left\|\left\|f^{(n)}\right\|\right\|_{L_2(x_0,b)}}{((n-1)!)\sqrt{2n - 1}\sqrt{2n}}. \tag{19.6}$

Proof.   By (19.3) we have

$$\|f\left(y\right)\| = \frac{1}{(n-1)!} \left\|\int_{x_0}^y \left(y - t\right)^{n-1} f^{(n)}\left(t\right) dt\right\|$$

$$\leq \frac{1}{(n-1)!} \int_{x_0}^y \left(y - t\right)^{n-1} \left\|f^{(n)}\left(t\right)\right\| dt$$

$$\leq \frac{1}{(n-1)!} \left(\int_{x_0}^{y} (y-t)^{p(n-1)} \, dt \right)^{1/p} \left(\int_{x_0}^{y} \left\| f^{(n)}(t) \right\|^{q} \, dt \right)^{1/q}$$

$$\leq \frac{1}{(n-1)!} \frac{(y-x_0)^{\left(n-1+\frac{1}{p}\right)}}{(p(n-1)+1)^{1/p}} \left\| \left\| \left\| f^{(n)} \right\| \right\| \right\|_{L_q(x_0,b)}. \tag{19.7}$$

Thus

$$\| f(y) \|^{\nu} \leq \frac{1}{((n-1)!)^{\nu}} \frac{(y-x_0)^{\nu\left(n-1+\frac{1}{p}\right)}}{(p(n-1)+1)^{\nu/p}} \left\| \left\| \left\| f^{(n)} \right\| \right\| \right\|_{L_q(x_0,b)}^{\nu}, \tag{19.8}$$

and

$$\int_{x_0}^{b} \| f(y) \|^{\nu} \, dy \leq \frac{1}{((n-1)!)^{\nu}} \frac{(b-x_0)^{\nu\left(n-1+\frac{1}{p}\right)+1} \left\| \left\| \left\| f^{(n)} \right\| \right\| \right\|_{L_q(x_0,b)}^{\nu}}{\left(\nu\left(n-1+\frac{1}{p}\right)+1 \right)(p(n-1)+1)^{\nu/p}}, \tag{19.9}$$

proving the claim. $\qquad \square$

For the next result we need

Theorem 19.2. (see [230], p. 90) *Let $f(t)$ be an integrable function on an interval $[a,b]$, with values in a Banach space X. Then the mean value of $f(t)$ on $[a,b]$ belongs to the closed convex hull of the set of values of $f(t)$ on $[a,b]$.*

That is we have that

$$\frac{1}{b-a} \int_{a}^{b} f(t) \, dt = \frac{1}{b-a} \lim_{d(\Pi) \to 0} \sum_{k=0}^{n-1} f(\tau_k) \Delta t_k, \tag{19.10}$$

where

$$\Pi := \{ a = t_0 \leq \tau_0 \leq t_1 \leq \tau_1 \leq t_2 \leq \ldots \leq t_{n-1} \leq \tau_{n-1} \leq t_n = b \}, \tag{19.11}$$

with

$$d(\Pi) := \max \{ \Delta t_0, \Delta t_1, \ldots, \Delta t_{n-1} \}, \tag{19.12}$$

$$\Delta t_k := t_{k+1} - t_k.$$

Next we give Neumann–Poincaré like inequalities.

Theorem 19.3. *Here $f \in C^1([a,b], X)$, X a Banach space, $p, q > 1 : \frac{1}{p} + \frac{1}{q} = 1$, $\nu > 0$.*

One has

$$1) \quad \left\| \left\| \left\| f - \frac{1}{b-a} \int_{a}^{b} f(t) \, dt \right\| \right\| \right\|_{L_\nu(a,b)} \leq (b-a)^{\left(\frac{1}{p}+\frac{1}{\nu}\right)} \left\| \left\| \left\| f' \right\| \right\| \right\|_{L_q(a,b)}. \tag{19.13}$$

When $\nu = q$ we obtain

$$2) \quad \left\| \left\| \left\| f - \frac{1}{b-a} \int_{a}^{b} f(t) \, dt \right\| \right\| \right\|_{L_q(a,b)} \leq (b-a) \left\| \left\| \left\| f' \right\| \right\| \right\|_{L_q(a,b)}. \tag{19.14}$$

When $\nu = p = q = 2$ we have

$$3)\quad \left\|\left\| f - \frac{1}{b-a}\int_a^b f(t)\,dt \right\|\right\|_{L_2(a,b)} \le (b-a)\,\left\|\,\|f'\|\,\right\|_{L_2(a,b)}, \tag{19.15}$$

equivalently,

$$\int_a^b \left\| f(s) - \frac{1}{b-a}\int_a^b f(t)\,dt \right\|^2 ds \le (b-a)^2 \left(\int_a^b \|f'(s)\|^2\,ds \right). \tag{19.15*}$$

Proof. Since $f \in C^1\left([a,b],X\right)$ we have by (19.2) that

$$f(y) = f(x) + \int_x^y f'(t)\,dt, \quad \forall x,y \in [a,b]. \tag{19.16}$$

So by (19.16) we have

$$f(s) = f(\tau_k) + \int_{\tau_k}^s f'(r)\,dr, \tag{19.17}$$

$\forall s \in [a,b]$ and $k = 0,1,\ldots,n-1$.

Hence

$$f(s)\frac{\Delta t_k}{b-a} = f(\tau_k)\frac{\Delta t_k}{b-a} + \frac{\Delta t_k}{b-a}\int_{\tau_k}^s f'(r)\,dr, \tag{19.18}$$

and

$$f(s)\frac{1}{b-a}\sum_{k=0}^{n-1}\Delta t_k = \frac{1}{b-a}$$

$$\cdot \sum_{k=0}^{n-1} f(\tau_k)\,\Delta t_k + \frac{1}{b-a}\sum_{k=0}^{n-1}\Delta t_k \int_{\tau_k}^s f'(r)\,dr. \tag{19.19}$$

Therefore we observe that

$$f(s) - \frac{1}{b-a}\int_a^b f(t)\,dt = \frac{1}{b-a}\sum_{k=0}^{n-1} f(\tau_k)\,\Delta t_k$$

$$-\frac{1}{b-a}\int_a^b f(t)\,dt + \frac{1}{b-a}\sum_{k=0}^{n-1}\Delta t_k \int_{\tau_k}^s f'(r)\,dr, \tag{19.20}$$

and

$$\left\| f(s) - \frac{1}{b-a}\int_a^b f(t)\,dt \right\|$$

$$\le \left\| \frac{1}{b-a}\sum_{k=0}^{n-1} f(\tau_k)\,\Delta t_k - \frac{1}{b-a}\int_a^b f(t)\,dt \right\|$$

$$+\frac{1}{b-a}\sum_{k=0}^{n-1}\Delta t_k\left\|\int_{\tau_k}^{s}f'\left(r\right)dr\right\|$$

$$\leq\left\|\frac{1}{b-a}\sum_{k=0}^{n-1}f\left(\tau_k\right)\Delta t_k-\frac{1}{b-a}\int_{a}^{b}f\left(t\right)dt\right\|$$

$$+\frac{1}{b-a}\sum_{k=0}^{n-1}\Delta t_k\left|\int_{\tau_k}^{s}\left\|f'\left(r\right)\right\|dr\right|$$

$$\leq\left\|\frac{1}{b-a}\sum_{k=0}^{n-1}f\left(\tau_k\right)\Delta t_k-\frac{1}{b-a}\int_{a}^{b}f\left(t\right)dt\right\|+\int_{a}^{b}\left\|f'\left(r\right)\right\|dr. \qquad (19.21)$$

That is we proved

$$\left\|f\left(s\right)-\frac{1}{b-a}\int_{a}^{b}f\left(t\right)dt\right\|$$

$$\leq\left\|\frac{1}{b-a}\sum_{k=0}^{n-1}f\left(\tau_k\right)\Delta t_k-\frac{1}{b-a}\int_{a}^{b}f\left(t\right)dt\right\|+\int_{a}^{b}\left\|f'\left(r\right)\right\|dr,\quad\forall s\in\left[a,b\right], \quad (19.22)$$

for any partition Π.

Taking the limit in both sides of (19.22) as $d\left(\Pi\right)\to 0$ we obtain by (19.10) that

$$\left\|f\left(s\right)-\frac{1}{b-a}\int_{a}^{b}f\left(t\right)dt\right\|\leq\int_{a}^{b}\left\|f'\left(r\right)\right\|dr$$

$$\leq\left(b-a\right)^{1/p}\left(\int_{a}^{b}\left\|f'\left(r\right)\right\|^{q}dr\right)^{1/q},\quad\forall s\in\left[a,b\right]. \qquad (19.23)$$

Consequently we find

$$\left\|f\left(s\right)-\frac{1}{b-a}\int_{a}^{b}f\left(t\right)dt\right\|^{\nu}\leq\left(b-a\right)^{\nu/p}\left(\int_{a}^{b}\left\|f'\left(r\right)\right\|^{q}dr\right)^{\nu/q},\quad\forall s\in\left[a,b\right].$$

$$(19.24)$$

Finally we have

$$\int_{a}^{b}\left\|f\left(s\right)-\frac{1}{b-a}\int_{a}^{b}f\left(t\right)dt\right\|^{\nu}ds\leq\left(b-a\right)^{\frac{\nu}{p}+1}\left(\int_{a}^{b}\left\|f'\left(r\right)\right\|^{q}dr\right)^{\nu/q}, \qquad (19.25)$$

proving the claim. $\qquad\square$

Corollary 19.4. (to Theorem 19.3). *Additionally assume that $\int_a^b f(t)\,dt = 0$. One has*

$$1)\ \||f\||_{L_\nu(a,b)} \le (b-a)^{\frac{1}{p}+\frac{1}{\nu}} \||f'\||_{L_q(a,b)}. \tag{19.26}$$

When $\nu = q$ we obtain

$$2)\ \||f\||_{L_q(a,b)} \le (b-a) \||f'\||_{L_q(a,b)}. \tag{19.27}$$

When $\nu = p = q = 2$ we have

$$3)\ \||f\||_{L_2(a,b)} \le (b-a) \||f'\||_{L_2(a,b)}, \tag{19.28}$$

equivalently

$$\int_a^b \|f(s)\|^2\,ds \le (b-a)^2 \left(\int_a^b \|f'(s)\|^2\,ds \right). \tag{19.28*}$$

We continue with

Theorem 19.5. *Let $y \in [a, x_0]$, $x_0 > a$. Assume $f^{(k)}(x_0) = 0$, $k = 0, 1, \ldots, n-1$. Then*

$$1)\ \||f\||_{L_\nu(a,x_0)} \le \frac{(x_0-a)^{\left(n-1+\frac{1}{p}+\frac{1}{\nu}\right)} \||f^{(n)}\||_{L_q(a,x_0)}}{((n-1)!)\,(p(n-1)+1)^{1/p}\left(\nu\left(n-1+\frac{1}{p}\right)+1\right)^{1/\nu}}. \tag{19.29}$$

When $\nu = q$ we derive

$$2)\ \||f\||_{L_q(a,x_0)} \le \frac{(x_0-a)^n \||f^{(n)}\||_{L_q(a,x_0)}}{((n-1)!)\,(p(n-1)+1)^{1/p}(qn)^{1/q}}. \tag{19.30}$$

And for $\nu = q = p = 2$ we have

$$3)\ \||f\||_{L_2(a,x_0)} \le \frac{(x_0-a)^n \||f^{(n)}\||_{L_2(a,x_0)}}{((n-1)!)\,\sqrt{2n-1}\sqrt{2n}}. \tag{19.31}$$

Proof. By (19.3) we have

$$\|f(y)\| = \frac{1}{(n-1)!} \left\| \int_y^{x_0} (t-y)^{n-1} f^{(n)}(t)\,dt \right\|$$

$$\le \frac{1}{(n-1)!} \int_y^{x_0} (t-y)^{n-1} \left\| f^{(n)}(t) \right\| dt$$

$$\le \frac{1}{(n-1)!} \left(\int_y^{x_0} (t-y)^{p(n-1)}\,dt \right)^{1/p} \left(\int_y^{x_0} \left\| f^{(n)}(t) \right\|^q dt \right)^{1/q}$$

$$\le \frac{1}{(n-1)!} \frac{(x_0-y)^{\left(n-1+\frac{1}{p}\right)}}{(p(n-1)+1)^{1/p}} \||f^{(n)}\||_{L_q(a,x_0)} \tag{19.32}$$

Hence

$$\|f(y)\|^{\nu} \le \frac{1}{((n-1)!)^{\nu}} \frac{(x_0-y)^{\nu\left(n-1+\frac{1}{p}\right)}}{(p(n-1)+1)^{\nu/p}} \left\|\left\|\left\|f^{(n)}\right\|\right\|\right\|^{\nu}_{L_q(a,x_0)}, \tag{19.33}$$

and

$$\int_a^{x_0} \|f(y)\|^{\nu}\, dy \le \frac{1}{((n-1)!)^{\nu}}$$

$$\cdot \frac{(x_0-a)^{\left(\nu\left(n-1+\frac{1}{p}\right)+1\right)} \left\|\left\|\left\|f^{(n)}\right\|\right\|\right\|^{\nu}_{L_q(a,x_0)}}{(p(n-1)+1)^{\nu/p}\left(\nu\left(n-1+\frac{1}{p}\right)+1\right)}, \tag{19.34}$$

proving the claim. $\qquad\square$

We continue with L_1 results.

Proposition 19.6. *Let $y \in [x_0,b]$, $x_0 < b$. Suppose $f^{(k)}(x_0) = 0$, $k = 0,1,\ldots,n-1$.*

Then

$$1)\ \left\|\left\|\left\|f\right\|\right\|\right\|_{L_{\nu}(x_0,b)} \le \frac{(b-x_0)^{\left(n-1+\frac{1}{\nu}\right)} \left\|\left\|\left\|f^{(n)}\right\|\right\|\right\|_{L_1(x_0,b)}}{(n-1)!\,(\nu(n-1)+1)^{1/\nu}}. \tag{19.35}$$

When $\nu = 1$ we get

$$2)\ \left\|\left\|\left\|f\right\|\right\|\right\|_{L_1(x_0,b)} \le \frac{(b-x_0)^n \left\|\left\|\left\|f^{(n)}\right\|\right\|\right\|_{L_1(x_0,b)}}{n!}. \tag{19.36}$$

Proof. By (19.3), (19.7) we have

$$\|f(y)\| \le \frac{1}{(n-1)!}\int_{x_0}^y (y-t)^{n-1}\left\|f^{(n)}(t)\right\| dt$$

$$\le \frac{(y-x_0)^{n-1}}{(n-1)!}\int_{x_0}^b \left\|f^{(n)}(t)\right\| dt = \frac{(y-x_0)^{n-1}}{(n-1)!} \left\|\left\|\left\|f^{(n)}\right\|\right\|\right\|_{L_1(x_0,b)}. \tag{19.37}$$

That is

$$\|f(y)\|^{\nu} \le \frac{(y-x_0)^{\nu(n-1)}}{((n-1)!)^{\nu}} \left\|\left\|\left\|f^{(n)}\right\|\right\|\right\|^{\nu}_{L_1(x_0,b)}, \tag{19.38}$$

and

$$\int_{x_0}^b \|f(y)\|^{\nu}\, dy \le \frac{(b-x_0)^{\nu(n-1)+1} \left\|\left\|\left\|f^{(n)}\right\|\right\|\right\|^{\nu}_{L_1(x_0,b)}}{((n-1)!)^{\nu}(\nu(n-1)+1)}, \tag{19.39}$$

proving the claim. $\qquad\square$

The countercase of the last result follows.

Proposition 19.7. *Let $y \in [a, x_0]$, $x_0 > a$. Suppose $f^{(k)}(x_0) = 0$, $k = 0, 1, \ldots, n - 1$.*

Then

$$1) \quad \|\|f\|\|_{L_\nu(a,x_0)} \leq \frac{(x_0 - a)^{\left(n - 1 + \frac{1}{\nu}\right)} \|\|f^{(n)}\|\|_{L_1(a,x_0)}}{((n-1)!)(\nu(n-1)+1)^{1/\nu}}. \tag{19.40}$$

When $\nu = 1$ we get

$$2) \quad \|\|f\|\|_{L_1(a,x_0)} \leq \frac{(x_0 - a)^n \|\|f^{(n)}\|\|_{L_1(a,x_0)}}{n!}. \tag{19.41}$$

Proof. By (19.3) and (19.32) we have

$$\|f(y)\| \leq \frac{1}{(n-1)!} \int_y^{x_0} (t - y)^{n-1} \left\|f^{(n)}(t)\right\| dt$$

$$\leq \frac{(x_0 - y)^{n-1}}{(n-1)!} \int_a^{x_0} \left\|f^{(n)}(t)\right\| dt$$

$$= \frac{(x_0 - y)^{n-1}}{(n-1)!} \|\|f^{(n)}\|\|_{L_1(a,x_0)}. \tag{19.42}$$

That is

$$\|f(y)\|^\nu \leq \frac{(x_0 - y)^{\nu(n-1)}}{((n-1)!)^\nu} \|\|f^{(n)}\|\|_{L_1(a,x_0)}^\nu, \tag{19.43}$$

and

$$\int_a^{x_0} \|f(y)\|^\nu \, dy \leq \frac{(x_0 - a)^{(\nu(n-1)+1)} \|\|f^{(n)}\|\|_{L_1(a,x_0)}^\nu}{((n-1)!)^\nu (\nu(n-1)+1)}, \tag{19.44}$$

proving the claim. $\square$

We continue with Sobolev like inequalities

Theorem 19.8. *Let $y \in [x_0, b]$, $x_0 < b$. Assume $f^{(k)}(x_0) = 0$, $k = 0, 1, \ldots, n - 1$.*
Then

$$1) \quad \|\|f\|\|_{L_\nu(x_0,b)} \leq \frac{M_\nu}{n} \left(\sum_{i=1}^n \|\|f^{(i)}\|\|_{L_q(x_0,b)}\right), \tag{19.45}$$

where

$$M_\nu := \max_{1 \leq i \leq n} \left\{ \frac{(b - x_0)^{\left(i - 1 + \frac{1}{p} + \frac{1}{\nu}\right)}}{((i-1)!)(p(i-1)+1)^{1/p} \left(\nu\left(i - 1 + \frac{1}{p}\right) + 1\right)^{1/\nu}} \right\}. \tag{19.46}$$

When $\nu = q$ we derive

$$2) \quad \|\|f\|\|_{L_q(x_0,b)} \leq \frac{M_q}{n} \left(\sum_{i=1}^n \|\|f^{(i)}\|\|_{L_q(x_0,b)}\right), \tag{19.47}$$

where

$$M_q := \max_{1 \le i \le n} \left\{ \frac{(b - x_0)^i}{((i-1)!)\,(p(i-1)+1)^{1/p}\,(qi)^{1/q}} \right\}. \tag{19.48}$$

For $\nu = q = p = 2$ we have

$$3) \quad \||f\||_{L_2(x_0,b)} \le \frac{M_2}{n} \left(\sum_{i=1}^{n} \||f^{(i)}\||_{L_2(x_0,b)} \right), \tag{19.49}$$

where

$$M_2 := \max_{1 \le i \le n} \left\{ \frac{(b - x_0)^i}{((i-1)!)\,\sqrt{2i-1}\,\sqrt{2i}} \right\}. \tag{19.50}$$

Proof. By assumption of Theorem 19.1 we have that $f \in C^i([a,b], X)$, for all $i = 1, \ldots, n$, and that $f^{(k)}(x_0) = 0$, $k = 0, 1, \ldots, i-1$; for all $i = 1, \ldots, n$, $n \in \mathbb{N}$. Thus by (19.4) we find

$$\||f\||_{L_\nu(x_0,b)} \le \frac{(b - x_0)^{\left(i-1+\frac{1}{p}+\frac{1}{\nu}\right)} \||f^{(i)}\||_{L_q(x_0,b)}}{((i-1)!)\,(p(i-1)+1)^{1/p}\,\left(\nu\left(i-1+\frac{1}{p}\right)+1\right)^{1/\nu}}$$

$$\le M_\nu \||f^{(i)}\||_{L_q(x_0,b)} \tag{19.51}$$

for all $i = 1, \ldots, n$.

So that

$$n \||f\||_{L_\nu(x_0,b)} \le M_\nu \left(\sum_{i=1}^{n} \||f^{(i)}\||_{L_q(x_0,b)} \right), \tag{19.52}$$

proving the claim. $\qquad\square$

We continue with

Theorem 19.9. *Let $y \in [a, x_0]$, $x_0 > a$. Assume $f^{(k)}(x_0) = 0$, $k = 0, 1, \ldots, n-1$.*
Then

$$1) \quad \||f\||_{L_\nu(a,x_0)} \le \frac{\Delta_\nu}{n} \left(\sum_{i=1}^{n} \||f^{(i)}\||_{L_q(a,x_0)} \right), \tag{19.53}$$

where

$$\Delta_\nu := \max_{1 \le i \le n} \left\{ \frac{(x_0 - a)^{\left(i-1+\frac{1}{p}+\frac{1}{\nu}\right)}}{((i-1)!)\,(p(i-1)+1)^{1/p}\,\left(\nu\left(i-1+\frac{1}{p}\right)+1\right)^{1/\nu}} \right\}. \tag{19.54}$$

When $\nu = q$ we obtain

$$2) \quad \||f\||_{L_q(a,x_0)} \le \frac{\Delta_q}{n} \left(\sum_{i=1}^{n} \||f^{(i)}\||_{L_q(a,x_0)} \right), \tag{19.55}$$

where

$$\Delta_q := \max_{1 \le i \le n} \left\{ \frac{(x_0 - a)^i}{((i-1)!)\,(p\,(i-1)+1)^{1/p}\,(qi)^{1/q}} \right\}. \tag{19.56}$$

For $\nu = q = p = 2$ we have

$$3) \ \left\| \|f\| \right\|_{L_2(a,x_0)} \le \frac{\Delta_2}{n} \left(\sum_{i=1}^{n} \left\| \left\| f^{(i)} \right\| \right\|_{L_2(a,x_0)} \right), \tag{19.57}$$

where

$$\Delta_2 := \max_{1 \le i \le n} \left\{ \frac{(x_0 - a)^i}{((i-1)!)\,\sqrt{2i-1}\,\sqrt{2i}} \right\}. \tag{19.58}$$

Proof. Based on Theorem 19.5, similar to the proof of Theorem 19.8. $\qquad\square$

We continue with L_1 results.

Proposition 19.10. *Let $y \in [x_0, b]$, $x_0 < b$. Suppose $f^{(k)}(x_0) = 0$, $k = 0, 1, \ldots, n-1$.*

Then

$$1) \ \left\| \|f\| \right\|_{L_\nu(x_0,b)} \le \frac{\theta_\nu}{n} \left(\sum_{i=1}^{n} \left\| \left\| f^{(i)} \right\| \right\|_{L_1(x_0,b)} \right), \tag{19.59}$$

where

$$\theta_\nu := \max_{1 \le i \le n} \left\{ \frac{(b - x_0)^{\left(i-1+\frac{1}{\nu}\right)}}{((i-1)!)\,(\nu\,(i-1)+1)^{1/\nu}} \right\}. \tag{19.60}$$

When $\nu = 1$ we get

$$2) \ \left\| \|f\| \right\|_{L_1(x_0,b)} \le \frac{\theta_1}{n} \left(\sum_{i=1}^{n} \left\| \left\| f^{(i)} \right\| \right\|_{L_1(x_0,b)} \right), \tag{19.61}$$

where

$$\theta_1 := \max_{1 \le i \le n} \left\{ \frac{(b - x_0)^i}{i!} \right\}. \tag{19.62}$$

Proof. Based on Proposition 19.6. We have from (19.35) that

$$\left\| \|f\| \right\|_{L_\nu(x_0,b)} \le \frac{(b - x_0)^{\left(i-1+\frac{1}{\nu}\right)} \left\| \left\| f^{(i)} \right\| \right\|_{L_1(x_0,b)}}{((i-1)!)\,(\nu\,(i-1)+1)^{1/\nu}} \tag{19.63}$$

for all $i = 1, \ldots, n$.

Consequently we see that

$$\left\| \|f\| \right\|_{L_\nu(x_0,b)} \le \theta_\nu \left\| \left\| f^{(i)} \right\| \right\|_{L_1(x_0,b)}, \tag{19.64}$$

all $i = 1, \ldots, n$.

Thus

$$n \left\| \|f\| \right\|_{L_\nu(x_0,b)} \le \theta_\nu \left(\sum_{i=1}^{n} \left\| \left\| f^{(i)} \right\| \right\|_{L_1(x_0,b)} \right), \tag{19.65}$$

proving the claim. $\qquad\square$

Similarly we give

Proposition 19.11. *Let $y \in [a, x_0]$, $x_0 > a$. Assume $f^{(k)}(x_0) = 0$, $k = 0, 1, \ldots, n-1$.*

Then

$$1) \quad \|\|f\|\|_{L_\nu(a,x_0)} \leq \frac{T_\nu}{n} \left(\sum_{i=1}^{n} \|\|f^{(i)}\|\|_{L_1(a,x_0)} \right), \tag{19.66}$$

where

$$T_\nu := \max_{1 \leq i \leq n} \left\{ \frac{(x_0 - a)^{\left(i-1+\frac{1}{\nu}\right)}}{((i-1)!)\,(\nu(i-1)+1)^{1/\nu}} \right\}. \tag{19.67}$$

When $\nu = 1$ we get

$$2) \quad \|\|f\|\|_{L_1(a,x_0)} \leq \frac{T_1}{n} \left(\sum_{i=1}^{n} \|\|f^{(i)}\|\|_{L_1(a,x_0)} \right), \tag{19.68}$$

where

$$T_1 := \max_{1 \leq i \leq n} \left\{ \frac{(x_0 - a)^i}{i!} \right\}. \tag{19.69}$$

Proof. Based on Proposition 19.7 as in Proposition 19.10. $\qquad\square$

19.4 Applications

Here $X = \mathbb{C}$, $f \in C^n([a,b], \mathbb{C})$, $n \in \mathbb{N}$, $x_0, y \in [a,b]$. Furthermore suppose that $f^{(k)}(x_0) = 0$, $k = 0, 1, \ldots, n-1$; $p, q > 1 : \frac{1}{p} + \frac{1}{q} = 1$, $\nu > 0$.

First we give the following Poincaré like inequalities results.

Corollary 19.12. *Let $x_0 < b$. One has*

$$\|f\|_{L_\nu(x_0,b)} \leq \frac{(b - x_0)^{\left(n-1+\frac{1}{p}+\frac{1}{\nu}\right)} \left\|f^{(n)}\right\|_{L_q(x_0,b)}}{((n-1)!)\,(p(n-1)+1)^{1/p} \left(\nu\left(n-1+\frac{1}{p}\right)+1\right)^{1/\nu}}. \tag{19.70}$$

Proof. Use of (19.4). $\qquad\square$

We continue with

Corollary 19.13. *One has*

$$\left\|f - \frac{1}{b-a}\int_a^b f(t)\,dt\right\|_{L_\nu(a,b)} \leq (b-a)^{\left(\frac{1}{p}+\frac{1}{\nu}\right)} \|f'\|_{L_q(a,b)}. \tag{19.71}$$

Proof. Use of (19.13). $\qquad\square$

Corollary 19.14. *Let $x_0 > a$. One has*

$$\|f\|_{L_\nu(a,x_0)} \leq \frac{(x_0 - a)^{\left(n-1+\frac{1}{p}+\frac{1}{\nu}\right)} \left\|f^{(n)}\right\|_{L_q(a,x_0)}}{((n-1)!)\,(p\,(n-1)+1)^{1/p} \left(\nu\left(n-1+\frac{1}{p}\right)+1\right)^{1/\nu}}.$$
(19.72)

Proof. Use of (19.29). $\qquad\square$

We continue with L_1 Poincaré like results on $\mathbb{C}$.

Corollary 19.15. *Let $x_0 < b$. Then*

$$\|f\|_{L_\nu(x_0,b)} \leq \frac{(b - x_0)^{\left(n-1+\frac{1}{\nu}\right)} \left\|f^{(n)}\right\|_{L_1(x_0,b)}}{((n-1)!)\,(\nu\,(n-1)+1)^{1/\nu}}.$$
(19.73)

Proof. Use of (19.35). $\qquad\square$

Corollary 19.16. *Let $x_0 > a$. Then*

$$\|f\|_{L_\nu(a,x_0)} \leq \frac{(x_0 - a)^{\left(n-1+\frac{1}{\nu}\right)} \left\|f^{(n)}\right\|_{L_1(a,x_0)}}{((n-1)!)\,(\nu\,(n-1)+1)^{1/\nu}}.$$
(19.74)

Proof. Use of (19.40). $\qquad\square$

We finish with Sobolev like inequalities results on $\mathbb{C}$.

Corollary 19.17. *Let $x_0 < b$. Then*

$$\|f\|_{L_\nu(x_0,b)} \leq \frac{M_\nu}{n} \left(\sum_{i=1}^{n} \left\|f^{(i)}\right\|_{L_q(x_0,b)}\right),$$
(19.75)

where M_ν as in (19.46).

Proof. Use of (19.45). $\qquad\square$

Corollary 19.18. *Let $x_0 > a$. Then*

$$\|f\|_{L_\nu(a,x_0)} \leq \frac{\Delta_\nu}{n} \left(\sum_{i=1}^{n} \left\|f^{(i)}\right\|_{L_q(a,x_0)}\right),$$
(19.76)

where Δ_ν as in (19.54).

Proof. Use of (19.53). $\qquad\square$

At last we give L_1 Sobolev like results on $\mathbb{C}$.

Corollary 19.19. *Let $x_0 < b$. Then*

$$\|f\|_{L_\nu(x_0,b)} \leq \frac{\theta_\nu}{n} \left(\sum_{i=1}^{n} \left\|f^{(i)}\right\|_{L_1(x_0,b)}\right),$$
(19.77)

where θ_ν as in (19.60).

Proof. Use of (19.59). $\qquad\square$

Corollary 19.20. *Let $x_0 > a$. Then*

$$\|f\|_{L_\nu(a,x_0)} \leq \frac{T_\nu}{n} \left(\sum_{i=1}^{n} \left\| f^{(i)} \right\|_{L_1(a,x_0)} \right), \tag{19.78}$$

where T_ν as in (19.67).

Proof.   Use of (19.66).   □

Chapter 20

Poincaré Type Inequalities for Semigroups, Cosine and Sine Operator Functions

Here we present Poincaré type general L_p inequalities regarding Semigroups, Cosine and Sine Operator functions. We give applications. This chapter relies on [50].

20.1 Introduction

This chapter is motivated by the famous Poincaré inequality, see [2]: *Given a bounded domain $\Omega \subset \mathbb{R}^n$, $n \in \mathbb{N}$, it holds*

$$\|u\|_{L^p(\Omega)} \leq C \|\nabla u\|_{L^p(\Omega)}$$

for functions u with vanishing mean value over Ω, $1 \leq p \leq \infty$, under very general assumptions on Ω, where $\|\nabla u\|_{L^p(\Omega)}$ is defined as the L^p-norm of the euclidean norm of ∇u.

Especially in [2] is proved for a convex domain $\Omega \subset \mathbb{R}^n$ with diameter d that

$$\|u\|_{L^1(\Omega)} \leq \frac{d}{2} \|\nabla u\|_{L^1(\Omega)}$$

for any u with zero mean value on Ω, also the constant $1/2$ is optimal.

We are also motivated by [92], where the authors prove the following: *Let $1 < p < \infty$, $-\infty < a < b < \infty$. The best constant C (independent of a, b) for which the $1-$dimensional Poincaré inequality*

$$\left\| f - \frac{\int_a^b f(t)\, dt}{b - a} \right\|_{L^1([a,b])} \leq C (b - a)^{2 - \frac{1}{p}} \|f'\|_{L^p([a,b])}$$

holds for all Lipschitz continuous functions f, is $C = \frac{1}{2}(1 + p')^{-1/p'}$, where $p' > 1$: $\frac{1}{p} + \frac{1}{p'} = 1$.

So here we present Poincaré type inequalities for Semigroups and for Cosine and Sine Operator functions.

20.2 Semigroups Background

All this background comes from [79] (in general see also [147]).

243

Let X a real or complex Banach space with elements $f, g, \ldots$ having norm $\|f\|, \|g\|, \ldots$ and let $\varepsilon(X)$ be the Banach algebra of endomorphisms of X.

If $T \in \varepsilon(X)$, $\|T\|$ denotes the norm of T.

Definition 20.1. If $T(t)$ is an operator function on the non-negative real axis $0 \le t < \infty$ to the Banach algebra $\varepsilon(X)$ satisfying the following conditions:

$$\begin{cases} (i) \ \ T(t_1 + t_2) = T(t_1) T(t_2), & (t_1, t_2 \ge 0), \\ \quad\quad (ii) \ \ T(0) = I & (I = \text{identity operator}), \end{cases} \tag{20.1}$$

then $\{T(t); \ 0 \le t < \infty\}$ is called a one-parameter semi-group of operators in $\varepsilon(X)$.

The semi-group $\{T(t); \ 0 \le t < \infty\}$ is said to be of class C_0 if it satisfies the further property

$$(iii) \ \ s - \lim_{t \to 0+} T(t) f = f, \quad (f \in X) \tag{20.2}$$

referred to as the strong continuity of $T(t)$ at the origin.

In this chapter we shall assume that the family of bounded linear operators $\{T(t); \ 0 \le t < \infty\}$ mapping X to itself is a semi-group of class C_0, thus that all three conditions of the above definition are satisfied.

Proposition 20.2. *(a)* $\|T(t)\|$ *is bounded on every finite subinterval of* $[0, \infty)$.

(b) *For each* $f \in X$, *the vector-valued function* $T(t) f$ *on* $[0, \infty)$ *is strongly continuous.*

Definition 20.3. The infinitesimal generator A of the semi-group $\{T(t); 0 \le t < \infty\}$ is defined by

$$Af = s - \lim_{\tau \to 0+} A_\tau f, \quad A_\tau f = \frac{1}{\tau} [T(\tau) - I] \tag{20.3}$$

whenever the limit exists; the domain of A, in symbols $D(A)$, is the set of elements f for which the limit exists.

Proposition 20.4. *(a)* $D(A)$ *is a linear manifold in* X *and* A *is a linear operator.*

(b) *If* $f \in D(A)$, *then* $T(t) f \in D(A)$ *for each* $t \ge 0$ *and*

$$\frac{d}{dt} T(t) f = AT(t) f = T(t) Af \quad (t \ge 0); \tag{20.4}$$

furthermore,

$$T(t) f - f = \int_0^t T(u) Af \, du \quad (t > 0). \tag{20.5}$$

(c) $D(A)$ *is dense in* X, *i.e.* $\overline{D(A)} = X$, *and* A *is a closed operator.*

Definition 20.5. For $r = 0, 1, 2, \ldots$ the operator A^r is defined inductively by the relations $A^0 = I$, $A^1 = A$, and

$$D(A^r) = \{f; \ f \in D(A^{r-1}) \text{ and } A^{r-1} f \in D(A)\}$$

$$A^r f = A(A^{r-1} f) = s - \lim_{\tau \to 0+} A_\tau (A^{r-1} f) \quad (f \in D(A^r)). \tag{20.6}$$

For the operator A^r and its domain $D(A^r)$ we have the following

Proposition 20.6. *(a) $D(A^r)$ is a linear subspace in X and A^r is a linear operator.*
(b) If $f \in D(A^r)$, so does $T(t) f$ for each $t \geq 0$ and

$$\frac{d^r}{dt^r} T(t) f = A^r T(t) f = T(t) A^r f. \tag{20.7}$$

Moreover

$$\Delta_r(t) f := T(t) f - \sum_{k=0}^{r-1} \frac{t^k}{k!} A^k f = \frac{1}{(r-1)!} \int_0^t (t-u)^{r-1} T(u) A^r f \, du, \tag{20.8}$$

the Taylor's formula for semigroups.

(c) $D(A^r)$ is dense in X for $r = 1, 2, \ldots$; furthermore , $\bigcap\limits_{r=1}^{\infty} D(A^r)$ is dense in
X. A^r *is a closed operator.*

The integral in (20.8) is a vector valued Riemann integral, see [79], [164].

20.3 Poincaré Type Inequalities for Semigroups

Here always we consider $T(t)$, A as in the Semigroups Background and $f \in D(A^r)$, $r \in \mathbb{N}$. Also we consider $p, q > 1 : \frac{1}{p} + \frac{1}{q} = 1$, along with $a, \nu \in \mathbb{R} : a, \nu > 0$.

We present the first result.

Theorem 20.7. *One has*

$$1) \ \left\| \|\Delta_r(t) f\| \right\|_{L_\nu(0,a)}$$

$$\leq \frac{a^{r-1+\frac{1}{p}+\frac{1}{\nu}}}{(r-1)! \, (p(r-1)+1)^{1/p} \left(\nu\left(r-1+\frac{1}{p}\right)+1\right)^{1/\nu}} \left\| \|T(t) A^r f\| \right\|_{L_q(0,a)}. \tag{20.9}$$

When $q = \nu$ we have

$$2) \ \left\| \|\Delta_r(t) f\| \right\|_{L_q(0,a)}$$

$$\leq \frac{a^r}{(r-1)! \, (p(r-1)+1)^{1/p} (qr)^{1/q}} \left\| \|T(t) A^r f\| \right\|_{L_q(0,a)}. \tag{20.10}$$

Proof. By (20.8) we have

$$\|\Delta_r(t) f\| = \frac{1}{(r-1)!} \left\| \int_0^t (t-u)^{r-1} T(u) A^r f \, du \right\|$$

$$\leq \frac{1}{(r-1)!} \int_0^t (t-u)^{r-1} \|T(u) A^r f\| \, du$$

$$\leq \frac{1}{(r-1)!} \left(\int_0^t (t-u)^{p(r-1)} \, du \right)^{1/p} \left(\int_0^t \|T(u) A^r f\|^q \, du \right)^{1/q}$$

$$= \frac{t^{r-1+\frac{1}{p}}}{(r-1)!\,(p\,(r-1)+1)^{1/p}} \left(\int_0^t \|T(u)\,A^r f\|^q \, du \right)^{1/q}. \tag{20.11}$$

That is we get

$$\|\Delta_r(t)\,f\| \leq \frac{t^{r-1+\frac{1}{p}}}{(r-1)!\,(p\,(r-1)+1)^{1/p}} \left(\int_0^a \|T(u)\,A^r f\|^q \, du \right)^{1/q}. \tag{20.12}$$

Thus

$$\|\Delta_r(t)\,f\|^\nu \leq \frac{t^{\nu\left(r-1+\frac{1}{p}\right)}}{((r-1)!)^\nu\,(p\,(r-1)+1)^{\nu/p}} \, \|\|T(t)\,A^r f\|\|_{L_q(0,a)}^\nu, \tag{20.13}$$

and

$$\int_0^a \|\Delta_r(t)\,f\|^\nu \, dt \leq \frac{a^{\nu\left(r-1+\frac{1}{p}\right)+1} \, \|\|T(t)\,A^r f\|\|_{L_q(0,a)}^\nu}{((r-1)!)^\nu\,(p\,(r-1)+1)^{\nu/p}\left(\nu\left(r-1+\frac{1}{p}\right)+1\right)}, \tag{20.14}$$

proving the claim. $\qquad\qquad\qquad\qquad\qquad\qquad\qquad\qquad\qquad\qquad\quad \square$

We continue with

Corollary 20.8. (to Theorem 20.7)

Case of $p = q = 2$. We have

$$1) \quad \|\|\Delta_r(t)\,f\|\|_{L_\nu(0,a)}$$

$$\leq \frac{a^{r-\frac{1}{2}+\frac{1}{\nu}}}{(r-1)!\,\sqrt{(2r-1)}\left(\nu\left(r-\frac{1}{2}\right)+1\right)^{1/\nu}} \, \|\|T(t)\,A^r f\|\|_{L_2(0,a)}. \tag{20.15}$$

When $\nu = 2$ we derive

$$2) \quad \|\|\Delta_r(t)\,f\|\|_{L_2(0,a)} \leq \frac{a^r}{(r-1)!\,\sqrt{(2r-1)}\,\sqrt{2r}} \, \|\|T(t)\,A^r f\|\|_{L_2(0,a)}. \tag{20.16}$$

We treat next the L_1 case.

Theorem 20.9. *One has*

$$1) \quad \|\|\Delta_r(t)\,f\|\|_{L_\nu(0,a)} \leq \frac{a^{\left(r-1+\frac{1}{\nu}\right)} \, \|\|T(t)\,A^r f\|\|_{L_1(0,a)}}{(r-1)!\,(\nu\,(r-1)+1)^{1/\nu}}. \tag{20.17}$$

For $\nu = 1$ we get

$$2) \quad \|\|\Delta_r(t)\,f\|\|_{L_1(0,a)} \leq \frac{a^r \, \|\|T(t)\,A^r f\|\|_{L_1(0,a)}}{r!}. \tag{20.18}$$

Proof. We observe that

$$\|\Delta_r(t)\,f\| \leq \frac{1}{(r-1)!} \int_0^t (t-u)^{r-1} \|T(u)\,A^r f\| \, du$$

$$\leq \frac{t^{r-1}}{(r-1)!} \int_0^a \|T(u)\,A^r f\| \, du. \tag{20.19}$$

So that

$$\|\Delta_r(t)f\|^\nu \leq \frac{t^{\nu(r-1)}}{((r-1)!)^\nu}\,\||T(u)A^rf|\|^\nu_{L_1(0,a)} \tag{20.20}$$

and

$$\int_0^a \|\Delta_r(t)f\|^\nu \, dt \leq \frac{a^{\nu(r-1)+1}\,\||T(u)A^rf|\|^\nu_{L_1(0,a)}}{((r-1)!)^\nu\,(\nu(r-1)+1)}, \tag{20.21}$$

proving the claim. $\square$

We finish this section with

Application 20.10. (see also [80]) It is known the classical diffusion equation

$$\frac{\partial W}{\partial t} = \frac{\partial^2 W}{\partial x^2}, \quad -\infty < x < \infty,\ t > 0 \tag{20.22}$$

with initial condition

$$\lim_{t \to 0+} W(x,t) = f(x), \tag{20.23}$$

has under general conditions its solution given by

$$W(x,t,f) = [T(t)f](x) = \frac{1}{2\sqrt{\pi t}}\int_{-\infty}^{\infty} f(x+u)\,e^{-u^2/4t}du, \tag{20.24}$$

the so called Gauss–Weierstrass singular integral.

The infinitesimal generator of the semigroup $\{T(t)\,;\ 0 \leq t < \infty\}$ is $A = \partial^2/\partial x^2$ ([159], p. 578).

Here we suppose that f, $f^{(2k)}$, $k = 1,\ldots,r$, all belong to the Banach space $UCB(\mathbb{R})$, the space of bounded and uniformly continuous functions from $\mathbb{R}$ into itself, with norm

$$\|f\|_C := \sup_{x \in \mathbb{R}}|f(x)|. \tag{20.25}$$

Here we define

$$\overline{\Delta}_r(t)f(x) := W(x,t,f) - \sum_{k=0}^{r-1}\frac{t^k}{k!}f^{(2k)}(x), \quad \text{for all } x \in \mathbb{R}. \tag{20.26}$$

So by (20.9) we find

$$\||\overline{\Delta}_r(t)f\|_C\|_{L_\nu(0,a)}$$

$$\leq \frac{a^{r-1+\frac{1}{p}+\frac{1}{\nu}}}{(r-1)!\,(p(r-1)+1)^{1/p}\left(\nu\left(r-1+\frac{1}{p}\right)+1\right)^{1/\nu}}\,\||W\left(\cdot,t,f^{(2r)}\right)\|_C\|_{L_q(0,a)}. \tag{20.27}$$

We notice here that

$$\left|W\left(x,t,f^{(2r)}\right)\right| = \frac{1}{2\sqrt{\pi t}}\left|\int_{-\infty}^{\infty} f^{(2r)}(x+u)\,e^{-u^2/4t}du\right|$$

$$\leq \frac{1}{2\sqrt{\pi t}} \int_{-\infty}^{\infty} \left| f^{(2r)}\left(x+u\right) \right| e^{-u^2/4t} du$$

$$\leq \left\| f^{(2r)} \right\|_C \frac{1}{2\sqrt{\pi t}} \int_{-\infty}^{\infty} e^{-u^2/4t} du \tag{20.28}$$

$$= \left\| f^{(2r)} \right\|_C \cdot 1 = \left\| f^{(2r)} \right\|_C < \infty, \quad \text{for all } t \in \mathbb{R}.$$

That is

$$\left\| W\left(\cdot, t, f^{(2r)}\right) \right\|_C \leq \left\| f^{(2r)} \right\|_C . \tag{20.29}$$

Also by (20.10) we obtain

$$\left\| \left\| \overline{\Delta}_r\left(t\right) f \right\|_C \right\|_{L_q(0,a)}$$

$$\leq \frac{a^r}{(r-1)!\,(p\,(r-1)+1)^{1/p}\,(qr)^{1/q}} \left\| \left\| W\left(\cdot, t, f^{(2r)}\right) \right\|_C \right\|_{L_q(0,a)} . \tag{20.30}$$

By (20.15) we have $(p = q = 2)$

$$\left\| \left\| \overline{\Delta}_r\left(t\right) f \right\|_C \right\|_{L_\nu(0,a)}$$

$$\leq \frac{a^{r-\frac{1}{2}+\frac{1}{\nu}}}{(r-1)!\sqrt{2r-1}\left(\nu\left(r-\frac{1}{2}\right)+1\right)^{1/\nu}} \left\| \left\| W\left(\cdot, t, f^{(2r)}\right) \right\|_C \right\|_{L_2(0,a)} . \tag{20.31}$$

And by (20.16) $(\nu = 2)$ we see that

$$\left\| \left\| \overline{\Delta}_r\left(t\right) f \right\|_C \right\|_{L_2(0,a)}$$

$$\leq \frac{a^r}{(r-1)!\sqrt{2r-1}\sqrt{2r}} \left\| \left\| W\left(\cdot, t, f^{(2r)}\right) \right\|_C \right\|_{L_2(0,a)} . \tag{20.32}$$

We finish application with L_1 results.

By (20.17) we find

$$\left\| \left\| \overline{\Delta}_r\left(t\right) f \right\|_C \right\|_{L_\nu(0,a)}$$

$$\leq \frac{a^{\left(r-1+\frac{1}{\nu}\right)}}{(r-1)!\,(\nu\,(r-1)+1)^{1/\nu}} \left\| \left\| W\left(\cdot, t, f^{(2r)}\right) \right\|_C \right\|_{L_1(0,a)} , \tag{20.33}$$

and by (20.18) $(\nu = 1)$ we obtain

$$\left\| \left\| \overline{\Delta}_r\left(t\right) f \right\|_C \right\|_{L_1(0,a)} \leq \frac{a^r}{r!} \left\| \left\| W\left(\cdot, t, f^{(2r)}\right) \right\|_C \right\|_{L_1(0,a)} . \tag{20.34}$$

20.4 Cosine and Sine Operator Functions Background

(see [93], [147], [192], [191])

Let $(X, \|\cdot\|)$ be a real or complex Banach space. By definition, a cosine operator function is a family $\{C(t) \, ; \, t \in \mathbb{R}\}$ of bounded linear operators from X into itself, satisfying

$$(i) \ \ C(0) = I, \ \ I \text{ the identity operator;}$$

$$(ii) \ \ C(t+s) + C(t-s) = 2C(t)C(s), \quad \text{for all } t, s \in \mathbb{R}; \tag{20.35}$$

(the last product is composition)

$$(iii) \ \ C(\cdot)f \text{ is continuous an } \mathbb{R}, \text{ for all } f \in X.$$

Notice that

$$C(t) = C(-t), \quad \text{for all } t \in \mathbb{R}.$$

The associated sine operator function $S(\cdot)$ is defined by

$$S(t)f := \int_0^t C(s)\,f\,ds, \quad \text{for all } t \in \mathbb{R}, \text{ for all } f \in X. \tag{20.36}$$

The cosine operator function $C(\cdot)$ is such that $\|C(t)\| \leq M e^{\omega |t|}$, for some $M \geq 1$, $\omega \geq 0$, for all $t \in \mathbb{R}$, here $\|\cdot\|$ is the norm of the operator.

The infinitesimal generator A of $C(\cdot)$ is the operator from X into itself defined as

$$Af := \lim_{t \to 0+} \frac{2}{t^2} (C(t) - I) f \tag{20.37}$$

with domain $D(A)$. The operator A is closed and $D(A)$ is dense in X, i.e. $\overline{D(A)} = X$, and one has

$$\int_0^t S(s)\,f\,ds \in D(A) \text{ and } A \int_0^t S(s)\,f\,ds = C(t)f - f, \quad \text{for all } f \in X. \tag{20.38}$$

Also one has $A = C''(0)$, and $D(A)$ is the set of $f \in X$ such that $C(t)f$ is twice differentiable at $t = 0$; equivalently,

$$D(A) = \{f \in X : \ C(\cdot)f \in C^2(\mathbb{R}, X)\}. \tag{20.39}$$

If $f \in D(A)$, then $C(t)f \in D(A)$, and $C''(t)f = C(t)Af = AC(t)f$, for all $t \in \mathbb{R}$; $C'(0)f = 0$, see [140], [232].

We define $A^0 = I$, $A^2 = A \circ A, \ldots, A^n = A \circ A^{n-1}$, $n \in \mathbb{N}$. Let $f \in D(A^n)$, then $C(t)f \in C^{2n}(\mathbb{R}, X)$, and $C^{(2n)}(t)f = C(t)A^n f = A^n C(t)f$, for all $t \in \mathbb{R}$, and $C^{(2k-1)}(0)f = 0, 1 \leq k \leq n$, see [191].

For $f \in D(A^n)$, $t \in \mathbb{R}$, we have the cosine operator function's Taylor formula ([191], [192]) saying that

$$T_n(t)f := C(t)f - \sum_{k=0}^{n-1} \frac{t^{2k}}{(2k)!} A^k f = \int_0^t \frac{(t-s)^{2n-1}}{(2n-1)!} C(s) A^n f\,ds. \tag{20.40}$$

By integrating (20.40) we obtain the sine operator function's Taylor formula (see [45])

$$M_n(t) f := S(t) f - ft - \frac{t^3}{3!} Af - \ldots - \frac{t^{2n-1}}{(2n-1)!} A^{n-1} f$$

$$= \int_0^t \frac{(t-s)^{2n}}{(2n)!} C(s) A^n f \, ds, \quad \text{for all } t \in \mathbb{R}, \tag{20.41}$$

all $f \in D(A^n)$.

The integrals in (20.40) and (20.41) are vector valued Riemann integrals, see [79], [164].

20.5   Poincaré Type Inequalities for Cosine and Sine Operator Functions

Here always we consider $C(t)$, $S(t)$, A as in the Cosine and Sine operator functions Background and $f \in D(A^n)$, $n \in \mathbb{N}$. Also we consider $p, q > 1 : \frac{1}{p} + \frac{1}{q} = 1$, along with $a, \nu \in \mathbb{R} : \nu > 0$.

We present the following result

Theorem 20.11. *Let $a > 0$, $t \in [0, a]$. One has*

$$1) \ \left\| \|T_n(t) f\| \right\|_{L_\nu(0,a)} \leq \frac{a^{\left(2n-1+\frac{1}{p}+\frac{1}{\nu}\right)}}{(2n-1)! \left(p(2n-1)+1\right)^{1/p}}$$

$$\cdot \frac{\left\| \|C(t) A^n f\| \right\|_{L_q(0,a)}}{\left(\nu\left(2n-1+\frac{1}{p}\right)+1\right)^{1/\nu}}. \tag{20.42}$$

When $\nu = q$ we have

$$2) \ \left\| \|T_n(t) f\| \right\|_{L_q(0,a)} \leq \frac{a^{2n}}{(2n-1)! \left(p(2n-1)+1\right)^{1/p}}$$

$$\cdot \frac{\left\| \|C(t) A^n f\| \right\|_{L_q(0,a)}}{(2qn)^{1/q}}. \tag{20.43}$$

When $\nu = q = p = 2$ we derive

$$3) \ \left\| \|T_n(t) f\| \right\|_{L_2(0,a)} \leq \frac{a^{2n}}{2(2n-1)! \sqrt{n}\sqrt{4n-1}} \cdot \left\| \|C(t) A^n f\| \right\|_{L_2(0,a)}. \tag{20.44}$$

Proof.   By (20.40) we have

$$\|T_n(t) f\| = \frac{1}{(2n-1)!} \left\| \int_0^t (t-s)^{2n-1} C(s) A^n f \, ds \right\|$$

$$\leq \frac{1}{(2n-1)!} \int_0^t (t-s)^{2n-1} \|C(s) A^n f\| \, ds$$

$$\leq \frac{1}{(2n-1)!} \left(\int_0^t (t-s)^{p(2n-1)} \, ds \right)^{1/p} \left(\int_0^t \|C(s) \, A^n f\|^q \, ds \right)^{1/q}$$

$$= \frac{1}{(2n-1)!} \frac{t^{2n-1+\frac{1}{p}}}{(p(2n-1)+1)^{1/p}} \, \||C(s) \, A^n f\||_{L_q(0,t)} \, . \tag{20.45}$$

That is

$$\|T_n(t) f\| \leq \frac{1}{(2n-1)!} \cdot \frac{t^{2n-1+\frac{1}{p}}}{(p(2n-1)+1)^{1/p}} \, \||C(s) \, A^n f\||_{L_q(0,a)} \, . \tag{20.46}$$

Consequently

$$\|T_n(t) f\|^\nu \leq \frac{1}{((2n-1)!)^\nu} \cdot \frac{t^{\nu\left(2n-1+\frac{1}{p}\right)}}{(p(2n-1)+1)^{\nu/p}} \, \||C(s) \, A^n f\||^\nu_{L_q(0,a)} \, , \tag{20.47}$$

and thus

$$\int_0^a \|T_n(t) f\|^\nu \, dt \leq \frac{1}{((2n-1)!)^\nu}$$

$$\cdot \frac{a^{\nu\left(2n-1+\frac{1}{p}\right)+1}}{(p(2n-1)+1)^{\nu/p} \left(\nu\left(2n-1+\frac{1}{p}\right)+1\right)} \, \||C(s) \, A^n f\||^\nu_{L_q(0,a)} \, . \tag{20.48}$$

Therefore

$$\left(\int_0^a \|T_n(t) f\|^\nu \, dt \right)^{1/\nu} \leq \frac{1}{(2n-1)!}$$

$$\cdot \frac{a^{\left(2n-1+\frac{1}{p}+\frac{1}{\nu}\right)} \, \||C(s) \, A^n f\||_{L_q(0,a)}}{(p(2n-1)+1)^{1/p} \left(\nu\left(2n-1+\frac{1}{p}\right)+1\right)^{1/\nu}} \, , \tag{20.49}$$

proving the claim. $\qquad\square$

Next we give the counterpart of the last theorem.

Theorem 20.12. *Let $a < 0$, $t \in [a, 0]$.*
One has

$$1) \quad \||T_n(t) f\||_{L_\nu(a,0)}$$

$$\leq \frac{(-a)^{\left(2n-1+\frac{1}{p}+\frac{1}{\nu}\right)} \, \||C(t) \, A^n f\||_{L_q(a,0)}}{(2n-1)! \, (p(2n-1)+1)^{1/p} \left(\nu\left(2n-1+\frac{1}{p}\right)+1\right)^{1/\nu}} \, . \tag{20.50}$$

When $\nu = q$ we have

$$2) \quad \||T_n(t) f\||_{L_q(a,0)} \leq \frac{a^{2n} \, \||C(t) \, A^n f\||_{L_q(a,0)}}{(2n-1)! \, (p(2n-1)+1)^{1/p} \, (2qn)^{1/q}} \, . \tag{20.51}$$

When $\nu = q = p = 2$ we obtain

$$3) \ \||T_n(t) f\||_{L_2(a,0)} \leq \frac{a^{2n} \||C(t) A^n f\||_{L_2(a,0)}}{2(2n-1)!\sqrt{n}\sqrt{4n-1}}. \tag{20.52}$$

Proof. By (20.40) we have

$$\|T_n(t) f\| = \frac{1}{(2n-1)!} \left\| \int_0^t (t-s)^{2n-1} C(s) A^n f ds \right\|$$

$$= \frac{1}{(2n-1)!} \left\| \int_t^0 (t-s)^{2n-1} C(s) A^n f ds \right\|$$

$$\leq \frac{1}{(2n-1)!} \int_t^0 (s-t)^{2n-1} \|C(s) A^n f\| ds$$

$$\leq \frac{1}{(2n-1)!} \left(\int_t^0 (s-t)^{p(2n-1)} ds \right)^{1/p} \left(\int_t^0 \|C(s) A^n f\|^q ds \right)^{1/q}$$

$$= \frac{1}{(2n-1)!} \frac{(-t)^{\left(2n-1+\frac{1}{p}\right)}}{(p(2n-1)+1)^{1/p}} \||C(s) A^n f\||_{L_q(t,0)}. \tag{20.53}$$

That is

$$\|T_n(t) f\| \leq \frac{(-t)^{\left(2n-1+\frac{1}{p}\right)}}{(2n-1)! \, (p(2n-1)+1)^{1/p}} \||C(t) A^n f\||_{L_q(a,0)}. \tag{20.54}$$

Consequently

$$\|T_n(t) f\|^\nu \leq \frac{(-t)^{\nu\left(2n-1+\frac{1}{p}\right)}}{((2n-1)!)^\nu \, (p(2n-1)+1)^{\nu/p}} \||C(t) A^n f\||_{L_q(a,0)}^\nu \tag{20.55}$$

and hence

$$\int_a^0 \|T_n(t) f\|^\nu dt \leq \frac{(-a)^{\nu\left(2n-1+\frac{1}{p}\right)+1}}{\left(\nu\left(2n-1+\frac{1}{p}\right)+1\right)}$$

$$\cdot \frac{\||C(t) A^n f\||_{L_q(a,0)}^\nu}{((2n-1)!)^\nu \, (p(2n-1)+1)^{\nu/p}}. \tag{20.56}$$

Consequently it holds

$$\left(\int_a^0 \|T_n(t) f\|^\nu dt \right)^{1/\nu} \leq \frac{1}{(2n-1)!}$$

$$\cdot \frac{(-a)^{\left(2n-1+\frac{1}{p}+\frac{1}{\nu}\right)} \||C(t) A^n f\||_{L_q(a,0)}}{(p(2n-1)+1)^{1/p} \left(\nu\left(2n-1+\frac{1}{p}\right)+1\right)^{1/\nu}}, \tag{20.57}$$

proving the claim. $\qquad\square$

Next we treat the L_1 case

Theorem 20.13. *Let $a > 0$, $t \in [0, a]$.*

One has

$$1)\ \||T_n(t)f\|\|_{L_\nu(0,a)} \le \frac{a^{\left(2n-1+\frac{1}{\nu}\right)} \||C(t)A^n f\|\|_{L_1(0,a)}}{((2n-1)!)\,(\nu(2n-1)+1)^{1/\nu}}. \tag{20.58}$$

When $\nu = 1$ we obtain

$$2)\ \||T_n(t)f\|\|_{L_1(0,a)} \le \frac{a^{2n} \||C(t)A^n f\|\|_{L_1(0,a)}}{(2n)!}. \tag{20.59}$$

Proof. By (20.40) we have

$$\|T_n(t)f\| = \frac{1}{(2n-1)!}\left\|\int_0^t (t-s)^{2n-1} C(s) A^n f\, ds\right\|$$

$$\le \frac{1}{(2n-1)!}\int_0^t (t-s)^{2n-1} \|C(s) A^n f\|\, ds \le \frac{t^{2n-1}}{(2n-1)!}\int_0^a \|C(s) A^n f\|\, ds. \tag{20.60}$$

That is

$$\|T_n(t)f\| \le \frac{t^{2n-1}}{(2n-1)!} \||C(s) A^n f\|\|_{L_1(0,a)}. \tag{20.61}$$

and

$$\|T_n(t)f\|^\nu \le \frac{t^{\nu(2n-1)}}{((2n-1)!)^\nu} \||C(s) A^n f\|\|^\nu_{L_1(0,a)}. \tag{20.62}$$

Therefore

$$\int_0^a \|T_n(t)f\|^\nu\, dt \le \frac{a^{(\nu(2n-1)+1)} \||C(t) A^n f\|\|^\nu_{L_1(0,a)}}{((2n-1)!)^\nu\,(\nu(2n-1)+1)}. \tag{20.63}$$

Finally we find

$$\left(\int_0^a \|T_n(t)f\|^\nu\, dt\right)^{1/\nu} \le \frac{a^{\left(2n-1+\frac{1}{\nu}\right)} \||C(t) A^n f\|\|_{L_1(0,a)}}{((2n-1)!)\,(\nu(2n-1)+1)^{1/\nu}}, \tag{20.64}$$

proving the claim. $\qquad\square$

Next we give the countercase of the last result.

Theorem 20.14. *Let $a < 0$, $t \in [a, 0]$.*

One has

$$1)\ \||T_n(t)f\|\|_{L_\nu(a,0)} \le \frac{(-a)^{\left(2n-1+\frac{1}{\nu}\right)} \||C(t)A^n f\|\|_{L_1(a,0)}}{((2n-1)!)\,(\nu(2n-1)+1)^{1/\nu}}. \tag{20.65}$$

When $\nu = 1$ we derive

$$2)\ \||T_n(t)f\|\|_{L_1(a,0)} \le \frac{a^{2n} \||C(t)A^n f\|\|_{L_1(a,0)}}{(2n)!}. \tag{20.66}$$

Proof.   By (20.40) we have

$$\|T_n(t)f\| = \frac{1}{(2n-1)!}\left\|\int_t^0 (t-s)^{2n-1} C(s) A^n f\, ds\right\|$$

$$\leq \frac{1}{(2n-1)!}\int_t^0 (s-t)^{2n-1}\|C(s)A^n f\|\, ds$$

$$\leq \frac{(-t)^{2n-1}}{(2n-1)!}\int_a^0 \|C(s)A^n f\|\, ds. \tag{20.67}$$

That is we see that

$$\|T_n(t)f\| \leq \frac{(-t)^{2n-1}}{(2n-1)!}\, \|\|C(t)A^n f\|\|_{L_1(a,0)} \tag{20.68}$$

and

$$\|T_n(t)f\|^{\nu} \leq \frac{(-t)^{\nu(2n-1)}}{((2n-1)!)^{\nu}}\, \|\|C(t)A^n f\|\|^{\nu}_{L_1(a,0)}. \tag{20.69}$$

Consequently we obtain

$$\int_a^0 \|T_n(t)f\|^{\nu}\, dt \leq \frac{(-a)^{(\nu(2n-1)+1)}\, \|\|C(t)A^n f\|\|^{\nu}_{L_1(a,0)}}{((2n-1)!)^{\nu}\,(\nu(2n-1)+1)}. \tag{20.70}$$

Finally we observe that

$$\left(\int_a^0 \|T_n(t)f\|^{\nu}\, dt\right)^{1/\nu} \leq \frac{(-a)^{\left(2n-1+\frac{1}{\nu}\right)}\, \|\|C(t)A^n f\|\|_{L_1(a,0)}}{((2n-1)!)\,(\nu(2n-1)+1)^{1/\nu}}, \tag{20.71}$$

proving the claim. $\square$

Next we continue with Poincaré like inequalities involving the Sine operator function.

Theorem 20.15. *Let $a > 0$, $t \in [0,a]$.*
   One has

$$1)\quad \|\|M_n(t)f\|\|_{L_{\nu}(0,a)} \leq \frac{a^{\left(2n+\frac{1}{p}+\frac{1}{\nu}\right)}\, \|\|C(t)A^n f\|\|_{L_q(0,a)}}{((2n)!)\,(2pn+1)^{1/p}\left(\nu\left(2n+\frac{1}{p}\right)+1\right)^{1/\nu}}. \tag{20.72}$$

When $\nu = q$ we have

$$2)\quad \|\|M_n(t)f\|\|_{L_q(0,a)} \leq \frac{a^{2n}\, \|\|C(t)A^n f\|\|_{L_q(0,a)}}{((2n)!)\,(2pn+1)^{1/p}\,(q(2n+1))^{1/q}}. \tag{20.73}$$

When $\nu = q = p = 2$ we obtain

$$3)\quad \|\|M_n(t)f\|\|_{L_2(0,a)} \leq \frac{a^{2n}\, \|\|C(t)A^n f\|\|_{L_2(0,a)}}{((2n)!)\,\sqrt{4n+1}\,\sqrt{2(2n+1)}}. \tag{20.74}$$

Proof.   By (20.41) we have

$$\|M_n(t)f\| = \frac{1}{(2n)!}\left\|\int_0^t (t-s)^{2n}\,C(s)\,A^n f\,ds\right\|$$

$$\leq \frac{1}{(2n)!}\int_0^t (t-s)^{2n}\,\|C(s)\,A^n f\|\,ds$$

$$\leq \frac{1}{(2n)!}\left(\int_0^t (t-s)^{2pn}\,ds\right)^{1/p}\,\left\|\|C(t)\,A^n f\|\right\|_{L_q(0,a)}$$

$$= \frac{t^{2n+\frac{1}{p}}}{(2n)!\,(2pn+1)^{1/p}}\,\left\|\|C(t)\,A^n f\|\right\|_{L_q(0,a)}. \tag{20.75}$$

Consequently

$$\|M_n(t)f\|^{\nu} \leq \frac{t^{\nu\left(2n+\frac{1}{p}\right)}}{((2n)!)^{\nu}\,(2pn+1)^{\nu/p}}\,\left\|\|C(t)\,A^n f\|\right\|_{L_q(0,a)}^{\nu}, \tag{20.76}$$

and

$$\int_0^a \|M_n(t)f\|^{\nu}\,dt \leq \frac{a^{\left(\nu\left(2n+\frac{1}{p}\right)+1\right)}}{((2n)!)^{\nu}} \cdot \frac{\left\|\|C(t)\,A^n f\|\right\|_{L_q(0,a)}^{\nu}}{(2pn+1)^{\nu/p}\left(\nu\left(2n+\frac{1}{p}\right)+1\right)}, \tag{20.77}$$

proving the claim. $\square$

Next we give the counterpart of the last theorem.

Theorem 20.16. *Let $a < 0$, $t \in [a,0]$.*
One has

1) $\left\|\|M_n(t)f\|\right\|_{L_{\nu}(a,0)} \leq \dfrac{(-a)^{\left(2n+\frac{1}{p}+\frac{1}{\nu}\right)}\,\left\|\|C(t)\,A^n f\|\right\|_{L_q(a,0)}}{((2n)!)\,(2pn+1)^{1/p}\left(\nu\left(2n+\frac{1}{p}\right)+1\right)^{1/\nu}}. \tag{20.78}$

When $\nu = q$ we have

2) $\left\|\|M_n(t)f\|\right\|_{L_q(a,0)} \leq \dfrac{a^{2n}\,\left\|\|C(t)\,A^n f\|\right\|_{L_q(a,0)}}{((2n)!)\,(2pn+1)^{1/p}\,(q(2n+1))^{1/q}}. \tag{20.79}$

When $\nu = q = p = 2$ we get

3) $\left\|\|M_n(t)f\|\right\|_{L_2(a,0)} \leq \dfrac{a^{2n}\,\left\|\|C(t)\,A^n f\|\right\|_{L_2(a,0)}}{((2n)!)\,\sqrt{4n+1}\,\sqrt{2(2n+1)}}. \tag{20.80}$

Proof.   By (20.41) we have

$$\|M_n(t)f\| = \frac{1}{(2n)!}\left\|\int_t^0 (s-t)^{2n}\,C(s)\,A^n f\,ds\right\|$$

$$\leq \frac{1}{(2n)!}\int_t^0 (s-t)^{2n}\,\|C(s)\,A^n f\|\,ds$$

$$\leq \frac{1}{(2n)!} \left(\int_t^0 (s-t)^{2pn} \, ds \right)^{1/p} \||| C(t) A^n f \||\|_{L_q(a,0)}$$

$$= \frac{1}{(2n)!} \frac{(-t)^{2n+\frac{1}{p}}}{(2pn+1)^{1/p}} \||| C(t) A^n f \||\|_{L_q(a,0)} . \tag{20.81}$$

Therefore

$$\| M_n(t) f \|^\nu \leq \frac{(-t)^{\nu\left(2n+\frac{1}{p}\right)}}{((2n)!)^\nu (2pn+1)^{\nu/p}} \||| C(t) A^n f \||\|_{L_q(a,0)}^\nu , \tag{20.82}$$

and

$$\int_a^0 \| M_n(t) f \|^\nu \, dt \leq \frac{(-a)^{\left(\nu\left(2n+\frac{1}{p}\right)+1\right)}}{\left(\nu\left(2n+\frac{1}{p}\right)+1\right)} \cdot \frac{\||| C(t) A^n f \||\|_{L_q(a,0)}^\nu}{((2n)!)^\nu (2pn+1)^{\nu/p}}, \tag{20.83}$$

proving the claim. $\qquad\square$

It follows the corresponding L_1 treatment

Theorem 20.17. *Let $a > 0$, $t \in [0, a]$.*
One has

$$1) \ \||| M_n(t) f \||\|_{L_\nu(0,a)} \leq \frac{a^{\left(2n+\frac{1}{\nu}\right)} \||| C(t) A^n f \||\|_{L_1(0,a)}}{((2n)!) (2\nu n + 1)^{1/\nu}} . \tag{20.84}$$

When $\nu = 1$ we find

$$2) \ \||| M_n(t) f \||\|_{L_1(0,a)} \leq \frac{a^{(2n+1)} \||| C(t) A^n f \||\|_{L_1(0,a)}}{(2n+1)!} . \tag{20.85}$$

Proof. By (20.41) and (20.75) we have

$$\| M_n(t) f \| \leq \frac{1}{(2n)!} \int_0^t (t-s)^{2n} \| C(s) A^n f \| \, ds$$

$$\leq \frac{t^{2n}}{(2n)!} \||| C(t) A^n f \||\|_{L_1(0,a)} . \tag{20.86}$$

Hence

$$\| M_n(t) f \|^\nu \leq \frac{t^{2\nu n}}{((2n)!)^\nu} \||| C(t) A^n f \||\|_{L_1(0,a)}^\nu \tag{20.87}$$

and

$$\int_0^a \| M_n(t) f \|^\nu \, dt \leq \frac{a^{(2\nu n+1)} \||| C(t) A^n f \||\|_{L_1(0,a)}^\nu}{((2n)!)^\nu (2\nu n + 1)} , \tag{20.88}$$

proving the claim. $\qquad\square$

The counterpart of last theorem follows.

Theorem 20.18. *Let* $a < 0$, $t \in [a, 0]$.
One has

$$1) \quad |||M_n(t)f|||_{L_\nu(a,0)} \leq \frac{(-a)^{\left(2n+\frac{1}{\nu}\right)} |||C(t)A^n f|||_{L_1(a,0)}}{((2n)!)(2\nu n + 1)^{1/\nu}}. \tag{20.89}$$

When $\nu = 1$ *we derive*

$$2) \quad |||M_n(t)f|||_{L_1(a,0)} \leq \frac{(-a)^{(2n+1)} |||C(t)A^n f|||_{L_1(a,0)}}{(2n+1)!}. \tag{20.90}$$

Proof. By (20.41) and (20.81) we have

$$\|M_n(t)f\| \leq \frac{1}{(2n)!} \int_t^0 (s-t)^{2n} \|C(s)A^n f\| \, ds$$

$$\leq \frac{(-t)^{2n}}{(2n)!} |||C(t)A^n f|||_{L_1(a,0)}. \tag{20.91}$$

Thus

$$\|M_n(t)f\|^\nu \leq \frac{(-t)^{2\nu n}}{((2n)!)^\nu} |||C(t)A^n f|||^\nu_{L_1(a,0)} \tag{20.92}$$

and

$$\int_a^0 \|M_n(t)f\|^\nu \, dt \leq \frac{(-a)^{2\nu n + 1} |||C(t)A^n f|||^\nu_{L_1(a,0)}}{((2n)!)^\nu (2\nu n + 1)}, \tag{20.93}$$

proving the claim. $\square$

Application 20.19. (see [147], p. 121)
Let X be the Banach space of odd, $2\pi-$periodic real functions in the space of bounded uniformly continuous functions from $\mathbb{R}$ into itself: $BUC(\mathbb{R})$. Let $A := \frac{d^2}{dx^2}$ with $D(A^n) = \{f \in X : f^{(2k)} \in X, \ k = 1, \ldots, n\}$, $n \in \mathbb{N}$. A generates a Cosine function C^* given by

$$C^*(t)f(x) = \frac{1}{2}[f(x+t) + f(x-t)], \quad \forall x, t \in \mathbb{R}. \tag{20.94}$$

The corresponding Sine function S^* is given by

$$S^*(t)f(x) = \frac{1}{2}\left[\int_0^t f(x+s)\,ds + \int_0^t f(x-s)\,ds\right], \quad \forall x, t \in \mathbb{R}. \tag{20.95}$$

Here we consider $f \in D(A^n)$, $n \in \mathbb{N}$, as above. By (20.40) we obtain

$$T_n^*(t)f := \frac{1}{2}[f(\cdot + t) + f(\cdot - t)] - \sum_{k=0}^{n-1} \frac{t^{2k}}{(2k)!} f^{(2k)}$$

$$- \int_0^t \frac{(t-s)^{2n-1}}{2(2n-1)!} \left[f^{(2n)}(\cdot + s) + f^{(2n)}(\cdot - s)\right] ds, \quad \forall t \in \mathbb{R}. \tag{20.96}$$

By (20.41) we get

$$M_n^* (t) f := \frac{1}{2} \left[\int_0^t f (\cdot + s) \, ds + \int_0^t f (\cdot - s) \, ds \right] - \sum_{k=1}^n \frac{t^{2k-1}}{(2k-1)!} f^{(2(k-1))}$$

$$= \int_0^t \frac{(t - s)^{2n}}{2 \, (2n)!} \left[f^{(2n)} (\cdot + s) + f^{(2n)} (\cdot - s) \right] ds, \quad \forall t \in \mathbb{R}. \tag{20.97}$$

Let $g \in BUC (\mathbb{R})$, we define $\|g\| = \|g\|_\infty := \sup_{x \in \mathbb{R}} |g (x)| < \infty$.

Notice also that

$$\left\| \|C^* (s) A^n f\|_\infty \right\|_\infty = \left\| \left\| C^* (s) f^{(2n)} \right\|_\infty \right\|_\infty$$

$$= \frac{1}{2} \left\| \left\| \left(f^{(2n)} (\cdot + s) + f^{(2n)} (\cdot - s) \right) \right\|_\infty \right\|_\infty$$

$$\leq \frac{1}{2} \left[\left\| \left\| f^{(2n)} (\cdot + s) \right\|_\infty \right\|_\infty + \left\| \left\| f^{(2n)} (\cdot - s) \right\|_\infty \right\|_\infty \right] \leq \Theta < \infty, \tag{20.98}$$

where

$$\Theta := \left\| f^{(2n)} \right\|_\infty . \tag{20.99}$$

Let $a > 0$, $t \in [0, a]$. By (20.42) we derive

$$\left\| \|T_n^* (t) f\| \right\|_{L_\nu (0,a)} \leq \frac{a^{\left(2n - 1 + \frac{1}{p} + \frac{1}{\nu} \right)}}{(2n - 1)! \, (p \, (2n - 1) + 1)^{1/p}}$$

$$\cdot \frac{\left\| \left\| C^* (t) f^{(2n)} \right\| \right\|_{L_q (0,a)}}{\left(\nu \left(2n - 1 + \frac{1}{p} \right) + 1 \right)^{1/\nu}} . \tag{20.100}$$

Also by (20.72) we get

$$\left\| \|M_n^* (t) f\| \right\|_{L_\nu (0,a)} \leq \frac{a^{\left(2n + \frac{1}{p} + \frac{1}{\nu} \right)}}{((2n)!) \, (2pn + 1)^{1/p}}$$

$$\cdot \frac{\left\| \left\| C^* (t) f^{(2n)} \right\| \right\|_{L_q (0,a)}}{\left(\nu \left(2n + \frac{1}{p} \right) + 1 \right)^{1/\nu}} . \tag{20.101}$$

By (20.44) we find

$$\left\| \|T_n^* (t) f\| \right\|_{L_2 (0,a)} \leq \frac{a^{2n}}{2 \, (2n - 1)! \sqrt{n} \sqrt{4n - 1}}$$

$$\cdot \left\| \left\| C^* (t) f^{(2n)} \right\| \right\|_{L_2 (0,a)} . \tag{20.102}$$

By (20.74) we have

$$\left\|\left\|M_n^*\left(t\right)f\right\|\right\|_{L_2(0,a)} \le \frac{a^{2n}}{\left((2n)!\right)\sqrt{4n+1}}$$

$$\cdot \frac{\left\|\left\|C^*\left(t\right)f^{(2n)}\right\|\right\|_{L_2(0,a)}}{\sqrt{2\left(2n+1\right)}}. \tag{20.103}$$

Also by (20.58) we obtain

$$\left\|\left\|T_n^*\left(t\right)f\right\|\right\|_{L_\nu(0,a)} \le \frac{a^{\left(2n-1+\frac{1}{\nu}\right)}\left\|\left\|C^*\left(t\right)f^{(2n)}\right\|\right\|_{L_1(0,a)}}{\left((2n-1)!\right)\left(\nu\left(2n-1\right)+1\right)^{1/\nu}}, \tag{20.104}$$

and by (20.84) we observe that

$$\left\|\left\|M_n^*\left(t\right)f\right\|\right\|_{L_\nu(0,a)} \le \frac{a^{\left(2n+\frac{1}{\nu}\right)}\left\|\left\|C^*\left(t\right)f^{(2n)}\right\|\right\|_{L_1(0,a)}}{\left((2n)!\right)\left(2\nu n+1\right)^{1/\nu}}. \tag{20.105}$$

Let now $a<0$, $t\in[a,0]$. Then one by (20.50) has

$$\left\|\left\|T_n^*\left(t\right)f\right\|\right\|_{L_\nu(a,0)} \le \frac{(-a)^{\left(2n-1+\frac{1}{p}+\frac{1}{\nu}\right)}}{(2n-1)!\left(p\left(2n-1\right)+1\right)^{1/p}}$$

$$\cdot \frac{\left\|\left\|C^*\left(t\right)f^{(2n)}\right\|\right\|_{L_q(a,0)}}{\left(\nu\left(2n-1+\frac{1}{p}\right)+1\right)^{1/\nu}}. \tag{20.106}$$

and by (20.78) we get

$$\left\|\left\|M_n^*\left(t\right)f\right\|\right\|_{L_\nu(a,0)} \le \frac{(-a)^{\left(2n+\frac{1}{p}+\frac{1}{\nu}\right)}}{\left((2n)!\right)\left(2pn+1\right)^{1/p}}$$

$$\cdot \frac{\left\|\left\|C^*\left(t\right)f^{(2n)}\right\|\right\|_{L_q(a,0)}}{\left(\nu\left(2n+\frac{1}{p}\right)+1\right)^{1/\nu}}. \tag{20.107}$$

By (20.66) we find

$$\left\|\left\|T_n^*\left(t\right)f\right\|\right\|_{L_1(a,0)} \le \frac{a^{2n}\left\|\left\|C^*\left(t\right)f^{(2n)}\right\|\right\|_{L_1(a,0)}}{(2n)!}, \tag{20.108}$$

and by (20.90) we see that

$$\left\|\left\|M_n^*\left(t\right)f\right\|\right\|_{L_1(a,0)} \le \frac{(-a)^{(2n+1)}\left\|\left\|C^*\left(t\right)f^{(2n)}\right\|\right\|_{L_1(a,0)}}{(2n+1)!}. \tag{20.109}$$

Similarly one can apply the rest of the results here on Cosine and Sine Operator functions.

Chapter 21

Hardy–Opial Type Inequalities

Various L_p form Hardy–Opial type sharp integral inequalities are presented involving two functions. This chapter follows [32].

21.1 Results

Let $f \in L_p([a,b])$ and $g \in L_q([a,b])$, with p,q be such that $\frac{1}{p} + \frac{1}{q} = 1$. We consider the *generalized Hardy type operators* (for the basic Hardy operator see [204], p. 306),

$$T_g f(t) = \int_a^t g(s) f(s) ds,$$

and

$$T_g^* f(t) = \int_t^b g(s) f(s) ds.$$

We present here integral Hardy–Opial type sharp L_p-inequalities involving or related to T_g, T_g^*. Here $\mathcal{L} \int$ stands for Lebesque integral, $\mathcal{R} \int$ for Riemann integral and $(\mathcal{R} - \mathcal{S}) \int$ stands for Riemann–Stieltjes integral.

The first result follows.

Theorem 21.1. *Let $p, q > 1$ such that $\frac{1}{p} + \frac{1}{q} = 1$, with $f \in L_p([a,b])$, $g \in L_q([a,b])$, $x_0 \in [a,b]$ be fixed. Then*

$$\mathcal{L} \int_{x_0}^s \left(\mathcal{L} \int_{x_0}^w |g(t) f(t)| dt \right) |f(w)| dw$$

$$\leq 2^{-1/p} \left(\mathcal{R} \int_{x_0}^s \left(\mathcal{L} \int_{x_0}^w |g(t)|^q dt \right) dw \right)^{1/q} \left(\mathcal{L} \int_{x_0}^s |f(t)|^p dt \right)^{2/p}, \quad \text{all } x_0 \leq s \leq b.$$

$$(21.1)$$

Inequality (21.1) is sharp, that is attained when $f(t) = g(t) = c > 0$ a constant, for all $t \in [x_0, b]$.

261

Proof.   1) For $x_0 \leq w \leq b$, by Hölder's inequality we have

$$\mathcal{L}\int_{x_0}^{w}|g(t)f(t)|dt \leq \left(\mathcal{L}\int_{x_0}^{w}|g(t)|^q dt\right)^{1/q}\left(\mathcal{L}\int_{x_0}^{w}|f(t)|^p dt\right)^{1/p}$$

$$= \left(\mathcal{L}\int_{x_0}^{w}|g(t)|^q dt\right)^{1/q}(z(w))^{1/p}, \tag{21.2}$$

where

$$z(w) := \mathcal{L}\int_{x_0}^{w}|f(t)|^p dt \geq 0, \qquad z(x_0) = 0. \tag{21.3}$$

Here z is an absolutely continuous function. Hence we have a.e. that

$$z'(w) = |f(w)|^p \geq 0, \tag{21.4}$$

and

$$|f(w)| = (z'(w))^{1/p}, \quad \text{a.e.} \tag{21.5}$$

Therefore it holds

$$\left(\mathcal{L}\int_{x_0}^{w}|g(t)f(t)|dt\right)|f(w)| \leq \left(\mathcal{L}\int_{x_0}^{w}|g(t)|^q dt\right)^{1/q}(z(w)z'(w))^{1/p}, \tag{21.6}$$

a.e. on $[x_0, b]$.

Hence by integrating (21.6) over $[x_0, s]$, where $s \in [x_0, b]$, we obtain

$$\mathcal{L}\int_{x_0}^{s}\left(\mathcal{L}\int_{x_0}^{w}|g(t)f(t)|dt\right)|f(w)|dw$$

$$\leq \mathcal{L}\int_{x_0}^{s}\left(\mathcal{L}\int_{x_0}^{w}|g(t)|^q dt\right)^{1/q}(z(w)z'(w))^{1/p}dw \tag{21.7}$$

(using again Hölder's inequality)

$$\leq \left(\mathcal{R}\int_{x_0}^{s}\left(\mathcal{L}\int_{x_0}^{w}|g(t)|^q dt\right)dw\right)^{1/q}\left(\mathcal{L}\int_{x_0}^{s}z(w)z'(w)dw\right)^{1/p} \tag{21.8}$$

$$= \left(\mathcal{R}\int_{x_0}^{s}\left(\mathcal{L}\int_{x_0}^{w}|g(t)|^q dt\right)dw\right)^{1/q} \tag{21.9}$$

$$\left((\mathcal{R}-\mathcal{S})\int_{x_0}^{s}z(w)dz(w)\right)^{1/p} = \left(\mathcal{R}\int_{x_0}^{s}\left(\mathcal{L}\int_{x_0}^{w}|g(t)|^q dt\right)dw\right)^{1/q}\left(\frac{z^2(s)}{2}\right)^{1/p} \tag{21.10}$$

$$= 2^{-1/p}\left(\mathcal{R}\int_{x_0}^{s}\left(\mathcal{L}\int_{x_0}^{w}|g(t)|^q dt\right)dw\right)^{1/q}\left(\mathcal{L}\int_{x_0}^{s}|f(t)|^p dt\right)^{2/p}. \tag{21.11}$$

That is proving inequality (21.1).

2) The sharpness follows. We see that

$$\text{L.H.S.}(21.1) = \text{R.H.S.}(21.1) = \frac{c^3(s-x_0)^2}{2}, \tag{21.12}$$

establishing attainability of (21.1). $\square$

We also give

Theorem 21.2. *Let* $p, q > 1$ *such that* $\frac{1}{p} + \frac{1}{q} = 1$, *with* $f \in L_p([a,b])$, $g \in L_q([a,b])$, $x_0 \in [a,b]$ *be fixed. Then*

$$\mathcal{L} \int_s^{x_0} \left(\mathcal{L} \int_w^{x_0} |g(t)f(t)|dt \right) |f(w)|dw$$

$$\leq 2^{-1/p} \left(\mathcal{R} \int_s^{x_0} \left(\mathcal{L} \int_w^{x_0} |g(t)|^q dt \right) dw \right)^{1/q} \left(\mathcal{L} \int_s^{x_0} |f(t)|^p dt \right)^{2/p},$$

all $a \leq s \leq x_0$. $\qquad\qquad\qquad\qquad\qquad\qquad\qquad\qquad\qquad\qquad\qquad\qquad$ (21.13)

Inequality (21.13) is sharp, that is attained when $f(t) = g(t) = c > 0$ *a constant, for all* $t \in [a, x_0]$.

Proof. 1) For $a \leq w \leq x_0$, by Hölder's inequality we have

$$\mathcal{L} \int_w^{x_0} |g(t)f(t)|dt \leq \left(\mathcal{L} \int_w^{x_0} |g(t)|^q dt \right)^{1/q} \left(\mathcal{L} \int_w^{x_0} |f(t)|^p dt \right)^{1/p} \quad (21.14)$$

$$= \left(\mathcal{L} \int_w^{x_0} |g(t)|^q dt \right)^{1/q} (z(w))^{1/p}, \quad (21.15)$$

where

$$z(w) := \mathcal{L} \int_w^{x_0} |f(t)|^p dt \geq 0, \qquad z(x_0) = 0. \quad (21.16)$$

That is

$$z(w) = \mathcal{L} \int_a^{x_0} |f(t)|^p dt - \mathcal{L} \int_a^{w} |f(t)|^p dt. \quad (21.17)$$

Here z is an absolutely continuous function.

Hence we have a.e. that

$$z'(w) = -|f(w)|^p \leq 0, \quad (21.18)$$

and a.e. that

$$-z'(w) = |f(w)|^p \geq 0, \quad (21.19)$$

and a.e. that

$$|f(w)| = (-z'(w))^{1/p}. \quad (21.20)$$

Therefore it holds

$$\left(\mathcal{L} \int_w^{x_0} |g(t)f(t)|dt \right) |f(w)| \leq \left(\mathcal{L} \int_w^{x_0} |g(t)|^q dt \right)^{1/q} (z(w)(-z'(w))^{1/p}, \quad (21.21)$$

a.e. on $[a, x_0]$. Integrating inequality (21.21) over $[s, x_0]$, where $s \in [a, x_0]$, we derive

$$\mathcal{L} \int_s^{x_0} \left(\mathcal{L} \int_w^{x_0} |g(t)f(t)|dt \right) |f(w)|dw$$

$$\leq \mathcal{L} \int_s^{x_0} \left(\mathcal{L} \int_w^{x_0} |g(t)|^q dt \right)^{1/q} \left(z(w)(-z'(w)) \right)^{1/p} dw \qquad (21.22)$$

(again using Hölder's inequality)

$$\leq \left(\mathcal{L} \int_s^{x_0} \left(\mathcal{L} \int_w^{x_0} |g(t)|^q dt \right) dw \right)^{1/q} \left(\mathcal{L} \int_s^{x_0} z(w)(-z'(w))dw \right)^{1/p} \qquad (21.23)$$

$$= \left(\mathcal{R} \int_s^{x_0} \left(\mathcal{L} \int_w^{x_0} |g(t)|^q dt \right) dw \right)^{1/q} \left(-(\mathcal{R} - S) \int_s^{x_0} z(w)dz(w) \right)^{1/p} \qquad (21.24)$$

$$= \left(\mathcal{R} \int_s^{x_0} \left(\mathcal{L} \int_w^{x_0} |g(t)|^q dt \right) dw \right)^{1/q} \left(\frac{z^2(s)}{2} \right)^{1/p} \qquad (21.25)$$

$$= 2^{-1/p} \left(\mathcal{R} \int_s^{x_0} \left(\mathcal{L} \int_w^{x_0} |g(t)|^q dt \right) dw \right)^{1/q} \left(\mathcal{L} \int_s^{x_0} |f(t)|^p dt \right)^{2/p}. \qquad (21.26)$$

That is proving (21.13).

2) The sharpness follows. We see that

$$\text{L.H.S.}(21.13) = \text{R.H.S.}(21.13) = \frac{c^3(x_0 - s)^2}{2}, \qquad (21.27)$$

establishing attainability of (21.13). $\square$

We have

Corollary 21.3. *Let $p, q > 1$ such that $\frac{1}{p} + \frac{1}{q} = 1$, with $f \in L_p([a,b])$, $g \in L_q([a,b])$. Then*

$$\|fg\|_1 \|f\|_1 \leq 2^{-1/p} \|f\|_p^2 \left[\left(\mathcal{R} \int_a^b \left(\mathcal{L} \int_a^w |g(t)|^q dt \right) dw \right)^{1/q} \right.$$

$$\left. + \left(\mathcal{R} \int_a^b \left(\mathcal{L} \int_w^b |g(t)|^q dt \right) dw \right)^{1/q} \right] \leq (b-a)^{1/q} \|f\|_p^2 \|g\|_q. \qquad (21.28)$$

Inequality is sharp, attained when $f = g = c > 0$.

Proof. We apply Theorem 21.1 for $x_0 = a$ and $s = b$. We have

$$\mathcal{L} \int_a^b \left(\mathcal{L} \int_a^w |g(t)f(t)|dt \right) |f(w)|dw$$

$$\leq 2^{-1/p} \left(\mathcal{R} \int_a^b \left(\mathcal{L} \int_a^w |g(t)|^q dt \right) dw \right)^{1/q} \left(\mathcal{L} \int_a^b |f(t)|^p dt \right)^{2/p}. \qquad (21.29)$$

We also apply Theorem 21.2 for $x_0 = b$, $s = a$. We find

$$\mathcal{L} \int_a^b \left(\mathcal{L} \int_w^b |g(t)f(t)|dt \right) |f(w)|dw$$

$$\leq 2^{-1/p} \left(\mathcal{R} \int_a^b \left(\mathcal{L} \int_w^b |g(t)|^q dt \right) dw \right)^{1/q} \left(\mathcal{L} \int_a^b |f(t)|^p dt \right)^{2/p}. \qquad (21.30)$$

We add (21.29) and (21.30) to observe that

$$\mathcal{L}\int_a^b\left(\mathcal{L}\int_a^w |g(t)f(t)|dt + \mathcal{L}\int_w^b |g(t)f(t)|dt\right)|f(w)|dw$$

$$\leq 2^{-1/p}\|f\|_p^2\left[\left(\mathcal{R}\int_a^b\left(\mathcal{L}\int_a^w |g(t)|^q dt\right)dw\right)^{1/q}\right.$$

$$\left.+\left(\mathcal{R}\int_a^b\left(\mathcal{L}\int_w^b |g(t)|^q dt\right)dw\right)^{1/q}\right]. \tag{21.31}$$

But it holds

$$\text{L.H.S.}(21.31) = \|fg\|_1\|f\|_1, \tag{21.32}$$

proving first inequality in (21.28).

Since $1/q < 1$ we get

$$\text{R.H.S.}(21.31) \leq 2^{-1/p}\|f\|_p^2 2^{1-\frac{1}{q}}\left[\mathcal{R}\int_a^b\left(\mathcal{L}\int_a^w |g(t)|^q dt\right)dw\right.$$

$$\left.+\mathcal{R}\int_a^b\left(\mathcal{L}\int_w^b |g(t)|^q dt\right)dw\right]^{1/q} \tag{21.33}$$

$$= \|f\|_p^2\left[\mathcal{R}\int_a^b\left(\mathcal{L}\int_a^b |g(t)|^q dt\right)dw\right]^{1/q} \tag{21.34}$$

$$= \|f\|_p^2\|g\|_q(b-a)^{1/q}, \tag{21.35}$$

proving second part of (21.28). Sharpness is obvious, by all three parts being equal to $c^3(b-a)^2$, when $f = g = c > 0$. $\qquad\square$

Next comes

Corollary 21.4. *Let $f, g \in L_2([0,1])$. Then*

$$\mathcal{L}\int_0^s\left(\mathcal{L}\int_0^w |g(t)f(t)|dt\right)|f(w)|dw$$

$$\leq \frac{\sqrt{2}}{2}\left(\mathcal{R}\int_0^s\left(\mathcal{L}\int_0^w g^2(t)dt\right)dw\right)^{1/2}\left(\mathcal{L}\int_0^s f^2(t)dt\right), all\ 0 \leq s \leq 1. \tag{21.36}$$

Inequality (21.36) is sharp, that is attained when $f(t) = g(t) = 1$, all $t \in [0,1]$.

Proof. Apply Theorem 21.1 for $p = q = 2$ on $[0,1]$, and $x_0 = 0$. $\qquad\square$

We obtain the following Opial type (see [75], [6], p. 8) basic inequality which implies Opial's inequality [195].

Corollary 21.5. *Let $f \in C^1([0,1])$ such that $f(0) = 0$. Then*

$$\int_0^s |f(t)f'(t)|dt \leq \frac{s}{2}\int_0^s (f'(t))^2 dt, \quad \forall \varepsilon \subset [0,1]. \tag{21.37}$$

Inequality (21.37) is sharp, namely it is attained by $f(t) = t$.

Proof. We apply (21.36) for $g(t) = 1$, $\forall t \in [0,1]$, and in the place of f we plug in f'. We then have by (21.36) that

$$\int_0^s |f(t)f'(t)|dt$$

$$= \int_0^s \left| \int_0^w f'(t)dt \right| |f'(w)|dw \leq \int_0^s \left(\int_0^w |f'(t)|dt \right) |f'(w)|dw \qquad (21.38)$$

$$\leq \frac{\sqrt{2}}{2} \left(\int_0^s w\,dw \right)^{1/2} \left(\int_0^s (f'(t))^2 dt \right) \qquad (21.39)$$

$$= \frac{s}{2} \int_0^s (f'(t))^2 dt. \qquad (21.40)$$

That is proving (21.37). Sharpness now is obvious. $\qquad \square$

The extreme case follows.

Proposition 21.6. *Let $f \in L_\infty([a,b])$, $g \in L_1([a,b])$, $x_0 \in [a,b]$ be fixed. Then*
i)

$$\mathcal{L}\int_{x_0}^s \left(\mathcal{L}\int_{x_0}^w |g(t)f(t)|dt \right) |f(w)|dw \leq \|f\|_\infty^2 \left(\mathcal{R}\int_{x_0}^s \left(\mathcal{L}\int_{x_0}^w |g(t)|dt \right) dw \right),$$

$$(21.41)$$

for all $s \in [x_0, b]$.
Furthermore
ii)

$$\mathcal{L}\int_s^{x_0} \left(\mathcal{L}\int_w^{x_0} |g(t)f(t)|dt \right) |f(w)|dw \leq \|f\|_\infty^2 \left(\mathcal{R}\int_s^{x_0} \left(\mathcal{L}\int_w^{x_0} |g(t)|dt \right) dw \right),$$

$$(21.42)$$

for all $s \in [a, x_0]$.
Both (21.41), (21.42) are sharp, attained by $f(t) = g(t) = c > 0$.

Proof. Obvious. $\qquad \square$

To complete the chapter we present

Theorem 21.7. *Let $0 < q < 1$ and $p < 0$ such that $\frac{1}{p} + \frac{1}{q} = 1$, a fixed $x_0 \in [a,b]$ and $s \in (x_0, b]$. Assume that $g \in L_q([a,b])$, $f \in L_p([a,b])$, f is nowhere zero and $|f| \leq k$ a.e., $k > 0$. Then*

$$\mathcal{L}\int_{x_0}^s \left(\mathcal{L}\int_{x_0}^w |g(t)f(t)|dt \right) |f(w)|dw$$

$$\geq 2^{-1/p} \left(\mathcal{R}\int_{x_0}^s \left(\mathcal{L}\int_{x_0}^w |g(t)|^q dt \right) dw \right)^{1/q} \left(\mathcal{L}\int_{x_0}^s |f(t)|^p dt \right)^{2/p}. \qquad (21.43)$$

Inequality (21.43) is sharp, namely it is attained when $f(t) = g(t) = c > 0$, for all $t \in [x_0, b]$.

Proof. Let $x_0 < w \le s$. Here $|f| > 0$ and $|f| \le k$ a.e., $k > 0$. Hence $\frac{1}{k} \le \frac{1}{|f|}$ a.e., $-p > 0$ and $k^p \le |f|^p$ a.e. Therefore

$$z(w) := \mathcal{L} \int_{x_0}^{w} |f(x)|^p dx \ge k^p (w - x_0) > 0, \ z(x_0) = 0, \tag{21.44}$$

and when $w_2 > w_1$ we derive

$$z(w_2) - z(w_1) = \mathcal{L} \int_{w_1}^{w_2} |f|^p dx \ge k^p (w_2 - w_1) > 0. \tag{21.45}$$

That is proving that $z(w)$ is strictly increasing on $[x_0, s]$, and z also is an absolutely continuous function there.

Next, by reverse Hölder's inequality we obtain

$$\mathcal{L} \int_{x_0}^{w} |g(t)f(t)| dt \ge \left(\mathcal{L} \int_{x_0}^{w} |g(t)|^q dt \right)^{1/q} \left(\mathcal{L} \int_{x_0}^{w} |f(t)|^p dt \right)^{1/p} \tag{21.46}$$

$$= \left(\mathcal{L} \int_{x_0}^{w} |g(t)|^q dt \right)^{1/q} (z(w))^{1/p}, \text{ for } x_0 < w \le s. \tag{21.47}$$

It holds a.e. on $[x_0, s]$ that

$$z'(w) = |f(w)|^p > 0$$

and

$$|f(w)| = (z'(w))^{1/p}, \quad \text{a.e. on } [x_0, s]. \tag{21.48}$$

Therefore

$$\left(\mathcal{L} \int_{x_0}^{w} |g(t)f(t)| dt \right) |f(w)|$$

$$\ge \left(\mathcal{L} \int_{x_0}^{w} |g(t)|^q dt \right)^{1/q} (z(w))^{1/p} (z'(w))^{1/p}, \quad \text{a.e. on } (x_0, s]. \tag{21.49}$$

Take now $x_0 < \theta \le w \le s$ and $\theta < s$. We observe that

$$\mathcal{L} \int_{x_0}^{s} \left(\mathcal{L} \int_{x_0}^{w} |g(t)f(t)| dt \right) |f(w)| dw = \lim_{\theta \downarrow x_0} \mathcal{L} \int_{\theta}^{s} \left(\mathcal{L} \int_{x_0}^{w} |g(t)f(t)| dt \right) |f(w)| dw$$

$$\ge \lim_{\theta \downarrow x_0} \left[\mathcal{L} \int_{\theta}^{s} \left(\mathcal{L} \int_{x_0}^{w} |g(t)|^q dt \right)^{1/q} (z(w)z'(w))^{1/p} dw \right] \tag{21.50}$$

(again by reverse Hölder's inequality)

$$\geq \lim_{\theta \downarrow x_0} \left(\mathcal{R} \int_\theta^s \left(\mathcal{L} \int_{x_0}^w |g(t)|^q dt \right) dw \right)^{1/q} \lim_{\theta \downarrow x_0} \left(\mathcal{L} \int_\theta^s z(w) z'(w) dw \right)^{1/p} \quad (21.51)$$

$$= \left(\mathcal{R} \int_{x_0}^s \left(\mathcal{L} \int_{x_0}^w |g(t)|^q dt \right) dw \right)^{1/q} \lim_{\theta \downarrow x_0} \left((\mathcal{R} - \mathcal{S}) \int_\theta^s z(w) dz(w) \right)^{1/p} \quad (21.52)$$

$$= \left(\mathcal{R} \int_{x_0}^s \left(\mathcal{L} \int_{x_0}^w |g(t)|^q dt \right) dw \right)^{1/q} \lim_{\theta \downarrow x_0} \left(\frac{z^2(s) - z^2(\theta)}{2} \right)^{1/p}$$

$$= 2^{-1/p} \left(\mathcal{R} \int_{x_0}^s \left(\mathcal{L} \int_{x_0}^w |g(t)|^q dt \right) dw \right)^{1/q} (z(s))^{2/p}$$

$$= 2^{-1/p} \left(\mathcal{R} \int_{x_0}^s \left(\mathcal{L} \int_{x_0}^w |g(t)|^q dt \right) dw \right)^{1/q} \left(\mathcal{L} \int_{x_0}^s |f(t)|^p dt \right)^{2/p}, \quad (21.53)$$

proving (21.43).

Sharpness same as in Theorem 21.1. $\qquad \square$

The counterpart of Theorem 21.7 follows.

Theorem 21.8. *Let $0 < q < 1$ and $p < 0$ such that $\frac{1}{p} + \frac{1}{q} = 1$, a fixed $x_0 \in [a, b]$ and $s \in [a, x_0)$. Assume that $g \in L_q([a, b])$, $f \in L_p([a, b])$, f is nowhere zero and $|f| \leq k$ a.e., $k > 0$. Then*

$$\mathcal{L} \int_s^{x_0} \left(\mathcal{L} \int_w^{x_0} |g(t) f(t)| dt \right) |f(w)| dw$$

$$\geq 2^{-1/p} \left(\mathcal{R} \int_s^{x_0} \left(\mathcal{L} \int_w^{x_0} |g(t)|^q dt \right) dw \right)^{1/q} \left(\mathcal{L} \int_s^{x_0} |f(t)|^p dt \right)^{2/p}. \quad (21.54)$$

Inequality (21.54) is sharp, that is attained when $f(t) = g(t) = c > 0$ a constant, for all $t \in [a, x_0]$.

Proof. Let $s \leq w < x_0$. Here again $k^p \leq |f|^p$ a.e. Therefore

$$z(w) := \mathcal{L} \int_w^{x_0} |f(t)|^p dt \geq k^p (x_0 - w) > 0, \qquad z(x_0) = 0, \quad (21.55)$$

and when $w_2 < w_1$ we find

$$z(w_2) - z(w_1) = \mathcal{L} \int_{w_2}^{w_1} |f(t)|^p dt \geq k^p (w_1 - w_2) > 0. \quad (21.56)$$

That is proving that $z(w)$ is strictly decreasing on $[s, x_0]$, and z also is an absolutely continuous function there.

Next, by reverse Hölder's inequality we obtain

$$\mathcal{L} \int_w^{x_0} |g(t) f(t)| dt \geq \left(\mathcal{L} \int_w^{x_0} |g(t)|^q dt \right)^{1/q} \left(\mathcal{L} \int_w^{x_0} |f(t)|^p dt \right)^{1/p}$$

$$= \left(\mathcal{L} \int_w^{x_0} |g(t)|^q dt \right)^{1/q} (z(w))^{1/p}, \text{ for } s \leq w < x_0. \quad (21.57)$$

It holds a.e. on $[s, x_0]$ that

$$z'(w) = -|f(w)|^p < 0, \tag{21.58}$$

i.e.

$$-z'(w) = |f(w)|^p > 0, \tag{21.59}$$

and

$$|f(w)| = (-z'(w))^{1/p}, \tag{21.60}$$

a.e. on $[s, x_0]$. Therefore

$$\left(\mathcal{L} \int_w^{x_0} |g(t)f(t)|dt \right) |f(w)| \geq \left(\mathcal{L} \int_w^{x_0} |g(t)|^q dt \right)^{1/q} \left(z(w)(-z'(w)) \right)^{1/p},$$
$$\text{a.e. on } [s, x_0). \tag{21.61}$$

Take now $s \leq w \leq \theta < x_0$ and $s < \theta$. We see that

$$\mathcal{L} \int_s^{x_0} \left(\mathcal{L} \int_w^{x_0} |g(t)f(t)|dt \right) |f(w)|dw$$

$$= \lim_{\theta \uparrow x_0} \left(\mathcal{L} \int_s^\theta \left(\mathcal{L} \int_w^{x_0} |g(t)f(t)|dt \right) |f(w)|dw \right)$$

$$\geq \lim_{\theta \uparrow x_0} \left[\mathcal{L} \int_s^\theta \left(\mathcal{L} \int_w^{x_0} |g(t)|^q dt \right)^{1/q} \left(z(w)(-z'(w)) \right)^{1/p} dw \right] \tag{21.62}$$

(again by reverse Hölder's inequality)

$$\geq \lim_{\theta \uparrow x_0} \left(\mathcal{R} \int_s^\theta \left(\mathcal{L} \int_w^{x_0} |g(t)|^q dt \right) dw \right)^{1/q} \lim_{\theta \uparrow x_0} \left(\mathcal{L} \int_s^\theta z(w)(-z'(w)) dw \right)^{1/p} \tag{21.63}$$

$$= \left(\mathcal{R} \int_s^{x_0} \left(\mathcal{L} \int_w^{x_0} |g(t)|^q dt \right) dw \right)^{1/q} \lim_{\theta \uparrow x_0} \left(-(\mathcal{R} - \mathcal{S}) \int_s^\theta z(w) dz(w) \right)^{1/p} \tag{21.64}$$

$$= \left(\mathcal{R} \int_s^{x_0} \left(\mathcal{L} \int_w^{x_0} |g(t)|^q dt \right) dw \right)^{1/q} \lim_{\theta \uparrow x_0} \left(\frac{z^2(s) - z^2(\theta)}{2} \right)^{1/p} \tag{21.65}$$

$$= 2^{-1/p} \left(\mathcal{R} \int_s^{x_0} \left(\mathcal{L} \int_w^{x_0} |g(t)|^q dt \right) dw \right)^{1/q} (z(s))^{2/p} \tag{21.66}$$

$$= 2^{-1/p} \left(\mathcal{R} \int_s^{x_0} \left(\mathcal{L} \int_w^{x_0} |g(t)|^q dt \right) dw \right)^{1/q} \left(\mathcal{L} \int_s^{x_0} |f(t)|^p dt \right)^{2/p}, \tag{21.67}$$

that is proving (21.54).

Sharpness is obvious. $\qquad\square$

We give

Corollary 21.9. *Let $p, q > 1$ such that $\frac{1}{p} + \frac{1}{q} = 1$, with $f \in L_p([a,b])$, $x_0 \in [a,b]$ be fixed. Then*

$$\mathcal{L} \int_{x_0}^{s} \left(\mathcal{L} \int_{x_0}^{w} e^t |f(t)| dt \right) |f(w)| dw$$

$$\leq 2^{-1/p} \left[\frac{1}{q} \left[\frac{1}{q} (e^{qs} - e^{qx_0}) - e^{qx_0}(s - x_0) \right] \right]^{1/q} \left(\mathcal{L} \int_{x_0}^{s} |f(t)|^p dt \right)^{2/p},$$

$$(21.68)$$

for all $x_0 \leq s \leq b$.

Proof. By (21.1) we have for all $x_0 \leq s \leq b$ that

$$\mathcal{L} \int_{x_0}^{s} \left(\mathcal{L} \int_{x_0}^{w} e^t |f(t)| dt \right) |f(w)| dw$$

$$\leq 2^{-1/p} \left(\int_{x_0}^{s} \left(\int_{x_0}^{w} e^{qt} dt \right) dw \right)^{1/q} \left(\mathcal{L} \int_{x_0}^{s} |f(t)|^p dt \right)^{2/p} \qquad (21.69)$$

$$= 2^{-1/p} \left(\frac{1}{q} \left[\frac{1}{q} (e^{qs} - e^{qx_0}) - e^{qx_0}(s - x_0) \right] \right)^{1/q} \left(\mathcal{L} \int_{x_0}^{s} |f(t)|^p dt \right)^{2/p}, (21.70)$$

proving (21.68). $\qquad\square$

We finish the chapter with

Corollary 21.10. *Let $p, q > 1$ such that $\frac{1}{p} + \frac{1}{q} = 1$, with $g \in L_q([a,b])$, here $a > 0$, and let $r \neq 0$. Then*

$$\mathcal{L} \int_{a}^{s} \frac{1}{w^r} \left(\mathcal{L} \int_{a}^{w} \frac{|g(t)|}{t^r} dt \right) dw$$

$$\leq 2^{-1/p} \left(\frac{s^{1-rp} - a^{1-rp}}{1 - rp} \right)^{2/p} \left(\mathcal{R} \int_{a}^{s} \left(\mathcal{L} \int_{a}^{w} |g(t)|^q dt \right) dw \right)^{1/q},$$

$$(21.71)$$

for all $a \leq s \leq b$.

Proof. From (21.1) for $f(t) = \frac{1}{t^r}$, we obtain

$$\mathcal{L} \int_{a}^{s} \frac{1}{w^r} \left(\mathcal{L} \int_{a}^{w} \frac{|g(t)|}{t^r} dt \right) dw$$

$$\leq 2^{-1/p} \left(\mathcal{R} \int_{a}^{s} \left(\mathcal{L} \int_{a}^{w} |g(t)|^q dt \right) dw \right)^{1/q} \left(\int_{a}^{s} \frac{1}{t^{rp}} dt \right)^{2/p},$$

all $a \leq s \leq b$. $\qquad (21.72)$

But

$$\int_{a}^{s} \frac{1}{t^{rp}} dt = \frac{s^{1-rp} - a^{1-rp}}{1 - rp}, \qquad (21.73)$$

proving (21.71). $\qquad\square$

Chapter 22

A Basic Sharp Integral Inequality

A sharp multidimensional integral type inequality is presented involving n-th order ($n \in \mathbb{N}$) mixed partial derivatives. This is subject to some basic boundary condition satisfied by the involved multivariate function. This chapter is based on [60].

22.1   Introduction

This chapter is motivated and inspired by [16] and [123] about Ostrowski type inequalities; see also Ostrowski's paper [196], 1938. Though the presented results are quite different from Ostrowski type inequalities, the working spirit is the same. T. Apostol's book [62] was used as a reference for some basic facts related to integration and differentiation.

22.2   Results

We present

Theorem 22.1. *Let* $f \in C^n(B)$, $n \in \mathbb{N}$, *where* $B = [a_1, b_1] \times \cdots \times [a_n, b_n]$, a_j, $b_j \in \mathbb{R}$, *with* $a_j < b_j$, $j = 1, ..., n$. *Denote by* ∂B *the boundary of the box* B. *Assume that* $f(x) = 0$, *for all* $x = (x_1, ..., x_n) \in \partial B$ *(in other words we suppose that* $f(\cdots, a_j, \cdots) = f(\cdots, b_j, \cdots) = 0$, *for all* $j = 1, ..., n$*). Then*

$$\int_B |f(x_1, ..., x_n)| \, dx_1 \cdots dx_n \leq \frac{m(B)}{2^n} \int_B \left| \frac{\partial^n f(x_1, ..., x_n)}{\partial x_1 \cdots \partial x_n} \right| dx_1 \cdots dx_n, \quad (22.1)$$

where $m(B) = \prod_{j=1}^{n}(b_j - a_j)$ *is the* n-*th dimensional volume (i.e. the Lebesgue measure) of* B.

Theorem 22.2. *Inequality* (22.1) *is sharp (in the sense that equality can be asymptotically attained).*

Proof of Theorem 22.1. Let $(x_1, ..., x_n) \in B$, i.e. $a_j \leq x_j \leq b_j$, for all $j = 1, ..., n$.

271

The assumptions give

$$f(x_1, ..., x_n) = \int_{a_1}^{x_1} \cdots \int_{a_n}^{x_n} \frac{\partial^n f(s_1, ..., s_n)}{\partial s_1 \cdots \partial s_n} ds_1 \cdots ds_n,$$

and

$$f(x_1, ..., x_n) = (-1)^n \int_{x_1}^{b_1} \cdots \int_{x_n}^{b_n} \frac{\partial^n f(s_1, ..., s_n)}{\partial s_1 \cdots \partial s_n} ds_1 \cdots ds_n.$$

More generally, if we introduce the intervals

$$I_{j,0} = [a_j, x_j] \quad \text{and} \quad I_{j,1} = [x_j, b_j], \qquad j = 1, ..., n,$$

we have

$$f(x_1, ..., x_n) = (-1)^{\varepsilon_1 + \cdots + \varepsilon_n} \int_{I_{1,\varepsilon_1}} \cdots \int_{I_{n,\varepsilon_n}} \frac{\partial^n f(s_1, ..., s_n)}{\partial s_1 \cdots \partial s_n} ds_1 \cdots ds_n, \qquad (22.2)$$

where each ε_j can be either 0 or 1. Adding up (22.2) for all 2^n choices of $(\varepsilon_1, ..., \varepsilon_n)$ we derive

$$2^n f(x_1, ..., x_n) = \sum_{\varepsilon_1, ..., \varepsilon_n} (-1)^{\varepsilon_1 + \cdots + \varepsilon_n} \int_{I_{1,\varepsilon_1}} \cdots \int_{I_{n,\varepsilon_n}} \frac{\partial^n f(s_1, ..., s_n)}{\partial s_1 \cdots \partial s_n} ds_1 \cdots ds_n.$$

$$(22.3)$$

Next by taking absolute values in (22.3) and using basic properties of integrals (noting that the 2^n "sub-boxes" $I_{1,\varepsilon_1} \times \cdots \times I_{n,\varepsilon_n}$ form a partition of B) and the subadditivity property of the absolute value we find that

$$|f(x_1, ..., x_n)| \leq \frac{1}{2^n} \int_B \left| \frac{\partial^n f(x_1, ..., x_n)}{\partial x_1 \cdots \partial x_n} \right| dx_1 \cdots dx_n, \qquad (22.4)$$

true for all $(x_1, ..., x_n) \in B$. Inequality (22.4) has by itself its merits.

Finally by integrating (22.4) over B we establish the result. $\qquad \square$

Proof of Theorem 22.2. Without loss of generality, to establish optimality of (22.1) it is enough to prove sharpness of the following inequality

$$\int_0^1 |f(x)| \, dx \leq \frac{1}{2} \int_0^1 |f'(x)| \, dx \qquad (22.5)$$

(this is obtained from (22.1) by taking $n = 1$ and $[a_1, b_1] = [0, 1]$). Clearly here it is assumed that $f \in C^1[0, 1]$ with $f(0) = f(1) = 0$.

Let $0 < \varepsilon < 1$. Consider the function

$$f_\varepsilon(x) = \begin{cases} \frac{2\varepsilon x - x^2}{\varepsilon^2}, & 0 \leq x \leq \varepsilon, \\ 1, & \varepsilon < x \leq 1 - \varepsilon, \\ \frac{2\varepsilon(1-x) - (1-x)^2}{\varepsilon^2}, & 1 - \varepsilon < x \leq 1. \end{cases}$$

Clearly $f_\varepsilon(x) \geq 0$ on $[0, 1]$, and $f_\varepsilon(0) = f_\varepsilon(1) = 0$ (in fact $f_\varepsilon(x) = f_\varepsilon(1 - x)$, that is $f_\varepsilon(x)$ is symmetric about $x = 1/2$). Furthermore

$$f_\varepsilon'(x) = \begin{cases} \frac{2(\varepsilon - x)}{\varepsilon^2}, & 0 \leq x \leq \varepsilon, \\ 0, & \varepsilon < x \leq 1 - \varepsilon, \\ \frac{2(1 - x - \varepsilon)}{\varepsilon^2}, & 1 - \varepsilon < x \leq 1, \end{cases}$$

hence f'_ε is continuous on $[0, 1]$.

Next we calculate and find for f_ε that the right hand side of (22.5) is 1, while the left hand side of (22.5) is $1 - (2/3)\varepsilon$, i.e. (22.5) becomes

$$1 - \frac{2}{3}\varepsilon \leq 1.$$

Therefore, by letting $\varepsilon \searrow 0$, we see that equality in (22.5) is asymptotically attained.

$\square$

Remark 22.3. Notice that $\lim_{\varepsilon \searrow 0} f_\varepsilon(x) = 1$ (pointwise) for all $x \in (0, 1)$. In this sense equality in (22.4) is also asymptotically attainable.

Remark 22.4. There is no function other than the trivial $f(x_1, ..., x_n) \equiv 0$ that makes (22.1) an equality. Indeed, if there were such an f, then, by continuity (22.4) should be an equality for all $(x_1, ..., x_n) \in B$, thus f should be constant which is impossible due to the boundary condition, unless, of course $f \equiv 0$.

Remark 22.5. The function f of Theorem 22.1 can be complex valued.

Remark 22.6. The case $n = 2$ is quite interesting: Let $f \in C^2([a, b] \times [c, d])$, $a < b$, $c < d$, with $f(a, \cdot) = f(b, \cdot) = f(\cdot, c) = f(\cdot, d) = 0$. Then (22.1) becomes

$$\int_c^d \int_a^b |f(x, y)| \, dx dy \leq \frac{(b - a)(d - c)}{4} \int_c^d \int_a^b |\partial_{xy} [f(x, y)]| \, dx dy.$$

Notice that the operator $2\partial_{xy}$ becomes the wave operator $\partial_{yy} - \partial_{xx}$ after a $45°$-rotation of the axes x and y.

Remark 22.7. It is a curious fact that inequality (22.1) fails badly in a variety of domains. For example, given $a, b > 0$, let $D_k \subset \mathbb{R}^2$, $k = 1, 2, 3, ...$, be the domain

$$D_k = \left\{ (x, y) \in \mathbb{R}^2 : \frac{x^{2k}}{a^{2k}} + \frac{y^{2k}}{b^{2k}} \leq 1 \right\}.$$

Notice that $D_k \subset B = [-a, a] \times [-b, b]$ and we can make D_k as close to B as we wish, by taking k sufficiently large. However, if f_k is the polynomial

$$f_k(x, y) = \frac{x^{2k}}{a^{2k}} + \frac{y^{2k}}{b^{2k}} - 1,$$

then, of course, $f_k \in C^2(D_k)$ and $f_k = 0$ on ∂D_k, but

$$\frac{\partial^2 f_k(x, y)}{\partial x \partial y} \equiv 0,$$

hence (22.1) fails completely if D_k is the domain of integration!

We finish chapter with

Theorem 22.8. *Let $f \in C^{n+2}(B)$, $n \in \mathbb{N}$, $B = [a_1, b_1] \times \ldots \times [a_n, b_n]$, $a_j, b_j \in \mathbb{R}$, with $a_j < b_j, j = 1, \ldots, n$. Assume that*

$$\frac{\partial^2 f}{\partial x_j^2}(\cdots, a_i, \cdots) = \frac{\partial^2 f}{\partial x_j^2}(\cdots, b_i, \cdots) = 0,$$

for all $i, j = 1, \cdots, n$. Then

$$\int_B \left| \nabla^2 f(x_1, \ldots, x_n) \right| dx_1, dx_2 \ldots dx_n$$

$$\leq \frac{(\Pi_{i=1}^n (b_i - a_i))}{2^n} \left(\sum_{i=1}^n \left(\int_B \left| \frac{\partial^{n+2} f(x_1, \ldots, x_n)}{\partial x_1 \ldots \partial^3 x_i \ldots \partial x_n} \right| dx_1 \ldots dx_n \right) \right). \quad (22.6)$$

Proof.   We observe the following

$$\left| \nabla^2 f(x_1, \ldots, x_n) \right| = \left| \sum_{i=1}^n \frac{\partial^2 f(x_1, \ldots, x_n)}{\partial x_i^2} \right| \leq \sum_{i=1}^n \left| \frac{\partial^2 f(x_1, \ldots, x_n)}{\partial x_i^2} \right| \quad (22.7)$$

$$\leq (by \quad (22.4), \quad \text{set instead of f the function } \tfrac{\partial^2 f}{\partial x_i^2})$$

$$\sum_{i=1}^n \frac{1}{2^n} \left(\int_B \left| \frac{\partial^n \left(\frac{\partial^2 f}{\partial x_i^2} \right)(x_1, \ldots, x_n)}{\partial x_1 \ldots \partial x_n} \right| dx_1 \ldots dx_n \right)$$

$$= \frac{1}{2^n} \sum_{i=1}^n \left(\int_B \left| \frac{\partial^{n+2} f(x_1, \ldots, x_n)}{\partial x_1 \ldots \partial x_i^3 \ldots \partial x_n} \right| dx_1 \ldots dx_n \right). \quad (22.8)$$

So we get true that

$$\left| \nabla^2 f(x_1, \ldots, x_n) \right| \leq \frac{1}{2^n} \sum_{i=1}^n \left(\int_B \left| \frac{\partial^{n+2} f(x_1, \ldots, x_n)}{\partial x_1 \ldots \partial x_i^3 \ldots \partial x_n} \right| dx_1 \ldots dx_n \right), \quad (22.9)$$

$\forall\, (x_1, \ldots, x_n) \in \mathbb{B}$. Integrating (22.9) over B we get (22.6). $\quad\square$

Chapter 23

Estimates of the Remainder in Taylor's Formula

Estimates of the remainder in Taylor's formula are given. This chapter relies on [55].

23.1 Introduction

The following theorem is well known as Taylor's formula or Taylor's theorem with the integral remainder.

Theorem 23.1. *Let $f : [a, b] \to \mathbb{R}$ and n a positive integer. If f is such that $f^{(n)}$ is absolutely continuous on $[a, b]$, $x_0 \in (a, b)$, then for all $x \in (a, b)$ we have*

$$f(x) = T_n(f; x_0, x) + R_n(f; x_0, x), \tag{23.1}$$

where $T_n(f; x_0, \cdot)$ is Taylor's polynomial of degree n, i.e.,

$$T_n(f; x_0, x) = \sum_{k=0}^{n} \frac{f^{(k)}(x_0)(x - x_0)^k}{k!} \tag{23.2}$$

(note that $f^{(0)} = f$ and $0! = 1$) and the remainder can be given by

$$R_n(f; x_0, x) = \frac{1}{n!} \int_{x_0}^{x} (x - t)^n f^{(n+1)}(t)\, dt. \tag{23.3}$$

For a mapping $g : [a, b] \to \mathbb{R}$ and two arbitrary points $x_0, x \in (a, b)$, define

$$\|g\|_{[x_0, x]; p} := \left| \int_{x_0}^{x} |g(t)|^p \, dt \right|^{\frac{1}{p}}, \quad p \geq 1$$

and

$$\|g\|_{[x_0, x]; \infty} := \operatorname*{ess\ sup}_{\substack{t \in [x_0, x] \\ (t \in [x, x_0])}} |g(t)|.$$

Using Hölder's inequality, we may state the following corollary.

275

Corollary 23.2. *With the above assumptions, we have*

$$|R_n (f; x_0, x)|$$

$$\leq \begin{cases} \frac{|x-x_0|^n}{n!} \left\| f^{(n+1)} \right\|_{[x_0,x],1} & \text{if } f^{(n+1)} \in L_1 [a,b]; \\[3ex] \frac{|x-x_0|^{n+\frac{1}{q}}}{n!(nq+1)^{\frac{1}{q}}} \left\| f^{(n+1)} \right\|_{[x_0,x],p} & \text{if } f^{(n+1)} \in L_q [a,b], \\[1ex] & \qquad p > 1, \; \frac{1}{p} + \frac{1}{q} = 1; \\[2ex] \frac{|x-x_0|^{n+1}}{(n+1)!} \left\| f^{(n+1)} \right\|_{[x_0,x],\infty} & \text{if } f^{(n+1)} \in L_\infty [a,b]. \end{cases}$$

$$(23.4)$$

For some applications of (23.4) for particular functions, see [101].

23.2   Some New Bounds for the Remainder

The following simple result comes from G.A. Anastassiou in [16].

Lemma 23.3. *Suppose that the mapping* $f : [a,b] \to \mathbb{R}$ *is such that* $f^{(n-1)}$ *is absolutely continuous on* $[a,b]$ *and* $x_0 \in (a,b)$. *Then for all* $x \in (a,b)$ *the remainder* $R_n (f; x_0, x)$ *in (23.1) can be represented by*

$$R_n (f; x_0, x) = \frac{1}{(n-1)!} \int_{x_0}^{x} \left[f^{(n)} (t) - f^{(n)} (x_0) \right] (x - t)^{n-1} dt, \quad n \geq 1. \quad (23.5)$$

Proof.   We apply Taylor's formula with the integral remainder for $n-1$ obtaining

$$f (x) = T_{n-1} (f; x_0, x) + \frac{1}{(n-1)!} \int_{x_0}^{x} (x - t)^{n-1} f^{(n)} (t) \, dt$$

$$= T_n (f; x_0, x) - \frac{(x - x_0)^n}{n!} f^{(n)} (x_0) + \frac{1}{(n-1)!} \int_{x_0}^{x} (x - t)^{n-1} f^{(n)} (t) \, dt$$

$$= T_n (f; x_0, x) + \frac{1}{(n-1)!} \int_{x_0}^{x} \left[f^{(n)} (t) - f^{(n)} (x_0) \right] (x - t)^{n-1} dt,$$

which produces the representation (23.5). $\square$

The following theorem holds.

Theorem 23.4. *Suppose that* f, x_0 *and* x *are as in Lemma 23.3. Then we have* *the bounds*

$$|R_n\left(f;x_0,x\right)|$$

$$\leq \begin{cases} \dfrac{|x-x_0|^{n-1}}{n!}\left\|f^{(n)}-f^{(n)}\left(x_0\right)\right\|_{[x_0,x],1} & \text{if } f^{(n)}\in L_1\left[a,b\right]; \\[4mm] \dfrac{|x-x_0|^{n-1+\frac{1}{q}}}{(n-1)!\left[(n-1)q+1\right]^{\frac{1}{q}}}\left\|f^{(n)}-f^{(n)}\left(x_0\right)\right\|_{[x_0,x],p} & \text{if } f^{(n)}\in L_q\left[a,b\right], \\ & \quad p>1,\ \frac{1}{p}+\frac{1}{q}=1; \\[4mm] \dfrac{|x-x_0|^{n}}{n!}\left\|f^{(n)}-f^{(n)}\left(x_0\right)\right\|_{[x_0,x],\infty} & \text{if } f^{(n)}\in L_\infty\left[a,b\right]. \end{cases}$$

$$(23.6)$$

Proof.   We have

$$|R_n\left(f;x_0,x\right)| \leq \frac{1}{(n-1)!}\left|\int_{x_0}^{x}\left[f^{(n)}\left(t\right)-f^{(n)}\left(x_0\right)\right]\left(x-t\right)^{n-1}dt\right|$$

$$\leq \frac{1}{(n-1)!}\left|\int_{x_0}^{x}\left|f^{(n)}\left(t\right)-f^{(n)}\left(x_0\right)\right||x-t|^{n-1}dt\right| := M\left(x_0,x\right).$$

If $f^{(n)}\in L_1\left[a,b\right]$, then

$$M\left(x_0,x\right) \leq \frac{1}{(n-1)!}\sup_{\substack{t\in[x_0,x] \\ (t\in[x,x_0])}}|x-t|^{n-1}\left|\int_{x_0}^{x}\left|f^{(n)}\left(t\right)-f^{(n)}\left(x_0\right)\right|dt\right|$$

$$= \frac{1}{n!}|x-x_0|^{n-1}\left\|f^{(n)}-f^{(n)}\left(x_0\right)\right\|_{[x_0,x],1}$$

and the first inequality in (23.6) is proved.

Using Hölder's integral inequality, we have, for $f^{(n)}\in L_p\left[a,b\right]$, that

$$M\left(x_0,x\right) \leq \frac{1}{(n-1)!}\left|\int_{x_0}^{x}\left|f^{(n)}\left(t\right)-f^{(n)}\left(x_0\right)\right|^{p}dt\right|^{\frac{1}{p}}\left|\int_{x_0}^{x}|x-t|^{(n-1)q}dt\right|^{\frac{1}{q}}$$

$$= \frac{1}{(n-1)!}\left\|f^{(n)}-f^{(n)}\left(x_0\right)\right\|_{[x_0,x],p}\left[\frac{|x-x_0|^{(n-1)q+1}}{(n-1)q+1}\right]^{\frac{1}{q}}$$

$$= \frac{1}{(n-1)!\left[(n-1)q+1\right]^{\frac{1}{q}}}|x-x_0|^{n-1+\frac{1}{q}}\left\|f^{(n)}-f^{(n)}\left(x_0\right)\right\|_{[x_0,x],p}$$

and the second inequality in (23.6) is proved.

Finally, we have for $f^{(n)}\in L_\infty\left[a,b\right]$, that

$$M\left(x_0,x\right) \leq ess\sup_{\substack{t\in[x_0,x] \\ (t\in[x,x_0])}}\left|f^{(n)}\left(t\right)-f^{(n)}\left(x_0\right)\right|\frac{1}{(n-1)!}\left|\int_{x_0}^{x}|x-t|^{n-1}dt\right|$$

$$= \frac{1}{n!}|x-x_0|^{n}\left\|f^{(n)}-f^{(n)}\left(x_0\right)\right\|_{[x_0,x],\infty}$$

and the theorem is proved. $\qquad\square$

The following result for Hölder type mappings also holds.

Theorem 23.5. *Suppose that the mapping $f : [a, b] \to \mathbb{R}$ is such that $f^{(n)}$ is of $H - r -$ Hölder type. That is,*

$$\left| f^{(n)}(t) - f^{(n)}(s) \right| \le H \left| t - s \right|^r \quad \text{for all } t, s \in (a, b) \tag{23.7}$$

and $H > 0$ is given. Then we have the inequality:

$$\left| R_n(f; x_0, x) \right| \le \frac{HB(r+1, n)}{(n-1)!} \left| x - x_0 \right|^{r+n}, \tag{23.8}$$

where $B(\cdot, \cdot)$ is Euler's beta function.

Proof. As $f^{(n)}$ is of $H - r -$ Hölder type, we may write

$$\left| R_n(f; x_0, x) \right| \le \frac{1}{(n-1)!} \left| \int_{x_0}^{x} \left| f^{(n)}(t) - f^{(n)}(x_0) \right| \left| x - t \right|^{n-1} dt \right| \tag{23.9}$$

$$\le \frac{H}{(n-1)!} \left| \int_{x_0}^{x} \left| t - x_0 \right|^r \left| x - t \right|^{n-1} dt \right| =: N(x_0, x).$$

Assume that $x_0 \le x$. Then

$$N(x_0, x) = \int_{x_0}^{x} (t - x_0)^r (x - t)^{n-1} dt = (x - x_0)^{r+n-1+1} \int_0^1 t^r (1 - t)^{n-1} dt$$

$$= (x - x_0)^{r+n} B(r+1, n).$$

A similar equality can be obtained if $x < x_0$. Consequently, in general

$$N(x_0, x) = \left| x - x_0 \right|^{r+n} B(r+1, n)$$

and then, by (23.9), we deduce (23.8). $\qquad\qquad\square$

Corollary 23.6. *Suppose that the mapping $f : [a, b] \to \mathbb{R}$ is such that $f^{(n)}$ is $L -$ Lipschitzian on $[a, b]$, i.e.,*

$$\left| f^{(n)}(t) - f^{(n)}(s) \right| \le L \left| t - s \right| \quad \text{for all } t, s \in (a, b), \tag{23.10}$$

where $L > 0$ is given. Then we have the inequality

$$\left| R_n(f; x_0, x) \right| \le \frac{L \left| x - x_0 \right|^{n+1}}{(n+1)!}. \tag{23.11}$$

Proof. For $r = 1$ we have

$$B(2, n) = \int_0^1 t^{n-1} (1 - t) \, dt = \frac{1}{n(n+1)}.$$

Using (23.8), we deduce (23.11). $\qquad\qquad\square$

We can now state the following result as well.

Theorem 23.7. *Let f, x_0 and x be as in Theorem 23.1. Then the remainder $R_n(f; x_0, x)$ satisfies the bound*
$$|R_n(f; x_0, x)|$$
$$\leq \begin{cases} \frac{1}{(n-1)!} \left| \int_{x_0}^{x} |x-t|^{n-1} |t-x_0| \left\| f^{(n+1)} \right\|_{[x_0,t],\infty} dt \right| & \text{if } f^{(n+1)} \in L_\infty[a,b]; \\[2mm] \frac{1}{(n-1)!} \left| \int_{x_0}^{x} |x-t|^{n-1} |t-x_0|^{\frac{1}{q}} \left\| f^{(n+1)} \right\|_{[x_0,t],p} dt \right| & \text{if } f^{(n+1)} \in L_q[a,b], \\ & \qquad p > 1, \; \frac{1}{p} + \frac{1}{q} = 1; \\[2mm] \frac{1}{(n-1)!} \left| \int_{x_0}^{x} |x-t|^{n-1} \left\| f^{(n+1)} \right\|_{[x_0,t],1} dt \right| & \text{if } f^{(n+1)} \in L_1[a,b]. \end{cases}$$
$$(23.12)$$

Proof. As $f^{(n)}$ is absolutely continuous on $[a,b]$ we may write that
$$f^{(n)}(t) - f^{(n)}(x_0) = \int_{x_0}^{t} f^{(n+1)}(u) \, du$$
and then, by (23.5), we have the representation:
$$R_n(f; x_0, x) = \frac{1}{(n-1)!} \int_{x_0}^{x} \left(\int_{x_0}^{t} f^{(n+1)}(u) \, du \right) (x-t)^{n-1} \, dt. \qquad (23.13)$$
By (23.13) we may write:
$$|R_n(f; x_0, x)| \leq \frac{1}{(n-1)!} \left| \int_{x_0}^{x} \left| \int_{x_0}^{t} f^{(n+1)}(u) \, du \right| |x-t|^{n-1} \, dt \right|. \qquad (23.14)$$
Now, if $f^{(n+1)} \in L_\infty[a,b]$, then
$$\left| \int_{x_0}^{t} f^{(n+1)}(u) \, du \right| \leq |t-x_0| \left\| f^{(n+1)} \right\|_{[x_0,t],\infty}.$$
Also, by Hölder's integral inequality we have (for $p > 1$, $\frac{1}{p} + \frac{1}{q} = 1$) that
$$\left| \int_{x_0}^{t} f^{(n+1)}(u) \, du \right| \leq |t-x_0|^{\frac{1}{q}} \left| \int_{x_0}^{t} \left| f^{(n+1)}(u) \right|^p du \right|^{\frac{1}{p}}$$
$$= |t-x_0|^{\frac{1}{q}} \left\| f^{(n+1)} \right\|_{[x_0,t],p}$$
and
$$\left| \int_{x_0}^{t} f^{(n+1)}(u) \, du \right| \leq \left| \int_{x_0}^{t} \left| f^{(n+1)}(u) \right| du \right| = \left\| f^{(n+1)} \right\|_{[x_0,t],1}.$$
Consequently, we have
$$\left| \int_{x_0}^{t} f^{(n+1)}(u) \, du \right| \leq \begin{cases} |t-x_0| \left\| f^{(n+1)} \right\|_{[x_0,t],\infty} \\[2mm] |t-x_0|^{\frac{1}{q}} \left\| f^{(n+1)} \right\|_{[x_0,t],p} \\[2mm] \left\| f^{(n+1)} \right\|_{[x_0,t],1}. \end{cases} \qquad (23.15)$$
Using (23.14) and (23.15), we easily deduce (23.12). $\qquad \square$

23.3   Some Further Bounds of the Remainder

Let us consider the Chebychev functional defined by

$$T\left(g,h\right) := \frac{1}{b-a} \int_a^b h\left(x\right) g\left(x\right) dx - \frac{1}{\left(b-a\right)^2} \int_a^b h\left(x\right) dx \int_a^b g\left(x\right) dx, \qquad (23.16)$$

where $h, g : [a, b] \to \mathbb{R}$ are measurable on $[a, b]$ and the involved integrals exist on $[a, b]$.

The following identity which can be proved by direct computation is well known as Korkine's identity:

$$T\left(g,h\right) = \frac{1}{2\left(b-a\right)^2} \int_a^b \int_a^b \left(h\left(x\right) - h\left(y\right)\right)\left(g\left(x\right) - g\left(y\right)\right) dx dy. \qquad (23.17)$$

The following lemma holds.

Lemma 23.8. *Suppose that the mapping $f : [a, b] \to \mathbb{R}$ and x_0, x are as in Lemma 23.3. Then we have the representation*

$$R_n\left(f; x_0, x\right)$$
$$= \left[\left[f^{(n-1)}; x_0, x\right] - f^{(n)}\left(x_0\right)\right] \frac{\left(x - x_0\right)^n}{n!} + \frac{1}{2\left(n-1\right)!\left(x - x_0\right)}$$
$$\times \int_{x_0}^x \int_{x_0}^x \left(f^{(n)}\left(t\right) - f^{(n)}\left(s\right)\right)\left(\left(x - t\right)^{n-1} - \left(x - s\right)^{n-1}\right) dt ds.$$

$$(23.18)$$

Proof.   Applying Korkine's identity, we may write:

$$\frac{1}{x - x_0} \int_{x_0}^x \left(f^{(n)}\left(t\right) - f^{(n)}\left(x_0\right)\right)\left(x - t\right)^{n-1} dt$$
$$-\frac{1}{\left(x - x_0\right)^2} \int_{x_0}^x \left(f^{(n)}\left(t\right) - f^{(n)}\left(x_0\right)\right) dt \cdot \frac{1}{x - x_0} \int_{x_0}^x \left(x - t\right)^{n-1} dt$$
$$= \frac{1}{2\left(x - x_0\right)^2} \int_{x_0}^x \int_{x_0}^x \left(f^{(n)}\left(t\right) - f^{(n)}\left(s\right)\right)\left(\left(x - t\right)^{n-1} - \left(x - s\right)^{n-1}\right) dt ds$$

which is clearly equivalent to:

$$\int_{x_0}^x \left(f^{(n)}\left(t\right) - f^{(n)}\left(x_0\right)\right)\left(x - t\right)^{n-1} dt$$
$$= \left[f^{(n-1)}\left(x\right) - f^{(n)}\left(x_0\right) - \left(x - x_0\right) f^{(n)}\left(x_0\right)\right] \frac{\left(x - x_0\right)^{n-1}}{n}$$
$$+\frac{1}{2\left(x - x_0\right)} \int_{x_0}^x \int_{x_0}^x \left(f^{(n)}\left(t\right) - f^{(n)}\left(s\right)\right)\left(\left(x - t\right)^{n-1} - \left(x - s\right)^{n-1}\right) dt ds$$

from where we get (23.18). $\square$

The following theorem holds.

Theorem 23.9. *Assume that the mapping $f : [a,b] \to \mathbb{R}$ has the property that $f^{(n)} \in L_2[a,b]$ and $x_0, x \in (a,b)$. Then we have the inequality:*

$$|R_n(f; x_0, x)|$$
$$\leq \left| \left[f^{(n-1)}; x_0, x \right] - f^{(n)}(x_0) \right| \frac{|x - x_0|^n}{n!}$$
$$+ \frac{(n-1)|x - x_0|^n}{n!\sqrt{2n-1}} \left[\frac{1}{x - x_0} \left\| f^{(n)} \right\|_{[x_0,x];2}^2 - \left(\left[f^{(n-1)}; x_0, x \right] \right)^2 \right]^{\frac{1}{2}},$$

$$(23.19)$$

where

$$\left\| f^{(n)} \right\|_{[x_0,x];2} = \left| \int_{x_0}^{x} \left| f^{(n)}(t) \right|^2 dt \right|^{\frac{1}{2}}.$$

Proof. We have, by (23.18), that

$$|R_n(f; x_0, x)|$$
$$\leq \left| \left[f^{(n-1)}; x_0, x \right] - f^{(n)}(x_0) \right| \frac{|x - x_0|^n}{n!} + \frac{1}{2(n-1)! |x - x_0|}$$
$$\times \left| \int_{x_0}^{x} \int_{x_0}^{x} \left(f^{(n)}(t) - f^{(n)}(s) \right) \left[(x-t)^{n-1} - (x-s)^{n-1} \right] dtds \right|.$$

$$(23.20)$$

Using the Cauchy-Buniakowski-Schwartz inequality, we have

$$\left| \int_{x_0}^{x} \int_{x_0}^{x} \left(f^{(n)}(t) - f^{(n)}(s) \right) \left[(x-t)^{n-1} - (x-s)^{n-1} \right] dtds \right|$$
$$\leq \left| \int_{x_0}^{x} \int_{x_0}^{x} \left(f^{(n)}(t) - f^{(n)}(s) \right)^2 dtds \right|^{\frac{1}{2}}$$
$$\left| \int_{x_0}^{x} \int_{x_0}^{x} \left[(x-t)^{n-1} - (x-s)^{n-1} \right]^2 dtds \right|^{\frac{1}{2}}$$
$$= 2 \left[(x - x_0) \int_{x_0}^{x} \left[f^{(n)}(t) \right]^2 dt - \left(\int_{x_0}^{x} f^{(n)}(t) \, dt \right)^2 \right]^{\frac{1}{2}}$$
$$\times \left[(x - x_0) \int_{x_0}^{x} (x-t)^{2(n-1)} dt - \left(\int_{x_0}^{x} (x-t)^{n-1} dt \right)^2 \right]^{\frac{1}{2}}$$
$$= 2 \left[(x - x_0) \left\| f^{(n)} \right\|_{[x_0,x];2}^2 - \left(f^{(n-1)}(x) - f^{(n-1)}(x_0) \right)^2 \right]^{\frac{1}{2}}$$
$$\times \left[(x - x_0) \frac{(x - x_0)^{2(n-1)+1}}{2(n-1)+1} - \left(\frac{(x - x_0)^n}{n} \right)^2 \right]^{\frac{1}{2}}$$

$$= 2\left|x - x_0\right| \left[\frac{1}{x - x_0} \left\|f^{(n)}\right\|_{[x_0,x];2}^2 - \left(\left[f^{(n-1)}; x_0, x\right]\right)^2\right]^{\frac{1}{2}}$$

$$\times \left[\frac{(x - x_0)^{2n}}{2n - 1} - \frac{(x - x_0)^{2n}}{n^2}\right]^{\frac{1}{2}}$$

$$= 2\left|x - x_0\right| \left[\frac{1}{x - x_0} \left\|f^{(n)}\right\|_{[x_0,x];2}^2 - \left(\left[f^{(n-1)}; x_0, x\right]\right)^2\right]^{\frac{1}{2}}$$

$$\times \left|x - x_0\right|^n \left[\frac{n^2 - 2n + 1}{n^2 (2n - 1)}\right]^{\frac{1}{2}}$$

$$= 2\left|x - x_0\right|^{n+1} \left[\frac{1}{x - x_0} \left\|f^{(n)}\right\|_{[x_0,x];2}^2 - \left(\left[f^{(n-1)}; x_0, x\right]\right)^2\right]^{\frac{1}{2}} \frac{n - 1}{n\sqrt{2n - 1}}$$

$$= \frac{2(n - 1)\left|x - x_0\right|^{n+1}}{n\sqrt{2n - 1}} \left[\frac{1}{x - x_0} \left\|f^{(n)}\right\|_{[x_0,x];2}^2 - \left(\left[f^{(n-1)}; x_0, x\right]\right)^2\right]^{\frac{1}{2}}$$

and then

$$\frac{1}{2(n - 1)!} \left|\int_{x_0}^{x} \int_{x_0}^{x} \left(f^{(n)}(t) - f^{(n)}(s)\right) \left[(x - t)^{n-1} - (x - s)^{n-1}\right] dt ds\right|$$

$$\leq \frac{(n - 1)\left|x - x_0\right|^n}{n!\sqrt{2n - 1}} \left[\frac{1}{x - x_0} \left\|f^{(n)}\right\|_{[x_0,x];2}^2 - \left(\left[f^{(n-1)}; x_0, x\right]\right)^2\right]^{\frac{1}{2}}.$$

Using (23.20), we deduce (23.19). $\qquad\qquad\square$

23.4 Some Inequalities for Special Cases

In this section we assume that $x_0, x \in (a, b)$ and $x \geq x_0$.

The following theorem holds.

Theorem 23.10. *Let* $f : [a, b] \to \mathbb{R}$ *be such that* $f^{(n)}$ *is monotonic nondecreasing (nonincreasing) on* $[x_0, x]$. *Then we have the inequality:*

$$f(x) \leq (\geq) T_n(f; x_0, x) + \left[\left[f^{(n-1)}; x_0, x\right] - f^{(n)}(x_0)\right] \frac{(x - x_0)^n}{n!} \tag{23.21}$$

or, equivalently,

$$f(x) \leq (\geq) T_{n-1}(f; x_0, x) + \frac{(x - x_0)^n}{n!} \left[f^{(n-1)}; x_0, x\right]. \tag{23.22}$$

Proof. We use the following Chebychev inequality

$$T(g, h) \geq 0 \ (\leq 0) \tag{23.23}$$

provided that (g, h) are synchronous (asynchronous), i.e., we recall that the mappings (g, h) are synchronous (asynchronous) if

$$(g(x) - g(y))(h(x) - h(y)) \geq 0 \ (\leq 0) \quad \text{for all } x, y \in [a, b]. \tag{23.24}$$

As the mapping $h(t) := (x - t)^{n-1}$ is monotonic nonincreasing on $[x_0, x]$, then we have, for $f^{(n)}$ nondecreasing,

$$T\left(f^{(n)}, (x - \cdot)^{n-1}\right) \leq 0$$

and then, by (23.18) we conclude that

$$R_n\left(f; x_0, x\right) \leq \left[\left[f^{(n-1)}; x_0, x\right] - f^{(n)}\left(x_0\right)\right] \frac{(x - x_0)^n}{n!}. \tag{23.25}$$

The case when $f^{(n)}$ is monotonic nonincreasing goes likewise and we omit the details.

The following refinement of Chebychev's inequality is known (see for example [129])

$$T(g, h) \geq \max\left\{|T(g, |h|)|, |T(|g|, h)|, |T(|g|, |h|)|\right\} \geq 0. \tag{23.26}$$

Using (23.9), we may improve (23.21) as follows. $\qquad\square$

Theorem 23.11. *Let $f : [a, b] \to \mathbb{R}$ be such that $f^{(n)}$ is monotonic nonincreasing on $[x_0, x]$. Then we have the inequality:*

$$f(x) - T_{n-1}(f; x_0, x) - \left[f^{(n-1)}; x_0, x\right] \frac{(x - x_0)^n}{n!}$$

$$\geq \frac{1}{(n-1)!} \left| \int_{x_0}^{x} \left|f^{(n)}(t)\right| (x - t)^{n-1} \, dt - \frac{1}{n} (x - x_0)^{n-1} \int_{x_0}^{x} \left|f^{(n)}(t)\right| \, dt \right| \geq 0. \tag{23.27}$$

Proof. Apply inequality (23.26) for $g = f^{(n)}$, $h = (x - \cdot)^{n-1}$ to derive

$$T\left(f^{(n)}, (x - \cdot)^{n-1}\right) \geq \max\left\{|A_1|, |B_1|, |C_1|\right\}, \tag{23.28}$$

where

$$A_1 = T\left(f^{(n)}, \left|(x - \cdot)^{n-1}\right|\right) = T\left(f^{(n)}, (x - \cdot)^{n-1}\right)$$

$$B_1 = T\left(\left|f^{(n)}\right|, (x - \cdot)^{n-1}\right) = \frac{1}{x - x_0} \int_{x_0}^{x} \left|f^{(n)}(t)\right| (x - t)^{n-1} \, dt$$

$$- \frac{1}{(x - x_0)^2} \int_{x_0}^{x} \left|f^{(n)}(t)\right| \, dt \cdot \int_{x_0}^{x} (x - t)^{n-1} \, dt$$

$$= \frac{1}{x - x_0} \int_{x_0}^{x} \left|f^{(n)}(t)\right| (x - t)^{n-1} \, dt - \frac{1}{(x - x_0)^2} \int_{x_0}^{x} \left|f^{(n)}(t)\right| \, dt \cdot \frac{(x - x_0)^n}{n}$$

and

$$C_1 = T\left(\left|f^{(n)}\right|, \left|(x - \cdot)^{n-1}\right|\right) = T\left(\left|f^{(n)}\right|, (x - \cdot)^{n-1}\right) = B_1.$$

Now, using the fact that (see Lemma 23.8)

$$f(x) - T_n(f; x_0, x) - \left[\left[f^{(n-1)}; x_0, x\right] - f^{(n)}(x_0)\right] \frac{(x - x_0)}{n!} \tag{23.29}$$

$$= \frac{(x - x_0)^n}{(n-1)!} T\left(f^{(n)}, (x - \cdot)^{n-1}\right)$$

then by the inequality (23.28), we may deduce (23.27). $\qquad\square$

The following theorem also holds.

Theorem 23.12. *Let $f : [a, b] \to \mathbb{R}$ be such that $f^{(n)}$ is convex (concave) on $[x_0, x]$. Then we have the inequality:*

$$f(x) - T_n(f; x_0, x) \underset{(\leq)}{\geq} \frac{1}{(n+1)!} f^{(n-1)}(x_0)(x - x_0)^{n+1}. \tag{23.30}$$

Proof. As $f^{(n)}$ is convex (concave) on $[x_0, x]$, we may write that

$$f^{(n)}(t) - f^{(n)}(x_0) \underset{(\leq)}{\geq} f^{(n-1)}(x_0)(t - x_0), \quad t \in [x_0, x]$$

which implies that

$$\left[f^{(n)}(t) - f^{(n)}(x_0) \right](x - t)^{n-1} \underset{(\leq)}{\geq} f^{(n-1)}(x_0)(t - x_0)(x - t)^{n-1}, \quad t \in [x_0, x].$$

Integrating over t on $[x_0, x]$ and using the representation (23.1), we may get

$$R_n(f; x_0, x) \geq (\leq) \frac{1}{(n-1)!} \int_{x_0}^{x} f^{(n-1)}(x_0)(t - x_0)(x - t)^{n-1} \, dt$$

$$= \frac{1}{(n-1)!} f^{(n-1)}(x_0) \int_{x_0}^{x} (t - x_0)(x - t)^{n-1} \, dt$$

$$= \frac{1}{(n-1)!} f^{(n-1)}(x_0)(x - x_0)^{n+1} B(2, n)$$

$$= \frac{1}{(n+1)!} f^{(n-1)}(x_0)(x - x_0)^{n+1}$$

and the inequality (23.30) is proved. $\qquad\square$

23.5   Taylor-Multivariate Case Estimates

Let Q be a compact convex subset of $\mathbb{R}^k$, $k \geq 2$; $\mathbf{z} := (z_1, \ldots, z_k)$, $\mathbf{x}_0 := (x_{01}, \ldots, x_{0k}) \in Q$.

Let $f : Q \to \mathbb{R}$ be such that all partial derivatives of order $(n-1)$ are coordinate-wise absolutely continuous functions, $n \in \mathbb{N}$. Also, $f \in C^{n-1}(Q)$. Each nth order partial derivative is denoted by $f_\alpha := \frac{\partial^\alpha f}{\partial x^\alpha}$, where $\alpha := (\alpha_1, \ldots, \alpha_k)$, $\alpha_i \in \mathbb{Z}^+$, $i = 1, \ldots, k$ and $|\alpha| := \sum_{i=1}^{k} \alpha_i = n$. Consider $g_z(t) := f(x_0 + t(z - x_0))$, $t \geq 0$. Then

$$g_z^{(j)}(t)$$

$$= \left[\left(\sum_{i=1}^{k} (z_i - x_{0i}) \frac{\partial}{\partial x_i} \right)^j f \right] (x_{01} + t(z_1 - x_{01}), \ldots, x_{0k} + t(z_k - x_{0k})), \tag{23.31}$$

for all $j = 0, 1, 2, \ldots, n - 1$.

Note that $g_z^{(n)}(t)$ is given in a similar way.

Example 23.13. Let $n = k = 2$. Then

$$g_z(t) = f(x_{01} + t(z_1 - x_{01}),\ x_{02} + t(z_2 - x_{02})),\ t \in \mathbb{R}$$

and

$$g_z'(t) = (z_1 - x_{01})\frac{\partial f}{\partial x_1}(x_0 + t(z - x_0)) + (z_2 - x_{02})\frac{\partial f}{\partial x_2}(x_0 + t(z - x_0)).$$

In addition,

$$g_z''(t) = (z_1 - x_{01})\left(\frac{\partial f}{\partial x_1}(x_0 + t(z - x_0))\right)' + (z_2 - x_{02}) \cdot \left(\frac{\partial f}{\partial x_2}(x_0 + t(z - x_0))\right)'$$

$$= (z_1 - x_{01})\left\{(z_1 - x_{01})\frac{\partial f^2}{\partial x_1^2}(\cdot) + (z_2 - x_{02})\frac{\partial f^2}{\partial x_2 \partial x_1}(\cdot)\right\}$$

$$+ (z_2 - x_{02})\left\{(z_1 - x_{01})\frac{\partial f^2}{\partial x_1 \partial x_2}(\cdot) + (z_2 - x_{02})\frac{\partial f^2}{\partial x_2^2}(\cdot)\right\}.$$

Hence,

$$g_z''(t) = (z_1 - x_{01})^2\frac{\partial f^2}{\partial x_1^2}(\cdot) + (z_1 - x_{01})(z_2 - x_{02})\frac{\partial f^2}{\partial x_2 \partial x_1}(\cdot)$$

$$+ (z_1 - x_{01})(z_2 - x_{02})\frac{\partial f^2}{\partial x_1 \partial x_2}(\cdot) + (z_2 - x_{02})^2\frac{\partial f^2}{\partial x_2^2}(\cdot).$$

Similarly, we derive the case for $n, k \in \mathbb{N}$ for $g_z^{(n)}(t)$.

Notice that, if $\|f_\alpha\|_{[x_0, z]}$ exists for all α such that $|\alpha| = n$, then $\left\|g_z^{(n)}\right\|_{[0,1]}$ also exists; $\|\cdot\|$ is any type of p−norm $(p \in [1, \infty])$.

Therefore, we derive the multivariate Taylor Theorem:

Theorem 23.14. *With the above assumptions, we have*

$$f(z_1, \ldots, z_k) = g_z(1) = \sum_{j=0}^{n}\frac{g_z^{(j)}(0)}{j!} + R_n(z, 0), \tag{23.32}$$

where

$$R_n(z, 0) := \int_0^1\left(\int_0^{t_1}\cdots\left(\int_0^{t_{n-1}}\left(g_z^{(n)}(t_n) - g_z^{(n)}(0)\right)dt_n\right)\cdots\right)dt_1, \tag{23.33}$$

or

$$R_n(z, 0) = \frac{1}{(n-1)!}\int_0^1(1 - \theta)^{n-1}\left(g_z^{(n)}(\theta) - g_z^{(n)}(0)\right)d\theta. \tag{23.34}$$

A simpler form is

$$f(z_1, \ldots, z_k) = g_z(1) = \sum_{j=0}^{n-1}\frac{g_z^{(j)}(0)}{j!} + \tilde{R}_n(z, 0), \tag{23.35}$$

where

$$\tilde{R}_n(z,0) := \int_0^1 \left(\int_0^{t_1} \cdots \left(\int_0^{t_{n-1}} g_z^{(n)}(t_n)\, dt_n \right) \cdots \right) dt_1, \qquad (23.36)$$

or

$$\tilde{R}_n(z,0) = \frac{1}{(n-1)!} \int_0^1 (1-\theta)^{n-1} g_z^{(n)}(\theta)\, d\theta. \qquad (23.37)$$

Notice that $g_z(0) = f(x_0)$.

For a mapping $f : Q \to \mathbb{R}$, $z, x_0 \in Q$, $Q \subset \mathbb{R}^k$ compact and convex, we define

$$\|f\|_{[x_0,z];p([z,x_0])} = \left| \int_{[x_0,z]([z,x_0])} |f(y)|^p \, dy \right|^{\frac{1}{p}}, \quad p \geq 1. \qquad (23.38)$$

Here $\int_{[x_0,z]([z,x_0])}$ is a kth multiple integral. Also,

$$\|f\|_{[x_0,z];\infty} := ess \sup_{y \in [x_0,z](y \in [z,x_0])} |f(y)|, \qquad (23.39)$$

where $[x_0,z] \equiv [z,x_0]$ are line segments in Q.

We first find estimates for $\tilde{R}_n(z,0)$ as in (23.37).

Remark 23.15. Let $\|\cdot\|$ be any norm on the functions from Q to $\mathbb{R}$. Let $\|f_\alpha^*\|_{[x_0,z]} = \max\limits_{|\alpha|=n} \|f_\alpha\|_{[x_0,z]}$. Then

$$\left\| g_z^{(n)}(t) \right\|_{[0,1]}$$

$$= \left\| \left[\left(\sum_{i=1}^k (z_i - x_{0i}) \frac{\partial}{\partial x_i} \right)^n f \right] (x_{01} + t(z_1 - x_{01}), \ldots, x_{0k} + t(z_k - x_{0k})) \right\|_{[0,1]}$$

$$\leq \left(\sum_{i=1}^k |z_i - x_{0i}| \right)^n \cdot \|f_\alpha^*\|_{[x_0,z]},$$

that is,

$$\left\| g_z^{(n)}(t) \right\|_{[0,1]} \leq \left(\|z - x_0\|_{l_1} \right)^n \cdot \|f_\alpha^*\|_{[x_0,z]}. \qquad (23.40)$$

Here $\|\cdot\|_{[0,1]}$, $\|\cdot\|_{[x_0,z]}$ may be any kind of $p-$norm ($p \in [1,\infty]$).

We now state the first result in estimating the remainder $\tilde{R}_n(z,0)$.

Theorem 23.16. *With the above assumptions, we have*

$$\left| \tilde{R}_n(z,0) \right| \leq \begin{cases} \frac{1}{(n-1)!} \left\| g_z^{(n)} \right\|_{L_1[0,1]} & \text{if } g_z^{(n)} \in L_1[0,1]; \\[2ex] \frac{1}{(n-1)!(pn-p+1)^{\frac{1}{p}}} \left\| g_z^{(n)} \right\|_{L_q[0,1]} & \text{if } g_z^{(n)} \in L_q[0,1], \\[1ex] & \qquad \frac{1}{p} + \frac{1}{q} = 1, \, p > 1; \\[2ex] \frac{1}{n!} \left\| g_z^{(n)} \right\|_{[0,1];\infty} & \text{if } g_z^{(n)} \in L_\infty[0,1]. \end{cases}$$

Proof.   We have

$$\left|\tilde{R}_n\left(z,0\right)\right| \le \frac{1}{(n-1)!} \int_0^1 (1-\theta)^{n-1} \left|g_z^{(n)}\left(\theta\right)\right| d\theta$$

$$\le \frac{1}{(n-1)!} \int_0^1 \left|g_z^{(n)}\left(\theta\right)\right| d\theta = \frac{1}{(n-1)!} \left\|g_z^{(n)}\left(t\right)\right\|_{L_1[0,1]}.$$

That is

$$\left|\tilde{R}_n\left(z,0\right)\right| \le \frac{\left\|g_z^{(n)}\right\|_{L_1[0,1]}}{(n-1)!}, \tag{23.41}$$

given that $g_z^{(n)} \in L_1\left[0,1\right]$, the last is implied by all $f_\alpha \in L_1\left[x_0,z\right]$.

Again we observe that

$$\left|\tilde{R}_n\left(z,0\right)\right| \le \frac{1}{(n-1)!} \int_0^1 (1-\theta)^{n-1} \left|g_z^{(n)}\left(\theta\right)\right| d\theta$$

$$\le \frac{1}{(n-1)!} \left(\int_0^1 \left((1-\theta)^{n-1}\right)^p d\theta\right)^{\frac{1}{p}} \left(\int_0^1 \left|g_z^{(n)}\left(\theta\right)\right|^q d\theta\right)^{\frac{1}{q}}$$

$$= \frac{1}{(n-1)!} \left(\int_0^1 (1-\theta)^{pn-p} d\theta\right)^{\frac{1}{p}} \left\|g_z^{(n)}\right\|_{L_q[0,1]}$$

$$= \frac{1}{(n-1)!\,(pn-p+1)^{\frac{1}{p}}} \left\|g_z^{(n)}\right\|_{L_q[0,1]}.$$

That is,

$$\left|\tilde{R}_n\left(z,0\right)\right| \le \frac{\left\|g_z^{(n)}\right\|_{L_q[0,1]}}{(n-1)!\,(pn-p+1)^{\frac{1}{p}}}, \tag{23.42}$$

where $p,q > 1$, $\frac{1}{p} + \frac{1}{q} = 1$; $g_z^{(n)} \in L_q\left[0,1\right]$, the last is implied by all $f_\alpha \in L_q\left[x_0,z\right]$.

Also it holds

$$\left|\tilde{R}_n\left(z,0\right)\right| \le \frac{1}{(n-1)!} \int_0^1 (1-\theta)^{n-1} \left|g_z^{(n)}\left(\theta\right)\right| d\theta$$

$$\le \frac{1}{(n-1)!} \left(\int_0^1 (1-\theta)^{n-1} d\theta\right) \left\|g_z^{(n)}\right\|_{[0,1];\infty}$$

$$= \frac{\left\|g_z^{(n)}\right\|_{[0,1];\infty}}{n!},$$

that is

$$\left|\tilde{R}_n\left(z,0\right)\right| \le \frac{\left\|g_z^{(n)}\right\|_{[0,1];\infty}}{n!}, \quad \text{if } g_z^{(n)} \in L_\infty\left[0,1\right], \tag{23.43}$$

the last is implied by all $f_\alpha \in L_\infty\left[x_0,z\right]$. $\quad\sqcup$

Remark 23.17. Observe that

$$g_z^{(n)}(0) = \left[\left(\sum_{i=1}^{k}(z_i - x_{0i})\frac{\partial}{\partial x_i}\right)^n f\right](x_0),$$

where $x_0 := (x_{01}, \ldots, x_{0k})$.

Similarly, we find

$$|R_n(z,0)| \le \frac{\left\|g_z^{(n)}(t) - g_z^{(n)}(0)\right\|_{L_1[0,1]}}{(n-1)!} \tag{23.44}$$

given that $g_z^{(n)} \in L_1[0,1]$ which holds when all $f_\alpha \in L_1[x_0, z]$.

Also

$$|R_n(z,0)| \le \frac{\left\|g_z^{(n)}(t) - g_z^{(n)}(0)\right\|_{L_q[0,1]}}{(n-1)!\,(pn-p+1)^{\frac{1}{p}}}, \tag{23.45}$$

where $p, q > 1$, $\frac{1}{p} + \frac{1}{q} = 1$; $g_z^{(n)} \in L_q[0,1]$, when all $f_\alpha \in L_q[x_0, z]$.

Furthermore we obtain

$$|R_n(z,0)| \le \frac{\left\|g_z^{(n)}(t) - g_z^{(n)}(0)\right\|_{[0,1];\infty}}{n!}, \quad \text{if } g_z^{(n)} \in L_\infty[0,1], \tag{23.46}$$

when all $f_\alpha \in L_\infty[x_0, z]$.

Suppose now that (for all α such that $|\alpha| = n$)

$$|f_\alpha(x) - f_\alpha(y)| \le L \cdot \|x - y\|_{l_1}^{\beta}, \quad 0 < \beta \le 1, \tag{23.47}$$

for all $x, y \in Q$, $L > 0$, where $\|\cdot\|_{l_1}$ is the l_1 norm in $\mathbb{R}^k$. Here f_α is any partial derivative of order n. Then clearly (for all α such that $|\alpha| = n$)

$$|f_\alpha(x_0 + t(z - x_0)) - f_\alpha(x_0)| \le L \cdot t^{\beta} \cdot \|z - x_0\|_{l_1}^{\beta}, \tag{23.48}$$

where $\|z - x_0\|_{l_1} = \sum_{i=1}^{k}|z_i - x_{0i}|$.

Thus, if $z \ne x_0$, then for at least one $i \in \{1, \ldots, k\}$, we have $z_i \ne x_{0i}$, i.e., $\|z - x_0\|_{l_1} \ne 0$.

So, without loss of generality *assume* that $z \ne x_0$, which implies that $\|z - x_0\|_{l_1} \ne 0$. Hence by (23.33)

$$|R_n(z,0)|$$

$$\le \int_0^1 \left(\int_0^{t_1} \cdots \left(\int_0^{t_{n-1}}\left(\sum_{|\alpha|=n}\frac{n!\prod_{i=1}^{k}|z_i - x_{0i}|^{\alpha_i}}{\alpha_1!\cdots\alpha_k!}L\,\|z - x_0\|_{l_1}^{\beta}\,t_n^{\beta}\right)dt_n\right)\cdots\right)dt_1$$

$$= \sum_{|\alpha|=n}\frac{n!\prod_{i=1}^{k}|z_i - x_{0i}|^{\alpha_i}}{\alpha_1!\cdots\alpha_k!}\cdot L\cdot\|z - x_0\|_{l_1}^{\beta}\cdot\int_0^1\left(\int_0^{t_1}\cdots\left(\int_0^{t_{n-1}}t_n^{\beta}dt_n\right)\cdots\right)dt_1$$

$$= L\left(\sum_{i=1}^{k}|z_i - x_{0i}|\right)^n\|z - x_0\|_{l_1}^{\beta}\frac{1}{\prod_{j=1}^{n}(\beta + j)} = L\frac{\|z - x_0\|_{l_1}^{\beta+n}}{\prod_{j=1}^{n}(\beta + j)}.$$

Consequently, we state the following result.

Theorem 23.18. *Let f_α satisfy (23.47), then*

$$|R_n(z,0)| \le L \cdot \frac{\|z - x_0\|_{l_1}^{\beta+n}}{\prod_{j=1}^{n}(\beta+j)}, \quad \text{for all } z \in Q, \ |\alpha| = n. \tag{23.49}$$

Another matter to discuss:
We have

$$f(z_1, \ldots, z_k) = \sum_{j=0}^{n-1} \frac{g_z^{(j)}(0)}{j!} + R_{n-1}(z,0), \tag{23.50}$$

where

$$R_{n-1}(z,0) = \frac{1}{(n-2)!} \int_0^1 (1-\theta)^{n-2} \left(g_z^{(n-1)}(\theta) - g_z^{(n-1)}(0) \right) d\theta$$

$$= \frac{1}{(n-2)!} \int_0^1 (1-\theta)^{n-2} \left(\int_0^\theta g_z^{(n)}(u) \, du \right) d\theta. \tag{23.51}$$

Furthermore

$$|R_{n-1}(z,0)| \le \frac{1}{(n-2)!} \int_0^1 (1-\theta)^{n-2} \left(\int_0^\theta \left| g_z^{(n)}(u) \right| du \right) d\theta. \tag{23.52}$$

Now, if all $f_\alpha \in L_\infty[x_0, z]$, then $g_z^{(n)} \in L_\infty[0,1]$, and thus

$$\int_0^\theta \left| g_z^{(n)}(u) \right| du \le \theta \left\| g_z^{(n)} \right\|_{L_\infty[0,\theta]}.$$

Let $p, q > 1$ such that $\frac{1}{p} + \frac{1}{q} = 1$, then

$$\left| \int_0^\theta \left| g_z^{(n)}(u) \right| du \right| \le \left(\int_0^\theta 1^q du \right)^{\frac{1}{q}} \left(\int_0^\theta \left| g_z^{(n)}(u) \right|^p du \right)^{\frac{1}{p}}$$

$$= \theta^{\frac{1}{q}} \left\| g_z^{(n)} \right\|_{L_p[0,\theta]}$$

where $g_z^{(n)} \in L_p[0,1]$, when all $f_\alpha \in L_p[x_0, z]$.

Also,

$$\int_0^\theta \left| g_z^{(n)}(u) \right| du \le \left\| g_z^{(n)} \right\|_{L_1[0,\theta]},$$

where $g_z^{(n)} \in L_1[0,1]$, when all $f_\alpha \in L_1[x_0, z]$.

Thus we state the following result.

Theorem 23.19. *With the above assumptions, we have*

$$|R_{n-1}(z,0)|$$

$$\leq \begin{cases} \frac{1}{(n-2)!} \int_0^1 (1-\theta)^{n-2} \cdot \theta \cdot \left\| g_z^{(n)} \right\|_{L_\infty[0,\theta]} d\theta; & \text{if } g_z^{(n)} \in L_\infty[0,1], \\ & \text{all } f_\alpha \in L_\infty[x_0, z]; \\ \frac{1}{(n-2)!} \int_0^1 (1-\theta)^{n-2} \cdot \theta^{\frac{1}{q}} \cdot \left\| g_z^{(n)} \right\|_{L_p[0,\theta]} d\theta; & \text{if } g_z^{(n)} \in L_p[0,1], \ \frac{1}{p} + \frac{1}{q} = 1, \\ & p, q > 1, \ \text{all } f_\alpha \in L_p[x_0, z]; \\ \frac{1}{(n-2)!} \int_0^1 (1-\theta)^{n-2} \left\| g_z^{(n)} \right\|_{L_1[0,\theta]} d\theta; & \text{if } g_z^{(n)} \in L_1[0,1], \\ & \text{all } f_\alpha \in L_1[x_0, z]. \end{cases}$$

$$(23.53)$$

Remark 23.20.

a) Using Lemma 23.8, we have the representation

$$R_n(z, 0) = \left[\left(g_z^{(n-1)}(1) - g_z^{(n-1)}(0) \right) - g_z^{(n)}(0) \right] \frac{1}{n!}$$
$$+ \frac{1}{2(n-1)!} \int_0^1 \int_0^1 \left(g_z^{(n)}(t) - g_z^{(n)}(s) \right) \left((1-t)^{n-1} - (1-s)^{n-1} \right) dt ds.$$

$$(23.54)$$

Next, suppose that $g_z^{(n)} \in L_2[-1, 2]$. By Theorem 23.9 we get $(0, 1 \in (-1, 2))$:

$$|R_n(z, 0)| \leq \frac{\left| \left(g_z^{(n-1)}(1) - g_z^{(n-1)}(0) \right) - g_z^{(n)}(0) \right|}{n!}$$
$$+ \frac{(n-1)}{n! \sqrt{2n-1}} \left[\left\| g_z^{(n)} \right\|_{[0,1];2}^2 \right.$$
$$\left. - \left(g_z^{(n-1)}(1) - g_z^{(n-1)}(0) \right)^2 \right]^{\frac{1}{2}},$$

$$(23.55)$$

where

$$\left\| g_z^{(n)} \right\|_{[0,1];2} := \left(\int_0^1 \left| g_z^{(n)}(t) \right|^2 dt \right)^{\frac{1}{2}}.$$

b) By Theorem 23.10, assuming that $g_z^{(n)}$ is nondecreasing (nonincreasing) over $[0, 1]$, we have

$$f(z_1, \ldots, z_k)$$
$$\leq (\geq) \sum_{j=0}^n \frac{g_z^{(j)}(0)}{j!} + \frac{\left[\left(g_z^{(n-1)}(1) - g_z^{(n-1)}(0) \right) - g_z^{(n)}(0) \right]}{n!}.$$

$$(23.56)$$

That is

$$f(z_1, \ldots, z_k)$$
$$\leq (\geq) \sum_{j=0}^{n-1} \frac{g_z^{(j)}(0)}{j!} + \frac{\left(g_z^{(n-1)}(1) - g_z^{(n-1)}(0) \right)}{n!}.$$

$$(23.57)$$

c) By Theorem 23.11, assuming that $g_z^{(n)}$ is monotonic increasing on $[0,1]$, we find that

$$f(z_1,\ldots,z_k) - \sum_{j=0}^{n-1} \frac{g_z^{(j)}(0)}{j!} - \frac{\left(g_z^{(n-1)}(1) - g_z^{(n-1)}(0)\right)}{n!}$$

$$\geq \frac{1}{(n-1)!} \left| \int_0^1 \left|g_z^{(n)}(t)\right|(1-t)^{n-1}\,dt - \frac{1}{n}\int_0^1 \left|g_z^{(n)}(t)\right|\,dt \right|. \qquad (23.58)$$

d) At last, by Theorem 23.12, for $g_z^{(n)}$ convex (concave) on $[0,1]$, we derive

$$f(z_1,\ldots,z_k) - \sum_{j=0}^{n} \frac{g_z^{(j)}(0)}{j!} \geq (\leq)\frac{g_z^{(n-1)}(0)}{(n+1)!}. \qquad (23.59)$$

Chapter 24

The Distributional Taylor Formula

We present a distributional Taylor formula with precise integral remainder. We give applications using it and estimates for the remainder. This chapter relies on [51].

24.1 Introduction and Background

This chapter is motivated by the important works of R. Estrada-R. P. Kanwal (1990) [135]; (1992), [136]; (1993), [137]; and A. Durán-R. Estrada-R. P. Kanwal (1996), [134]; B. Stankovic (1996), [234], and R. Estrada-R. P. Kanwal (2002), Chapter 3, section 3.2, [138]; I. M. Gel'fand-G. E. Shilov (1964), Vol I, pp. 146-151, and pp. 331-345, [146]. It is also motivated by the seminal works of B. Ziemian (1988), [251]; (1988), [253]; (1989), [252].

All the above authors either give distributional Taylor type asymptotic expansions, or distributional Taylor formulae where the remainder is not precisely specified and it is rather vague. Other times Taylor expansions are only for the delta function distribution.

Author's work with S. S. Dragomir (2001), [55], was also another inspiration for this chapter.

Other stimulating works are the ones by V. I. Burenkov (1974), [77] and (1998), [78], see Chapter 3, both for Sobolev's integral representation. But this chapter would have been impossible without the existence of the excellent monograph "Analysis" by E. H. Lieb-M. Loss, 2001, [178].

We are based mainly on the following result from last.

Theorem 24.1. ([178], p. 143, Fundamental theorem of Calculus for distributions) *Let* $f \in W_{loc}^{1,1}(\mathbb{R}^n)$. *Then, for each* $y \in \mathbb{R}^n$ *and almost every* $x \in \mathbb{R}^n$,

$$f(x+y) - f(x) = \int_0^1 y \cdot \nabla f(x+ty)\, dt. \tag{24.1}$$

Before using (24.1) we would like to give an equivalent result to Theorem 24.1 and some useful definitions, however for full details on basic distribution theory we use we refer the reader to [178], Ch. 6 pp. 135–169.

293

So we give the equivalent to Theorem 24.1 result.

Theorem 24.2. *Let $f \in W^{1,1}_{loc}(\mathbb{R}^n)$. Then, almost every $x, z \in \mathbb{R}^n$ we have,*

$$f(z) - f(x) = \int_0^1 (z - x) \cdot \nabla f(x + t(z - x))\, dt. \tag{24.2}$$

Proof. Easily we see that (24.1) is equivalent to (24.2). $\qquad\qquad\square$

Remark 24.3. Notice that (24.1) is true for any $y \in A \subseteq \mathbb{R}^n$ and almost all $x \in A$ such that $x + y \in A$, where A is a compact and convex subset of $\mathbb{R}^n$. The null set of x's (24.1) does not hold depends on y, f. Similarly (24.2) holds for almost every $x, z \in A$, where A is as above.

In both cases $f \in W^{1,1}_{loc}(\mathbb{R}^n)$.

The following terms and definitions are taken from [178], pp. 136–144.

Let Ω be an open, nonempty subset of $\mathbb{R}^n$, $n \geq 2$, $C^\infty_c(\Omega)$ denotes the space of all infinitely differentiable, complex-valued functions whose support is compact and in Ω. The space of test functions is denoted by $D(\Omega)$. The dual space of $D(\Omega)$ is denoted by $D'(\Omega)$ which is all continuous linear functionals $T : D(\Omega) \to \mathbb{C}$, we call T's distributions.

By $L^p_{loc}(\Omega)$ we denote the space of locally p^{th}–power summable functions, $1 \leq p \leq \infty$. Such functions are Borel measurable functions defined on all of Ω and values in $\mathbb{C}$ and with the property $\|f\|_{L^p(K)} < \infty$, for every compact set $K \subset \Omega$. Clearly $L^p_{loc}(\Omega) \supset L^p(\Omega)$ and, if $r > p$, we have $L^p_{loc}(\Omega) \supset L^r_{loc}(\Omega)$.

We must mention

Theorem 24.4. (Functions are uniquely determined by distributions, see [178], p. 138) *Let $\Omega \subset \mathbb{R}^n$ be open and let $f, g \in L^1_{loc}(\Omega)$. Assume that the distributions defined by f and g are equal, i.e.*

$$\int_\Omega f\Phi = \int_\Omega g\Phi \tag{24.3}$$

for all $\Phi \in D(\Omega)$. Then $f(x) = g(x)$ for almost every $x \in \Omega$.

By ∇f we denote the distributional gradient of f, that is the n–tuple $(\partial_1 f, \partial_2 f, \ldots, \partial_n f)$, where $f \in L^1_{loc}(\Omega)$ –functions are an important class of distributions.

We denote by $W^{1,1}_{loc}(\Omega)$ the class of functions from $L^1_{loc}(\Omega)$ whose distributional (weak) first derivatives are also in $L^1_{loc}(\Omega)$.

We further define, $1 \leq p \leq \infty$,

$$W^{1,p}_{loc}(\Omega) := \{f : \Omega \to \mathbb{C} : f \in L^p_{loc}(\Omega) \text{ and} \partial_i f, \text{ as a distribution in } D'(\Omega), \text{ is a}$$

$$L^p_{loc}(\Omega) - \text{function for } i = 1, \ldots, n\}. \tag{24.4}$$

We have $W^{1,p}_{loc}(\Omega) \supset W^{1,r}_{loc}(\Omega)$ if $r > p$. We can also define the Sobolev space $W^{1,p}(\Omega) \subset W^{1,p}_{loc}(\Omega)$:

$$W^{1,p}(\Omega) := \{f : \Omega \to \mathbb{C} : f \text{ and}$$

$$\partial_i f \in L^p(\Omega), \; i = 1, \ldots, n\}. \tag{24.5}$$

Similarly we define $W_{loc}^{m,p}(\Omega)$, and $W^{m,p}(\Omega)$, $m > 1$. In these the Borel measurable functions $f : \Omega \to \mathbb{C}$ and all of its partial distributional derivatives up to order m belong to $L_{loc}^p(\Omega)$, and $L^p(\Omega)$, respectively. Clearly $W^{m,p}(\Omega) \subset W_{loc}^{m,p}(\Omega)$.

We end this section with the known basic and useful results, usually given as exercises. We put them together.

Theorem 24.5. *Let $f \in W_{loc}^{m,1}(\Omega)$, Ω is open subset in $\mathbb{R}^n$. Then the distributional partial derivative $D_{r_1,\ldots,r_k} f(x)$, $x \in \Omega$, remains unchanged almost everywhere, when the indices $r_1, \ldots, r_k$ are permuted, each r_i is a positive integer $\leq m$. There are $\dbinom{k+n-1}{k}$ distinct distributional partial derivatives of order k in n dimensions, however there is a total of n^k distributional partial derivatives of order $k \leq m$.*

We demonstrate the first part of the last theorem.

Example 24.6. Let $f \in W_{loc}^{3,1}(\Omega)$. We see that

$$\int_\Omega f \frac{\partial^3 \Phi}{\partial x \partial y \partial z} = - \int_\Omega \frac{\partial f}{\partial x} \frac{\partial^2 \Phi}{\partial y \partial z} = \int_\Omega \frac{\partial^2 f}{\partial y \partial x} \frac{\partial \Phi}{\partial z}$$

$$= - \int_\Omega \frac{\partial^3 f}{\partial z \partial y \partial x} \Phi, \; \forall \Phi \in D(\Omega). \tag{24.6}$$

Similarly we obtain

$$\int_\Omega f \frac{\partial^3 \Phi}{\partial x \partial z \partial y} = - \int_\Omega \frac{\partial f}{\partial x} \frac{\partial^2 \Phi}{\partial z \partial y}$$

$$= \int_\Omega \frac{\partial^2 f}{\partial z \partial x} \frac{\partial \Phi}{\partial y} = - \int_\Omega \frac{\partial^3 f}{\partial y \partial z \partial x} \Phi, \; \forall \Phi \in D(\Omega). \tag{24.7}$$

However

$$\frac{\partial^3 \Phi}{\partial x \partial y \partial z} = \frac{\partial^3 \Phi}{\partial x \partial z \partial y},$$

thus

$$\int_\Omega f \frac{\partial^3 \Phi}{\partial x \partial y \partial z} = \int_\Omega f \frac{\partial^3 \Phi}{\partial x \partial z \partial y}, \tag{24.8}$$

hence

$$\int_\Omega \frac{\partial^3 f}{\partial z \partial y \partial x} \Phi = \int_\Omega \frac{\partial^3 f}{\partial y \partial z \partial x} \Phi, \; \forall \Phi \in D(\Omega). \tag{24.9}$$

Therefore by Theorem 24.4 we derive

$$\frac{\partial^3 f}{\partial z \partial y \partial x} = \frac{\partial^3 f}{\partial y \partial z \partial x}, \; \text{a.e. on } \Omega, \tag{24.10}$$

etc.

24.2   Main Results

We present the main results.

We start with

Theorem 24.7. *Let* $f \in W_{loc}^{m,1}(\mathbb{R}^n)$; $m \in \mathbb{N}$, $n \in \mathbb{N} - \{1\}$. *Then for every* $y \in \mathbb{R}^n$ *and almost every* $x \in \mathbb{R}^n$ *we have*

$$f(x+y) - f(x) = \sum_{l=1}^{m-1} \left\{ \sum_{j_1,j_2,\ldots,j_l=1}^{n} (\partial_{j_1\ldots j_l}^{l} f(x)) \left(\prod_{i=1}^{l} y_{j_i} \right) \right\}$$

$$+ \sum_{j_1,\ldots,j_m=1}^{n} \left(\prod_{i=1}^{m} y_{j_i} \right) \int_{[0,1]^m} \partial_{j_1\ldots j_m}^{m} f \left(x + \prod_{r=1}^{m} t_r y \right) dt_1 \ldots dt_m, \qquad (24.11)$$

where $y := (y_1, \ldots, y_n)$, $x := (x_1, \ldots, x_n)$.

Equality (24.11) is true for any $y \in A$ and almost every $x \in A$ such that $x+y \in A$, where A is a compact and convex subset of $\mathbb{R}^n$.

The above theorem is equivalent to

Theorem 24.8. *Let* $f \in W_{loc}^{m,1}(\mathbb{R}^n)$; $m \in \mathbb{N}$, $n \in \mathbb{N} - \{1\}$. *Then for almost every* $x, z \in \mathbb{R}^n$ *we have*

$$f(z) - f(x) = \sum_{l=1}^{m-1} \left\{ \sum_{j_1,\ldots,j_l=1}^{n} (\partial_{j_1\ldots j_l}^{l} f(x)) \left(\prod_{i=1}^{l} (z_{j_i} - x_{j_i}) \right) \right\}$$

$$+ \sum_{j_1,\ldots,j_m=1}^{n} \left(\prod_{i=1}^{m} (z_{j_i} - x_{j_i}) \right) \int_{[0,1]^m} \partial_{j_1\ldots j_m}^{m} f \left(x + \prod_{r=1}^{m} t_r (z - x) \right) dt_1 \ldots dt_m,$$

$$(24.12)$$

where $z := (z_1, \ldots, z_n)$.

Equality (24.12) is true for almost every $x, z \in A$, where A is a compact and convex subset of $\mathbb{R}^n$.

Another useful equivalent form of Theorem 24.7 follows.

Theorem 24.9. *Let* $f \in W_{loc}^{m,1}(\mathbb{R}^n)$; $m \in \mathbb{N}$, $n \in \mathbb{N} - \{1\}$. *Then for every* $y \in \mathbb{R}^n$ *and almost every* $x \in \mathbb{R}^n$ *we have*

$$f(x+y) - f(x) = \sum_{l=1}^{m} \left\{ \sum_{j_1,\ldots,j_l=1}^{n} (\partial_{j_1\ldots j_l}^{l} f(x)) \left(\prod_{i=1}^{l} y_{j_i} \right) \right\}$$

$$+ \sum_{j_1,\ldots,j_m=1}^{n} \left(\prod_{i=1}^{m} y_{j_i} \right) \int_{[0,1]^m} \left(\partial_{j_1\ldots j_m}^{m} f \left(x + \prod_{r=1}^{m} t_r y \right) - \partial_{j_1\ldots j_m}^{m} f(x) \right) dt_1 \ldots dt_m.$$

$$(24.13)$$

Above (24.13) is also true for any $y \in A$, almost every $x \in A : x + y \in A$, where A is compact, convex subset of $\mathbb{R}^n$.

Similarly, Theorem 24.8 is equivalent to

Theorem 24.10. *Let* $f \in W_{loc}^{m,1}(\mathbb{R}^n)$; $m \in \mathbb{N}$, $n \in \mathbb{N} - \{1\}$. *Then for almost every* $x, z \in \mathbb{R}^n$ *we have*

$$f(z) - f(x) = \sum_{l=1}^{m} \left\{ \sum_{j_1,\ldots,j_l=1}^{n} \left(\partial_{j_1 \ldots j_l}^{l} f(x) \right) \left(\prod_{i=1}^{l} (z_{j_i} - x_{j_i}) \right) \right\}$$

$$+ \sum_{j_1,\ldots,j_m=1}^{n} \left(\prod_{i=1}^{m} (z_{j_i} - x_{j_i}) \right) \int_{[0,1]^m} \left(\partial_{j_1 \ldots j_m}^{m} f \left(x + \prod_{r=1}^{m} t_r (z - x) \right) \right.$$

$$\left. - \partial_{j_1 \ldots j_m}^{m} f(x) \right) dt_1 \ldots dt_m. \tag{24.14}$$

The last (24.14) is true for almost every $x, z \in A$, where A is a compact, convex subset of $\mathbb{R}^n$.

We demonstrate the validity of Theorem 24.7 by giving the following.

Proposition 24.11. *Let* $f \in W_{loc}^{2,1}(\mathbb{R}^n)$, $n \geq 2$. *Then for every* $y \in \mathbb{R}^n$ *and almost every* $x \in \mathbb{R}^n$ *we have*

$$f(x+y) - f(x) = y \cdot \nabla f(x)$$

$$+ \sum_{j_1=1}^{n} \sum_{j_2=1}^{n} y_{j_1} y_{j_2} \int_0^1 \int_0^1 \partial_{j_2} \partial_{j_1} f(x + t_2 t_1 y) \, dt_2 dt_1. \tag{24.15}$$

Proof. By Theorem 24.1 we have

$$f(x+y) - f(x) = \sum_{j_1=1}^{n} y_{j_1} \int_0^1 \partial_{j_1} f(x + t_1 y) \, dt_1. \tag{24.16}$$

Let $\Phi \in D(\mathbb{R}^n)$, then

$$\int_{\mathbb{R}^n} \Phi(x) \left[f(x+y) - f(x) \right] dx =$$

$$\int_{\mathbb{R}^n} \Phi(x) \left\{ \sum_{j_1=1}^{n} y_{j_1} \int_0^1 \partial_{j_1} f(x + t_1 y) \, dt_1 \right\} dx. \tag{24.17}$$

Next we apply (24.17) for each $\partial_{j_1} f$, $j_1 = 1, \ldots, n$, to obtain

$$\int_{\mathbb{R}^n} \Phi(x) \left[\partial_{j_1} f(x + t_1 y) - \partial_{j_1} f(x) \right] dx =$$

$$\int_{\mathbb{R}^n} \Phi(x) \left\{ \sum_{j_2=1}^{n} y_{j_2} \int_0^1 \partial_{j_2} \partial_{j_1} f(x + t_2 t_1 y) \, dt_2 \right\} dx. \tag{24.18}$$

That is

$$\int_{\mathbb{R}^n} \Phi(x)\,\partial_{j_1} f(x + t_1 y)\,dx =$$

$$\int_{\mathbb{R}^n} \Phi(x) \left[\partial_{j_1} f(x) \sum_{j_2=1}^{n} y_{j_2} \int_0^1 \partial_{j_2 j_1}^2 f(x + t_2 t_1 y)\,dt_2 \right] dx. \tag{24.19}$$

Here all integrands are integrable and we can apply Fubini's theorem. Therefore we derive

$$\int_{\mathbb{R}^n} \Phi(x) \left(\int_0^1 \partial_{j_1} f(x + t_1 y)\,dt_1 \right) dx$$

$$= \int_{\mathbb{R}^n} \Phi(x) \left[\partial_{j_1} f(x) + \sum_{j_2=1}^{n} y_{j_2} \int_0^1 \int_0^1 \partial_{j_2 j_1}^2 f(x + t_2 t_1 y)\,dt_2 dt_1 \right] dx. \tag{24.20}$$

Next we apply (24.20) into (24.17) to find

$$\int_{\mathbb{R}^n} \Phi(x) \left[f(x + y) - f(x) \right] dx$$

$$\overset{(24.17)}{=} \sum_{j_1=1}^{n} y_{j_1} \left(\int_{\mathbb{R}^n} \Phi(x) \left\{ \int_0^1 \partial_{j_1} f(x + t_1 y)\,dt_1 \right\} dx \right)$$

$$\overset{(24.20)}{=} \sum_{j_1=1}^{n} y_{j_1} \left(\int_{\mathbb{R}^n} \Phi(x) \left[\partial_{j_1} f(x) + \sum_{j_2=1}^{n} y_{j_2} \right. \right.$$

$$\left. \left. \int_0^1 \int_0^1 \partial_{j_2 j_1}^2 f(x + t_2 t_1 y)\,dt_2 dt_1 \right] dx \right). \tag{24.21}$$

So we find from (24.21) that

$$\int_{\mathbb{R}^n} \Phi(x) \left[f(x + y) - f(x) \right] dx$$

$$= \int_{\mathbb{R}^n} \Phi(x) \left[\sum_{j_1=1}^{n} y_{j_1} \partial_{j_1} f(x) + \sum_{j_1=1}^{n} \sum_{j_2=1}^{n} y_{j_1} y_{j_2} \right.$$

$$\left. \int_0^1 \int_0^1 \partial_{j_2 j_1}^2 f(x + t_2 t_1 y)\,dt_2 dt_1 \right] dx, \tag{24.22}$$

for every $\Phi \in D(\mathbb{R}^n)$.

Using now Theorem 24.4 on (24.22) we get (24.15). $\qquad\square$

We continue with the next step of the above established procedure of iteration.

Proposition 24.12. *Let* $f \in W_{loc}^{3,1}\left(\mathbb{R}^n\right)$, $n \geq 2$. *Then for every* $y \in \mathbb{R}^n$ *and almost every* $x \in \mathbb{R}^n$, *we have*

$$f\left(x+y\right) - f\left(x\right) = y \cdot \nabla f\left(x\right) + \sum_{j_1,j_2=1}^{n} y_{j_1} y_{j_2} \partial_{j_2 j_1}^2 f\left(x\right) + \sum_{j_1,j_2,j_3=1}^{n} y_{j_1} y_{j_2} y_{j_3}$$

$$+ \sum_{j_1,j_2,j_3=1}^{n} y_{j_1} y_{j_2} y_{j_3} \int_{[0,1]^3} \partial_{j_3 j_2 j_1}^3 f\left(x + t_3 t_2 t_1 y\right) dt_3 dt_2 dt_1. \tag{24.23}$$

Proof. Here we plug into (24.17) $\partial_{j_2}\partial_{j_1} f$, for all j_1, $j_2 \in \left\{1,\ldots,n\right\}$, to obtain

$$\int_{\mathbb{R}^n} \Phi\left(x\right) \left[\partial_{j_2}\partial_{j_1} f\left(x + t_2 t_1 y\right) - \partial_{j_2}\partial_{j_1} f\left(x\right)\right] dx$$

$$= \int_{\mathbb{R}^n} \Phi\left(x\right) \left\{\sum_{j_3=1}^{n} y_{j_3} \int_0^1 \partial_{j_3 j_2 j_1}^3 f\left(x + t_3 t_2 t_1 y\right) dt_3\right\} dx. \tag{24.24}$$

That is

$$\int_{\mathbb{R}^n} \Phi\left(x\right) \partial_{j_2 j_1}^2 f\left(x + t_2 t_1 y\right) dx$$

$$= \int_{\mathbb{R}^n} \Phi\left(x\right) \left[\partial_{j_2 j_1}^2 f\left(x\right) + \left\{\sum_{j_3=1}^{n} y_{j_3} \int_0^1 \partial_{j_3 j_2 j_1}^3 f\left(x + t_3 t_2 t_1 y\right) dt_3\right\}\right] dx. \tag{24.25}$$

By Fubini theorem we derive

$$\int_{\mathbb{R}^n} \Phi\left(x\right) \left(\int_{[0,1]^2} \partial_{j_2 j_1}^2 f\left(x + t_2 t_1 y\right) dt_2 dt_1\right) dx =$$

$$\int_{\mathbb{R}^n} \Phi\left(x\right) \left[\partial_{j_2 j_1}^2 f\left(x\right) + \sum_{j_3=1}^{n} y_{j_3} \int_{[0,1]^3} \partial_{j_3 j_2 j_1}^3 f\left(x + t_3 t_2 t_1 y\right) dt_3 dt_2 dt_1\right] dx. \tag{24.26}$$

Next we put (24.26) into (24.22) to get

$$\int_{\mathbb{R}^n} \Phi\left(x\right) \left[f\left(x+y\right) - f\left(x\right)\right] dx = \int_{\mathbb{R}^n} \Phi\left(x\right) \left[y \cdot \nabla f\left(x\right) + \sum_{j_1,j_2=1}^{n} y_{j_1} y_{j_2}\right.$$

$$\left[\partial_{j_2 j_1}^2 f\left(x\right) + \sum_{j_3=1}^{n} y_{j_3} \int_{[0,1]^3} \partial_{j_3 j_2 j_1}^3 f\left(x + t_3 t_2 t_1 y\right) dt_3 dt_2 dt_1\right] dx$$

$$= \int_{\mathbb{R}^n} \Phi\left(x\right) \left[y \cdot \nabla f\left(x\right) + \sum_{j_1,j_2=1}^{n} y_{j_1} y_{j_2} \partial_{j_2 j_1}^2 f\left(x\right) +\right.$$

$$\left.\sum_{j_1,j_2,j_3=1}^{n} y_{j_1} y_{j_2} y_{j_3} \int_{[0,1]^3} \partial_{j_3 j_2 j_1}^3 f\left(x + t_3 t_2 t_1 y\right) dt_3 dt_2 dt_1\right] dx, \tag{24.27}$$

true for every $\Phi \in D\left(\mathbb{R}^n\right)$. By Theorem 24.4 we derive (24.23). $\qquad\square$

We continue recursively as above but now on $W_{loc}^{4,1}\left(\mathbb{R}^n\right)$ to establish

Proposition 24.13. *Let* $f \in W_{loc}^{4,1}\left(\mathbb{R}^n\right)$, $n \geq 2$. *Then for every* $y \in \mathbb{R}^n$ *and almost every* $x \in \mathbb{R}^n$ *we derive*

$$f\left(x+y\right)-f\left(x\right) = y\cdot\nabla f\left(x\right)+ \sum_{j_1,j_2=1}^{n} y_{j_1}y_{j_2}\partial_{j_1 j_2}^2 f\left(x\right)+ \sum_{j_1,j_2,j_3=1}^{n} y_{j_1}y_{j_2}y_{j_3}\partial_{j_1 j_2 j_3}^3 f\left(x\right) +$$

$$\sum_{j_1,j_2,j_3,j_4=1}^{n} y_{j_1}y_{j_2}y_{j_3}y_{j_4} \int_{[0,1]^4} \partial_{j_1 j_2 j_3 j_4}^4 f\left(x+t_1 t_2 t_3 t_4 y\right) dt_1 dt_2 dt_3 dt_4. \qquad (24.28)$$

Proof.   Now we plug into (24.17) $\partial_{j_3 j_2 j_1}^3 f$, for all $j_1, j_2, j_3 \in \{1,\ldots,n\}$, to obtain

$$\int_{\mathbb{R}^n} \Phi\left(x\right) \left[\partial_{j_3 j_2 j_1}^3 f\left(x+t_3 t_2 t_1 y\right) - \partial_{j_3 j_2 j_1}^3 f\left(x\right)\right] dx$$

$$= \int_{\mathbb{R}^n} \Phi\left(x\right) \left\{ \sum_{j_4=1}^{n} y_{j_4} \int_0^1 \partial_{j_4 j_3 j_2 j_1}^4 f\left(x+t_4 t_3 t_2 t_1 y\right) dt_4 \right\} dx. \qquad (24.29)$$

That is we have

$$\int_{\mathbb{R}^n} \Phi\left(x\right) \partial_{j_3 j_2 j_1}^3 f\left(x+t_3 t_2 t_1 y\right) dx$$

$$= \int_{\mathbb{R}^n} \Phi\left(x\right) \left[\partial_{j_3 j_2 j_1}^3 f\left(x\right) + \sum_{j_4=1}^{n} y_{j_4} \int_0^1 \partial_{j_4 j_3 j_2 j_1}^4 f\left(x+t_4 t_3 t_2 t_1 y\right) dt_4\right] dx. \qquad (24.30)$$

Using again Fubini's theorem we find

$$\int_{\mathbb{R}^n} \Phi\left(x\right) \left(\int_{[0,1]^3} \partial_{j_3 j_2 j_1}^3 f\left(x+t_3 t_2 t_1 y\right) dt_3 dt_2 dt_1\right) dx$$

$$= \int_{\mathbb{R}^n} \Phi\left(x\right) \left[\partial_{j_3 j_2 j_1}^3 f\left(x\right) + \sum_{j_4=1}^{n} y_{j_4} \int_{[0,1]^4} \partial_{j_4 j_3 j_2 j_1}^4 f\left(x+t_4 t_3 t_2 t_1 y\right) dt_4 dt_3 dt_2 dt_1\right] dx.$$
$$\qquad (24.31)$$

Finally we put (24.31) into (24.27) to derive

$$\int_{\mathbb{R}^n} \Phi\left(x\right) \left[f\left(x+y\right) - f\left(x\right)\right] dx = \int_{\mathbb{R}^n} \Phi\left(x\right) \left[y \cdot \nabla f\left(x\right) + \sum_{j_1,j_2=1}^{n} y_{j_1}y_{j_2}\partial_{j_2 j_1}^2 f\left(x\right)\right.$$

$$+ \sum_{j_1,j_2,j_3=1}^{n} y_{j_1}y_{j_2}y_{j_3} \left\{\partial_{j_3 j_2 j_1}^3 f\left(x\right)\right.$$

$$\left.\left.+ \sum_{j_4=1}^{n} y_{j_4} \int_{[0,1]^4} \partial_{j_4 j_3 j_2 j_1}^4 f\left(x+t_4 t_3 t_2 t_1 y\right) dt_4 dt_3 dt_2 dt_1 \right\}\right] dx, \qquad (24.32)$$

for every $\Phi \in D\left(\mathbb{R}^n\right)$.

Using Theorems 24.4 and 24.5 and Fubini's theorem we establish (24.28). $\qquad\square$

Clearly by mathematical induction we establish

Theorem 24.14. *Let $f \in W_{loc}^{m,1}(\mathbb{R}^n)$, where $m \in \mathbb{N}$, $n \in \mathbb{N}-\{1\}$. Then for every $y \in \mathbb{R}^n$ and almost every $x \in \mathbb{R}^n$ we have*

$$f(x+y) - f(x) = y \cdot \nabla f(x) + \sum_{j_1,j_2=1}^{n} y_{j_1} y_{j_2} \partial^2_{j_1 j_2} f(x)$$

$$+ \sum_{j_1,j_2,j_3=1}^{n} y_{j_1} y_{j_2} y_{j_3} \partial^3_{j_1 j_2 j_3} f(x) + \ldots + \sum_{j_1,j_2,\ldots,j_{m-1}=1}^{n} y_{j_1} y_{j_2} \cdots y_{j_{m-1}} \partial^{m-1}_{j_1 j_2 \ldots j_{m-1}} f(x)$$

$$+ \sum_{j_1,j_2,j_3,\ldots,j_m=1}^{n} y_{j_1} y_{j_2} \cdots y_{j_m} \int_{[0,1]^m} \partial^m_{j_1 j_2 \ldots j_m} f(x + t_1 \ldots t_m y) \, dt_1 dt_2 \ldots dt_m.$$

$$(24.33)$$

Writing (24.33) into a compact form we have (24.11).

Remark 24.15. (on Theorem 24.14)

1) Let $f \in W_{loc}^{3,1}(\mathbb{R}^3)$, for every $y \in \mathbb{R}^3$ and almost every $x \in \mathbb{R}^3$ we get

$$f(x+y) - f(x) = y \cdot \nabla f(x) + \sum_{j_1,j_2=1}^{3} y_{j_1} y_{j_2} \partial^2_{j_1 j_2} f(x)$$

$$+ \sum_{j_1,j_2,j_3=1}^{3} y_{j_1} y_{j_2} y_{j_3} \int_{[0,1]^3} \partial^3_{j_1 j_2 j_3} f(x + t_1 t_2 t_3 y) \, dt_1 dt_2 dt_3. \qquad (24.34)$$

2) Let $f \in W_{loc}^{4,1}(\mathbb{R}^4)$, for every $y \in \mathbb{R}^4$ and almost every $x \in \mathbb{R}^4$ it holds

$$f(x+y) - f(x) = y \cdot \nabla f(x) + \sum_{j_1,j_2=1}^{4} y_{j_1} y_{j_2} \partial^2_{j_1 j_2} f(x)$$

$$+ \sum_{j_1,j_2,j_3=1}^{4} y_{j_1} y_{j_2} y_{j_3} \partial^3_{j_1 j_2 j_3} f(x) + \sum_{j_1,j_2,j_3,j_4=1}^{4} y_{j_1} y_{j_2} y_{j_3} y_{j_4}$$

$$\cdot \int_{[0,1]^4} \partial^4_{j_1 j_2 j_3 j_4} f(x + t_1 t_2 t_3 t_4 y) \, dt_1 dt_2 dt_3 dt_4. \qquad (24.35)$$

3) Let $f \in W_{loc}^{5,1}(\mathbb{R}^4)$, for every $y \in \mathbb{R}^4$ and almost every $x \in \mathbb{R}^4$ we derive

$$f(x+y) - f(x) = y \cdot \nabla f(x) + \sum_{j_1,j_2=1}^{4} y_{j_1} y_{j_2} \partial^2_{j_1 j_2} f(x)$$

$$+ \sum_{j_1,j_2,j_3=1}^{4} y_{j_1} y_{j_2} y_{j_3} \partial^3_{j_1 j_2 j_3} f(x) + \sum_{j_1,j_2,j_3,j_4=1}^{4} y_{j_1} y_{j_2} y_{j_3} y_{j_4} \partial^4_{j_1 j_2 j_3 j_4} f(x)$$

$$+ \sum_{j_1,j_2,j_3,j_4,j_5=1}^{4} y_{j_1} y_{j_2} y_{j_3} y_{j_4} y_{j_5}$$

$$\cdot \int_{[0,1]^5} \partial^5_{j_1 j_2 j_3 j_4 j_5} f\left(x + t_1 t_2 t_3 t_4 t_5 y\right) dt_1 dt_2 dt_3 dt_4 dt_5. \tag{24.36}$$

4) Let $f \in W^{6,1}_{loc}\left(\mathbb{R}^4\right)$, for every $y \in \mathbb{R}^4$ and almost every $x \in \mathbb{R}^4$ it holds

$$f\left(x + y\right) - f\left(x\right) = y \cdot \nabla f\left(x\right) + \sum_{j_1,j_2=1}^{4} y_{j_1} y_{j_2} \partial^2_{j_1 j_2} f\left(x\right) +$$

$$\sum_{j_1,j_2,j_3=1}^{4} y_{j_1} y_{j_2} y_{j_3} \partial^3_{j_1 j_2 j_3} f\left(x\right) + \sum_{j_1,j_2,j_3,j_4=1}^{4} y_{j_1} y_{j_2} y_{j_3} y_{j_4} \partial^4_{j_1 j_2 j_3 j_4} f\left(x\right)$$

$$+ \sum_{j_1,j_2,j_3,j_4,j_5=1}^{4} y_{j_1} y_{j_2} y_{j_3} y_{j_4} y_{j_5} \partial^5_{j_1 j_2 j_3 j_4 j_5} f\left(x\right)$$

$$+ \sum_{j_1,j_2,j_3,j_4,j_5,j_6=1}^{4} y_{j_1} y_{j_2} y_{j_3} y_{j_4} y_{j_5} y_{j_6}$$

$$\cdot \int_{[0,1]^6} \partial^6_{j_1 j_2 j_3 j_4 j_5 j_6} f\left(x + \prod_{l=1}^{6} t_l y\right) dt_1 \ldots dt_6. \tag{24.37}$$

Notice here that for example

$$\partial^5_{32234} f = \frac{\partial^5 f}{\partial x_3 \partial x_2^2 \partial x_3 \partial x_4}, \quad \partial^5_{44444} f = \frac{\partial^5 f}{\partial x_4^5}, \text{ etc.}$$

24.3   Applications

1) Let $f \in W^{m,1}\left(\mathbb{R}^n\right)$; $m \in \mathbb{N}$, $n \in \mathbb{N} - \{1\}$, $y \in \mathbb{R}^n$. Consider the convolution operator

$$\left(Lf\right)\left(y\right) := \int_{\mathbb{R}^n} f\left(x + y\right) dx. \tag{24.38}$$

Using (24.11) and Fubini's theorem we obtain

$$\left(Lf\right)\left(y\right) - \left(Lf\right)\left(0\right) = \int_{\mathbb{R}^n} f\left(x + y\right) dx - \int_{\mathbb{R}^n} f\left(x\right) dx$$

$$= \sum_{l=1}^{m-1} \sum_{j_1,\ldots,j_l=1}^{n} \left(\prod_{i=1}^{l} y_{j_i}\right) \int_{\mathbb{R}^n} \partial^l_{j_1 \ldots j_l} f\left(x\right) dx$$

$$+ \sum_{j_1,\ldots,j_m=1}^{n} \left(\prod_{i=1}^{m} y_{j_i}\right) \int_{[0,1]^m} \left(\int_{\mathbb{R}^n} \partial_{j_1\ldots j_m}^m f\left(x + \prod_{r=1}^{m} t_r y\right) dx\right) dt_1 \ldots dt_m. \quad (24.39)$$

Using (24.13) we find

$$(Lf)(y) - (Lf)(0) = \sum_{l=1}^{m} \left[\sum_{j_1,\ldots,j_l=1}^{n} \left(\prod_{i=1}^{l} y_{j_i}\right) (L(\partial_{j_1\ldots j_l}^l f))(0)\right] + \sum_{j_1,\ldots,j_m=1}^{n} \left(\prod_{i=1}^{m} y_{j_i}\right)$$

$$\cdot \int_{[0,1]^m} \left[(L(\partial_{j_1\ldots j_m}^m f)) \left(\prod_{r=1}^{m} t_r y\right) - (L(\partial_{j_1\ldots j_m}^m f))(0)\right] dt_1 \ldots dt_m. \quad (24.40)$$

2) Let nonempty B measurable subset of $\mathbb{R}^n$, with $Vol(B) \neq 0$. Let $f \in W_{loc}^{m,1}(\mathbb{R}^n)$; $m \in \mathbb{N}$, $n \geq 2$, $y \in \mathbb{R}^n$, then it holds

$$\frac{\int_B f(x+y)\,dx}{Vol(B)} - \frac{\int_B f(x)\,dx}{Vol(B)} = \sum_{l=1}^{m-1} \sum_{j_1,\ldots,j_l=1}^{n} \left(\prod_{i=1}^{l} y_{j_i}\right) \frac{\int_B \partial_{j_1\ldots j_l}^l f(x)\,dx}{Vol(B)}$$

$$+ \sum_{j_1,\ldots,j_m=1}^{n} \left(\prod_{i=1}^{m} y_{j_i}\right) \int_{[0,1]^m} \left(\frac{\int_B \partial_{j_1\ldots j_m}^m f\left(x + \prod_{r=1}^{m} t_r y\right) dx}{Vol(B)}\right) dt_1 \ldots dt_m. \quad (24.41)$$

3) Let $f \in W_{loc}^{m,1}(\mathbb{R}^n)$, $m \in \mathbb{N}$, $n \geq 2$. Then for $y \in \mathbb{R}^n$ and almost all $x \in \mathbb{R}^n$, by (24.11) we derive

$$T := \left| f(x+y) - f(x) - \sum_{l=1}^{m-1} \left\{\sum_{j_1,\ldots,j_l=1}^{n} (\partial_{j_1\ldots j_l}^l f(x)) \left(\prod_{i=1}^{l} y_{j_i}\right)\right\}\right|$$

$$\leq \sum_{j_1,\ldots,j_m=1}^{n} \left(\prod_{i=1}^{m} |y_{j_i}|\right) \int_{[0,1]^m} \left|\partial_{j_1\ldots j_m}^m f\left(x + \prod_{r=1}^{m} t_r y\right)\right| dt_1 \ldots dt_m < \infty, \quad (24.42)$$

by $\partial_{j_1\ldots j_m}^m f\left(x + \prod_{r=1}^{m} t_r y\right)$ being in $L_1([0,1]^m)$, as a function of $t_1,\ldots,t_m$.

If additionally we suppose that

$$\partial_{j_1\ldots j_m}^m f\left(x + \prod_{r=1}^{m} t_r y\right) \in L_\infty([0,1]^m),$$

then by (24.42) we derive

$$T \leq \sum_{j_1,\ldots,j_m=1}^{n} \left(\prod_{i=1}^{m} |y_{j_i}|\right) \|\partial_{j_1\ldots j_m}^m f(x + \ldots y)\|_{\infty,[0,1]^m}. \quad (24.43)$$

4) Let $a, b \in \mathbb{R}^n$, we say $a \leq b$ iff $a_i \leq b_i$, for all $i = 1,\ldots,n$. Let $h \in \mathbb{R}^n$, $h \geq 0$, $f \in W^{m,1}(\mathbb{R}^n)$, $m \in \mathbb{N}$, $n \in \mathbb{N} - \{1\}$. We define the first L_1 modulus of continuity of f as follows:

$$\omega_1(f,\delta)_1 := \sup_{0 \leq h \leq \delta} \int_{\mathbb{R}^n} |f(x+h) - f(x)|\,dx, \quad (24.44)$$

$0 \leq \delta \in \mathbb{R}^n$, i.e. $h_i \leq \delta_i$, $i = 1, \ldots, n$.

By (24.13) we obtain (for almost every $x \in \mathbb{R}^n$)

$$|f(x+h) - f(x)| \leq \sum_{l=1}^{m} \left\{ \sum_{j_1,\ldots,j_l=1}^{n} |\partial_{j_1\ldots j_l}^{l} f(x)| \left(\prod_{i=1}^{l} \delta_{j_i} \right) \right\}$$

$$+ \sum_{j_1,\ldots,j_m=1}^{n} \left(\prod_{i=1}^{m} \delta_{j_i} \right) \int_{[0,1]^m} \left| \partial_{j_1\ldots j_m}^{m} f\left(x + \prod_{r=1}^{m} t_r h \right) - \partial_{j_1\ldots j_m}^{m} f(x) \right| dt_1 \ldots dt_m.$$

$$(24.45)$$

Consequently we have

$$\int_{\mathbb{R}^n} |f(x+h) - f(x)| \, dx \leq \sum_{l=1}^{m} \left\{ \sum_{j_1,\ldots,j_l=1}^{n} \left(\int_{\mathbb{R}^n} |\partial_{j_1\ldots j_l}^{l} f(x)| \, dx \right) \left(\prod_{i=1}^{l} \delta_{j_i} \right) \right\}$$

$$+ \sum_{j_1,\ldots,j_m=1}^{n} \left(\prod_{i=1}^{m} \delta_{j_i} \right)$$

$$\cdot \int_{[0,1]^m} \int_{\mathbb{R}^n} \left| \partial_{j_1\ldots j_m}^{m} f\left(x + \prod_{r=1}^{m} t_r h \right) - \partial_{j_1\ldots j_m}^{m} f(x) \right| dx \, dt_1 \ldots dt_m. \qquad (24.46)$$

We have derived the conclusion

$$\omega_1(f, \delta)_1 \leq \sum_{l=1}^{m} \left\{ \sum_{j_1,\ldots,j_l=1}^{n} \|\partial_{j_1\ldots j_l}^{l} f\|_{L_1(\mathbb{R}^n)} \left(\prod_{i=1}^{l} \delta_{j_i} \right) \right\}$$

$$+ \sum_{j_1,\ldots,j_m=1}^{n} \left(\prod_{i=1}^{m} \delta_{j_i} \right) \int_{[0,1]^m} \omega_1 \left(\partial_{j_1\ldots j_m}^{m} f, \prod_{r=1}^{m} t_r \delta \right)_1 dt_1 \ldots dt_m. \qquad (24.47)$$

We finish the chapter with

5) Let $f \in W_{loc}^{m,1}(\mathbb{R}^n)$; $m \in \mathbb{N}$, $n \geq 2$, $y_0 \in \mathbb{R}^n$ be fixed. Then for almost every $x_0 \in \mathbb{R}^n$ with the property $\partial_{j_1\ldots j_l}^{l} f(x_0) = 0$, for all $l = 0, 1, \ldots, m-1$, and all $j_1, \ldots, j_l \in \{1, \ldots, n\}$, by (24.11) it holds

$$f(x_0 + y_0) =$$

$$\sum_{j_1,\ldots,j_m=1}^{n} \left(\prod_{i=1}^{m} y_{0j_i} \right) \int_{[0,1]^m} \partial_{j_1\ldots j_m}^{m} f\left(x_0 + \prod_{r=1}^{m} t_r y_0 \right) dt_1 \ldots dt_m. \qquad (24.48)$$

Chapter 25

Chebyshev–Grüss Type Inequalities Using Euler Type and Fink Identities

In this chapter we present Chebyshev–Grüss type univariate inequalities by using the generalized Euler type and Fink identities. The results involve functions f, g, $f^{(n)}$, $g^{(n)}$, $n \in \mathbb{N}$, and are with respect to $\| \cdot \|_p$, $1 \le p \le \infty$. This chapter relies on [41].

25.1 Background

Here we mention the following inspiring and motivating results.

Theorem 25.1 (Čebyšev, 1882, [88]). *Let $f, g \colon [a, b] \to \mathbb{R}$ absolutely continuous functions. If $f', g' \in L_\infty([a, b])$, then*

$$\left| \frac{1}{b-a} \int_a^b f(x)g(x)dx - \left(\frac{1}{b-a} \int_a^b f(x)dx \right) \left(\frac{1}{b-a} \int_a^b g(x)dx \right) \right|$$

$$\le \frac{1}{12}(b-a)^2 \|f'\|_\infty \|g'\|_\infty. \tag{25.1}$$

Also we mention

Theorem 25.1* (Grüss, 1935, [150]). *Let f, g integrable functions from $[a, b]$ into $\mathbb{R}$, such that $m \le f(x) \le M$, $\rho \le g(x) \le \sigma$, for all $x \in [a, b]$, where m, M, $\rho, \sigma \in \mathbb{R}$. Then*

$$\left| \frac{1}{b-a} \int_a^b f(x)g(x)dx - \left(\frac{1}{b-a} \int_a^b f(x)dx \right) \left(\frac{1}{b-a} \int_a^b g(x)dx \right) \right|$$

$$\le \frac{1}{4}(M - m)(\sigma - \rho). \tag{25.1*}$$

Let $B_k(x)$, $k \ge 0$, the Bernoulli polynomials, $B_k := B_k(0)$, $k \ge 0$, the Bernoulli numbers, and $B_k^*(x)$, $k \ge 0$, are the periodic functions of period one, related to the Bernoulli polynomials as

$$B_k^*(x) = B_k(x), \quad 0 \le x < 1, \tag{25.2}$$

305

and

$$B_k^*(x+1) = B_k^*(x), \quad x \in \mathbb{R}. \tag{25.3}$$

Some basic properties of Bernoulli polynomials follow (see [1, 23.1]). We have

$$B_0(x) = 1, \quad B_1(x) = x - \frac{1}{2}, \quad B_2(x) = x^2 - x + \frac{1}{6},$$

and

$$B_k'(x) = kB_{k-1}(x), \quad k \in \mathbb{N} \tag{25.4}$$

$$B_k(x+1) - B_k(x) = kx^{k-1}, \quad k \geq 0. \tag{25.5}$$

Clearly $B_0^* = 1$, B_1^* is a discontinuous function with a jump of -1 at each integer, and B_k^*, $k \geq 2$, is a continuous function. Notice that $B_k(0) = B_k(1) = B_k$, $k \geq 2$.

We need the general

Theorem 25.2 (see [35]). *Let $f\colon [a,b] \to \mathbb{R}$ be such that $f^{(n-1)}$, $n \geq 1$, is a continuous function and $f^{(n)}(x)$ exists and is finite for all but a countable set of x in (a,b) and that $f^{(n)} \in L_1([a,b])$. Then for every $x \in [a,b]$ we have*

$$f(x) = \frac{1}{b-a} \int_a^b f(t)dt + \sum_{k=1}^{n-1} \frac{(b-a)^{k-1}}{k!} B_k\left(\frac{x-a}{b-a}\right) [f^{(k-1)}(b) - f^{(k-1)}(a)]$$

$$+ \frac{(b-a)^{n-1}}{n!} \int_{[a,b]} \left[B_n\left(\frac{x-a}{b-a}\right) - B_n^*\left(\frac{x-t}{b-a}\right) \right] f^{(n)}(t)dt. \tag{25.6}$$

The sum in (25.6) when $n = 1$ is zero.

If $f^{(n-1)}$ is just absolutely continuous then (25.6) is valid again. Formula (25.6) is a *generalized Euler type identity*, see also [98], [171].

We need also Fink's identity.

Theorem 25.3 (Fink, [141]). *Let $a, b \in \mathbb{R}$, $f\colon [a,b] \to \mathbb{R}$, $n \geq 1$, $f^{(n-1)}$ is absolutely continuous on $[a,b]$. Then*

$$f(x) = \frac{n}{b-a} \int_a^b f(t)dt - \sum_{k=1}^{n-1} \left(\frac{n-k}{k!}\right) \left(\frac{f^{(k-1)}(a)(x-a)^k - f^{(k-1)}(b)(x-b)^k}{b-a} \right)$$

$$+ \frac{1}{(n-1)!(b-a)} \int_a^b (x-t)^{n-1} k(t,x) f^{(n)}(t)dt, \tag{25.7}$$

where

$$k(t,x) := \begin{cases} t - a, & a \leq t \leq x \leq b, \\ t - b, & a \leq x < t \leq b. \end{cases} \tag{25.8}$$

When $n = 1$ the sum $\sum\limits_{k=1}^{n-1}$ in (25.7) is zero.

In this chapter based on Theorems 25.2 and 25.3 we present Čebyšev–Grüss type inequalities, see (25.1), (25.1)*, involving f, g, $f^{(n)}$, $g^{(n)}$, $n \in \mathbb{N}$, with respect to $\|\cdot\|_p$, $1 \leq p \leq \infty$, see Theorems 25.4 and 25.6 and Corollaries 25.5 and 25.7.

25.2 Main Results

We present the first main result based on generalized Euler type identity.

Theorem 25.4. *Let $f, g \colon [a, b] \to \mathbb{R}$ be such that $f^{(n-1)}$, $g^{(n-1)}$, $n \in \mathbb{N}$, be continuous functions and $f^{(n)}$, $g^{(n)}$ exist and are finite for all but a countable set of $x \in (a, b)$ and that $f^{(n)}, g^{(n)} \in L_1([a, b])$. Denote*

$$T_{n-1}^{f}(x) := \sum_{k=1}^{n-1} \frac{(b-a)^{k-1}}{k!} B_k\left(\frac{x-a}{b-a}\right)\left(f^{(k-1)}(b) - f^{(k-1)}(a)\right), \qquad (25.9)$$

$$(T_0^{f}(x) = 0),$$

$$T_{n-1}^{g}(x) := \sum_{k=1}^{n-1} \frac{(b-a)^{k-1}}{k!} B_k\left(\frac{x-a}{b-a}\right)\left(g^{(k-1)}(b) - g^{(k-1)}(a)\right), \qquad (25.10)$$

$$(T_0^{g}(x) = 0),$$

and

$$\Delta_{(f,g)} := \int_a^b f(x)g(x)dx - \frac{1}{b-a}\left(\int_a^b f(x)dx\right)\left(\int_a^b g(x)dx\right)$$

$$- \frac{1}{2}\int_a^b \left(f(x)T_{n-1}^{g}(x) + g(x)T_{n-1}^{f}(x)\right)dx. \qquad (25.11)$$

1) If $f^{(n)}, g^{(n)} \in L_\infty([a, b])$, then

$$|\Delta_{(f,g)}| \leq \frac{(b-a)^n}{2n!}\left(\int_a^b \left(\sqrt{\frac{(n!)^2}{(2n)!}|B_{2n}| + B_n^2\left(\frac{x-a}{b-a}\right)}\right)dx\right)$$

$$\times \left[\|f\|_\infty \|g^{(n)}\|_\infty + \|g\|_\infty \|f^{(n)}\|_\infty\right]. \qquad (25.12)$$

2) If $f^{(n)}, g^{(n)} \in L_p([a, b])$, where $p, q > 1$ such that $\frac{1}{p} + \frac{1}{q} = 1$, then

$$|\Delta_{(f,g)}| \leq \frac{(b-a)^{n-\frac{1}{p}}}{2n!}\left[\int_a^b \left(\int_0^1 \left|B_n(t) - B_n\left(\frac{x-a}{b-a}\right)\right|^q dt\right)dx\right]^{1/q}$$

$$\times \left\{\|f\|_p \|g^{(n)}\|_p + \|g\|_p \|f^{(n)}\|_p\right\}. \qquad (25.13)$$

When $p = q = 2$, it holds

$$|\Delta_{(f,g)}| \leq \frac{(b-a)^{n-\frac{1}{2}}}{2n!}\left[\int_a^b \left(\frac{(n!)^2}{(2n)!}|B_{2n}| + B_n^2\left(\frac{x-a}{b-a}\right)\right)dx\right]^{1/2}$$

$$\times \left\{\|f\|_2 \|g^{(n)}\|_2 + \|g\|_2 \|f^{(n)}\|_2\right\}. \qquad (25.14)$$

3) With respect to $\|\cdot\|_1$ it holds

$$|\Delta_{(f,g)}| \leq \frac{(b-a)^n}{2n!}\left\|\sqrt{\frac{(n!)^2}{(2n)!}|B_{2n}| + B_n^2\left(\frac{x-a}{b-a}\right)}\right\|_\infty$$

$$\times \left\{\|f\|_1 \|g^{(n)}\|_\infty + \|g\|_1 \|f^{(n)}\|_\infty\right\}. \qquad (25.15)$$

Proof. We have by Theorem 25.2 that

$$f(x) = \frac{1}{b-a}\int_a^b f(t)dt + T_{n-1}^f(x) + \mathcal{R}_n^f(x), \tag{25.16}$$

where

$$T_{n-1}^f(x) := \sum_{k=1}^{n-1} \frac{(b-a)^{k-1}}{k!} B_k\left(\frac{x-a}{b-a}\right)\left(f^{(k-1)}(b) - f^{(k-1)}(a)\right), \tag{25.17}$$

$(T_0^f(x) = 0)$, and

$$\mathcal{R}_n^f(x) := -\frac{(b-a)^{n-1}}{n!}\int_a^b \left[B_n^*\left(\frac{x-t}{b-a}\right) - B_n\left(\frac{x-a}{b-a}\right)\right] f^{(n)}(t)dt, \ \forall x \in [a,b]. \tag{25.18}$$

Similarly we derive

$$g(x) = \frac{1}{b-a}\int_a^b g(t)dt + T_{n-1}^g(x) + \mathcal{R}_n^g(x), \tag{25.19}$$

where

$$T_{n-1}^g(x) := \sum_{k=1}^{n-1} \frac{(b-a)^{k-1}}{k!} B_k\left(\frac{x-a}{b-a}\right)\left(g^{(k-1)}(b) - g^{(k-1)}(a)\right), \tag{25.20}$$

$(T_0^g(x) = 0)$, and

$$\mathcal{R}_n^g(x) := -\frac{(b-a)^{n-1}}{n!}\int_a^b \left[B_n^*\left(\frac{x-t}{b-a}\right) - B_n\left(\frac{x-a}{b-a}\right)\right] g^{(n)}(t)dt, \ \forall x \in [a,b]. \tag{25.21}$$

Then

$$f(x)g(x) = \frac{g(x)}{b-a}\int_a^b f(t)dt + g(x)T_{n-1}^f(x) + g(x)\mathcal{R}_n^f(x), \tag{25.22}$$

and

$$g(x)f(x) = \frac{f(x)}{b-a}\int_a^b g(t)dt + f(x)T_{n-1}^g(x) + f(x)\mathcal{R}_n^g(x), \ \forall x \in [a,b]. \tag{25.23}$$

So that

$$\int_a^b f(x)g(x)dx = \frac{1}{b-a}\left(\int_a^b g(x)dx\right)\left(\int_a^b f(x)dx\right)$$
$$+ \int_a^b g(x)T_{n-1}^f(x)dx + \int_a^b g(x)\mathcal{R}_n^f(x)dx, \tag{25.24}$$

and

$$\int_a^b f(x)g(x)dx = \frac{1}{b-a}\left(\int_a^b f(x)dx\right)\left(\int_a^b g(x)dx\right)$$
$$+ \int_a^b f(t)T_{n-1}^g(x)dx + \int_a^b f(x)\mathcal{R}_n^g(x)dx. \tag{25.25}$$

That is we get

$$\int_a^b f(x)g(x)dx - \frac{1}{b-a}\left(\int_a^b f(x)dx\right)\left(\int_a^b g(x)dx\right)$$

$$= \int_a^b g(x)T_{n-1}^f(x)dx + \int_a^b g(x)\mathcal{R}_n^f(x)dx$$

$$= \int_a^b f(x)T_{n-1}^g(x)dx + \int_a^b f(x)\mathcal{R}_n^g(x)dx. \tag{25.26}$$

Therefore we have

$$\int_a^b f(x)g(x)dx - \frac{1}{b-a}\left(\int_a^b f(x)dx\right)\left(\int_a^b g(x)dx\right)$$

$$= \frac{1}{2}\left\{\int_a^b \left(f(x)T_{n-1}^g(x) + g(x)T_{n-1}^f(x)\right)dx\right.$$

$$\left. + \int_a^b \left(f(x)\mathcal{R}_n^g(x) + g(x)\mathcal{R}_n^f(x)\right)dx\right\}, \tag{25.27}$$

and

$$\Delta_{(f,g)} := \int_a^b f(x)g(x)dx - \frac{1}{b-a}\left(\int_a^b f(x)dx\right)\left(\int_a^b g(x)dx\right)$$

$$- \frac{1}{2}\int_a^b \left(f(x)T_{n-1}^g(x) + g(x)T_{n-1}^f(x)\right)dx$$

$$= \frac{1}{2}\int_a^b \left(f(x)\mathcal{R}_n^g(x) + g(x)\mathcal{R}_n^f(x)\right)dx. \tag{25.28}$$

1) We estimate $\Delta_{(f,g)}$ with respect to $\|\cdot\|_\infty$. We have

$$|\Delta_{(f,g)}| \le \frac{1}{2}\left[\|f\|_\infty \int_a^b |\mathcal{R}_n^g(x)|dx + \|g\|_\infty \int_a^b |\mathcal{R}_n^f(x)|dx\right]. \tag{25.29}$$

But we see that

$$|\mathcal{R}_n^f(x)| \le \frac{(b-a)^{n-1}}{n!}\int_a^b \left|B_n^*\left(\frac{x-t}{b-a}\right) - B_n\left(\frac{x-a}{b-a}\right)\right||f^{(n)}(t)|dt$$

$$\le \frac{(b-a)^{n-1}}{n!}\|f^{(n)}\|_\infty \int_a^b \left|B_n^*\left(\frac{x-t}{b-a}\right) - B_n\left(\frac{x-a}{b-a}\right)\right|dt$$

$$= \frac{\|f^{(n)}\|_\infty}{n!}(b-a)^n \int_0^1 \left|B_n(t) - B_n\left(\frac{x-a}{b-a}\right)\right|dt$$

$$\le \frac{\|f^{(n)}\|_\infty}{n!}(b-a)^n \left(\int_0^1 \left(B_n(t) - B_n\left(\frac{x-a}{b-a}\right)\right)^2 dt\right)^{1/2}. \tag{25.30}$$

That is

$$|\mathcal{R}_n^f(x)| \le \frac{\|f^{(n)}\|_\infty}{n!}(b-a)^n \left(\int_0^1 \left(B_n(t) - B_n\left(\frac{x-a}{b-a}\right)\right)^2 dt\right)^{1/2}. \tag{25.31}$$

Using (25.31) into (25.29) we get

$$|\Delta_{(f,g)}| \le \frac{(b-a)^n}{2n!} \left[\|f\|_\infty \|g^{(n)}\|_\infty + \|g\|_\infty \|f^{(n)}\|_\infty \right]$$

$$\times \left(\int_a^b \left(\int_0^1 \left(B_n(t) - B_n\left(\frac{x-a}{b-a}\right) \right)^2 dt \right)^{1/2} dx \right). \qquad (25.32)$$

Finally by [98], p. 352, we find

$$|\Delta_{(f,g)}| \le \frac{(b-a)^n}{2n!} \left[\|f\|_\infty \|g^{(n)}\|_\infty + \|g\|_\infty \|f^{(n)}\|_\infty \right]$$

$$\times \left(\int_a^b \sqrt{\frac{(n!)^2}{(2n)!} |B_{2n}| + B_n^2\left(\frac{x-a}{b-a}\right)} \, dx \right), \qquad (25.33)$$

proving the claim.

2) We estimate $\Delta_{(f,g)}$ with respect to $\|\cdot\|_p$. Let $p, q > 1$: $\frac{1}{p} + \frac{1}{q} = 1$. Then

$$|\Delta_{(f,g)}| \le \frac{1}{2} \left[\|f\|_p \|\mathcal{R}_n^g\|_q + \|g\|_p \|\mathcal{R}_n^f\|_q \right]. \qquad (25.34)$$

Notice that

$$|\mathcal{R}_n^f(x)|^q \le \frac{(b-a)^{q(n-1)}}{(n!)^q} \left(\int_a^b \left| B_n^*\left(\frac{x-t}{b-a}\right) - B_n\left(\frac{x-a}{b-a}\right) \right| |f^{(n)}(t)| dt \right)^q$$

$$\le \frac{(b-a)^{q(n-1)}}{(n!)^q} \left(\int_a^b \left| B_n^*\left(\frac{x-t}{b-a}\right) - B_n\left(\frac{x-a}{b-a}\right) \right|^q dt \right) \|f^{(n)}\|_p^q. \quad (25.35)$$

Furthermore it holds that

$$|\mathcal{R}_n^f(x)|^q \le \frac{\|f^{(n)}\|_p^q}{(n!)^q} (b-a)^{q(n-1)+1} \left(\int_0^1 \left| B_n(t) - B_n\left(\frac{x-a}{b-a}\right) \right|^q dt \right). \qquad (25.36)$$

Also it holds

$$\|\mathcal{R}_n^f(x)\|_q \le \frac{\|f^{(n)}\|_p}{n!} (b-a)^{n-\frac{1}{p}} \left(\int_a^b \left(\int_0^1 \left| B_n(t) - B_n\left(\frac{x-a}{b-a}\right) \right|^q dt \right) dx \right)^{1/q}.$$
$$(25.37)$$

Hence by using (25.37) into (25.34) we get

$$|\Delta_{(f,g)}| \le \frac{(b-a)^{n-\frac{1}{p}}}{2n!} \left[\int_a^b \left(\int_0^1 \left| B_n(t) - B_n\left(\frac{x-a}{b-a}\right) \right|^q dt \right) dx \right]^{1/q}$$

$$\times \left\{ \|f\|_p \|g^{(n)}\|_p + \|g\|_p \|f^{(n)}\|_p \right\}, \qquad (25.38)$$

proving the claim.

When $p = q = 2$, then

$$|\Delta_{(f,g)}| \le \frac{(b-a)^{n-\frac{1}{2}}}{2n!} \left[\int_a^b \left(\int_0^1 \left(B_n(t) - B_n\left(\frac{x-a}{b-a}\right) \right)^2 dt \right) dx \right]^{1/2}$$

$$\times \left\{ \|f\|_2 \|g^{(n)}\|_2 + \|g\|_2 \|f^{(n)}\|_2 \right\}. \qquad (25.39)$$

Consequently by [98], p. 352 we derive

$$|\Delta_{(f,g)}| \le \frac{(b-a)^{n-\frac{1}{2}}}{2n!} \left[\int_a^b \left(\frac{(n!)^2}{(2n)!}|B_{2n}| + B_n^2\left(\frac{x-a}{b-a}\right) \right) dx \right]^{1/2}$$
$$\times \left\{ \|f\|_2\|g^{(n)}\|_2 + \|g\|_2\|f^{(n)}\|_2 \right\}. \tag{25.40}$$

3) We estimate $\Delta_{(f,g)}$ with respect to $\|\cdot\|_1$. We observe that

$$|\Delta_{(f,g)}| \le \frac{1}{2}\left(\int_a^b |f(x)|\,|\mathcal{R}_n^g(x)|dx + \int_a^b |g(x)|\,|\mathcal{R}_n^f(x)|dx \right) \tag{25.41}$$

$$\le \frac{1}{2}\left(\|f\|_1\|\mathcal{R}_n^g(x)\|_\infty + \|g\|_1\|\mathcal{R}_n^f(x)\|_\infty \right). \tag{25.42}$$

Next we observe that

$$\|\mathcal{R}_n^f(x)\|_\infty = \frac{(b-a)^{n-1}}{n!}\left\| \int_a^b \left[B_n^*\left(\frac{x-t}{b-a}\right) - B_n\left(\frac{x-a}{b-a}\right) \right] f^{(n)}(t)dt \right\|_\infty. \tag{25.43}$$

But we see that

$$|\mathcal{R}_n^f(x)| \le \frac{(b-a)^{n-1}}{n!} \int_a^b \left| B_n^*\left(\frac{x-t}{b-a}\right) - B_n\left(\frac{x-a}{b-a}\right) \right| |f^{(n)}(t)|dt$$

$$\le \frac{(b-a)^n}{n!}\|f^{(n)}\|_\infty \left(\int_0^1 \left| B_n(t) - B_n\left(\frac{x-a}{b-a}\right) \right| dt \right). \tag{25.44}$$

That is

$$\|\mathcal{R}_n^f(x)\|_\infty \le \frac{(b-a)^n}{n!}\|f^{(n)}\|_\infty \left\| \int_0^1 \left| B_n(t) - B_n\left(\frac{x-a}{b-a}\right) \right| dt \right\|_\infty \tag{25.45}$$
$$\text{(by [98], p. 352)}$$

$$\le \frac{(b-a)^n}{n!}\|f^{(n)}\|_\infty \left\| \sqrt{\frac{(n!)^2}{(2n)!}|B_{2n}| + B_n^2\left(\frac{x-a}{b-a}\right)} \right\|_\infty. \tag{25.46}$$

Using (25.46) into (25.42) we find

$$|\Delta_{(f,g)}| \le \frac{(b-a)^n}{2n!} \left\| \sqrt{\frac{(n!)^2}{(2n)!}|B_{2n}| + B_n^2\left(\frac{x-a}{b-a}\right)} \right\|_\infty$$
$$\times \left\{ \|f\|_1\|g^{(n)}\|_\infty + \|g\|_1\|f^{(n)}\|_\infty \right\}, \tag{25.47}$$

proving the claim. $\qquad\square$

We give

Corollary 25.5. *Let $f, g \in C([a,b])$ and f', g' exist and are finite for all but a countable set of $x \in (a,b)$ and that $f', g' \in L_1([a,b])$.*

1) *If $f', g' \in L_\infty([a,b])$, then*

$$\left| \frac{1}{b-a} \int_a^b f(x)g(x)dx - \frac{1}{(b-a)^2} \left(\int_a^b f(x)dx \right) \left(\int_a^b g(x)dx \right) \right|$$

$$\leq \frac{(b-a)}{48} \{ 4\sqrt{3} - \ln[3a + 2\sqrt{3}(b-a) - 3b]$$

$$+ \ln[-3a + 2\sqrt{3}(b-a) + 3b] \} \left[\|f\|_\infty \|g'\|_\infty + \|g\|_\infty \|f'\|_\infty \right]. \quad (25.48)$$

2) *If $f', g' \in L_p([a,b])$, where $p, q > 1$ such that $\frac{1}{p} + \frac{1}{q} = 1$, then*

$$\left| \frac{1}{b-a} \int_a^b f(x)g(x)dx - \frac{1}{(b-a)^2} \left(\int_a^b f(x)dx \right) \left(\int_a^b g(x)dx \right) \right|$$

$$\leq 2^{-1}(b-a)^{-1/p} \left[\int_a^b \left(\int_0^1 \left| t - \left(\frac{x-a}{b-a} \right) \right|^q dt \right) dx \right]^{1/q}$$

$$\times \left\{ \|f\|_p \|g'\|_p + \|g\|_p \|f'\|_p \right\}. \quad (25.49)$$

When $p = q = 2$, it holds

$$\left| \frac{1}{b-a} \int_a^b f(x)g(x)dx - \frac{1}{(b-a)^2} \left(\int_a^b f(x)dx \right) \left(\int_a^b g(x)dx \right) \right|$$

$$\leq \frac{1}{2\sqrt{6}} \{ \|f\|_2 \|g'\|_2 + \|g\|_2 \|f'\|_2 \}. \quad (25.50)$$

3) *With respect to $\| \cdot \|_1$ it holds*

$$\left| \frac{1}{b-a} \int_a^b f(x)g(x)dx - \frac{1}{(b-a)^2} \left(\int_a^b f(x)dx \right) \left(\int_a^b g(x)dx \right) \right|$$

$$\leq \frac{1}{2\sqrt{3}} \{ \|f\|_1 \|g'\|_\infty + \|g\|_1 \|f'\|_\infty \}. \quad (25.51)$$

Proof.   By Theorem 25.4 for $n = 1$, etc. $\square$

Next we present the second main result here based on Fink's identity.

Theorem 25.6. *Let $f, g \colon [a,b] \to \mathbb{R}$, $n \in \mathbb{N}$, $f^{(n-1)}$, $g^{(n-1)}$ are absolutely continuous on $[a,b]$. Denote*

$$F_{n-1}^f(x) := \sum_{k=1}^{n-1} \left(\frac{n-k}{k!} \right) \left(\frac{f^{(k-1)}(b)(x-b)^k - f^{(k-1)}(a)(x-a)^k}{b-a} \right), \quad (25.52)$$

$(F_0^f(x) = 0)$,

$$F_{n-1}^g(x) := \sum_{k=1}^{n-1} \left(\frac{n-k}{k!} \right) \left(\frac{g^{(k-1)}(b)(x-b)^k - g^{(k-1)}(a)(x-a)^k}{b-a} \right), \quad (25.53)$$

$(F_0^g(x) = 0)$, *and*

$$\Delta_{(f,g)} := \int_a^b f(x)g(x)dx - \frac{n}{b-a}\left(\int_a^b f(x)dx\right)\left(\int_a^b g(x)dx\right)$$
$$- \frac{1}{2}\left[\int_a^b \left(g(x)F_{n-1}^f(x) + f(x)F_{n-1}^g(x)\right)dx\right]. \tag{25.54}$$

1) *If* $f^{(n)}, g^{(n)} \in L_\infty([a,b])$, *then*

$$|\Delta_{(f,g)}| \le \frac{(b-a)^{n+1}}{(n+2)!}\left[\|f\|_\infty\|g^{(n)}\|_\infty + \|g\|_\infty\|f^{(n)}\|_\infty\right]. \tag{25.55}$$

2) *If* $f^{(n)}, g^{(n)} \in L_p([a,b])$, *where* $p, q > 1$ *such that* $\frac{1}{p} + \frac{1}{q} = 1$, *then*

$$|\Delta_{(f,g)}| \le 2^{-1/p}(qn+2)^{-1/q}\left(B(q(n-1)+1, q+1)\right)^{1/q}\frac{(b-a)^{n-1+\frac{2}{q}}}{(n-1)!}$$
$$\times \left[\|f\|_p\|g^{(n)}\|_p + \|g\|_p\|f^{(n)}\|_p\right]. \tag{25.56}$$

When $p = q = 2$, *it holds*

$$|\Delta_{(f,g)}| \le \frac{(b-a)^n}{(n-1)!2\sqrt{n(n+1)(4n^2-1)}}\left[\|f\|_2\|g^{(n)}\|_2 + \|g\|_2\|f^{(n)}\|_2\right]. \tag{25.57}$$

3) *With respect to* $\|\cdot\|_1$ *it holds*

$$|\Delta_{(f,g)}| \le \frac{(b-a)^n}{2(n+1)!}\left(\|f\|_1\|g^{(n)}\|_\infty + \|g\|_1\|f^{(n)}\|_\infty\right). \tag{25.58}$$

Proof. Since $f\colon [a,b] \to \mathbb{R}$ has $f^{(n-1)}$ absolutely continuous over $[a,b]$, by Theorem 25.3 we have that

$$f(x) = \frac{n}{b-a}\int_a^b f(t)dt + F_{n-1}^f(x) + \mathcal{R}_n^f(x), \tag{25.59}$$

where

$$F_{n-1}^f(x) := \sum_{k=1}^{n-1}\left(\frac{n-k}{k!}\right)\left(\frac{f^{(k-1)}(b)(x-b)^k - f^{(k-1)}(a)(x-a)^k}{b-a}\right), \tag{25.60}$$

and

$$\mathcal{R}_n^f(x) := \frac{1}{(n-1)!(b-a)}\int_a^b (x-t)^{n-1}k(t,x)f^{(n)}(t)dt, \tag{25.61}$$

with

$$k(t,x) := \begin{cases} t-a, & a \le t \le x \le b, \\ t-b, & a \le x < t \le b. \end{cases} \tag{25.62}$$

Let f, g as above, i.e.

$$g(x) = \frac{n}{b-a}\int_a^b g(t)dt + F_{n-1}^g(x) + \mathcal{R}_n^g(x). \tag{25.63}$$

Then

$$f(x)g(x) = \frac{n}{b-a}g(x)\int_a^b f(t)dt + g(x)F_{n-1}^f(x) + g(x)\mathcal{R}_n^f(x),$$

$$f(x)g(x) = \frac{n}{b-a}f(x)\int_a^b g(t)dt + f(x)F_{n-1}^g(x) + f(x)\mathcal{R}_n^g(x). \tag{25.64}$$

Then by integrating we obtain

$$\int_a^b f(x)g(x)dx = \frac{n}{b-a}\left(\int_a^b g(x)dx\right)\left(\int_a^b f(x)dx\right)$$

$$+ \int_a^b g(x)F_{n-1}^f(x)dx + \int_a^b g(x)\mathcal{R}_n^f(x), \tag{25.65}$$

$$\int_a^b f(x)g(x)dx = \frac{n}{b-a}\left(\int_a^b f(x)dx\right)\left(\int_a^b g(x)dx\right)$$

$$+ \int_a^b f(x)F_{n-1}^g(x)dx + \int_a^b f(x)\mathcal{R}_n^g(x)dx. \tag{25.66}$$

That is

$$\int_a^b f(x)g(x)dx - \frac{n}{b-a}\left(\int_a^b f(x)dx\right)\left(\int_a^b g(x)dx\right)$$

$$= \int_a^b g(x)F_{n-1}^f(x)dx + \int_a^b g(x)\mathcal{R}_n^f(x)dx$$

$$= \int_a^b f(x)F_{n-1}^g(x)dx + \int_a^b f(x)\mathcal{R}_n^g(x)dx. \tag{25.67}$$

Adding the last we derive

$$\Delta_{(f,g)} := \int_a^b f(x)g(x)dx - \frac{n}{b-a}\left(\int_a^b f(x)dx\right)\left(\int_a^b g(x)dx\right)$$

$$- \frac{1}{2}\left[\int_a^b \left(g(x)F_{n-1}^f(x) + f(x)F_{n-1}^g(x)\right)dx\right]$$

$$= \frac{1}{2}\left[\int_a^b \left(f(x)\mathcal{R}_n^g(x) + g(x)\mathcal{R}_n^f(x)\right)dx\right]. \tag{25.68}$$

Next we estimate $\Delta_{(f,g)}$.

1) Estimate with respect to $\|\cdot\|_\infty$. We have that

$$|\Delta_{(f,g)}| \le \frac{1}{2}\left[\|f\|_\infty \int_a^b |\mathcal{R}_n^g(x)|dx + \|g\|_\infty \int_a^b |\mathcal{R}_n^f(x)|dx\right]. \tag{25.69}$$

But it holds

$$\int_a^b |\mathcal{R}_n^f(x)|dx \le \frac{\|f^{(n)}\|_\infty}{(n-1)!(b-a)}\int_a^b \left(\int_a^b |x-t|^{n-1}|k(t,x)|dt\right)dx,$$

$$\int_a^b |\mathcal{R}_n^g(x)|dx \le \frac{\|g^{(n)}\|_\infty}{(n-1)!(b-a)}\int_a^b \left(\int_a^b |x-t|^{n-1}|k(t,x)|dt\right)dx. \tag{25.70}$$

Consequently we find

$$|\Delta_{(f,g)}| \le \frac{1}{2(n-1)!(b-a)} \left(\int_a^b \left(\int_a^b |x-t|^{n-1}|k(t,x)|dt \right) dx \right)$$
$$\times \left[\|f\|_\infty \|g^{(n)}\|_\infty + \|g\|_\infty \|f^{(n)}\|_\infty \right]. \tag{25.71}$$

We have that

$$\int_a^b |x-t|^{n-1}|k(t,x)|dt = \frac{1}{n(n+1)}\left[(x-a)^{n+1} + (b-x)^{n+1} \right], \tag{25.72}$$

and

$$\int_a^b \left(\int_a^b |x-t|^{n-1}|k(t,x)|dt \right) dx = \frac{2(b-a)^{n+2}}{n(n+1)(n+2)}, \tag{25.73}$$

giving us

$$|\Delta_{(f,g)}| \le \frac{(b-a)^{n+1}}{(n+2)!}\left[\|f\|_\infty \|g^{(n)}\|_\infty + \|g\|_\infty \|f^{(n)}\|_\infty \right], \tag{25.74}$$

that is proving the claim.

2) Case of $\|\cdot\|_p$: Let $p, q > 1$ such that $\frac{1}{p} + \frac{1}{q} = 1$. Then

$$|\Delta_{(f,g)}| \le \frac{1}{2}\left[\|f\|_p \|\mathcal{R}_n^g\|_q + \|g\|_p \|\mathcal{R}_n^f\|_q \right]. \tag{25.75}$$

But it holds

$$\|\mathcal{R}_n^f\|_q = \frac{1}{(n-1)!(b-a)} \left(\int_a^b \left| \int_a^b (x-t)^{n-1}k(t,x)f^{(n)}(t)dt \right|^q dx \right)^{1/q}$$

$$\le \frac{1}{(n-1)!(b-a)} \left(\int_a^b \left(\int_a^b |x-t|^{n-1}|k(t,x)|\,|f^{(n)}(t)|dt \right)^q dx \right)^{1/q} \tag{25.76}$$

$$\le \frac{1}{(n-1)!(b-a)} \left(\int_a^b \left[\left(\int_a^b |x-t|^{q(n-1)}|k(t,x)|^q dt \right)^{1/q} \|f^{(n)}\|_p \right]^q dx \right)^{1/q}.$$

That is

$$\|\mathcal{R}_n^f\|_q \le \frac{\|f^{(n)}\|_p}{(n-1)!(b-a)} \left(\int_a^b \left(\int_a^b |x-t|^{q(n-1)}|k(t,x)|^q dt \right) dx \right)^{1/q}. \tag{25.77}$$

We find that

$$\int_a^b |x-t|^{q(n-1)}|k(t,x)|^q dt$$
$$= B\big(q(n-1)+1, q+1\big)\left[(x-a)^{qn+1} + (b-x)^{qn+1} \right], \; \forall x \in [a,b], \tag{25.78}$$

and

$$\left(\int_a^b \left(\int_a^b |x-t|^{q(n-1)}|k(t,x)|^q dt \right) dx \right)^{1/q}$$
$$= \big(B(q(n-1)+1, q+1)\big)^{1/q} 2^{1/q}(qn+2)^{-1/q}(b-a)^{n+\frac{2}{q}}. \tag{25.79}$$

Consequently by (25.79) and (25.77) it holds

$$\|\mathcal{R}_n^f\|_q \le \frac{\|f^{(n)}\|_p}{(n-1)!}\big[2(qn+2)^{-1}B(q(n-1)+1,q+1)\big]^{1/q}(b-a)^{n-1+\frac{2}{q}}. \tag{25.80}$$

We conclude by (25.80) and (25.75) that

$$|\Delta_{(f,g)}| \le 2^{-1/p}(qn+2)^{-1/q}\big(B(q(n-1)+1,q+1)\big)^{1/q}$$
$$\times \frac{(b-a)^{n-1+\frac{2}{q}}}{(n-1)!}\big[\|f\|_p\|g^{(n)}\|_p + \|g\|_p\|f^{(n)}\|_p\big], \tag{25.81}$$

proving the claim.

When $p = q = 2$ we find

$$|\Delta_{(f,g)}| \le 2^{-1}\big(n(n+1)(2n-1)(2n+1)\big)^{-1/2}$$
$$\times \frac{(b-a)^n}{(n-1)!}\big[\|f\|_2\|g^{(n)}\|_2 + \|g\|_2\|f^{(n)}\|_2\big]. \tag{25.82}$$

3) Case of $\|\cdot\|_1$. We have

$$|\Delta_{(f,g)}| \le \frac{1}{2}\big(\|f\|_1\|\mathcal{R}_n^g\|_\infty + \|g\|_1\|\mathcal{R}_n^f\|_\infty\big). \tag{25.83}$$

It holds

$$|\mathcal{R}_n^f(x)| \le \frac{\|f^{(n)}\|_\infty}{(n-1)!(b-a)}\int_a^b |x-t|^{n-1}|k(t,x)|dt, \tag{25.84}$$

and

$$\|\mathcal{R}_n^f\|_\infty \le \frac{\|f^{(n)}\|_\infty}{(n-1)!(b-a)}\left\|\int_a^b |x-t|^{n-1}|k(t,x)|dt\right\|_\infty$$
$$= \frac{\|f^{(n)}\|_\infty}{(n+1)!(b-a)}\|(x-a)^{n+1}+(b-x)^{n+1}\|_\infty$$
$$= \frac{\|f^{(n)}\|_\infty}{(n+1)!}(b-a)^n. \tag{25.85}$$

That is

$$\|\mathcal{R}_n^f\|_\infty \le \|f^{(n)}\|_\infty\frac{(b-a)^n}{(n+1)!}. \tag{25.86}$$

Finally by (25.86) and (25.83) we obtain

$$|\Delta_{(f,g)}| \le \frac{(b-a)^n}{2(n+1)!}\big(\|f\|_1\|g^{(n)}\|_\infty + \|g\|_1\|f^{(n)}\|_\infty\big), \tag{25.87}$$

proving the claim. $\qquad\square$

We end chapter with

Corollary 25.7. *Let $f,g\colon [a,b] \to \mathbb{R}$ be absolutely continuous.*

1) *If $f', g' \in L_\infty([a,b])$, then*

$$\left| \frac{1}{b-a} \int_a^b f(x)g(x)dx - \frac{1}{(b-a)^2}\left(\int_a^b f(x)dx \right)\left(\int_a^b g(x)dx \right) \right|$$

$$\leq \frac{(b-a)}{6}\left[\|f\|_\infty \|g'\|_\infty + \|g\|_\infty \|f'\|_\infty \right]. \tag{25.88}$$

2) *If $f', g' \in L_p([a,b])$, where $p, q > 1$ such that $\frac{1}{p} + \frac{1}{q} = 1$, then*

$$\left| \frac{1}{b-a} \int_a^b f(x)g(x)dx - \frac{1}{(b-a)^2}\left(\int_a^b f(x)dx \right)\left(\int_a^b g(x)dx \right) \right|$$

$$\leq 2^{-\frac{1}{p}}(q+2)^{-\frac{1}{q}}(q+1)^{-\frac{1}{q}}(b-a)^{\frac{2}{q}-1}\left[\|f\|_p \|g'\|_p + \|g\|_p \|f'\|_p \right]. \tag{25.89}$$

When $p = q = 2$, it holds

$$\left| \frac{1}{b-a} \int_a^b f(x)g(x)dx - \frac{1}{(b-a)^2}\left(\int_a^b f(x)dx \right)\left(\int_a^b g(x)dx \right) \right|$$

$$\leq \frac{1}{2\sqrt{6}}\left[\|f\|_2 \|g'\|_2 + \|g\|_2 \|f'\|_2 \right]. \tag{25.90}$$

3) *With respect to $\| \cdot \|_1$ it holds*

$$\left| \frac{1}{b-a} \int_a^b f(x)g(x)dx - \frac{1}{(b-a)^2}\left(\int_a^b f(x)dx \right)\left(\int_a^b g(x)dx \right) \right|$$

$$\leq \frac{1}{4}\left(\|f\|_1 \|g'\|_\infty + \|g\|_1 \|f'\|_\infty \right). \tag{25.91}$$

Proof. By Theorem 25.6 for $n = 1$, etc. $\qquad\square$

Chapter 26

Grüss Type Multivariate Integral Inequalities

Grüss type multidimensional integral inequalities are presented involving two functions of any number of independent variables. This chapter is based on [27].

26.1 Introduction

One of the most famous integral inequalities was given by Grüss [150] in 1935 and it can be stated as follows (see [187, p. 296]),

$$\left| \frac{1}{b-a} \int_a^b f(x)g(x)\,dx - \left(\frac{1}{b-a} \int_a^b f(x)\,dx \right)\left(\frac{1}{b-a} \int_a^b g(x)\,dx \right) \right|$$
$$\leq \frac{1}{4}(M-m)(N-n),$$

where f and g are integrable functions on $[a,b]$ and satisfy the conditions

$$m \leq f(x) \leq M, \qquad n \leq g(x) \leq N,$$

for all $x \in [a,b]$, where m, M, n, N are given real numbers.

A great deal of attention has been given to the above inequality and many articles related to various generalizations, extensions, and variants of it have appeared in the literature; see Chapter X of the book [187] by Mitrinović, Pečarić, and Fink, where more references are given.

Here we give the multivariate analog of Grüss inequality for as many as possible independent variables, given in two variations. Grüss inequalities for functions of two variables were first given by B. Pachpatte in [200]. His paper is one of our motivations. Next we mention one of his results. We follow the notations of [200] exactly.

Here $\mathbb{R}$ denotes the set of real numbers and $\Delta = [a,b] \times [c,d]$, a, b, c, $d \in \mathbb{R}$. We denote by $G(\Delta)$ the set of continuous functions $z \colon \Delta \to \mathbb{R}$ for which $D_2 D_1 z(x,y) = \frac{\partial^2 z(x,y)}{\partial y \partial x}$ exists and is continuous on Δ and belong to $L_\infty(\Delta)$. For any function $z(x,y) \in L_\infty(\Delta)$, we define $\|z\|_\infty = \sup_{(x,y)\in\Delta} |z(x,y)|$.

319

We need the following notation

$$k = (b - a)(d - c),$$

$$H_1(x) = \left[\frac{1}{4}(b - a)^2 + \left(x - \frac{a + b}{2}\right)^2\right],$$

$$H_2(y) = \left[\frac{1}{4}(d - c)^2 + \left(y - \frac{c + d}{2}\right)^2\right],$$

$$F(x, y) = \left[(d - c)\int_a^b f(t, y)\, dt + (b - a)\int_c^d f(x, s)\, ds\right],$$

$$G(x, y) = \left[(d - c)\int_a^b g(t, y)\, dt + (b - a)\int_c^d g(x, s)\, ds\right],$$

$$M(x, y) = |g(x, y)|\, \|D_2 D_1 f\|_\infty + |f(x, y)|\, \|D_2 D_1 g\|_\infty,$$

for $f, g \in G(\Delta)$.

Then we have

Theorem 26.1 ([200]). *Let $f, g \in G(\Delta)$. It holds*

$$\left| \frac{1}{k}\int_a^b \int_c^d f(x, y)g(x, y)\, dy\, dx \right.$$

$$+ \left(\frac{1}{k}\int_a^b \int_c^d f(x, y)\, dy\, dx\right)\left(\frac{1}{k}\int_a^b \int_c^d g(x, y)\, dy\, dx\right)$$

$$\left. - \frac{1}{2k^2}\int_a^b \int_c^d (g(x, y)F(x, y) + f(x, y)G(x, y))\, dy\, dx \right|$$

$$\leq \frac{1}{2k^2}\int_a^b \int_c^d M(x, y)H_1(x)H_2(y)\, dy\, dx.$$

Another motivation for this work is the approach in [60].

26.2   Auxiliary Result

We need the following generalized Montgomery type identity.

Theorem 26.2 ([23]). *Let $f : \underset{i=1}{\overset{n}{\times}} [a_i, b_i] \to \mathbb{R}$ be a continuous mapping on $\underset{i=1}{\overset{n}{\times}} [a_i, b_i]$, and $\frac{\partial^n f(x_1, \ldots, x_n)}{\partial x_1 \cdots \partial x_n}$ exists on $\underset{i=1}{\overset{n}{\times}} [a_i, b_i]$ and is integrable. Let also $(x_1, \ldots, x_n) \in \underset{i=1}{\overset{n}{\times}} [a_i, b_i]$ be fixed. We define the kernels $p_i : [a_i, b_i]^2 \to \mathbb{R}$:*

$$p_i(x_i, s_i) := \begin{cases} s_i - a_i, & s_i \in [a_i, x_i], \\ s_i - b_i, & s_i \in (x_i, b_i], \end{cases}$$

for all $i = 1, \ldots, n$.

Then

$$\theta_{1,n} := \int_{\underset{i=1}{\overset{n}{\times}} [a_i,b_i]} \prod_{i=1}^{n} p_i(x_i, s_i) \frac{\partial^n f(s_1, \ldots, s_n)}{\partial s_1 \cdots \partial s_n} ds_1 \cdots ds_n$$

$$= \left\{ \left(\prod_{i=1}^{n}(b_i - a_i) \right) \cdot f(x_1, \ldots, x_n) \right\}$$

$$- \left[\sum_{i=1}^{\binom{n}{1}} \left(\prod_{\substack{j=1 \\ j \neq i}}^{n}(b_j - a_j) \int_{a_i}^{b_i} f(x_1, \ldots, s_i, \ldots, x_n) ds_i \right) \right]$$

$$+ \left[\sum_{\ell=1}^{\binom{n}{2}} \left(\prod_{\substack{k=1 \\ k \neq i,j}}^{n}(b_k - a_k) \int_{a_i}^{b_i} \int_{a_j}^{b_j} f(x_1, \ldots, s_i, \ldots, s_j, \ldots, x_n) ds_i ds_j \right)_{(\ell)} \right]$$

$$- + \cdots - + \cdots + (-1)^{n-1}$$

$$\cdot \left[\sum_{j=1}^{\binom{n}{n-1}} (b_j - a_j) \int_{\underset{\substack{i=1 \\ i \neq j}}{\overset{n}{\times}} [a_i,b_i]} f(s_1, \ldots, x_j, \ldots, s_n) ds_1 \cdots \widehat{ds_j} \cdots ds_n \right]$$

$$+ (-1)^n \int_{\underset{i=1}{\overset{n}{\times}} [a_i,b_i]} f(s_1, \ldots, s_n) ds_1 \cdots ds_n =: \theta_{2,n}. \tag{26.1}$$

The above ℓ counts all the (i,j)'s, $i < j$ and $i, j = 1, \ldots, n$. Also $\widehat{ds_j}$ means ds_j is missing.

26.3 Main Results

We present

Theorem 26.3. *Let $f, g \in C^n(B)$, $n \in \mathbb{N}$, where $B := \overset{n}{\underset{i=1}{\times}} [a_i, b_i]$, $a_i, b_i \in \mathbb{R}$, with $a_i < b_i$, $i = 1, \ldots, n$. Denote by ∂B the boundary of the box B. Suppose that $f(x) = g(x) = 0$, for all $x = (x_1, \ldots, x_n) \in \partial B$ (in other words we assume that*

$$f(\ldots, a_i, \ldots) = g(\ldots, a_i, \ldots) = f(\ldots, b_i, \ldots)$$
$$= g(\ldots, b_i, \ldots) = 0,$$

for all $i = 1, \ldots, n$). Call $V_n := \prod_{i=1}^{n} (b_i - a_i)$. Then

$$\frac{1}{V_n} \int_B |f(x_1, \ldots, x_n)| \, |g(x_1, \ldots, x_n)| \, dx_1 \cdots dx_n$$

$$\leq \frac{1}{2^{n+1}} \left[\int_B |g(x_1, \ldots, x_n)| \left\| \frac{\partial^n f}{\partial s_1 \cdots \partial s_n} \right\|_\infty \right.$$

$$\left. + |f(x_1, \ldots, x_n)| \left\| \frac{\partial^n g}{\partial s_1 \cdots \partial s_n} \right\|_\infty \, dx_1 \cdots dx_n \right]. \qquad (26.2)$$

Proof.   Let $(x_1, \ldots, x_n) \in B$, i.e., $a_i \leq x_i \leq b_i$, for all $i = 1, \ldots, n$. By the assumptions we obtain

$$f(x_1, \ldots, x_n) = \int_{a_1}^{x_1} \cdots \int_{a_n}^{x_n} \frac{\partial^n f(s_1, \ldots, s_n)}{\partial s_1 \cdots \partial s_n} ds_1 \cdots ds_n,$$

and

$$f(x_1, \ldots, x_n) = (-1)^n \int_{x_1}^{b_1} \cdots \int_{x_n}^{b_n} \frac{\partial^n f(s_1, \ldots, s_n)}{\partial s_1 \cdots \partial s_n} ds_1 \cdots ds_n.$$

In general we introduce the subintervals

$$I_{i,0} = [a_i, x_i] \qquad \text{and} \qquad I_{i,1} = [x_i, b_i], \quad i = 1, \ldots, n.$$

Then we find that

$$f(x_1, \ldots, x_n) = (-1)^{\varepsilon_1 + \cdots + \varepsilon_n} \int_{I_{1,\varepsilon_1}} \cdots \int_{I_{n,\varepsilon_n}} \frac{\partial^n f(s_1, \ldots, s_n)}{\partial s_1 \cdots \partial s_n} ds_1 \cdots ds_n, \quad (26.3)$$

where each ε_i can be either 0 or 1. Adding up (26.3) for all 2^n choices for $(\varepsilon_1, \ldots, \varepsilon_n)$ we derive

$$2^n f(x_1, \ldots, x_n) = \sum_{\varepsilon_1, \ldots, \varepsilon_n} (-1)^{\varepsilon_1 + \cdots + \varepsilon_n} \int_{I_{1,\varepsilon_1}}$$

$$\cdots \int_{I_{n,\varepsilon_n}} \frac{\partial^n f(s_1, \ldots, s_n)}{\partial s_1 \cdots \partial s_n} ds_1 \cdots ds_n. \qquad (26.4)$$

Next by taking absolute values in (26.4) and using basic properties of integrals (noticing that the 2^n "sub-boxes" $I_{1,\varepsilon_1} \times \cdots \times I_{n,\varepsilon_n}$ form a partition of B) and the subadditivity property of the absolute value we find that

$$|f(x_1, \ldots, x_n)| \leq \frac{1}{2^n} \int_B \left| \frac{\partial^n f(x_1, \ldots, x_n)}{\partial x_1 \cdots \partial x_n} \right| dx_1 \cdots dx_n$$

$$\leq \frac{V_n}{2^n} \left\| \frac{\partial^n f}{\partial x_1 \cdots \partial x_n} \right\|_\infty.$$

That is we have that

$$|f(x_1, \ldots, x_n)| \leq \frac{V}{2^n} \left\| \frac{\partial^n f}{\partial x_1 \cdots \partial x_n} \right\|_\infty, \qquad (26.5)$$

is true for all $(x_1, \ldots, x_n) \in B$. Similarly it holds

$$|g(x_1, \ldots, x_n)| \leq \frac{V_n}{2^n} \left\| \frac{\partial^n g}{\partial x_1 \cdots \partial x_n} \right\|_\infty, \tag{26.6}$$

true for all $(x_1, \ldots, x_n) \in B$. Therefore

$$|f(x_1, \ldots, x_n)| \, |g(x_1, \ldots, x_n)| \leq \frac{V_n}{2^n} |g(x_1, \ldots, x_n)| \left\| \frac{\partial^n f}{\partial x_1 \cdots \partial x_n} \right\|_\infty, \tag{26.7}$$

and

$$|f(x_1, \ldots, x_n)| \, |g(x_1, \ldots, x_n)| \leq \frac{V_n}{2^n} |f(x_1, \ldots, x_n)| \left\| \frac{\partial^n g}{\partial x_1 \cdots \partial x_n} \right\|_\infty. \tag{26.8}$$

Hence by adding (26.7) and (26.8) we see that

$$|f(x_1, \ldots, x_n)| \, |g(x_1, \ldots, x_n|$$
$$\leq \frac{V_n}{2^{n+1}} \left(|g(x_1, \ldots, x_n)| \left\| \frac{\partial^n f}{\partial x_1 \cdots \partial x_n} \right\|_\infty \right.$$
$$\left. + |f(x_1, \ldots, x_n)| \left\| \frac{\partial^n g}{\partial x_1 \cdots \partial x_n} \right\|_\infty \right), \tag{26.9}$$

is true for all $(x_1, \ldots, x_n) \in B$. Integrating (26.9) over B we obtain (26.2). $\qquad \square$

Remark 26.4. Inequality (26.9) has by itself its own merits.

We also give

Theorem 26.5. *Consider the class of functions* $G := \left\{ f \colon \underset{i=1}{\overset{n}{\times}} [a_i, b_i] \to \mathbb{R} \text{ continu-} \right.$
ous, $n \in \mathbb{N}$: *the partial* $\frac{\partial^n f}{\partial x_1 \cdots \partial x_n}$ *exists on* $\underset{i=1}{\overset{n}{\times}} [a_i, b_i]$ *and belongs to* $L_\infty \left(\underset{i=1}{\overset{n}{\times}} [a_i, b_i] \right)$
with norm $\| \cdot \|_\infty \}$. *Here* $(x_1, \ldots, x_n) \in \underset{i=1}{\overset{n}{\times}} [a_i, b_i]$. *Let* $f, g \in G$. *Denote*
$V_n := \prod_{i=1}^n (b_i - a_i)$, *and*

$$H_j(x_j) := \frac{(x_j - a_j)^2 + (b_j - x_j)^2}{2}, \quad j = 1, \ldots, n.$$

Set

$$F_1(x_1, \ldots, x_n) := \sum_{i=1}^{\binom{n}{1}} \left(\prod_{\substack{j=1 \\ j \neq i}}^n (b_j - a_j) \int_{a_i}^{b_i} f(x_1, \ldots, s_i, \ldots, x_n) ds_i \right),$$

$$G_1(x_1, \ldots, x_n) := \sum_{i=1}^{\binom{n}{1}} \left(\prod_{\substack{j=1 \\ j \neq i}}^n (b_j - a_j) \int_{a_i}^{b_i} g(x_1, \ldots, s_i, \ldots, x_n) ds_i \right);$$

$$F_2(x_1, \ldots, x_n) := - \left[\sum_{\ell=1}^{\binom{n}{2}} \left(\prod_{\substack{k=1 \\ k \neq i,j}}^n (b_k - a_k) \int_{a_i}^{b_i} \int_{a_j}^{b_j} \right. \right.$$
$$\left. \left. \cdot f(x_1, \ldots, s_i, \ldots, s_j, \ldots, x_n) ds_i ds_j \right)_{(\ell)} \right],$$

where ℓ counts all (i,j)'s, $i < j$, and $i, j = 1, \ldots, n$, also

$$G_2(x_1, \ldots, x_n) := -\left[\sum_{\ell=1}^{\binom{n}{2}} \left(\prod_{\substack{k=1 \\ k \neq i,j}}^{n} (b_k - a_k) \int_{a_i}^{b_i} \int_{a_j}^{b_j} \right.\right.$$

$$\left.\left. \cdot\, g(x_1, \ldots, s_i, \ldots, s_j, \ldots, x_n) ds_i ds_j \right)_{(\ell)} \right];$$

$$\cdots\cdots\cdots\cdots\cdots\cdots ;$$

$$F_{n-1}(x_1, \ldots, x_n) := (-1)^n \left[\sum_{j=1}^{\binom{n}{n-1}} (b_j - a_j) \int_{\substack{n \\ \times \\ i=1 \\ i \neq j}}^{} [a_i, b_i] \right.$$

$$\left. \cdot\, f(s_1, \ldots, x_j, \ldots, s_n) ds_1 \cdots \widehat{ds_j} \cdots ds_n \right],$$

where $\widehat{ds_j}$ means ds_j is missing, also

$$G_{n-1}(x_1, \ldots, x_n) := (-1)^n \left[\sum_{j=1}^{\binom{n}{n-1}} (b_j - a_j) \int_{\substack{n \\ \times \\ i=1 \\ i \neq j}}^{} [a_i, b_i] \right.$$

$$\left. \cdot\, g(s_1, \ldots, x_j, \ldots, s_n) ds_1 \cdots \widehat{ds_j} \cdots ds_n \right].$$

Also define

$$M_n(x_1, \ldots, x_n) := |g(x_1, \ldots, x_n)| \left\| \frac{\partial^n f}{\partial s_1 \cdots \partial s_n} \right\|_{\infty}$$

$$+ |f(x_1, \ldots, x_n)| \left\| \frac{\partial^n g}{\partial s_1 \cdots \partial s_n} \right\|_{\infty}.$$

Then

$$\Gamma_n := \left| \frac{1}{V_n} \int_{\substack{n \\ \times \\ i=1}}^{} [a_i, b_i] f(x_1, \ldots, x_n) g(x_1, \ldots, x_n) dx_1 \cdots dx_n \right.$$

$$+ (-1)^n \left(\frac{1}{V_n} \int_{\substack{n \\ \times \\ i=1}}^{} [a_i, b_i] f(x_1, \ldots, x_n) dx_1 \cdots dx_n \right)$$

$$\cdot \left(\frac{1}{V_n} \int_{\substack{n \\ \times \\ i=1}}^{} [a_i, b_i] g(x_1, \ldots, x_n) dx_1 \cdots dx_n \right)$$

$$- \frac{1}{2V_n^2} \left(\int_{\substack{n \\ \times \\ i=1}}^{} [a_i, b_i] \left[g(x_1, \ldots, x_n) \sum_{j=1}^{n-1} F_j(x_1, \ldots, x_n) \right. \right.$$

$$+ f(x_1, \ldots, x_n) \sum_{j=1}^{n-1} G_j(x_1, \ldots, x_n) \bigg] dx_1 \cdots dx_n \bigg) \bigg|$$

$$\leq \frac{1}{2V_n^2} \int_{\substack{n \\ \times \\ i=1}}^{} M_n(x_1, \ldots, x_n) \left(\prod_{j=1}^{n} H_j(x_j) \right) dx_1 \cdots dx_n. \qquad (26.10)$$

Proof. We define the kernels $p_i \colon [a_i, b_i]^2 \to \mathbb{R}$:

$$p_i(x_i, s_i) := \begin{cases} s_i - a_i, \, s_i \in [a_i, x_i], \\ s_i - b_i, \, s_i \in (x_i, b_i], \end{cases}$$

for all $i = 1, \ldots, n$.

We also have that

$$\int_{[a_j, b_j]} |p_j(x_j, s_j)| ds_j = H_j(x_j), \quad j = 1, \ldots, n.$$

From Theorem 26.5 ([23]) we get

$$V_n f(x_1, \ldots, x_n) = \sum_{j=1}^{n-1} F_j(x_1, \ldots, x_n) + (-1)^{n+1} \int_{\substack{n \\ \times \\ i=1}}^{} f(s_1, \ldots, s_n) ds_1 \cdots ds_n$$

$$+ \int_{\substack{n \\ \times [a_i, b_i] \\ i=1}}^{} \prod_{i=1}^{n} p_i(x_i, s_i) \frac{\partial^n f(s_1, \ldots, s_n)}{\partial s_1 \cdots \partial s_n} ds_1 \cdots ds_n. \qquad (26.11)$$

And also

$$V_n g(x_1, \ldots, x_n) = \sum_{j=1}^{n-1} G_j(x_1, \ldots, x_n) + (-1)^{n+1} \int_{\substack{n \\ \times \\ i=1}}^{} g(s_1, \ldots, s_n) ds_1 \cdots ds_n$$

$$+ \int_{\substack{n \\ \times [a_i, b_i] \\ i=1}}^{} \prod_{i=1}^{n} p_i(x_i, s_i) \frac{\partial^n g(s_1, \ldots, s_n)}{\partial s_1 \cdots \partial s_n} ds_1 \cdots ds_n. \qquad (26.12)$$

Next we multiply (26.11) by $g(x_1, \ldots, x_n)$ and (26.12) by $f(x_1, \ldots, x_n)$ and we add the resulting identities, to find

$$2V_n f(x_1, \ldots, x_n) g(x_1, \ldots, x_n)$$

$$= \bigg[g(x_1, \ldots, x_n) \sum_{j=1}^{n-1} F_j(x_1, \ldots, x_n) + f(x_1, \ldots, x_n) \sum_{j=1}^{n-1} G_j(x_1, \ldots, x_n) \bigg]$$

$$+ A_n(x_1, \ldots, x_n) + B_n(x_1, \ldots, x_n). \qquad (26.13)$$

Here we have

$$A_n(x_1, \ldots, x_n) := (-1)^{n+1} \bigg[g(x_1, \ldots, x_n) \cdot \int_{\substack{n \\ \times [a_i, b_i] \\ i=1}}^{} f(s_1, \ldots, s_n) ds_1 \cdots ds_n$$

$$+ f(x_1, \ldots, x_n) \int_{\substack{n \\ \times [a_i, b_i] \\ i=1}}^{} g(s_1, \ldots, s_n) ds_1 \cdots ds_n \bigg], \qquad (26.14)$$

and

$$B_n(x_1,\ldots,x_n) := \left[g(x_1,\ldots,x_n) \int_{\underset{i=1}{\overset{n}{\times}} [a_i,b_i]} \prod_{i=1}^{n} p_i(x_i,s_i) \cdot \frac{\partial^n f(s_1,\ldots,s_n)}{\partial x_1 \cdots \partial s_n} ds_1 \cdots ds_n \right.$$

$$\left. + f(x_1,\ldots,x_n) \int_{\underset{i=1}{\overset{n}{\times}} [a_i,b_i]} \prod_{i=1}^{n} p_i(x_i,s_i) \cdot \frac{\partial^n g(s_1,\ldots,s_n)}{\partial s_1 \cdots \partial s_n} ds_1 \cdots ds_n \right].$$

$$(26.15)$$

We see that

$$|B_n(x_1,\ldots,x_n)| \le M_n(x_1,\ldots,x_n) \left(\prod_{j=1}^{n} H_j(x_j) \right). \qquad (26.16)$$

Next we integrate (26.13) over $\underset{i=1}{\overset{n}{\times}} [a_i,b_i]$ and we derive

$$\frac{1}{V_n} \int_{\underset{i=1}{\overset{n}{\times}} [a_i,b_i]} f(x_1,\ldots,x_n) g(x_1,\ldots,x_n) dx_1 \cdots dx_n$$

$$= \frac{1}{2V_n^2} \left(\int_{\underset{i=1}{\overset{n}{\times}} [a_i,b_i]} \left[g(x_1,\ldots,x_n) \sum_{j=1}^{n-1} F_j(x_1,\ldots,x_n) \right. \right.$$

$$\left. + f(x_1,\ldots,x_n) \sum_{j=1}^{n-1} G_j(x_1,\ldots,x_n) \right] dx_1 \cdots dx_n \Bigg)$$

$$+ (-1)^{n+1} \left[\left(\frac{1}{V_n} \int_{\underset{i=1}{\overset{n}{\times}} [a_i,b_i]} f(x_1,\ldots,x_n) dx_1 \cdots dx_n \right) \right.$$

$$\left. \cdot \left(\frac{1}{V_n} \int_{\underset{i=1}{\overset{n}{\times}} [a_i,b_i]} g(x_1,\ldots,x_n) dx_1 \cdots dx_n \right) \right]$$

$$+ \frac{1}{2V_n^2} \int_{\underset{i=1}{\overset{n}{\times}} [a_i,b_i]} B_n(x_1,\ldots,x_n) dx_1 \cdots dx_n. \qquad (26.17)$$

Consequently we obtain

$$\Gamma_n \le \frac{1}{2V_n^2} \int_{\underset{i=1}{\overset{n}{\times}} [a_i,b_i]} |B_n(x_1,\ldots,x_n)| dx_1 \cdots dx_n. \qquad (26.18)$$

At last using (26.18) along with (26.16) we have established (26.10). $\qquad\square$

Remark 26.6. From (26.13), (26.14), (26.16) we get

$$\left| 2V_n f(x_1,\ldots,x_n) g(x_1,\ldots,x_n) - g(x_1,\ldots,x_n) \right.$$

$$\left. \cdot \sum_{j=1}^{n-1} F_j(x_1,\ldots,x_n) - f(x_1,\ldots,x_n) \sum_{j=1}^{n-1} G_j(x_1,\ldots,x_n) - A_n(x_1,\ldots,x_n) \right|$$

$$\leq M_n(x_1,\ldots,x_n)\left(\prod_{j=1}^{n} H_j(x_j)\right), \text{ for } (x_1,\ldots,x_n) \in \underset{i=1}{\overset{n}{\times}}[a_i,b_i]. \qquad (26.19)$$

We also give

Corollary 26.7 (to Theorem 26.3). *Consider the class of functions* $\mathcal{F} := \{f \in C^n(B),$ *where* $n \in \mathbb{N},$ $B := \underset{i=1}{\overset{n}{\times}}[a_i,b_i]$ *such that* $f(x) = 0,$ *for all* $x = (x_1,\ldots,x_n) \in \partial B$ *(the boundary of* B*)*$\}.$ *Let* $f \in \mathcal{F}.$ *Let also* $g \in C^n(B).$ *Put* $V_n := \prod_{i=1}^{n}(b_i - a_i).$ *Then*

$$\frac{1}{V_n}\int_B |f(x_1,\ldots,x_n)|\,|g(x_1,\ldots,x_n)|\,dx_1\cdots dx_n$$

$$\leq \frac{1}{2^n}\int_B \left|\frac{\partial^n(fg)}{\partial x_1\cdots\partial x_n}(x_1,\ldots,x_n)\right| dx_1\cdots dx_n. \qquad (26.20)$$

Proof. Totally the same way as in the proof of Theorem 26.3 we find

$$|f(x_1,\ldots,x_n)| \leq \frac{1}{2^n}\int_B \left|\frac{\partial^n f(x_1,\ldots,x_n)}{\partial x_1\cdots\partial x_n}\right| dx_1\cdots dx_n, \quad \forall f \in \mathcal{F}.$$

Integrating over B the last one we obtain

$$\int_B |f(x_1,\ldots,x_n)|\,dx_1\cdots dx_n \leq \frac{V_n}{2^n}\int_B \left|\frac{\partial^n f(x_1,\ldots,x_n)}{\partial x_1\cdots\partial x_n}\right| dx_1\cdots dx_n, \quad (26.21)$$

true for any $f \in \mathcal{F}.$

The inequality (26.21) was also proved in [60]. Clearly here $f \cdot g \in \mathcal{F}.$
Finally applying (26.21) for $f \cdot g$ we establish (26.20). $\qquad\qquad \square$

Finally we have

Corollary 26.8 (to Theorem 26.5). *Case of* $n = 3.$ *Here we consider the class of functions* $G^* := \{f: \underset{i=1}{\overset{3}{\times}}[a_i,b_i] \to \mathbb{R}$ *continuous:* $\frac{\partial^3 f}{\partial x_1 \partial x_2 \partial x_3}$ *exists on* $\underset{i=1}{\overset{3}{\times}}[a_i,b_i]$ *and belongs to* $L_\infty\left(\underset{i=1}{\overset{3}{\times}}[a_i,b_i]\right)$ *with norm* $\|\cdot\|_\infty\}.$ *Let* $(x_1,x_2,x_3) \in \underset{i=1}{\overset{3}{\times}}[a_i,b_i]$ *and* $f,g \in G^*.$ *Denote* $V_3 := \prod_{i=1}^{3}(b_i - a_i),$ *and*

$$H_j(x_j) := \frac{(x_j - a_j)^2 + (b_j - x_j)^2}{2}, \quad j - 1,\ldots,n.$$

Put

$$F_1(x_1, x_2, x_3) := \sum_{i=1}^{3} \left(\prod_{\substack{j=1 \\ j \neq i}}^{3} (b_j - a_j) \int_{a_i}^{b_i} f(x_1, s_i, x_3) ds_i \right),$$

$$G_1(x_1, x_2, x_3) := \sum_{i=1}^{3} \left(\prod_{\substack{j=1 \\ j \neq i}}^{3} (b_j - a_j) \int_{a_i}^{b_i} g(x_1, s_i, x_3) ds_i \right);$$

$$F_2(x_1, x_2, x_3) := -\left[(b_3 - a_3) \int_{a_1}^{b_1} \int_{a_2}^{b_2} f(s_1, s_2, x_3) ds_1 ds_2 \right.$$

$$+ (b_2 - a_2) \int_{a_1}^{b_1} \int_{a_3}^{b_3} f(s_1, x_2, s_3) ds_1 ds_3$$

$$\left. + (b_1 - a_1) \int_{a_2}^{b_2} \int_{a_3}^{b_3} f(x_1, s_2, s_3) ds_2 ds_3 \right],$$

$$G_2(x_1, x_2, x_3) := -\left[(b_3 - a_3) \int_{a_1}^{b_1} \int_{a_2}^{b_2} g(s_1, s_2, x_3) ds_1 ds_2 \right.$$

$$+ (b_2 - a_2) \int_{a_1}^{b_1} \int_{a_3}^{b_3} g(s_1, x_2, s_3) ds_1 ds_3$$

$$\left. + (b_1 - a_1) \int_{a_2}^{b_2} \int_{a_3}^{b_3} g(x_1, s_2, s_3) ds_2 ds_3 \right];$$

$$M_3(x_1, x_2, x_3) := |g(x_1, x_2, x_3)| \left\| \frac{\partial^3 f}{\partial s_1 \partial s_2 \partial s_3} \right\|_{\infty}$$

$$+ |f(x_1, x_2, x_3)| \left\| \frac{\partial^3 g}{\partial s_1 \partial s_2 \partial s_3} \right\|_{\infty}.$$

Then

$$\left| \frac{1}{V_3} \int_{a_1}^{b_1} \int_{a_2}^{b_2} \int_{a_3}^{b_3} f(x_1, x_2, x_3) g(x_1, x_2, x_3) dx_1 dx_2 dx_3 \right.$$

$$- \left(\frac{1}{V_3} \int_{a_1}^{b_1} \int_{a_2}^{b_2} \int_{a_3}^{b_3} f(x_1, x_2, x_3) dx_1 dx_2 dx_3 \right)$$

$$\left(\frac{1}{V_3} \int_{a_1}^{b_1} \int_{a_2}^{b_2} \int_{a_3}^{b_3} g(x_1, x_2, x_3) dx_1 dx_2 dx_3 \right)$$

$$- \frac{1}{2V_3^2} \int_{a_1}^{b_1} \int_{a_2}^{b_2} \int_{a_3}^{b_3} \left[g(x_1, x_2, x_3)(F_1(x_1, x_2, x_3) \right.$$

$$\left. + F_2(x_1, x_2, x_3)) + f(x_1, x_2, x_3)(G_1(x_1, x_2, x_3) + G_2(x_1, x_2, x_3)) \right] dx_1 dx_2 dx_3 \Bigg|$$

$$\leq \frac{1}{2V_3^2} \int_{a_1}^{b_1} \int_{a_2}^{b_2} \int_{a_3}^{b_3} M_3(x_1, x_2, x_3) \left(\prod_{j=1}^{3} H_j(x_j) \right) dx_1 dx_2 dx_3. \quad (26.22)$$

Comment 26.9. Theorem 26.5 clearly generalizes Theorem 26.1 of [200], that is for $n = 2$ the corresponding inequalities coincide.

Chapter 27

Chebyshev–Grüss Type Inequalities on Spherical Shells and Balls

We present Chebyshev–Grüss type inequalities over spherical shells and balls by extending some basic univariate results of Pachpatte. This chapter is based on [46].

27.1   Introduction

For two absolutely continuous functions $f, g \colon [a, b] \to \mathbb{R}$ consider the functional

$$T(f, g) = \frac{1}{b - a} \int_a^b f(x)g(x)dx - \left(\frac{1}{b - a} \int_a^b f(x)dx \right) \left(\frac{1}{b - a} \int_a^b g(x)dx \right), \quad (27.1)$$

where the involved integrals exist.

In 1882, Chebyshev [88] proved that if $f', g' \in L_\infty[a, b]$, then

$$|T(f, g)| \leq \frac{1}{12}(b - a)^2 \|f'\|_\infty \|g'\|_\infty. \quad (27.2)$$

In 1935, Grüss [150] showed that

$$|T(f, g)| \leq \frac{1}{4}(M - \theta)(\Gamma - \Delta), \quad (27.3)$$

provided θ, M, Δ, Γ are real numbers satisfying the condition $-\infty < \theta \leq M < \infty$, $-\infty < \Delta \leq \Gamma < \infty$ for $x \in [a, b]$.

The purpose of this chapter is to extend above fundamental results over spherical shells and balls in $\mathbb{R}^N$, $N \geq 1$. For that we need the following machinery from Pachpatte [206] which also motivates this chapter.

Let $f \colon [a, b] \to \mathbb{R}$ be differentiable on $[a, b]$ and $f' \colon [a, b] \to \mathbb{R}$ is integrable on $[a, b]$. Let $w \colon [a, b] \to [0, \infty)$ be some probability density function, that is, an integrable function satisfying $\int_a^b w(t)dt = 1$, and $W(t) = \int_a^t w(x)dx$ for $t \in [a, b]$, $W(t) = 0$ for $t < a$, and $W(t) = 1$ for $t > b$. In [213] Pečarić has given the following weighted extension of the Montgomery identity:

$$f(x) = \int_a^b w(t)f(t)dt + \int_a^b P_w(x, t)f'(t)dt, \quad (27.4)$$

where $P_w(x, t)$ is the weighted Peano kernel defined by

$$P_w(x, t) = \begin{cases} W(t), & a \leq t \leq x, \\ W(t) - 1, & x < t \leq b. \end{cases} \quad (27.5)$$

331

For some suitable functions $w, f, g \colon [a, b] \to \mathbb{R}$, we define

$$T(w, f, g) = \int_a^b w(x)f(x)g(x)dx - \left(\int_a^b w(x)f(x)dx\right)\left(\int_a^b w(x)g(x)dx\right), \quad (27.6)$$

and $\|\cdot\|_\infty$ the usual Lebesgue norm on $L_\infty[a,b]$ that is, $\|h\|_\infty := \operatorname{ess\,sup}_{t\in[a,b]} |h(t)|$ for $h \in L_\infty[a,b]$. Pachpatte Theorems 2.1, 2.2 ([206]) follow.

Theorem 27.1. *Let $f, g \colon [a,b] \to \mathbb{R}$ be differentiable on $[a,b]$ and $f', g' \colon [a,b] \to \mathbb{R}$ are integrable on $[a,b]$. Let $w \colon [a,b] \to [0,\infty)$ be an integrable function satisfying $\int_a^b w(t)dt = 1$. Then*

$$|T(w,f,g)| \leq \|f'\|_\infty \|g'\|_\infty \int_a^b w(x)H^2(x)dx, \qquad (27.7)$$

where

$$H(x) = \int_a^b |P_w(x,t)|dt \qquad (27.8)$$

for $x \in [a,b]$ and $P_w(x,t)$ is the weighted Peano kernel given by (27.5).

Theorem 27.2. *Let f, g, f', g', w be as in Theorem 27.1. Then*

$$|T(w,f,g)| \leq \frac{1}{2}\int_a^b w(x)[|g(x)|\,\|f'\|_\infty + |f(x)|\,\|g'\|_\infty]H(x)dx, \qquad (27.9)$$

where $H(x)$ is defined by (27.8).

27.2 Main Results

We make

Remark 27.3. Let A be a *spherical shell* $\subseteq \mathbb{R}^N$, $N \geq 1$, i.e. $A := B(0, R_2) - \overline{B(0, R_1)}$, $0 < R_1 < R_2$. Here the ball $B(0, R) := \{x \in \mathbb{R}^N : |x| < R\}$, $R > 0$, where $|\cdot|$ is the Euclidean norm, also $S^{N-1} := \{x \in \mathbb{R}^N : |x| = 1\}$ is the unit sphere in $\mathbb{R}^N$ with surface area $\omega_N := \frac{2\pi^{N/2}}{\Gamma\left(\frac{N}{2}\right)}$. For $x \in \mathbb{R}^N - \{0\}$ one can write uniquely $x = r\omega$, where $r > 0$, $\omega \in S^{N-1}$.

Let $F, G \in C^1(\overline{A})$. First we suppose that F, G are *radial*, i.e. $F(x) = f(r)$, $G(x) = g(r)$, where $r = |x|$, $R_1 \leq r \leq R_2$. Of course $f, g \in C^1([R_1, R_2])$.

We notice that

$$\frac{N}{R_2^N - R_1^N}\int_{R_1}^{R_2} s^{N-1}ds = 1,$$

i.e. $w(s) := \frac{Ns^{N-1}}{R_2^N - R_1^N}$, $R_1 \leq s \leq R_2$, is a probability density function.

Thus

$$W(s) := \int_{R_1}^s w(\tau)d\tau = \frac{s^N - R_1^N}{R_2^N - R_1^N}, \quad \text{for } s \in [R_1, R_2],$$

also $W(s) = 1$, for $s > R_2$; with $W(s) = 0$, for $s < R_1$. That is W is the associate distribution function here.

We introduce

$$P_w(r, s) = \begin{cases} W(s), & R_1 \le s \le r, \\ W(s) - 1, & r < s \le R_2. \end{cases} \tag{27.10}$$

Pečarić in [213], proved a weighted extension of Montgomery identity, see (27.4), which in our case is

$$f(r) = \int_{R_1}^{R_2} \left(\frac{N s^{N-1}}{R_2^N - R_1^N} \right) f(s) ds + \int_{R_1}^{R_2} P_w(r, s) f'(s) ds, \tag{27.11}$$

and

$$g(r) = \int_{R_1}^{R_2} \left(\frac{N s^{N-1}}{R_2^N - R_1^N} \right) g(s) ds + \int_{R_1}^{R_2} P_w(r, s) g'(s) ds. \tag{27.12}$$

Here we observe that

$$\mathrm{Vol}(A) = \frac{\omega_N (R_2^N - R_1^N)}{N}. \tag{27.13}$$

We denote by

$$\begin{aligned}
\tilde{T}(F, G) &:= \frac{\int_A F(x) G(x) dx}{\mathrm{Vol}(A)} - \frac{1}{(\mathrm{Vol}(A))^2} \left(\int_A F(x) dx \right) \left(\int_A G(x) dx \right) \\
&= \frac{N}{\omega_N (R_2^N - R_1^N)} \int_A F(x) G(x) dx \\
&\quad - \left(\frac{N}{\omega_N (R_2^N - R_1^N)} \right)^2 \left(\int_A F(x) dx \right) \left(\int_A G(x) dx \right),
\end{aligned} \tag{27.14}$$

the *Chebyshev functional* in this setting.

We notice that

$$\frac{1}{\omega_N} \int_A F(x) dx = \frac{1}{\omega_N} \int_{S^{N-1}} \left(\int_{R_1}^{R_2} f(r) r^{N-1} dr \right) d\omega = \int_{R_1}^{R_2} f(r) r^{N-1} dr. \tag{27.15}$$

Similarly we find

$$\frac{1}{\omega_N} \int_A G(x) dx = \int_{R_1}^{R_2} g(r) r^{N-1} dr, \tag{27.16}$$

and

$$\frac{1}{\omega_N} \int_A F(x) G(x) dx = \int_{R_1}^{R_2} f(r) g(r) r^{N-1} dr. \tag{27.17}$$

Consequently we have

$$\begin{aligned}
\tilde{T}(F, G) &= \left(\frac{N}{R_2^N - R_1^N} \right) \int_{R_1}^{R_2} f(r) g(r) r^{N-1} dr \\
&\quad - \left(\frac{N}{R_2^N - R_1^N} \right)^2 \left(\int_{R_1}^{R_2} f(r) r^{N-1} dr \right) \left(\int_{R_1}^{R_2} g(r) r^{N-1} dr \right) =: T(w, f, g),
\end{aligned} \tag{27.18}$$

as in Pachpatte [206], see (27.6).

By Theorem 2.1 of Pachpatte [206], see Theorem 27.1, we obtain that

$$|\tilde{T}(F,G)| \le \|f'\|_\infty \|g'\|_\infty \left(\frac{N}{R_2^N - R_1^N}\right) \int_{R_1}^{R_2} r^{N-1} H^2(r)dr, \tag{27.19}$$

where

$$H(r) := \int_{R_1}^{R_2} |P_w(r,s)|ds = \int_{R_1}^{r} W(s)ds + \int_{r}^{R_2} (1 - W(s))ds.$$

That is

$$H(r) = \left(\frac{1}{R_2^N - R_1^N}\right) \left[\left(\frac{2r^{N+1} + N(R_1^{N+1} + R_2^{N+1})}{N+1}\right) - r(R_1^N + R_2^N) \right], \tag{27.20}$$

$r \in [R_1, R_2]$.

In general it holds

$$\left\|\frac{\partial F}{\partial r}\right\|_\infty \le \|\nabla F\|_\infty,$$

$$\left\|\frac{\partial G}{\partial r}\right\|_\infty \le \|\nabla G\|_\infty, \tag{27.21}$$

with equality in the radial case.

So we derive that

$$|\tilde{T}(F,G)| \le \left\|\frac{\partial F}{\partial r}\right\|_\infty \left\|\frac{\partial G}{\partial r}\right\|_\infty \frac{NI}{(R_2^N - R_1^N)^3(N+1)^2}, \tag{27.22}$$

where

$$I := \int_{R_1}^{R_2} r^{N-1} \left[2r^{N+1} + N(R_1^{N+1} + R_2^{N+1}) - r(N+1)(R_1^N + R_2^N)\right]^2 dr. \tag{27.23}$$

One calculates the above integral to find

$$\begin{aligned}
I = 4 &\left(\frac{R_2^{3N+2} - R_1^{3N+2}}{3N+2}\right) + N(R_1^{N+1} + R_2^{N+1})^2(R_2^N - R_1^N) \\
&+ \frac{(N+1)^2}{(N+2)}(R_1^N + R_2^N)^2(R_2^{N+2} - R_1^{N+2}) \\
&+ \left(\frac{4N}{2N+1}\right)(R_1^{N+1} + R_2^{N+1})(R_2^{2N+1} - R_1^{2N+1}) \\
&- 2(N+1)(R_1^N + R_2^N)(R_2^{2N+2} - R_1^{2N+2}).
\end{aligned} \tag{27.24}$$

We have established the first main result.

Theorem 27.4. *Let $F, G \in C^1(\overline{A})$ that are radial functions. Then*

$$|\tilde{T}(F,G)| \le \|\nabla F\|_\infty \|\nabla G\|_\infty \frac{NI}{(R_2^N - R_1^N)^3(N+1)^2}. \tag{27.25}$$

We continue from Remark 27.3.

Remark 27.5. By Theorem 2.2 of Pachpatte [206], see Theorem 27.2, under the same terms and assumptions. We get

$$|\tilde{T}(F,G)| \le \frac{1}{2}\left(\frac{N}{R_2^N - R_1^N}\right) \int_{R_1}^{R_2} r^{N-1}\big[|g(r)|\,\|f'\|_\infty + |f(r)|\,\|g'\|_\infty\big] H(r)dr. \tag{27.26}$$

We have derived the following result.

Theorem 27.6. *Let* $F, G \in C^1(\overline{A})$ *that are radial functions. Then*

$$\tilde{T}(F,G) \le \frac{1}{2\,\mathrm{Vol}(A)} \int_A \big[|G(x)|\,\|\nabla F\|_\infty + |F(x)|\,\|\nabla G\|_\infty\big] H(|x|)dx, \tag{27.27}$$

where

$$H(|x|) = \left(\frac{1}{R_2^N - R_1^N}\right)\left[\left(\frac{2|x|^{N+1} + N(R_1^{N+1} + R_2^{N+1})}{N+1}\right)\right.$$
$$\left. - |x|(R_1^N + R_2^N)\right], \quad x \in A. \tag{27.28}$$

We continue from Remark 27.5 to transfer our results over the ball.

Remark 27.7. Here we set $R := R_2$ and let $R > R_1 \to 0$. We suppose $F, G \in C^1(\overline{B(0,R)})$ that are radial. Inequality (27.25) is clearly true when $\|\nabla F\|_\infty$, $\|\nabla G\|_\infty$ are taken over $\overline{B(0,R)}$ which is a larger set containing $\overline{A}$. We consider here this modified form of (27.25).

Let now $0 < R_{1,n} \downarrow 0$, as $n \to \infty$, i.e. $B(0,R_{1,1}) \supset B(0,R_{1,2}) \supset \cdots \supset B(0,R_{1,n})\ldots$, with $\bigcap_{n=1}^{\infty} B(0,R_{1,n}) = \{0\}$. That is, as $R_1 \to 0$ then $B(0,R_1) \downarrow \{0\}$. Hence

$$\chi_{B(0,R_1)} \overset{R_1 \to 0}{\longrightarrow} \chi_{\{0\}}.$$

Let h be Lebesque integrable on $B(0,R)$. I.e. $|h\chi_{B(0,R_1)}| \le |h|$. Consequently

$$h\chi_{B(0,R_1)} \overset{R_1 \to 0}{\longrightarrow} h\chi_{\{0\}}, \quad \text{pointwise over } B(0,R).$$

Thus, by Dominated convergence Theorem we get

$$\mathcal{L}\int_{B(0,R)} h\chi_{B(0,R_1)}dx \overset{R_1 \to 0}{\longrightarrow} \mathcal{L}\int_{B(0,R)} h\chi_{\{0\}}dx.$$

That is

$$\int_{B(0,R_1)} hdx \overset{R_1 \to 0}{\longrightarrow} 0. \tag{27.29}$$

Applying (27.29) we obtain

$$\lim_{R_1 \to 0} \tilde{T}(F,G) = \frac{\int_{B(0,R)} F(x)G(x)dx}{\mathrm{Vol}(B(0,R))}$$
$$- \frac{1}{(\mathrm{Vol}(B(0,R)))^2}\left(\int_{B(0,R)} F(x)dx\right)\left(\int_{B(0,R)} G(x)dx\right). \tag{27.30}$$

We also notice that

$$\lim_{R_1 \to 0} (\text{R.H.S.}(27.25)) = \|\nabla F\|_\infty \|\nabla G\|_\infty \frac{NR^2}{(N+1)^2} \left[\frac{4}{(3N+2)} + \frac{(N+1)^2}{(N+2)} \right.$$
$$\left. + \frac{4N}{(2N+1)} - (N+2) \right]. \tag{27.31}$$

So based on (27.25) and the above we derive

Theorem 27.8. *Let* $F, G \in C^1(\overline{B(0,R)})$ *that are radial functions. Then*

$$\left| \frac{\int_{B(0,R)} F(x)G(x)dx}{\text{Vol}(B(0,R))} - \frac{1}{(\text{Vol}(B(0,R)))^2} \left(\int_{B(0,R)} F(x)dx \right) \left(\int_{B(0,R)} G(x)dx \right) \right|$$
$$\leq \|\nabla F\|_\infty \|\nabla G\|_\infty \frac{NR^2}{(N+1)^2} \left[\frac{4}{(3N+2)} + \frac{(N+1)^2}{(N+2)} + \frac{4N}{(2N+1)} - (N+2) \right]. \tag{27.32}$$

Similarly from (27.27) we obtain

Theorem 27.9. *Let* $F, G \in C^1(\overline{B(0,R)})$ *that are radial functions. Then*

$$\left| \frac{\int_{B(0,R)} F(x)G(x)dx}{\text{Vol}(B(0,R))} - \frac{1}{(\text{Vol}(B(0,R)))^2} \left(\int_{B(0,R)} F(x)dx \right) \left(\int_{B(0,R)} G(x)dx \right) \right|$$
$$\leq \frac{1}{2\,\text{Vol}(B(0,R))} \int_{B(0,R)} \left[|G(x)| \, \|\nabla F\|_\infty + |F(x)| \, \|\nabla G\|_\infty \right] H^*(|x|)dx, \tag{27.33}$$

where

$$H^*(|x|) = \frac{1}{R^N} \left[\left(\frac{2|x|^{N+1} + NR^{N+1}}{N+1} \right) - |x|R^N \right], \quad x \in B(0,R). \tag{27.34}$$

Next we treat not necessarily radial functions in our setting. We give

Theorem 27.10. *Let* $F, G \in C^1(\overline{A})$. *Then*

$$\frac{1}{\text{Vol}(A)} \left| \int_A F(x)G(x)dx - \frac{N}{(R_2^N - R_1^N)} \int_{S^{N-1}} \left[\left(\int_{R_1}^{R_2} F(r\omega)r^{N-1}dr \right) \right. \right.$$
$$\left. \left. \times \left(\int_{R_1}^{R_2} G(r\omega)r^{N-1}dr \right) \right] d\omega \right| \tag{27.35}$$
$$\leq \left\| \frac{\partial F}{\partial r} \right\|_\infty \left\| \frac{\partial G}{\partial r} \right\|_\infty \frac{NI}{(R_2^N - R_1^N)^3 (N+1)^2}$$
$$\leq \|\nabla F\|_\infty \|\nabla G\|_\infty \frac{NI}{(R_2^N - R_1^N)^3 (N+1)^2}, \tag{27.36}$$

where I *is given by (27.24).*

Proof. The functions $F(r\omega)$, $G(r\omega)$ are considered radial in r, $\forall \omega \in S^{N-1}$. Also there exist

$$\frac{\partial F(r\omega)}{\partial r}, \frac{\partial G(r\omega)}{\partial r} \in C([R_1, R_2]), \quad \forall \omega \in S^{N-1}.$$

As in (27.18), (27.19) and (27.22), (27.25) we get

$$\left| \left(\frac{N}{R_2^N - R_1^N} \right) \int_{R_1}^{R_2} F(r\omega)G(r\omega)r^{N-1}dr \right.$$

$$\left. - \left(\frac{N}{R_2^N - R_1^N} \right)^2 \left(\int_{R_1}^{R_2} F(r\omega)r^{N-1}dr \right) \left(\int_{R_1}^{R_2} G(r\omega)r^{N-1}dr \right) \right| \quad (27.37)$$

$$\leq \left\| \frac{\partial F(r\omega)}{\partial r} \right\|_{\infty,[R_1,R_2]} \left\| \frac{\partial G(r\omega)}{\partial r} \right\|_{\infty,[R_1,R_2]} \frac{NI}{(R_2^N - R_1^N)^3(N+1)^2}$$

$$\leq \left\| \frac{\partial F}{\partial r} \right\|_{\infty} \left\| \frac{\partial G}{\partial r} \right\|_{\infty} \frac{NI}{(R_2^N - R_1^N)^3(N+1)^2}. \quad (27.38)$$

Consequently we have

$$\left| \frac{N}{\omega_N(R_2^N - R_1^N)} \int_{S^{N-1}} \left(\int_{R_1}^{R_2} F(r\omega)G(r\omega)r^{N-1}dr \right) d\omega \right.$$

$$\left. - \frac{1}{\omega_N} \left(\frac{N}{R_2^N - R_1^N} \right)^2 \int_{S^{N-1}} \left[\left(\int_{R_1}^{R_2} F(r\omega)r^{N-1}dr \right) \left(\int_{R_1}^{R_2} G(r\omega)r^{N-1}dr \right) \right] d\omega \right|$$

$$\leq \left\| \frac{\partial F}{\partial r} \right\|_{\infty} \left\| \frac{\partial G}{\partial r} \right\|_{\infty} \frac{NI}{(R_2^N - R_1^N)^3(N+1)^2}. \quad (27.39)$$

But it holds

$$\int_{S^{N-1}} \left(\int_{R_1}^{R_2} F(r\omega)G(r\omega)r^{N-1}dr \right) d\omega = \int_A F(x)G(x)dx. \quad (27.40)$$

That is completing the proof. $\qquad \square$

Next we generalize Theorem 27.6.

Theorem 27.11. *Let* $F, G \in C^1(\overline{A})$. *Then*

$$\left| \int_A F(x)G(x)dx - \left(\frac{N}{R_2^N - R_1^N} \right) \int_{S^{N-1}} \left[\left(\int_{R_1}^{R_2} F(r\omega)r^{N-1}dr \right) \right. \right.$$

$$\left. \left. \times \left(\int_{R_1}^{R_2} G(r\omega)r^{N-1}dr \right) \right] d\omega \right| \quad (27.41)$$

$$\leq \frac{1}{2} \int_A \left[|G(x)| \left\| \frac{\partial F}{\partial r} \right\|_{\infty} + |F(x)| \left\| \frac{\partial G}{\partial r} \right\|_{\infty} \right] H(|x|)dx$$

$$\leq \frac{1}{2} \int_A \left[|G(x)| \, \|\nabla F\|_{\infty} + |F(x)| \, \|\nabla G\|_{\infty} \right] H(|x|)dx, \quad (27.42)$$

where H *is given by (27.20).*

Proof. Acting similarly as in the proof of Theorem 27.10 and by (27.26) we have

$$
\left| \left(\frac{N}{R_2^N - R_1^N} \right) \int_{R_1}^{R_2} F(r\omega)G(r\omega)r^{N-1}dr \right.
$$
$$
\left. - \left(\frac{N}{R_2^N - R_1^N} \right)^2 \left(\int_{R_1}^{R_2} F(r\omega)r^{N-1}dr \right) \left(\int_{R_1}^{R_2} G(r\omega)r^{N-1}dr \right) \right|
$$
$$
\leq \frac{1}{2} \left(\frac{N}{R_2^N - R_1^N} \right) \left[\int_{R_1}^{R_2} r^{N-1} \left[|G(r\omega)| \left\| \frac{\partial F}{\partial r} \right\|_\infty + |F(r\omega)| \left\| \frac{\partial G}{\partial r} \right\|_\infty \right] H(r)dr \right].
$$
$$(27.43)$$

Therefore we derive

$$
\left| \frac{N}{\omega_N(R_2^N - R_1^N)} \int_{S^{N-1}} \left(\int_{R_1}^{R_2} F(r\omega)G(r\omega)r^{N-1}dr \right) d\omega \right.
$$
$$
\left. - \frac{1}{\omega_N} \left(\frac{N}{R_2^N - R_1^N} \right)^2 \int_{S^{N-1}} \left[\left(\int_{R_1}^{R_2} F(r\omega)r^{N-1}dr \right) \left(\int_{R_1}^{R_2} G(r\omega)r^{N-1}dr \right) \right] d\omega \right|
$$
$$
\leq \frac{1}{2\omega_N} \left(\frac{N}{R_2^N - R_1^N} \right) \int_{S^{N-1}} \left[\int_{R_1}^{R_2} r^{N-1} \left[|G(r\omega)| \left\| \frac{\partial F}{\partial r} \right\|_\infty \right. \right.
$$
$$
\left. \left. + |F(\omega)| \left\| \frac{\partial G}{\partial r} \right\|_\infty \right] H(r)dr \right] d\omega,
$$
$$(27.44)$$

which proves the claim. $\qquad\square$

Next we give general results over the ball.

Theorem 27.12. *Let* $F, G \in C^1(\overline{B(0,R)})$. *Then*

$$
\frac{1}{\mathrm{Vol}(B(0,R))} \left| \int_{B(0,R)} F(x)G(x)dx \right.
$$
$$
\left. - \frac{N}{R^N} \int_{S^{N-1}} \left[\left(\int_0^R F(r\omega)r^{N-1}dr \right) \left(\int_0^R G(r\omega)r^{N-1}dr \right) \right] d\omega \right| \qquad (27.45)
$$
$$
\leq \left\| \frac{\partial F}{\partial r} \right\|_\infty \left\| \frac{\partial G}{\partial r} \right\|_\infty \frac{NR^2}{(N+1)^2} \left[\frac{4}{(3N+2)} + \frac{(N+1)^2}{(N+2)} + \frac{4N}{(2N+1)} - (N+2) \right]
$$
$$
\leq \|\nabla F\|_\infty \|\nabla G\|_\infty \frac{NR^2}{(N+1)^2} \left[\frac{4}{(3N+2)} + \frac{(N+1)^2}{(N+2)} + \frac{4N}{(2N+1)} - (N+2) \right].
$$
$$(27.46)$$

Proof. Here put $R := R_2$. The functions $F(r\omega)$, $G(r\omega)$ are considered radial in $r \in [0, R]$, $\forall \omega \in S^{N-1}$. Also there exist

$$
\frac{\partial F(r\omega)}{\partial r}, \ \frac{\partial G(r\omega)}{\partial r} \in C((0, R]), \quad \forall \omega \in S^{N-1}.
$$

Then as in (27.37) and (27.38) we obtain

$$\left| \left(\frac{N}{R^N - R_1^N} \right) \int_{R_1}^{R} F(r\omega)G(r\omega) r^{N-1} dr \right.$$

$$\left. - \left(\frac{N}{R^N - R_1^N} \right)^2 \left(\int_{R_1}^{R} F(r\omega) r^{N-1} dr \right) \left(\int_{R_1}^{R} G(r\omega) r^{N-1} dr \right) \right| \qquad (27.47)$$

$$\leq \left\| \frac{\partial F(r\omega)}{\partial r} \right\|_{\infty,[0,R]} \left\| \frac{\partial G(r\omega)}{\partial r} \right\|_{\infty,[0,R]} \frac{NI}{(R^N - R_1^N)^3 (N+1)^2}$$

$$\leq \left\| \frac{\partial F}{\partial r} \right\|_{\infty} \left\| \frac{\partial G}{\partial r} \right\|_{\infty} \frac{NI}{(R^N - R_1^N)^3 (N+1)^2}. \qquad (27.48)$$

Taking the limit as $R_1 \downarrow 0$ to both sides of (27.14), (27.48) we find

$$\left| \frac{N}{R^N} \int_0^R F(r\omega)G(r\omega) r^{N-1} dr - \left(\frac{N}{R^N} \right)^2 \left(\int_0^R F(r\omega) r^{N-1} dr \right) \left(\int_0^R G(r\omega) r^{N-1} dr \right) \right|$$

$$\leq \left\| \frac{\partial F}{\partial r} \right\|_{\infty} \left\| \frac{\partial G}{\partial r} \right\|_{\infty} \frac{NR^2}{(N+1)^2} \left[\frac{4}{(3N+2)} + \frac{(N+1)^2}{(N+2)} + \frac{4N}{(2N+1)} - (N+2) \right]. \qquad (27.49)$$

Consequently one obtains that

$$\left| \frac{N}{\omega_N R^N} \int_{S^{N-1}} \left(\int_0^R F(r\omega)G(r\omega) r^{N-1} dr \right) d\omega \right.$$

$$\left. - \frac{1}{\omega_N} \left(\frac{N}{R^N} \right)^2 \int_{S^{N-1}} \left[\left(\int_0^R F(r\omega) r^{N-1} dr \right) \left(\int_0^R G(r\omega) r^{N-1} dr \right) \right] d\omega \right|$$

$$\leq \left\| \frac{\partial F}{\partial r} \right\|_{\infty} \left\| \frac{\partial G}{\partial r} \right\|_{\infty} \frac{NR^2}{(N+1)^2} \left[\frac{4}{(3N+2)} + \frac{(N+1)^2}{(N+2)} + \frac{4N}{(2N+1)} - (N+2) \right]. \qquad (27.50)$$

But it holds

$$\int_{S^{N-1}} \left(\int_0^R F(r\omega)G(r\omega) r^{N-1} dr \right) d\omega = \int_{B(0,R)} F(x)G(x) dx. \qquad (27.51)$$

That completes the proof. $\qquad \square$

We end chapter with

Theorem 27.13. *Let $F, G \in C^1(\overline{B(0,R)})$. Then*

$$\left| \int_{B(0,R)} F(x)G(x) dx - \frac{N}{R^N} \int_{S^{N-1}} \left[\left(\int_0^R F(r\omega) r^{N-1} dr \right) \left(\int_0^R G(r\omega) r^{N-1} dr \right) \right] d\omega \right|$$

$$\leq \frac{1}{2} \int_{B(0,R)} \left[|G(x)| \left\| \frac{\partial F}{\partial r} \right\|_{\infty} + |F(x)| \left\| \frac{\partial G}{\partial r} \right\|_{\infty} \right] H^*(|x|) dx \qquad (27.52)$$

$$\leq \frac{1}{2} \int_{B(0,R)} [|G(x)| \|\nabla F\|_{\infty} + |F(x)| \|\nabla G\|_{\infty}] H^*(|x|) dx, \qquad (27.53)$$

where H^* is given by (27.34).

Proof. Here put $R := R_2$. The functions $F(r\omega)$, $G(r\omega)$ are considered radial in $r \in [0, R]$, $\forall \omega \in S^{N-1}$. Also there exist

$$\frac{\partial F(r\omega)}{\partial r}, \ \frac{\partial G(r\omega)}{\partial r} \in C((0, R]), \quad \forall \omega \in S^{N-1}.$$

Then as in (27.43) we derive

$$\left| \left(\frac{N}{R^N - R_1^N} \right) \int_{R_1}^R F(r\omega)G(r\omega)r^{N-1}dr \right.$$

$$\left. - \left(\frac{N}{R^N - R_1^N} \right)^2 \left(\int_{R_1}^R F(r\omega)r^{N-1}dr \right) \left(\int_{R_1}^R G(r\omega)r^{N-1}dr \right) \right|$$

$$\leq \frac{1}{2} \left(\frac{N}{R^N - R_1^N} \right) \left[\int_{R_1}^R r^{N-1} \left[|G(r\omega)| \left\| \frac{\partial F}{\partial r} \right\|_{\infty, \overline{B(0,R)}} \right. \right.$$

$$\left. \left. + |F(r\omega)| \left\| \frac{\partial G}{\partial r} \right\|_{\infty, \overline{B(0,R)}} \right] H(r)dr. \right. \tag{27.54}$$

Taking the limit as $R_1 \downarrow 0$ to both sides of (27.54) and after simplification, we find

$$\left| \int_0^R F(r\omega)G(r\omega)r^{N-1}dr - \left(\frac{N}{R^N} \right) \left(\int_0^R F(r\omega)r^{N-1}dr \right) \left(\int_0^R G(r\omega)r^{N-1}dr \right) \right|$$

$$\leq \frac{1}{2} \left[\int_0^R r^{N-1} \left[|G(r\omega)| \left\| \frac{\partial F}{\partial r} \right\|_\infty + |F(r\omega)| \left\| \frac{\partial G}{\partial r} \right\|_\infty \right] H^*(r)dr, \right. \tag{27.55}$$

where H^* is given by (27.34). Consequently one obtains

$$\left| \int_{S^{N-1}} \left(\int_0^R F(r\omega)G(r\omega)r^{N-1}dr \right) d\omega \right.$$

$$\left. - \left(\frac{N}{R^N} \right) \int_{S^{N-1}} \left[\left(\int_0^R F(r\omega)r^{N-1}dr \right) \left(\int_0^R G(r\omega)r^{N-1}dr \right) \right] d\omega \right|$$

$$\leq \frac{1}{2} \int_{S^{N-1}} \left[\int_0^R r^{N-1} \left[|G(r\omega)| \left\| \frac{\partial F}{\partial r} \right\|_\infty + |F(r\omega)| \left\| \frac{\partial G}{\partial r} \right\|_\infty \right] H^*(r)dr \right] d\omega, \tag{27.56}$$

which proves the claim. $\qquad\square$

Chapter 28

Multivariate Chebyshev–Grüss and Comparison of Integral Means Inequalities Using a Multivariate Euler Type Identity

In this chapter we present tight multivariate Chebyshev–Grüss and Comparison of Integral means inequalities by using a general multivariate Euler type identity. The results involve functions f, g and their high order single partials and are with respect to $\| \cdot \|_p$, $1 \le p \le \infty$. This chapter relies on [34].

28.1 Background

Here we mention the following motivating results.

Theorem 28.1. (Chebyshev, 1882, [88]) *Let* $f, g : [a, b] \to \mathbb{R}$ *absolutely continuous functions. If* $f', g' \in L_\infty([a, b])$, *then*

$$\left| \frac{1}{b-a} \int_a^b f(x)g(x)dx - \left(\frac{1}{b-a} \int_a^b f(x)dx \right) \left(\frac{1}{b-a} \int_a^b g(x)dx \right) \right|$$

$$\le \frac{1}{12}(b-a)^2 \|f'\|_\infty \|g'\|_\infty. \tag{28.1}$$

Also we mention

Theorem 28.2. (Grüss, 1935, [150]) *Let* f, g *integrable functions from* $[a, b]$ *into* $\mathbb{R}$, *such that* $m \le f(x) \le M$, $\rho \le g(x) \le \sigma$, *for all* $x \in [a, b]$, *where* $m, M, \rho, \sigma \in \mathbb{R}$. *Then*

$$\left| \frac{1}{b-a} \int_a^b f(x)g(x)dx - \left(\frac{1}{b-a} \int_a^b f(x)dx \right) \left(\frac{1}{b-a} \int_a^b g(x)dx \right) \right|$$

$$\le \frac{1}{4}(M - m)(\sigma - \rho). \tag{28.2}$$

Here $B_k(x)$, $k \ge 0$, are the Bernoulli polynomials, $B_k = B_k(0)$, $k \ge 0$, the Bernoulli numbers, and $B_k^*(x)$, $k \ge 0$, are periodic functions of period one, related to the Bernoulli polynomials as

$$B_k^*(x) - B_k(x), \quad 0 \le x < 1, \tag{28.3}$$

341

and

$$B_k^*(x+1) = B_k^*(x), \quad x \in \mathbb{R}. \tag{28.4}$$

Some basic properties of Bernoulli polynomials follow (see [1, 23.1]). We have

$$B_0(x) = 1, \quad B_1(x) = x - \frac{1}{2}, \quad B_2(x) = x^2 - x + \frac{1}{6}, \tag{28.5}$$

and

$$B_k'(x) = kB_{k-1}(x), \quad k \in \mathbb{N}, \tag{28.6}$$

$$B_k(x+1) - B_k(x) = kx^{k-1}, \quad k \geq 0. \tag{28.7}$$

Clearly $B_0^* = 1$, B_1^* is a discontinuous function with a jump of -1 at each integer, and B_k^*, $k \geq 2$, is a continuous function. Notice that $B_k(0) = B_k(1) = B_k$, $k \geq 2$.

We make

Assumption 28.3. Let f and the existing $\dfrac{\partial^l f}{\partial x_j^l}$, all $l = 1, \ldots, m$; $j = 1, \ldots, n$, be continuous real valued functions on $\prod_{i=1}^{n}[a_i, b_i]$; $m, n \in \mathbb{N}$, $a_i, b_i \in \mathbb{R}$.

In [29] we proved a general Multivariate Euler-type identity, see next.

Theorem 28.4. *All as in Assumption 28.3 for $m, n \in \mathbb{N}$, $x_i \in [a_i, b_i]$, $i = 1, 2, \ldots, n$. Then*

$$f(x_1, x_2, \ldots, x_n) - \frac{1}{\prod\limits_{i=1}^{n}(b_i - a_i)} \int_{\prod\limits_{i=1}^{n}[a_i, b_i]} f(s_1, \ldots, s_n) ds_1 \ldots ds_n$$

$$- \sum_{j=1}^{n} A_j^f = \sum_{j=1}^{n} B_j^f, \tag{28.8}$$

where for $j = 1, \ldots, n$ we have

$$A_j^f := A_j^f(x_j, x_{j+1}, \ldots, x_n) = \frac{1}{\left(\prod\limits_{i=1}^{j-1}(b_i - a_i)\right)} \left\{ \sum_{k=1}^{m-1} \frac{(b_j - a_j)^{k-1}}{k!} B_k\left(\frac{x_j - a_j}{b_j - a_j}\right) \right.$$

$$\times \left(\int_{\prod\limits_{i=1}^{j-1}[a_i, b_i]} \left(\frac{\partial^{k-1} f}{\partial x_j^{k-1}}(s_1, s_2, \ldots, s_{j-1}, b_j, x_{j+1}, \ldots, x_n) \right. \right.$$

$$\left. \left. \left. - \frac{\partial^{k-1} f}{\partial x_j^{k-1}}(s_1, s_2, \ldots, s_{j-1}, a_j, x_{j+1}, \ldots, x_n) \right) ds_1 \ldots ds_{j-1} \right) \right\} \tag{28.9}$$

and

$$B_j^f := B_j^f(x_j, x_{j+1}, \ldots, x_n) := \frac{(b_j - a_j)^{m-1}}{m! \left(\prod_{i=1}^{j-1} (b_i - a_i) \right)} \left\{ \int_{\prod_{i=1}^{j} [a_i, b_i]} \left(\left(B_m \left(\frac{x_j - a_j}{b_j - a_j} \right) \right. \right. \right.$$

$$\left. \left. \left. - B_m^* \left(\frac{x_j - s_j}{b_j - a_j} \right) \right) \frac{\partial^m f}{\partial x_j^m} (s_1, s_2, \ldots, s_j, x_{j+1}, \ldots, x_n) \right) ds_1 ds_2 \ldots ds_j \right\} \quad (28.10)$$

When $m = 1$ then $A_j^f = 0$, $j = 1, \ldots, n$.

A general set of suppositions follow.

Assumption 28.5. Here $m \in \mathbb{N}$, $j = 1, \ldots, n$. We assume

1) $f : \prod_{i=1}^{n} [a_i, b_i] \to \mathbb{R}$ is continuous.

2) $\dfrac{\partial^l f}{\partial x_j^l}$ are existing real valued functions for all $j = 1, \ldots, n$; $l = 1, \ldots, m - 2$.

3) For each $j = 1, \ldots, n$ we suppose that

$$\frac{\partial^{m-1} f}{\partial x_j^{m-1}} (x_1, \ldots, x_{j-1}, \cdot, x_{j+1}, \ldots, x_n)$$

is a continuous real valued function.

4) For each $j = 1, \ldots, n$ we suppose that

$$g_j(\cdot) := \frac{\partial^m f}{\partial x_j^m} (x_1, \ldots, x_{j-1}, \cdot, x_{j+1}, \ldots, x_n)$$

exists and is real valued with the possibility of being infinite only over an at most countable subset of (a_j, b_j).

5) Parts #3, #4 are true for all

$$(x_1, \ldots, x_{j-1}, x_{j+1}, \ldots, x_n) \in \prod_{\substack{i=1 \\ i \neq j}}^{n} [a_i, b_i].$$

6) The functions for $j = 2, \ldots, n$; $l = 1, \ldots, m - 2$,

$$q_j(\overbrace{\cdot, \cdot, \ldots, \cdot}^{j-1}) := \frac{\partial^l f}{\partial x_j^l} (\overbrace{\cdot, \cdot, \ldots, \cdot}^{j-1}, x_j, x_{j+1}, \ldots, x_n)$$

are continuous on $\prod_{i=1}^{j-1} [a_i, b_i]$, for each $(x_j, x_{j+1}, \ldots, x_n) \in \prod_{i=j}^{n} [a_i, b_i]$.

7) The functions for each $j = 1, \ldots, n$,

$$\varphi_j(\overbrace{\cdot, \cdot, \ldots, \cdot}^{j}) := \frac{\partial^m f}{\partial x_j^m} (\overbrace{\cdot, \cdot, \ldots, \cdot}^{j}, x_{j+1}, \ldots, x_n) \in L_1 \left(\prod_{i=1}^{j} [a_i, b_i] \right),$$

for any $(x_{j+1}, \ldots, x_n) \in \prod_{i=j+1}^{n} [a_i, b_i]$.

We give (see [29])

Theorem 28.6. *All as in Assumption 28.5. Then (28.8) is valid again.*

Some weaker general suppositions follow.

Assumption 28.7. Here $m \in \mathbb{N}$, $j = 1, \ldots, n$ and only the Parts #1, #2, #6, #7 of Assumption 28.5 remain the same. We further assume that for each $j = 1, \ldots, n$ and over $[a_j, b_j]$, the function

$$\frac{\partial^{m-1} f}{\partial x_j^{m-1}}(x_1, \ldots, x_{j-1}, \cdot, x_{j+1}, \ldots, x_n)$$

is absolutely continuous, and this is true for all

$$(x_1, \ldots, x_{j-1}, x_{j+1}, \ldots, x_n) \in \prod_{\substack{i=1 \\ i \neq j}}^{n} [a_i, b_i].$$

We give (see [29])

Theorem 28.8. *All as in Assumption 28.7. Then (28.8) is valid again.*

In this chapter based on Theorems 28.4, 28.6, 28.8 we present Chebyshev–Grüss type and comparison of integral means multivariate inequalities involving f, g and their high order single partials with respect to $\| \cdot \|_p$, $1 \leq p \leq \infty$.

28.2 Main Results

We present the first main result regarding Chebyshev–Grüss type multivariate inequalities, see also [27], [201].

Theorem 28.9. *Let f, g as in Assumptions 28.3 or 28.5 or 28.7.*
Here A_j^f, A_j^g as in (28.9), $j = 1, \ldots, n$. Denote by

$$\Delta_{(f,g)} := \int_{\prod_{i=1}^{n} [a_i, b_i]} f(\overrightarrow{r}) g(\overrightarrow{x}) d\overrightarrow{x}$$

$$- \frac{1}{\prod_{i=1}^{n}(b_i - a_i)} \left(\int_{\prod_{i=1}^{n} [a_i, b_i]} f(\overrightarrow{x}) d\overrightarrow{x} \right) \left(\int_{\prod_{i=1}^{n} [a_i, b_i]} g(\overrightarrow{x}) d\overrightarrow{x} \right)$$

$$- \frac{1}{2} \left[\int_{\prod_{i=1}^{n} [a_i, b_i]} \left[f(\overrightarrow{x}) \left(\sum_{j=1}^{n} A_j^g(x_j, \ldots, x_n) \right) \right. \right.$$

$$\left. \left. + g(\overrightarrow{x}) \left(\sum_{j=1}^{n} A_j^f(x_j, \ldots, x_n) \right) \right] d\overrightarrow{x} \right], \tag{28.11}$$

$$\overrightarrow{x} := (x_1, \ldots, x_n) \in \prod_{i=1}^{n} [a_i, b_i].$$

Then we have: 1) By assuming that $\dfrac{\partial^m g}{\partial x_j^m}, \dfrac{\partial^m f}{\partial x_j^m} \in L_\infty \left(\prod_{i=1}^{n} [a_i, b_i] \right)$*, for* $j = 1, \ldots, n$*, it holds*

$$|\Delta_{(f,g)}| \leq \frac{\prod_{i=1}^{n}(b_i - a_i)}{2m!} \left\{ \sum_{j=1}^{n} (b_j - a_j)^{m-1} \right.$$

$$\left(\int_{a_j}^{b_j} \left(\sqrt{\frac{(m!)^2}{(2m)!}|B_{2m}| + B_m^2 \left(\frac{x_j - a_j}{b_j - a_j} \right)} \right) dx_j \right)$$

$$\left. \left[\|f\|_\infty \left\| \frac{\partial^m g}{\partial x_j^m} \right\|_\infty + \|g\|_\infty \left\| \frac{\partial^m f}{\partial x_j^m} \right\|_\infty \right] \right\}. \tag{28.12}$$

2)Let $p_j, q_j > 1$: $\dfrac{1}{p_j} + \dfrac{1}{q_j} = 1$; $j = 1, \ldots, n$. *We further suppose that*

$$\frac{\partial^m g}{\partial x_j^m}, \frac{\partial^m f}{\partial x_j^m} \in L_{q_j} \left(\prod_{i=1}^{n} [a_i, b_i] \right).$$

It holds

$$|\Delta_{(f,g)}| \leq \frac{1}{2m!} \sum_{j=1}^{n} \left\{ \left\{ (b_j - a_j)^{m - \frac{1}{q_j}} \right. \right.$$

$$\left(\int_{a_j}^{b_j} \left(\int_0^1 \left| B_m \left(\frac{x_j - a_j}{b_j - a_j} \right) - B_m(t_j) \right|^{p_j} dt_j \right)^{q_j/p_j} dx_j \right)^{1/q_j} \right\}$$

$$\left. \left[\|f\|_{p_j} \left\| \frac{\partial^m g}{\partial x_j^m} \right\|_{q_j} + \|g\|_{p_j} \left\| \frac{\partial^m f}{\partial x_j^m} \right\|_{q_j} \right] \right\}. \tag{28.13}$$

When $p_j = q_j = 2$, *all* $j = 1, \ldots, n$, *it holds*

$$|\Delta_{(f,g)}| \leq \frac{1}{2m!} \sum_{j=1}^{n} \left\{ \left\{ (b_j - a_j)^{m - \frac{1}{2}} \right. \right.$$

$$\left(\int_{a_j}^{b_j} \left(\frac{(m!)^2}{(2m)!}|B_{2m}| + B_m^2 \left(\frac{x_j - a_j}{b_j - a_j} \right) \right) dx_j \right)^{1/2} \right\}$$

$$\left. \left[\|f\|_2 \left\| \frac{\partial^m g}{\partial x_j^m} \right\|_2 + \|g\|_2 \left\| \frac{\partial^m f}{\partial x_j^m} \right\|_2 \right] \right\}. \tag{28.14}$$

3) By assuming again that

$$\frac{\partial^m g}{\partial x_j^m}, \frac{\partial^m f}{\partial x_j^m} \in L_\infty \left(\prod_{i=1}^{n} [a_i, b_i] \right), \quad j = 1, \ldots, n,$$

it holds:

    i) Case of $m = 2r$, $r \in \mathbb{N}$, then

$$|\Delta_{(f,g)}| \leq \frac{1}{2(2r)!} \sum_{j=1}^{n} \left\{ (b_j - a_j)^{2r} \right.$$

$$\left[(1 - 2^{-2r})|B_{2r}| + \left\| 2^{-2r} B_{2r} - B_{2r}\left(\frac{x_j - a_j}{b_j - a_j} \right) \right\|_{\infty,[a_j,b_j]} \right]$$

$$\left. \left[\|f\|_1 \left\| \frac{\partial^{2r} g}{\partial x_j^{2r}} \right\|_\infty + \|g\|_1 \left\| \frac{\partial^{2r} f}{\partial x_j^{2r}} \right\|_\infty \right] \right\} \tag{28.15}$$

    ii) Case of $m = 2r + 1$, $r \in \mathbb{N}$, then

$$|\Delta_{(f,g)}| \leq \frac{1}{2(2r+1)!} \sum_{j=1}^{n} \left\{ (b_j - a_j)^{2r+1} \right.$$

$$\left[\frac{2(2r+1)!}{(2\pi)^{2r+1}(1 - 2^{-2r})} + \left\| B_{2r+1}\left(\frac{x_j - a_j}{b_j - a_j} \right) \right\|_{\infty,[a_j,b_j]} \right]$$

$$\left. \left[\|f\|_1 \left\| \frac{\partial^{2r+1} g}{\partial x_j^{2r+1}} \right\|_\infty + \|g\|_1 \left\| \frac{\partial^{2r+1} f}{\partial x_j^{2r+1}} \right\|_\infty \right] \right\}. \tag{28.16}$$

    iii) Case of $m = 1$, then

$$|\Delta_{(f,g)}| \leq \frac{1}{4} \sum_{j=1}^{n} \left\{ (b_j - a_j)(1 + b_j - a_j) \left[\|f\|_1 \left\| \frac{\partial g}{\partial x_j} \right\|_\infty + \|g\|_1 \left\| \frac{\partial f}{\partial x_j} \right\|_\infty \right] \right\}. \tag{28.17}$$

When $m = 1$ then $A_j^f = A_j^g = 0$, $j = 1, \ldots, n$.

Proof.   Since f, g are as in Assumptions 28.3 or 28.5 or 28.7 by (28.8) we obtain

$$f(x_1, x_2, \ldots, x_n) - \frac{1}{\prod\limits_{i=1}^{n}(b_i - a_i)} \int_{\prod\limits_{i=1}^{n} [a_i,b_i]} f(s_1, \ldots, s_n) ds_1 \ldots ds_n$$

$$- \sum_{j=1}^{n} A_j^f(x_j, x_{j+1}, \ldots, x_n) = \sum_{j=1}^{n} B_j^f(x_j, \ldots, x_n), \tag{28.18}$$

and

$$g(x_1, \ldots, x_n) - \frac{1}{\prod\limits_{i=1}^{n}(b_i - a_i)} \int_{\prod\limits_{i=1}^{n} [a_i,b_i]} g(s_1, \ldots, s_n) ds_1 \ldots ds_n$$

$$- \sum_{j=1}^{n} A_j^g(x_j, x_{j+1}, \ldots, x_n) = \sum_{j=1}^{n} B_j^g(x_j, \ldots, x_n). \tag{28.19}$$

Consequently it holds

$$f(x_1,\ldots,x_n)g(x_1,\ldots,x_n)$$

$$-\frac{1}{\prod\limits_{i=1}^{n}(b_i-a_i)}g(x_1,\ldots,x_n)\int_{\prod\limits_{i=1}^{n}[a_i,b_i]}f(s_1,\ldots,s_n)ds_1\ldots ds_n$$

$$-g(x_1,\ldots,x_n)\left(\sum_{j=1}^{n}A_j^f(x_j,\ldots,x_n)\right)$$

$$=g(x_1,\ldots,x_n)\left(\sum_{j=1}^{n}B_j^f(x_j,\ldots,x_n)\right),\tag{28.20}$$

and

$$f(x_1,\ldots,x_n)g(x_1,\ldots,x_n)$$

$$-\frac{1}{\prod\limits_{i=1}^{n}(b_i-a_i)}f(x_1,\ldots,x_n)\int_{\prod\limits_{i=1}^{n}[a_i,b_i]}g(s_1,\ldots,s_n)ds_1\ldots ds_n$$

$$-f(x_1,\ldots,x_n)\left(\sum_{j=1}^{n}A_j^g(x_j,\ldots,x_n)\right)$$

$$=f(x_1,\ldots,x_n)\left(\sum_{j=1}^{n}B_j^g(x_j,\ldots,x_n)\right).\tag{28.21}$$

By integrating the last identities we derive

$$\int_{\prod\limits_{i=1}^{n}[a_i,b_i]}f(x_1,\ldots,x_n)g(x_1,\ldots,x_n)dx_1\ldots dx_n$$

$$-\frac{1}{\prod\limits_{i=1}^{n}(b_i-a_i)}\left(\int_{\prod\limits_{i=1}^{n}[a_i,b_i]}g(x_1,\ldots,x_n)dx_1\ldots dx_n\right)$$

$$\left(\int_{\prod\limits_{i=1}^{n}[a_i,b_i]}f(x_1,\ldots,x_n)dx_1\ldots dx_n\right)$$

$$-\int_{\prod\limits_{i=1}^{n}[a_i,b_i]}g(x_1,\ldots,x_n)\left(\sum_{j=1}^{n}A_j^f(x_j,\ldots,x_n)\right)dx_1\ldots dx_n$$

$$=\int_{\prod\limits_{i=1}^{n}[a_i,b_i]}g(x_1,\ldots,x_n)\left(\sum_{j=1}^{n}B_j^f(x_j,\ldots,x_n)\right)dx_1\ldots dx_n,\tag{28.22}$$

and

$$\int_{\prod_{i=1}^{n}[a_i,b_i]} f(x_1,\ldots,x_n)g(x_1,\ldots,x_n)dx_1\ldots dx_n$$

$$-\frac{1}{\prod_{i=1}^{n}(b_i-a_i)}\left(\int_{\prod_{i=1}^{n}[a_i,b_i]} f(x_1,\ldots,x_n)dx_1\ldots dx_n\right)$$

$$\left(\int_{\prod_{i=1}^{n}[a_i,b_i]} g(x_1,\ldots,x_n)dx_1\ldots dx_n\right)$$

$$-\int_{\prod_{i=1}^{n}[a_i,b_i]} f(x_1,\ldots,x_n)\left(\sum_{j=1}^{n}A_j^g(x_j,\ldots,x_n)\right)dx_1\ldots dx_n$$

$$=\int_{\prod_{i=1}^{n}[a_i,b_i]} f(x_1,\ldots,x_n)\left(\sum_{j=1}^{n}B_j^g(x_j,\ldots,x_n)\right)dx_1\ldots dx_n. \tag{28.23}$$

Adding and dividing by 2 the latter two identities we get ($\overrightarrow{x} := (x_1,\ldots,x_n)$)

$$\Delta_{(f,g)} := \int_{\prod_{i=1}^{n}[a_i,b_i]} f(\overrightarrow{x})g(\overrightarrow{x})d\overrightarrow{x}$$

$$-\frac{1}{\prod_{i=1}^{n}(b_i-a_i)}\left(\int_{\prod_{i=1}^{n}[a_i,b_i]} f(\overrightarrow{x})d\overrightarrow{x}\right)\left(\int_{\prod_{i=1}^{n}[a_i,b_i]} g(\overrightarrow{x})d\overrightarrow{x}\right)$$

$$-\frac{1}{2}\left[\int_{\prod_{i=1}^{n}[a_i,b_i]}\left[f(\overrightarrow{x})\left(\sum_{j=1}^{n}A_j^g(x_j,\ldots,x_n)\right)\right.\right.$$

$$\left.\left.+g(\overrightarrow{x})\left(\sum_{j=1}^{n}A_j^f(x_j,\ldots,x_n)\right)\right]d\overrightarrow{x}\right.$$

$$=\frac{1}{2}\left[\int_{\prod_{i=1}^{n}[a_i,b_i]}\left[f(\overrightarrow{x})\left(\sum_{j=1}^{n}B_j^g(x_j,\ldots,x_n)\right)\right.\right.$$

$$\left.\left.+g(\overrightarrow{x})\left(\sum_{j=1}^{n}B_j^f(x_j,\ldots,x_n)\right)\right]d\overrightarrow{x}\right] := \Gamma. \tag{28.24}$$

That is

$$\Delta_{(f,g)} = \Gamma. \tag{28.25}$$

Next we estimate Γ.

1) Estimate with respect to $\|\cdot\|_\infty$.

We have

$$|\Gamma| \le \frac{1}{2} \sum_{j=1}^{n} \left[\|f\|_\infty \int_{\prod_{i=1}^{n}[a_i,b_i]} |B_j^g(x_j,\ldots,x_n)| d\overrightarrow{x} \right.$$

$$+ \|g\|_\infty \int_{\prod_{i=1}^{n}[a_i,b_i]} |B_j^f(x_j,\ldots,x_n)| d\overrightarrow{x} \Bigg]$$

$$= \frac{1}{2} \left\{ \sum_{j=1}^{n} \left(\prod_{i=1}^{j-1}(b_i - a_i) \right) \left[\|f\|_\infty \int_{\prod_{i=j}^{n}[a_i,b_i]} |B_j^g(x_j,\ldots,x_n)| dx_j \ldots dx_n \right. \right.$$

$$+ \|g\|_\infty \int_{\prod_{i=j}^{n}[a_i,b_i]} |B_j^f(x_j,\ldots,x_n)| dx_j \ldots dx_n \Bigg] \Bigg\} =: (*) \tag{28.26}$$

By Lemma 1 of [29], see there (58), we find

$$|B_j^f(x_j,\ldots,x_n)| \le \frac{(b_j - a_j)^m}{m!} \left(\sqrt{\frac{(m!)^2}{(2m)!}|B_{2m}| + B_m^2\left(\frac{x_j - a_j}{b_j - a_j} \right)} \right)$$

$$\times \left\| \frac{\partial^m f}{\partial x_j^m}(\overset{j}{\overbrace{\ldots}}, x_{j+1},\ldots,x_n) \right\|_{\infty, \prod_{i=1}^{j}[a_i,b_i]}, \tag{28.27}$$

for all $j = 1,\ldots,n$. Similarly for g.

Using (28.27) into (28.26) we derive

$$|\Gamma| \le (*) \le \frac{1}{2m!} \left\{ \sum_{j=1}^{n} \left(\prod_{i=1}^{j-1}(b_i - a_i) \right)(b_j - a_j)^m \right.$$

$$\left(\int_{a_j}^{b_j} \left(\sqrt{\frac{(m!)^2}{(2m)!}|B_{2m}| + B_m^2\left(\frac{x_j - a_j}{b_j - a_j} \right)} \right) dx_j \right)$$

$$\left[\|f\|_\infty \left(\int_{\prod_{i=j+1}^{n}[a_i,b_i]} \left\| \frac{\partial^m g}{\partial x_j^m}(\overset{j}{\overbrace{\ldots}}, x_{j+1},\ldots,x_n) \right\|_{\infty, \prod_{i=1}^{j}[a_i,b_i]} dx_{j+1} \ldots dx_n \right) \right. \tag{28.28}$$

$$+ \|g\|_\infty \left(\int_{\prod_{i=j+1}^{n}[a_i,b_i]} \left\| \frac{\partial^m f}{\partial x_j^m}(\overset{j}{\overbrace{\ldots}}, x_{j+1},\ldots,x_n) \right\|_{\infty, \prod_{i=1}^{j}[a_i,b_i]} dx_{j+1} \ldots dx_n \right) \Bigg] \Bigg\}$$

$$\le \frac{\prod_{i=1}^{n}(b_i - a_i)}{2m!} \left\{ \sum_{j=1}^{n}(b_j - a_j)^{m-1} \right.$$

$$\left(\int_{a_j}^{b_j} \left(\sqrt{ \frac{(m!)^2}{(2m)!} |B_{2m}| + B_m^2 \left(\frac{x_j - a_j}{b_j - a_j} \right) } \right) dx_j \right)$$

$$\left[\|f\|_\infty \left\| \frac{\partial^m g}{\partial x_j^m} \right\|_\infty + \|g\|_\infty \left\| \frac{\partial^m f}{\partial x_j^m} \right\|_\infty \right] \right\}, \tag{28.29}$$

proving the claim (28.12).

2) Estimate with respect to $\| \cdot \|_p$, $1 < p < \infty$. Here $p_j, q_j > 1$: $\dfrac{1}{p_j} + \dfrac{1}{q_j} = 1$; $j = 1, \ldots, n$.

We have

$$|\Gamma| \le \frac{1}{2} \sum_{j=1}^{n} \left[\|f\|_{p_j} \|B_j^g\|_{q_j} + \|g\|_{p_j} \|B_j^f\|_{q_j} \right]. \tag{28.30}$$

The assumption implies

$$\frac{\partial^m g}{\partial x_j^m} (\ldots, x_{j+1}, \ldots, x_n) \in L_{q_j} \left(\prod_{i=1}^{j} [a_i, b_i] \right),$$

for almost all $(x_{j+1}, \ldots, x_n) \in \displaystyle\prod_{i=j+1}^{n} [a_i, b_i]$, $j = 1, \ldots, n$. Then as in Lemma 2 of [29], see (62) there, we obtain

$$|B_j^g| \le \frac{(b_j - a_j)^{m - \frac{1}{q_j}}}{m!} \left(\prod_{i=1}^{j-1} (b_i - a_i) \right)^{-1/q_j} \left(\int_0^1 \left| B_m \left(\frac{x_j - a_j}{b_j - a_j} \right) - B_m(t_j) \right|^{p_j} dt_j \right)^{1/p_j}$$

$$\left\| \frac{\partial^m g}{\partial x_j^m} (\ldots, x_{j+1}, \ldots, x_n) \right\|_{q_j, \prod_{i=1}^{j} [a_i, b_i]}, \tag{28.31}$$

true for almost all $(x_{j+1}, \ldots, x_n) \in \displaystyle\prod_{i=j+1}^{n} [a_i, b_i]$, $j = 1, \ldots, n$.

Thus we derive

$$\left(\int_{\prod_{i=1}^{n} [a_i, b_i]} |B_j^g|^{q_j} d\vec{x} \right)^{1/q_j} \le \left(\frac{(b_j - a_j)^{m - \frac{1}{q_j}}}{m!} \left(\prod_{i=1}^{j-1} (b_i - a_i) \right)^{-1/q_j} \right)$$

$$\left(\int_{\prod_{i=1}^{n} [a_i, b_i]} \left\{ \left(\int_0^1 \left| B_m \left(\frac{x_j - a_j}{b_j - a_j} \right) - B_m(t_j) \right|^{p_j} dt_j \right)^{\frac{q_j}{p_j}} \right. \right.$$

$$\left. \left. \left\| \frac{\partial^m g}{\partial x_j^m} (\ldots, x_{j+1}, \ldots, x_n) \right\|_{q_j, \prod_{i=1}^{j} [a_i, b_i]}^{q_j} \right\} d\vec{x} \right)^{1/q_j} \tag{28.32}$$

$$= \left(\frac{(b_j - a_j)^{m - \frac{1}{q_j}}}{m!} \left(\prod_{i=1}^{j-1} (b_i - a_i) \right)^{-\frac{1}{q_j}} \right) \left(\prod_{i=1}^{j-1} (b_i - a_i) \right)^{1/q_j}$$

$$\left(\int_{a_j}^{b_j} \left(\int_0^1 \left| B_m \left(\frac{x_j - a_j}{b_j - a_j} \right) - B_m(t_j) \right|^{p_j} dt_j \right)^{q_j/p_j} dx_j \right)^{1/q_j}$$

$$\left(\int_{\prod_{i=j+1}^{n} [a_i, b_i]} \left\| \frac{\partial^m g}{\partial x_j^m} (\dots, x_{j+1}, \dots, x_n) \right\|_{q_j, \prod_{i=1}^{j} [a_i, b_i]}^{q_j} dx_{j+1} \dots dx_n \right)^{1/q_j}. \qquad (28.33)$$

That is we find

$$\|B_j^g\|_{q_j} \le \frac{(b_j - a_j)^{m - \frac{1}{q_j}}}{m!}$$

$$\left(\int_{a_j}^{b_j} \left(\int_0^1 \left| B_m \left(\frac{x_j - a_j}{b_j - a_j} \right) - B_m(t_j) \right|^{p_j} dt_j \right)^{q_j/p_j} dx_j \right)^{1/q_j} \left\| \frac{\partial^m g}{\partial x_j^m} \right\|_{q_j}, \qquad (28.34)$$

$j = 1, \dots, n$. Similarly for f.

Using (28.34) into (28.30) we find

$$|\Gamma| \le \frac{1}{2m!} \sum_{j=1}^{n} \left\{ \left\{ (b_j - a_j)^{m - \frac{1}{q_j}} \right. \right.$$

$$\left(\int_{a_j}^{b_j} \left(\int_0^1 \left| B_m \left(\frac{x_j - a_j}{b_j - a_j} \right) - B_m(t_j) \right|^{p_j} dt_j \right)^{q_j/p_j} dx_j \right)^{1/q_j} \Bigg\}$$

$$\left[\|f\|_{p_j} \left\| \frac{\partial^m g}{\partial x_j^m} \right\|_{q_j} + \|g\|_{p_j} \left\| \frac{\partial^m f}{\partial x_j^m} \right\|_{q_j} \right] \Bigg\}, \qquad (28.35)$$

proving the claim (28.13).

When $p_j = q_j = 2$, all $j = 1, \dots, n$, then

$$|\Gamma| \le \frac{1}{2m!} \sum_{j=1}^{n} \left\{ \left\{ (b_j - a_j)^{m - \frac{1}{2}} \right. \right.$$

$$\left(\int_{a_j}^{b_j} \left(\int_0^1 \left(B_m \left(\frac{x_j - a_j}{b_j - a_j} \right) - B_m(t_j) \right)^2 dt_j \right) dx_j \right)^{1/2} \Bigg\}$$

$$\left[\|f\|_2 \left\| \frac{\partial^m g}{\partial x_j^m} \right\|_2 + \|g\|_2 \left\| \frac{\partial^m f}{\partial x_j^m} \right\|_2 \right] \Bigg\} \qquad (28.36)$$

$$\text{(by [98], p. 352)}$$

$$= \frac{1}{2m!} \sum_{j=1}^{n} \left\{ \left\{ (b_j - a_j)^{m-\frac{1}{2}} \left(\int_{a_j}^{b_j} \left(\frac{(m!)^2}{(2m)!} |B_{2m}| + B_m^2 \left(\frac{x_j - a_j}{b_j - a_j} \right) \right) dx_j \right)^{1/2} \right\} \right.$$

$$\left. \left[\|f\|_2 \left\| \frac{\partial^m g}{\partial x_j^m} \right\|_2 + \|g\|_2 \left\| \frac{\partial^m f}{\partial x_j^m} \right\|_2 \right] \right\} \tag{28.37}$$

proving the claim (28.14).

3) Estimate with respect to $\| \cdot \|_1$.

We see that

$$|\Gamma| \le \frac{1}{2} \sum_{j=1}^{n} \left[\int_{\prod_{i=1}^{n} [a_i,b_i]} |f(\overrightarrow{x})| |B_j^g(x_j, \ldots, x_n)| d\overrightarrow{x} \right.$$

$$\left. + \int_{\prod_{i=1}^{n} [a_i,b_i]} |g(\overrightarrow{x})| |B_j^f(x_j, \ldots, x_n)| d\overrightarrow{x} \right] \tag{28.38}$$

$$\le \frac{1}{2} \sum_{j=1}^{n} \left[\|f\|_1 \|B_j^g(x_j, \ldots, x_n)\|_{\infty, \prod_{i=j}^{n} [a_i,b_j]} + \|g\|_1 \|B_j^f(x_j, \ldots, x_n)\|_{\infty, \prod_{i=j}^{n} [a_i,b_i]} \right]. \tag{28.39}$$

By Lemma 3 of [29], (70) there, we have

$$|B_j^f(x_j, x_{j+1}, \ldots, x_n)| \le \frac{(b_j - a_j)^{m-1}}{m! \left(\prod_{i=1}^{j-1} (b_i - a_i) \right)}$$

$$\left\| \frac{\partial^m f}{\partial x_j^m} (\ldots, x_{j+1}, \ldots, x_n) \right\|_{1, \prod_{i=1}^{j} [a_i,b_i]} \left\| B_m(t) - B_m \left(\frac{x_j - a_j}{b_j - a_j} \right) \right\|_{\infty, [0,1]}. \tag{28.40}$$

The special cases follow:

1)′ When $m = 2r$, $r \in \mathbb{N}$ we have (see again Lemma 3 of [29], (71) there)

$$|B_j^f(x_j, \ldots, x_n)| \le \frac{(b_j - a_j)^{2r-1}}{(2r)! \left(\prod_{i=1}^{j-1} (b_i - a_i) \right)}$$

$$\left\| \frac{\partial^{2r} f}{\partial x_j^{2r}} (\ldots, x_{j+1}, \ldots, x_n) \right\|_{1, \prod_{i=1}^{j} [a_i,b_i]} \left[(1 - 2^{-2r}) |B_{2r}| + \left| 2^{-2r} B_{2r} - B_{2r} \left(\frac{x_j - a_j}{b_j - a_j} \right) \right| \right]. \tag{28.41}$$

Consequently, when $m = 2r$, $r \in \mathbb{N}$, we have

$$\|B_j^f(x_j, \ldots, x_n)\|_{\infty, \prod_{i=j}^{n} [a_i,b_i]} \le \frac{(b_j - a_j)^{2r}}{(2r)!} \left\| \frac{\partial^{2r} f}{\partial x_j^{2r}} \right\|_\infty$$

$$\left[(1 - 2^{-2r})|B_{2r}| + \left\|2^{-2r}B_{2r} - B_{2r}\left(\frac{x_j - a_j}{b_j - a_j}\right)\right\|_{\infty,[a_j,b_j]}\right]. \tag{28.42}$$

2)$'$ When $m = 2r + 1$, $r \in \mathbb{N}$ (see again Lemma 3 of [29], (72) there)

$$|B_j^f(x_j, \ldots, x_n)| \leq \frac{(b_j - a_j)^{2r}}{(2r + 1)!\left(\prod\limits_{i=1}^{j-1}(b_i - a_i)\right)}$$

$$\left\|\frac{\partial^{2r+1}}{\partial x_j^{2r+1}}f(\ldots, x_{j+1}, \ldots, x_n)\right\|_{1,\prod\limits_{i=1}^{j}[a_i,b_i]}\left[\frac{2(2r+1)!}{(2\pi)^{2r+1}(1 - 2^{-2r})} + \left|B_{2r+1}\left(\frac{x_j - a_j}{b_j - a_j}\right)\right|\right].$$

$$\tag{28.43}$$

Consequently, when $m = 2r + 1$, $r \in \mathbb{N}$ we have

$$\|B_j^f(x_j, \ldots, x_n)\|_{\infty,\prod\limits_{i=j}^{n}[a_i,b_i]} \leq \frac{(b_j - a_j)^{2r+1}}{(2r + 1)!}\left\|\frac{\partial^{2r+1}f}{\partial x_j^{2r+1}}\right\|_{\infty}$$

$$\left[\frac{2(2r+1)!}{(2\pi)^{2r+1}(1 - 2^{-2r})} + \left\|B_{2r+1}\left(\frac{x_j - a_j}{b_j - a_j}\right)\right\|_{\infty,[a_j,b_j]}\right]. \tag{28.44}$$

3)$'$ When $m = 1$, we get again from Lemma 3 of [29], (73) there, that

$$|B_j^f(x_j, \ldots, x_n)| \leq \frac{1}{\prod\limits_{i=1}^{j-1}(b_i - a_i)}\left\|\frac{\partial f}{\partial x_j}(\ldots, x_{j+1}, \ldots, x_n)\right\|_{1,\prod\limits_{i=1}^{j}[a_i,b_i]}$$

$$\left[\frac{1}{2} + \left|x_j - \left(\frac{a_j + b_j}{2}\right)\right|\right]. \tag{28.45}$$

Thus it holds

$$\|B_j^f(x_j, \ldots, x_n)\|_{\infty,\prod\limits_{i=j}^{n}[a_i,b_i]} \leq \frac{(b_j - a_j)}{2}\left\|\frac{\partial f}{\partial x_j}\right\|_{\infty}[1 + b_j - a_j]. \tag{28.46}$$

Similar conclusions follow for g.

Using (28.42), (28.44) and (28.46) into (28.38)–(28.39) we derive:

1)$''$ Case of $m = 2r$, $r \in \mathbb{N}$, then

$$|\Gamma| \leq \frac{1}{2(2r)!}\sum_{j=1}^{n}\left\{(b_j - a_j)^{2r}\right.$$

$$\left[(1 - 2^{-2r})|B_{2r}| + \left\|2^{-2r}B_{2r} - B_{2r}\left(\frac{x_j - a_j}{b_j - a_j}\right)\right\|_{\infty,[a_j,b_j]}\right]$$

$$\left[\|f\|_1 \left\|\frac{\partial^{2r} g}{\partial x_j^{2r}}\right\|_\infty + \|g\|_1 \left\|\frac{\partial^{2r} f}{\partial x_j^{2r}}\right\|_\infty\right]\right\}. \tag{28.47}$$

2)″ Case of $m = 2r + 1$, $r \in \mathbb{N}$, then

$$|\Gamma| \le \frac{1}{2(2r+1)!} \sum_{j=1}^{n} \left\{ (b_j - a_j)^{2r+1} \right.$$

$$\left[\frac{2(2r+1)!}{(2\pi)^{2r+1}(1 - 2^{-2r})} + \left\|B_{2r+1}\left(\frac{x_j - a_j}{b_j - a_j}\right)\right\|_{\infty,[a_j,b_j]}\right]$$

$$\left[\|f\|_1 \left\|\frac{\partial^{2r+1} g}{\partial x_j^{2r+1}}\right\|_\infty + \|g\|_1 \left\|\frac{\partial^{2r+1} f}{\partial x_j^{2r+1}}\right\|_\infty\right]\right\}. \tag{28.48}$$

3)″ Case of $m = 1$, then

$$|\Gamma| \le \frac{1}{4} \sum_{j=1}^{n} \left\{ (b_j - a_j)(1 + b_j - a_j) \left[\|f\|_1 \left\|\frac{\partial g}{\partial x_j}\right\|_\infty + \|g\|_1 \left\|\frac{\partial f}{\partial x_j}\right\|_\infty\right]\right\}. \tag{28.49}$$

That is proving all the claims. $\qquad\qquad\qquad\qquad\qquad\qquad\qquad\quad \square$

When $m = 1$ our basic assumptions are simplified to

Assumption 28.10. Let f and the existing $\dfrac{\partial f}{\partial x_j}$, $j = 1, \ldots, n$, be continuous real valued functions on $\prod_{i=1}^{n}[a_i, b_i]$.

Assumption 28.11. Here $j = 1, \ldots, n$. We assume

1) $f : \prod_{i=1}^{n}[a_i, b_i] \to \mathbb{R}$ is continuous.

2) For each $j = 1, \ldots, n$, we suppose that

$$g_j(\cdot) := \frac{\partial f}{\partial x_j}(x_1, \ldots, x_{j-1}, \cdot, x_{j+1}, \ldots, x_n)$$

exists and is real valued with the possibility of being infinite only over an at most countable subset of (a_j, b_j).

3) Part #2 is true for all

$$(x_1, \ldots, x_{j-1}, x_{j+1}, \ldots, x_n) \in \prod_{\substack{i=1 \\ i \ne j}}^{n}[a_i, b_i].$$

4) The functions for each $j = 1, \ldots, n$,

$$\varphi_j(\overbrace{\cdots}^{j}) := \frac{\partial f}{\partial x_j}(\overbrace{\cdot, \cdot, \cdot}^{j}, x_{j+1}, \ldots, x_n) \in L_1\left(\prod_{i=1}^{j}[a_i, b_i]\right),$$

for any $(x_{j+1}, \ldots, x_n) \in \prod_{i=j+1}^{n} [a_i, b_i]$.

Assumption 28.12. Parts #1, #4 of Assumption 28.11 remain the same. We further suppose that for each $j = 1, \ldots, n$ and over $[a_j, b_j]$, the function

$$f(x_1, \ldots, x_{j-1}, \cdot, x_{j+1}, \ldots, x_n)$$

is absolutely continuous, and this is true for all

$$(x_1, \ldots, x_{j-1}, x_{j+1}, \ldots, x_n) \in \prod_{\substack{i=1 \\ i \neq j}}^{n} [a_i, b_i].$$

We give

Corollary 28.13. *Let f, g as in Assumptions 28.10 or 28.11 or 28.12. Denote by*

$$K_{(f,g)} := \frac{1}{\prod_{i=1}^{n}(b_i - a_i)} \int_{\prod_{i=1}^{n}[a_i,b_i]} f(\overrightarrow{x})g(\overrightarrow{x})dx$$

$$- \left(\frac{1}{\prod_{i=1}^{n}(b_i - a_i)} \int_{\prod_{i=1}^{n}[a_i,b_i]} f(\overrightarrow{x})d\overrightarrow{x} \right) \left(\frac{1}{\prod_{i=1}^{n}(b_i - a_i)} \int_{\prod_{i=1}^{n}[a_i,b_i]} g(\overrightarrow{x})d\overrightarrow{x} \right). \quad (28.50)$$

Then we have:

1) By assuming that $\dfrac{\partial g}{\partial x_j}, \dfrac{\partial f}{\partial x_j} \in L_\infty \left(\prod_{i=1}^{n}[a_i, b_i] \right)$, for $j = 1, \ldots, n$, it holds

$$|K_{(f,g)}| \leq \frac{1}{48} \left\{ \sum_{j=1}^{n} (b_j - a_j)\{4\sqrt{3} - \ln[3a_j + 2\sqrt{3}(b_j - a_j) - 3b_j] \right.$$

$$+ \ln[-3a_j + 2\sqrt{3}(b_j - a_j) + 3b_j]\}$$

$$\left. \left[\|f\|_\infty \left\| \frac{\partial g}{\partial x_j} \right\|_\infty + \|g\|_\infty \left\| \frac{\partial f}{\partial x_j} \right\|_\infty \right] \right\}. \quad (28.51)$$

2) Let $p_j, q_j > 1$: $\dfrac{1}{p_j} + \dfrac{1}{q_j} = 1$; $j = 1, \ldots, n$. Suppose also that

$$\frac{\partial g}{\partial x_j}, \frac{\partial f}{\partial x_j} \in L_{q_j} \left(\prod_{i=1}^{n}[a_i, b_i] \right).$$

It holds

$$|K_{(f,g)}| \leq \frac{1}{2 \left(\prod_{i=1}^{n}(b_i - a_i) \right)} \sum_{j=1}^{n} \left\{ \left\{ (b_j - a_j)^{1-\frac{1}{q_j}} \right. \right.$$

$$\left(\int_{a_j}^{b_j} \left(\int_0^1 \left| \frac{x_j - a_j}{b_j - a_j} - t_j \right|^{p_j} dt_j \right)^{q_j/p_j} dx_j \right)^{1/q_j} \right\}$$

$$\left[\|f\|_{p_j} \left\| \frac{\partial g}{\partial x_j} \right\|_{q_j} + \|g\|_{p_j} \left\| \frac{\partial f}{\partial x_j} \right\|_{q_j} \right] \right\}. \tag{28.52}$$

When $p_j = q_j = 2$, all $j = 1, \ldots, n$, it holds

$$|K_{(f,g)}| \le \frac{1}{48 \left(\prod_{i=1}^n (b_i - a_i) \right)} \sum_{j=1}^n \left\{ (b_j - a_j)^{3/2} \right.$$

$$\{ 4\sqrt{3} - \ln[3a_j + 2\sqrt{3}(b_j - a_j) - 3b_j] + \ln[-3a_j + 2\sqrt{3}(b_j - a_j) + 3b_j] \}$$

$$\left[\|f\|_2 \left\| \frac{\partial g}{\partial x_j} \right\|_2 + \|g\|_2 \left\| \frac{\partial f}{\partial x_j} \right\|_2 \right] \right\}. \tag{28.53}$$

3) Assume again

$$\frac{\partial g}{\partial x_j}, \frac{\partial f}{\partial x_j} \in L_\infty \left(\prod_{i=1}^n [a_i, b_i] \right), \quad j = 1, \ldots, n.$$

It holds

$$|K_{(f,g)}| \le \frac{1}{4} \sum_{j=1}^n \left\{ \frac{(1 + b_j - a_j)}{\prod_{\substack{i=1 \\ i \ne j}}^n [a_i, b_i]} \left[\|f\|_1 \left\| \frac{\partial g}{\partial x_j} \right\|_\infty + \|g\|_1 \left\| \frac{\partial f}{\partial x_j} \right\|_\infty \right] \right\}. \tag{28.54}$$

Proof.   By Theorem 28.9 and

$$\int_{a_j}^{b_j} \left(\sqrt{\frac{1}{12} + \left(\frac{x_j - a_j}{b_j - a_j} - \frac{1}{2} \right)^2} \right) dx_j = \frac{(b_j - a_j)}{24} \tag{28.55}$$

$$\{ 4\sqrt{3} - \ln[3a_j + 2\sqrt{3}(b_j - a_j) - 3b_j] + \ln[-3a_j + 2\sqrt{3}(b_j - a_j) + 3b_j] \},$$

$j = 1, \ldots, n.$ □

Remark 28.14. In the Assumptions 28.3 or 28.5 or 28.7. We further suppose that

$$\frac{\partial^l f}{\partial x_j^l}(\ldots, b_j, \ldots) = \frac{\partial^l f}{\partial x_j^l}(\ldots, a_j, \ldots), \tag{28.56}$$

for all $j = 1, \ldots, n$ and all $l = 0, 1, \ldots, m - 2$.

In that case $A_j^f = 0$, $j = 1, \ldots, n$, see (28.9). So if we consider f, g as above, we also find $A_j^g = 0$, $j = 1, \ldots, n$.

In that case, see (28.11), it collapses to

$$\Delta_{(f,g)} = \int_{\prod\limits_{i=1}^{n}[a_i,b_i]} f(\overrightarrow{x})g(\overrightarrow{x})d\overrightarrow{x}$$

$$-\frac{1}{\prod\limits_{i=1}^{n}(b_i-a_i)}\left(\int_{\prod\limits_{i=1}^{n}[a_i,b_i]} f(\overrightarrow{x})d\overrightarrow{x}\right)\left(\int_{\prod\limits_{i=1}^{n}[a_i,b_i]} g(\overrightarrow{x})d\overrightarrow{x}\right), \qquad (28.57)$$

that is simplified a lot, etc.

Remark 28.15. Let $[c_i,d_i] \subseteq [a_i,b_i]$, $i=1,\ldots,n$. Let μ be a probability measure on $\left(\prod\limits_{i=1}^{n}[c_i,d_i], \mathcal{P}\left(\prod\limits_{i=1}^{n}[c_i,d_i]\right)\right)$, $\mathcal{P}$ stands for the powerset.

Let f as in Assumptions 28.3 or 28.5 or 28.7. Then by Theorem 28.4, see (28.8), we have

$$M := \left| \int_{\prod\limits_{i=1}^{n}[c_i,d_i]} f(\overrightarrow{x})\mu(d\overrightarrow{x}) - \frac{1}{\prod\limits_{i=1}^{n}(b_i-a_i)} \int_{\prod\limits_{i=1}^{n}[a_i,b_i]} f(\overrightarrow{x})d\overrightarrow{x} \right.$$

$$\left. -\sum_{j=1}^{n}\int_{\prod\limits_{i=1}^{n}[c_i,d_i]} A_j^f d\mu \right| \qquad (28.58)$$

$$= \left|\sum_{j=1}^{n}\int_{\prod\limits_{i=1}^{n}[c_i,d_i]} B_j^f d\mu\right| \leq \sum_{j=1}^{n}\int_{\prod\limits_{i=1}^{n}[c_i,d_i]} |B_j^f|d\mu. \qquad (28.59)$$

When conditions (28.56) are valid then

$$M = \left| \int_{\prod\limits_{i=1}^{n}[c_i,d_i]} f(\overrightarrow{x})d\mu(\overrightarrow{x}) - \frac{1}{\prod\limits_{i=1}^{n}(b_i-a_i)} \int_{\prod\limits_{i=1}^{n}[a_i,b_i]} f(\overrightarrow{x})d\overrightarrow{x} \right|. \qquad (28.60)$$

When $m=1$ then again M is given by (28.60).

We present the second main result, now is regarding comparison of integral means, see also [30].

Theorem 28.16. *Let f as in Assumptions 28.3 or 28.5 or 28.7. Let μ and M as in Remark 28.15, see (28.58). Additionally assume that*

$$\frac{\partial^m f}{\partial x_j^m} \in L_\infty\left(\prod_{i=1}^{n}[a_i,b_i]\right), \quad j=1,\ldots,n; \ m\in\mathbb{N}.$$

Then

1)

$$M \leq \frac{1}{m!} \sum_{j=1}^{n} \left\{ (b_j - a_j)^m \left\| \frac{\partial^m f}{\partial x_j^m} \right\|_{\infty} \right.$$

$$\left. \left(\int_{\prod_{i=1}^{n} [c_i,d_i]} \left(\sqrt{\frac{(m!)^2}{(2m)!} |B_{2m}| + B_m^2 \left(\frac{x_j - a_j}{b_j - a_j} \right)} \right) d\mu(\overrightarrow{x}) \right) \right\} =: L_1. \qquad (28.61)$$

2) Let $p_j, q_j > 1$: $\dfrac{1}{p_j} + \dfrac{1}{q_j} = 1$, $j = 1, \ldots, n$, *it holds*

$$M \leq \frac{1}{m!} \sum_{j=1}^{n} \left\{ (b_j - a_j)^{m - \frac{1}{q_j}} \left(\prod_{i=1}^{j-1} (b_i - a_i) \right)^{-1/q_j} \right.$$

$$\left(\int_{\prod_{i=1}^{n} [c_i,d_i]} \left(\int_0^1 \left| B_m \left(\frac{x_j - a_j}{b_j - a_j} \right) - B_m(t_j) \right|^{p_j} dt_j \right) d\mu(\overrightarrow{x}) \right)^{1/p_j}$$

$$\left. \times \left(\int_{\prod_{i=1}^{n} [c_i,d_i]} \left\| \frac{\partial^m f}{\partial x_j^m} (\ldots, x_{j+1}, \ldots, x_n) \right\|_{q_j, \prod_{i=1}^{j} [a_i,b_i]}^{q_j} d\mu(\overrightarrow{x}) \right)^{1/q_j} \right\} =: L_2. \qquad (28.62)$$

When $p_j = q_j = 2$, $j = 1, \ldots, n$, *it holds*

$$M \leq \frac{1}{m!} \sum_{j=1}^{n} \left\{ (b_j - a_j)^{m - \frac{1}{2}} \left(\prod_{i=1}^{j-1} (b_i - a_i) \right)^{-1/2} \right.$$

$$\left(\int_{\prod_{i=1}^{n} [c_i,d_i]} \left(\frac{(m!)^2}{(2m)!} |B_{2m}| + B_m^2 \left(\frac{x_j - a_j}{b_j - a_j} \right) \right) d\mu(\overrightarrow{x}) \right)^{1/2}$$

$$\left. \times \left(\int_{\prod_{i=1}^{n} [c_i,d_i]} \left\| \frac{\partial^m f}{\partial x_j^m} (\ldots, x_{j+1}, \ldots, x_n) \right\|_{2, \prod_{i=1}^{j} [a_i,b_i]}^{2} d\mu(\overrightarrow{x}) \right)^{1/2} \right\} = L_3. \qquad (28.63)$$

3) (i) *Case* $m = 2r$, $r \in \mathbb{N}$, *it holds*

$$M \leq \frac{1}{(2r)!} \sum_{j=1}^{n} \left\{ \frac{(b_j - a_j)^{2r-1}}{\left(\prod_{i=1}^{j-1} (b_i - a_i) \right)} \right.$$

$$\left(\int_{\prod_{i=1}^{n} [c_i,d_i]} \left\| \frac{\partial^{2r} f}{\partial x_j^{2r}} (\ldots, x_{j+1}, \ldots, x_n) \right\|_{1, \prod_{i=1}^{j} [a_i,b_i]} d\mu(\overrightarrow{x}) \right)$$

$$\left[(1 - 2^{-2r})|B_{2r}| + \left\|2^{-2r}B_{2r} - B_{2r}\left(\frac{x_j - a_j}{b_j - a_j}\right)\right\|_\infty\right]\right\} =: L_4.$$
(28.64)

(ii) Case $m = 2r + 1$, $r \in \mathbb{N}$, it holds

$$M \leq \frac{1}{(2r + 1)!} \sum_{j=1}^n \left\{ \frac{(b_j - a_j)^{2r}}{\left(\prod_{i=1}^{j-1}(b_i - a_i)\right)} \right.$$

$$\left(\int_{\prod_{i=1}^n [c_i,d_i]} \left\|\frac{\partial^{2r+1} f}{\partial x_j^{2r+1}}(\ldots, x_{j+1}, \ldots, x_n)\right\|_{1,\prod_{i=1}^j [a_i,b_i]} d\mu(\overrightarrow{x})\right)$$

$$\times \left[\frac{2(2r+1)!}{(2\pi)^{2r+1}(1 - 2^{-2r})} + \left\|B_{2r+1}\left(\frac{x_j - a_j}{b_j - a_j}\right)\right\|_\infty\right]\right\} =: L_5.$$
(28.65)

(iii) When $m = 1$, it holds

$$M \leq \frac{1}{2} \sum_{j=1}^n \left\{ \frac{(1 + b_j - a_j)}{\prod_{i=1}^{j-1}(b_i - a_i)} \right.$$

$$\left(\int_{\prod_{i=1}^n [c_i,d_i]} \left\|\frac{\partial f}{\partial x_j}(\ldots, x_{j+1}, \ldots, x_n)\right\|_{1,\prod_{i=1}^j [a_i,b_i]} d\mu(\overrightarrow{x})\right)\right\} =: L_6.$$
(28.66)

Proof. We rely on (28.59).

1) Estimate with respect to $\|\cdot\|_\infty$. From (28.27) we obtain

$$|B_j^f(x_j, \ldots, x_n)| \leq \frac{(b_j - a_j)^m}{m!}\left\|\frac{\partial^m f}{\partial x_j^m}\right\|_\infty \left(\sqrt{\frac{(m!)^2}{(2m)!}|B_{2m}| + B_m^2\left(\frac{x_j - a_j}{b_j - a_j}\right)}\right),$$
(28.67)

all $j = 1, \ldots, n$, any $(x_j, \ldots, x_n) \in \prod_{i=j}^n [a_i, b_i]$.

Thus

$$\sum_{j=1}^n \int_{\prod_{i=1}^n [c_i,d_i]} |B_j^f| d\mu \leq \frac{1}{m!} \sum_{j=1}^n \left\{ (b_j - a_j)^m \left\|\frac{\partial^m f}{\partial x_j^m}\right\|_\infty \right.$$

$$\left(\int_{\prod_{i=1}^n [c_i,d_i]} \left(\sqrt{\frac{(m!)^2}{(2m)!}|B_{2m}| + B_m^2\left(\frac{x_j - a_j}{b_j - a_j}\right)}\right) d\mu(\overrightarrow{x})\right)\right\},$$
(28.68)

proving the claim (28.61).

2) Estimate with respect to $\|\cdot\|_{q_j}$. Here $p_j > 1$, $q_j > 1$: $\dfrac{1}{p_j} + \dfrac{1}{q_j} = 1$; $j = 1, \ldots, n$.

By Lemma 2 of [29], see (28.62) there, we derive

$$|B_j^f| \leq \frac{(b_j - a_j)^{m - \frac{1}{q_j}}}{m!} \left(\prod_{i=1}^{j-1} (b_i - a_i)^{-1/q_j} \left(\int_0^1 \left| B_m\left(\frac{x_j - a_j}{b_j - a_j} \right) - B_m(t_j) \right|^{p_j} dt_j \right)^{1/p_j} \right.$$

$$\left\| \frac{\partial^m f}{\partial x_j^m}(\ldots, x_{j+1}, \ldots, x_n) \right\|_{q_j, \prod\limits_{i=1}^{j} [a_i, b_i]} , \tag{28.69}$$

true for all $(x_j, x_{j+1}, \ldots, x_n) \in \prod\limits_{i=j}^{n} [a_i, b_i]$, $j = 1, \ldots, n$.

Hence we have

$$\sum_{j=1}^{n} \int_{\prod\limits_{i=1}^{n} [c_i, d_i]} |B_j^f| d\mu \leq \frac{1}{m!} \sum_{j=1}^{n} \left\{ (b_j - a_j)^{m - \frac{1}{q_j}} \left(\prod_{i=1}^{j-1} (b_i - a_i) \right)^{-1/q_j} \right.$$

$$\int_{\prod\limits_{i=1}^{n} [c_i, d_i]} \left[\left(\int_0^1 \left| B_m\left(\frac{x_j - a_j}{b_j - a_j} \right) - B_m(t_j) \right|^{p_j} dt_j \right)^{1/p_j} \right.$$

$$\left. \left\| \frac{\partial^m f}{\partial x_j^m}(\ldots, x_{j+1}, \ldots, x_n) \right\|_{q_j, \prod\limits_{i=1}^{j} [a_i, b_i]} \right] d\mu(\overrightarrow{x}) \right\} \tag{28.70}$$

$$\leq \frac{1}{m!} \sum_{j=1}^{n} \left\{ (b_j - a_j)^{m - \frac{1}{q_j}} \left(\prod_{i=1}^{j-1} (b_i - a_i) \right)^{-1/q_j} \right.$$

$$\left(\int_{\prod\limits_{i=1}^{n} [c_i, d_i]} \left(\int_0^1 \left| B_m\left(\frac{x_j - a_j}{b_j - a_j} \right) - B_m(t_j) \right|^{p_j} dt_j \right) d\mu(\overrightarrow{x}) \right)^{1/p_j}$$

$$\left. \times \left(\int_{\prod\limits_{i=1}^{n} [c_i, d_i]} \left(\left\| \frac{\partial^m f}{\partial x_j^m}(\ldots, x_{j+1}, \ldots, x_n) \right\|_{q_j, \prod\limits_{i=1}^{j} [a_i, b_i]} \right)^{q_j} d\mu(\overrightarrow{x}) \right)^{1/q_j} \right\}, \tag{28.71}$$

proving the claim (28.62).

When $p_j = q_j = 2$, $j = 1, \ldots, n$, it holds (see p.352 of [98])

$$\sum_{j=1}^{n} \int_{\prod\limits_{i=1}^{n} [c_i, d_i]} |B_j^f| d\mu \leq \frac{1}{m!} \sum_{j=1}^{n} \left\{ (b_j - a_j)^{m - \frac{1}{2}} \right.$$

$$\left(\prod_{i=1}^{j-1}(b_i - a_i)\right)^{-1/2}\left(\int_{\prod_{i=1}^{n}[c_i,d_i]}\left(\frac{(m!)^2}{(2m)!}|B_{2m}| + B_m^2\left(\frac{x_j - a_j}{b_j - a_j}\right)\right)d\mu(\overrightarrow{x})\right)^{1/2}$$

$$\times\left(\int_{\prod_{i=1}^{n}[c_i,d_i]}\left\|\frac{\partial^m f}{\partial x_j^m}(\ldots,x_{j+1},\ldots,x_n)\right\|_{2,\prod_{i=1}^{j}[a_i,b_i]}^2 d\mu(\overrightarrow{x})\right)^{1/2}\Bigg\},$$

proving the claim (28.63).

3) Estimate with respect to $\|\cdot\|_1$. The special cases follow:

(i) When $m = 2r$, $r \in \mathbb{N}$, by (28.4) we obtain

$$\sum_{j=1}^{n}\int_{\prod_{i=1}^{n}[c_i,d_i]}|B_j^f|d\mu \leq \frac{1}{(2r)!}\sum_{j=1}^{n}\left\{\frac{(b_j - a_j)^{2r-1}}{\left(\prod_{i=1}^{j-1}(b_i - a_i)\right)}\right.$$

$$\left(\int_{\prod_{i=1}^{n}[c_i,d_i]}\left\|\frac{\partial^{2r} f}{\partial x_j^{2r}}(\ldots,x_{j+1},\ldots,x_n)\right\|_{1,\prod_{i=1}^{j}[a_i,b_i]}d\mu(\overrightarrow{x})\right)\left[(1 - 2^{-2r})|B_{2r}|\right.$$

$$\left.\left.+\left\|2^{-2r}B_{2r} - B_{2r}\left(\frac{x_j - a_j}{b_j - a_j}\right)\right\|_\infty\right]\right\}. \tag{28.72}$$

(ii) When $m = 2r + 1$, $r \in \mathbb{N}$, then by (28.43) it holds

$$\sum_{j=1}^{n}\int_{\prod_{i=1}^{n}[c_i,d_i]}|B_j^f|d\mu \leq \frac{1}{(2r+1)!}\sum_{j=1}^{n}\left\{\frac{(b_j - a_j)^{2r}}{\left(\prod_{i=1}^{j-1}(b_i - a_i)\right)}\right.$$

$$\left(\int_{\prod_{i=1}^{n}[c_i,d_i]}\left\|\frac{\partial^{2r+1} f}{\partial x_j^{2r+1}}(\ldots,x_{j+1},\ldots,x_n)\right\|_{1,\prod_{i=1}^{j}[a_i,b_i]}d\mu(\overrightarrow{x})\right)$$

$$\left.\times\left[\frac{2(2r+1)!}{(2\pi)^{2r+1}(1 - 2^{-2r})} + \left\|B_{2r+1}\left(\frac{x_j - a_j}{b_j - a_j}\right)\right\|_\infty\right]\right\}. \tag{28.73}$$

(iii) When $m = 1$, from (28.45) it holds

$$\sum_{j=1}^{n}\int_{\prod_{i=1}^{n}[c_i,d_i]}|B_j^f|d\mu \leq \frac{1}{2}\sum_{j=1}^{m}\left\{\frac{(1 + b_j - a_j)}{\prod_{i=1}^{j-1}(b_i - a_i)}\right.$$

$$\left.\left(\int_{\prod_{i=1}^{n}[c_i,d_i]}\left\|\frac{\partial f}{\partial x_j}(\ldots,x_{j+1},\ldots,x_n)\right\|_{1,\prod_{i=1}^{j}[a_i,b_i]}d\mu(\overrightarrow{x})\right)\right\}, \tag{28.74}$$

proving all claims. $\qquad\square$

We give

Corollary 28.17. *All as in Theorem 28.16.*
1) Let $m = 2r$, $r \in \mathbb{N}$. Then

$$M \leq \min\{L_1, L_2, L_3, L_4\}. \tag{28.75}$$

2) Let $m = 2r + 1$, $r \in \mathbb{N}$. Then

$$M \leq \min\{L_1, L_2, L_3, L_5\}. \tag{28.76}$$

Next we compare means when $m = 1$.

Corollary 28.18. *Let f as in Assumptions 28.10 or 28.11 or 28.12. Let μ as in Remark 28.15. Additionally suppose that*

$$\frac{\partial f}{\partial x_j} \in L_\infty \left(\prod_{i=1}^{n} [a_i, b_i] \right), \quad j = 1, \dots, n.$$

Then

$$M = \left| \int_{\prod_{i=1}^{n} [c_i, d_i]} f(\overrightarrow{x}) d\mu(\overrightarrow{x}) - \frac{1}{\prod_{i=1}^{n}(b_i - a_i)} \int_{\prod_{i=1}^{n} [a_i, b_i]} f(\overrightarrow{x}) d\overrightarrow{x} \right|$$

$$\leq \sum_{j=1}^{n} \left\{ (b_j - a_j) \left\| \frac{\partial f}{\partial x_j} \right\|_\infty \right.$$

$$\left. \left(\int_{\prod_{i=1}^{n} [c_i, d_i]} \left(\sqrt{\frac{1}{12} + \left(\frac{x_j - a_j}{b_j - a_j} - \frac{1}{2} \right)^2} \right) d\mu(\overrightarrow{x}) \right) \right\} =: L_1'. \tag{28.77}$$

2) Let $p_j, q_j > 1$: $\dfrac{1}{p_j} + \dfrac{1}{q_j} = 1$, $j = 1 \dots, n$. It holds

$$M \leq \sum_{j=1}^{n} \left\{ (b_j - a_j)^{1 - \frac{1}{q_j}} \left(\prod_{i=1}^{j-1}(b_i - a_i) \right)^{-1/q_j} \right.$$

$$\left(\int_{\prod_{i=1}^{n} [c_i, d_i]} \left(\int_0^1 \left| \frac{x_j - a_j}{b_j - a_j} - t_j \right|^{p_j} dt_j \right) d\mu(\overrightarrow{x}) \right)^{1/p_j}$$

$$\left. \times \left(\int_{\prod_{i=1}^{n} [c_i, d_i]} \left\| \frac{\partial f}{\partial x_j}(\dots, x_{j+1}, \dots, x_n) \right\|_{q_j, \prod_{i=1}^{j} [a_i, b_i]}^{q_j} d\mu(\overrightarrow{x}) \right)^{1/q_j} \right\} =: L_2'. \tag{28.78}$$

When $p_j = q_j = 2$, $j = 1, \ldots, n$, it holds

$$M \leq \sum_{j=1}^{n} \left\{ \sqrt{b_j - a_j} \left(\prod_{i=1}^{j-1} (b_i - a_i) \right)^{-1/2} \right.$$

$$\left(\int_{\prod_{i=1}^{n} [c_i, d_i]} \left(\frac{1}{12} + \left(\frac{x_j - a_j}{b_j - a_j} - \frac{1}{2} \right)^2 \right) d\mu(\overrightarrow{x}) \right)^{1/2}$$

$$\times \left(\int_{\prod_{i=1}^{n} [c_i, d_i]} \left\| \frac{\partial f}{\partial x_j}(\ldots, x_{j+1}, \ldots, x_n) \right\|_{2, \prod_{i=1}^{j} [a_i, b_i]}^{2} d\mu(\overrightarrow{x}) \right)^{1/2} \right\} =: L_3'. \qquad (28.79)$$

Proof. By Theorem 28.16. $\qquad\qquad\square$

We end chapter with

Corollary 28.19. *All as in Corollary 28.18. Then*

$$M \leq \min\{L_1', L_2', L_3', L_6\}. \qquad (28.80)$$

Proof. By Corollary 28.18 and (28.66). $\qquad\qquad\square$

Chapter 29

Multivariate Fink Type Identity Applied to Multivariate Ostrowski, Comparison of Means and Grüss Inequalities

We present a general multivariate Fink type identity which is a representation formula for a multivariate function. Using it we derive general tight multivariate high order Ostrowski type, comparison of means and Grüss type inequalities. The estimates involve L_p norms, any $1 \le p \le \infty$. This chapter is based on [39].

29.1 Introduction

The article that motivates the most this chapter is of A. Fink (1992), see [141], where he presents among others a representation formulae for a univariate function. We generalize this to multi-dimension, see Theorem 29.6. Using this multivariate representation formula we derive multivariate Ostrowski type inequalities, see Theorems 29.9–29.12. We continue the comparison of general multivariate integral mean to Riemann integral average, see Theorem 29.15. Then we present multivariate Grüss type inequalities, see Theorems 29.17, 29.19, 29.20. At the end we apply the general results to the cases: of a two dimensional function involving partials of order one, see Corollaries 29.23, 29.25–29.33, and of a three-dimensional functions involving second order partials, see Corollaries 29.36, 29.38–29.46. The results here are also motivated by the works of Ostrowski [196], Grüss [150], and of the author [16], [17], [24], [38]. We would like to mention the following inspiring results.

Theorem 29.1 (Ostrowski, 1938, [196]). *Let $f \colon [a,b] \to \mathbb{R}$ be continuous on $[a,b]$ and differentiable on (a,b) whose derivative $f' \colon (a,b) \to \mathbb{R}$ is bounded on (a,b), i.e.,*
$$\|f'\|_\infty = \sup_{t\in(a,b)} |f'(t)| < +\infty. \ \textit{Then}$$

$$\left| \frac{1}{b-a} \int_a^b f(t)dt - f(x) \right| \le \left[\frac{1}{4} + \frac{\left(x - \frac{a+b}{2}\right)^2}{(b-a)^2} \right] (b-a)\|f'\|_\infty$$

for any $x \in [a,b]$. The constant $1/4$ is the best possible.

Theorem 29.2 (Grüss, 1935, [150]). *Let f,g integrable functions from $[a,b]$ into $\mathbb{R}$, such that $m \le f(x) \le M$, $\rho \le g(x) \le \sigma$, for all $x \in [a,b]$, where $m, M, \rho, \sigma \in \mathbb{R}$.*

Then

$$\left| \frac{1}{b-a} \int_a^b f(x)g(x)dx - \left(\frac{1}{b-a} \int_a^b f(x)dx \right) \left(\frac{1}{b-a} \int_a^b g(x)dx \right) \right|$$

$$\leq \frac{1}{4}(M-m)(\sigma-\rho).$$

29.2   Main Results

Here $\prod\limits_{i=1}^{m} [a_i, b_i] \subseteq \mathbb{R}^m$, $m, n \in \mathbb{N}$.

We mention

General Assumptions 29.3. Let $f \colon \prod\limits_{i=1}^{m} [a_i, b_i] \to \mathbb{R}$. We assume

1) for $j = 1, \ldots, m$ we have that

$$\frac{\partial^{n-1} f}{\partial x_j^{n-1}}(x_1, x_2, \ldots, x_{j-1}, s_j, x_{j+1}, \ldots, x_m)$$

is absolutely continuous in s_j on $[a_j, b_j]$, for every

$$(x_1, x_2, \ldots, x_{j-1}, x_{j+1}, \ldots, x_m) \in \prod\limits_{\substack{i=1 \\ i \neq j}}^{m} [a_i, b_i],$$

2) for $j = 1, \ldots, m$ we have that

$$\frac{\partial^n f}{\partial x_j^n}(s_1, \ldots, s_j, x_{j+1}, \ldots, x_m)$$

is continuous on $\prod\limits_{i=1}^{j} [a_i, b_i]$ for every

$$(x_{j+1}, \ldots, x_m) \in \prod\limits_{i=j+1}^{m} [a_i, b_i],$$

3) for each $j = 1, \ldots, m$, and for every $\ell = 1, \ldots, n-2$, we have that

$$\frac{\partial^\ell f}{\partial x_j^\ell}(s_1, s_2, \ldots, s_{j-1}, x_j, \ldots, x_m)$$

is continuous on $\prod\limits_{i=1}^{j-1} [a_i, b_i]$, for every $(x_j, \ldots, x_m) \in \prod\limits_{i=j}^{m} [a_j, b_j]$.

4) f is continuous on $\prod\limits_{i=1}^{m} [a_i, b_i]$.

We mention

Brief Assumptions 29.4. Let $f \colon \prod\limits_{i=1}^{m} [a_i, b_i] \to \mathbb{R}$ with $\frac{\partial^\ell f}{\partial x_i^\ell}$ for $\ell = 0, 1, \ldots, n$;

$i = 1, \ldots, m$, are continuous on $\prod\limits_{i=1}^{m} [a_i, b_i]$.

We mention

Definition 29.5. We put

$$k(s_i, x_i) := \begin{cases} s_i - a_i, & a_i \le s_i \le x_i \le b_i, \\ s_i - b_i, & a_i \le x_i < s_i \le b_i, \end{cases} \quad i = 1, \ldots, m.$$

We present the representation result of Fink type.

Theorem 29.6. *Let f as in General Assumptions 29.3 or Brief Assumptions 29.4,*

$$(x_1, \ldots, x_m) \in \prod_{j=1}^{m} [a_i, b_i].$$

Then

$$f(x_1, \ldots, x_m) = \frac{n^m}{\prod_{j=1}^{m} (b_j - a_j)} \int_{\prod_{j=1}^{m} [a_j, b_j]} f(s_1, \ldots, s_m) ds_1 \cdots ds_m + \sum_{i=1}^{m} T_i, \quad (29.1)$$

where for $i = 1, \ldots, m$ we set

$$T_i := T_i(x_i, \ldots, x_m) := \left(\frac{n^{i-1}}{\prod_{j=1}^{i} (b_j - a_j)} \right) \sum_{k=1}^{n-1} \left(\frac{n-k}{k!} \right)$$

$$\times \left\{ \int_{\prod_{j=1}^{i-1} [a_j, b_j]} \left(\frac{\partial^{k-1} f}{\partial x_i^{k-1}} (s_1, s_2, \ldots, s_{i-1}, b_i, x_{i+1}, \ldots, x_m)(x_i - b_i)^k \right. \right.$$

$$\left. - \frac{\partial^{k-1} f}{\partial x_i^{k-1}} (s_1, s_2, \ldots, s_{i-1}, a_i, x_{i+1}, \ldots, x_m)(x_i - a_i)^k \right) ds_1 ds_2 \cdots ds_{i-1} \Big\}$$

$$+ \left(\frac{n^{i-1}}{(n-1)! \prod_{j=1}^{i} (b_j - a_j)} \right) \left(\int_{\prod_{j=1}^{i} [a_j, b_j]} (x_i - s_i)^{n-1} k(s_i, x_i) \right.$$

$$\left. \times \frac{\partial^n f}{\partial x_i^n} (s_1, s_2, \ldots, s_i, x_{i+1}, \ldots, x_m) ds_1 ds_2 \cdots ds_i \right). \quad (29.2)$$

When $n = 1$ the sum $\sum_{k=1}^{n-1}$ in (29.2) is zero.

For the proof of Theorem 29.6 we need Fink's identity.

Theorem 29.7 (Fink, [141]). *Let any $a, b \in \mathbb{R}$, $f : [a, b] \to \mathbb{R}$, $n \ge 1$, $f^{(n-1)}$ is absolutely continuous on $[a, b]$. Then*

$$f(x) = \frac{n}{b - a} \int_a^b f(t) dt - \sum_{k=1}^{n-1} \left(\frac{n-k}{k!} \right) \left(\frac{f^{(k-1)}(a)(x-a)^k - f^{(k-1)}(b)(x-b)^k}{b - a} \right)$$

$$+ \frac{1}{(n-1)!(b-a)} \int_a^b (x-t)^{n-1} k(t, x) f^{(n)}(t) dt, \quad (29.3)$$

where

$$k(t, x) := \begin{cases} t - a, & a \leq t \leq x \leq b, \\ t - b, & a \leq x < t \leq b. \end{cases}$$

When $n = 1$ the sum $\sum_{k=1}^{n-1}$ in (29.3) is zero.

Proof of Theorem 29.6. Here we apply (29.3) repeatedly. We have

$$f(x_1, \ldots, x_m) = \frac{n}{b_1 - a_1} \int_{a_1}^{b_1} f(s_1, x_2, \ldots, x_m) ds_1 + T_1(x_1, \ldots, x_m), \qquad (29.4)$$

where

$$T_1(x_1, \ldots, x_m) := \frac{1}{(b_1 - a_1)} \sum_{k=1}^{n-1} \left(\frac{n-k}{k!} \right) \left(\frac{\partial^{k-1} f(b_1, x_2, \ldots, x_m)}{\partial x_1^{k-1}} (x_1 - b_1)^k \right.$$

$$\left. - \frac{\partial^{k-1} f}{\partial x_1^{k-1}} (a_1, x_2, \ldots, x_m)(x_1 - a_1)^k \right)$$

$$+ \frac{1}{(n-1)!(b_1 - a_1)} \int_{a_1}^{b_1} (x_1 - s_1)^{n-1} k(s_1, x_1) \frac{\partial^n f}{\partial x_1^n}(s_1, x_2, \ldots, x_m) ds_1. \qquad (29.5)$$

But it holds

$$f(s_1, x_2, \ldots, x_m) = \frac{n}{b_2 - a_2} \int_{a_2}^{b_2} f(s_1, s_2, x_3, \ldots, x_m) ds_2$$

$$+ \frac{1}{(b_2 - a_2)} \sum_{k=1}^{n-1} \left(\frac{n-k}{k!} \right) \left(\frac{\partial^{k-1} f}{\partial x_2^{k-1}}(s_1, b_2, x_3, \ldots, x_m)(x_2 - b_2)^k \right.$$

$$\left. - \frac{\partial^{k-1}}{\partial x_2^{k-1}} f(s_1, a_2, x_3, \ldots, x_m)(x_2 - a_2)^k \right)$$

$$+ \frac{1}{(n-1)!(b_2 - a_2)} \int_{a_2}^{b_2} (x_2 - s_2)^{n-1} k(s_2, x_2)$$

$$\times \frac{\partial^n f}{\partial x_2^n}(s_1, s_2, x_3, \ldots, x_m) ds_2. \qquad (29.6)$$

Combining (29.4) and (29.6) we obtain

$$f(x_1, x_2, \ldots, x_m) = \frac{n^2}{(b_1 - a_1)(b_2 - a_2)} \int_{a_1}^{b_1} \int_{a_2}^{b_2} f(s_1, s_2, x_3, \ldots, x_m) ds_1 ds_2$$

$$+ T_2(x_2, x_3, \ldots, x_m) + T_1(x_1, \ldots, x_m), \qquad (29.7)$$

where

$$T_2(x_2, x_3, \ldots, x_m) := \frac{n}{(b_1 - a_1)(b_2 - a_2)} \sum_{k=1}^{n-1} \left(\frac{n-k}{k!} \right)$$

$$\times \left(\int_{a_1}^{b_1} \left(\frac{\partial^{k-1} f}{\partial x_2^{k-1}}(s_1, b_2, x_3, \ldots, x_m)(x_2 - b_2)^k \right. \right.$$

$$\left. \left. - \frac{\partial^{k-1} f}{\partial x_2^{k-1}}(s_1, a_2, x_3, \ldots, x_m)(x_2 - a_2)^k \right) ds_1 \right)$$

$$+ \frac{n}{(n-1)!(b_1 - a_1)(b_2 - a_2)} \int_{a_1}^{b_1} \int_{a_2}^{b_2} (x_2 - s_2)^{n-1} k(s_2, x_2)$$

$$\times \frac{\partial^n f}{\partial x_2^n}(s_1, s_2, x_3, \ldots, x_m) ds_1 ds_2. \tag{29.8}$$

Next we observe that

$$f(s_1, s_2, x_3, \ldots, x_m) = \frac{n}{b_3 - a_3} \int_{a_3}^{b_3} f(s_1, s_2, s_3, x_4, \ldots, x_m) ds_3$$

$$+ \frac{1}{(b_3 - a_3)} \left(\sum_{k=1}^{n-1} \left(\frac{n-k}{k!} \right) \left(\frac{\partial^{k-1} f}{\partial x_3^{k-1}}(s_1, s_2, b_3, x_4, \ldots, x_m)(x_3 - b_3)^k \right. \right.$$

$$\left. \left. - \frac{\partial^{k-1} f}{\partial x_3^{k-1}}(s_1, s_2, a_3, x_4, \ldots, x_m)(x_3 - a_3)^k \right) \right)$$

$$+ \frac{1}{(n-1)!(b_3 - a_3)} \int_{a_3}^{b_3} (x_3 - s_3)^{n-1} k(s_3, x_3)$$

$$\times \frac{\partial^n f}{\partial x_3^n}(s_1, s_2, s_3, x_4, \ldots, x_m) ds_3. \tag{29.9}$$

Combining (29.7) and (29.9) we derive

$$f(x_1, x_2, \ldots, x_m) = \frac{n^3}{\prod\limits_{i=1}^{3}(b_i - a_i)} \int_{a_1}^{b_1} \int_{a_2}^{b_2} \int_{a_3}^{b_3} f(s_1, s_2, s_3, x_4, \ldots, x_m) ds_1 ds_2 ds_3$$

$$+ T_3(x_3, x_4, \ldots, x_m) + T_1 + T_2, \tag{29.10}$$

where

$$T_3(x_3, x_4, \ldots, x_m) := \left(\frac{n^2}{\prod\limits_{j=1}^{3}(b_j - a_j)} \right) \left(\sum_{k=1}^{n-1} \left(\frac{n-k}{k!} \right) \right)$$

$$\times \left(\int_{a_1}^{b_1} \int_{a_2}^{b_2} \left(\frac{\partial^{k-1} f}{\partial x_3^{k-1}}(s_1, s_2, b_3, x_4, \ldots, x_m)(x_3 - b_3)^k \right. \right.$$

$$\left. \left. - \frac{\partial^{k-1} f}{\partial x_3^{k-1}}(s_1, s_2, a_3, x_4, \ldots, x_m)(x_3 - a_3)^k \right) ds_1 ds_2 \right) \right)$$

$$+ \left(\frac{n^2}{(n-1)! \prod\limits_{j=1}^{3} (b_j - a_j)} \right) \left(\int_{\prod\limits_{j=1}^{3} [a_j,b_j]} (x_3 - s_3)^{n-1} k(s_3, x_3) \right.$$

$$\left. \times \frac{\partial^n f}{\partial x_3^n}(s_1, s_2, s_3, x_4, \ldots, x_m) ds_1 ds_2 ds_3 \right). \tag{29.11}$$

We also observe that

$$f(s_1, s_2, s_3, x_4, \ldots, x_m) = \frac{n}{b_4 - a_4} \int_{a_4}^{b_4} f(s_1, s_2, s_3, s_4, x_5, \ldots, x_m) ds_4$$

$$+ \frac{1}{(b_4 - a_4)} \sum_{k=1}^{n-1} \left(\frac{n-k}{k!} \right) \left(\frac{\partial^{k-1} f}{\partial x_4^{k-1}}(s_1, s_2, s_3, b_4, x_5, \ldots, x_m)(x_4 - b_4)^k \right.$$

$$\left. - \frac{\partial^{k-1} f}{\partial x_4^{k-1}}(s_1, s_2, s_3, a_4, x_5, \ldots, x_m)(x_4 - a_4)^k \right)$$

$$+ \frac{1}{(n-1)!(b_4 - a_4)} \int_{a_4}^{b_4} (x_4 - s_4)^{n-1} k(s_4, x_4)$$

$$\times \frac{\partial^n f}{\partial x_4^n}(s_1, s_2, s_3, s_4, x_5, \ldots, x_m) ds_4. \tag{29.12}$$

Combining (29.10) and (29.12) we obtain

$$f(x_1, x_2, \ldots, x_m)$$

$$= \frac{n^4}{\prod\limits_{j=1}^{4} (b_j - a_j)} \int_{\prod\limits_{j=1}^{4} [a_j,b_j]} f(s_1, s_2, s_3, s_4, x_5, \ldots, x_m) ds_1 ds_2 ds_3 ds_4$$

$$+ \sum_{j=1}^{4} T_j, \tag{29.13}$$

where

$$T_4(x_4, x_5, \ldots, x_m) := \left(\frac{n^3}{\prod\limits_{j=1}^{4} (b_j - a_j)} \right) \left\{ \sum_{k=1}^{n-1} \left(\frac{n-k}{k!} \right) \left[\int_{\prod\limits_{j=1}^{3} [a_j,b_j]} \right. \right.$$

$$\left(\frac{\partial^{k-1} f}{\partial x_4^{k-1}}(s_1, s_2, s_3, b_4, x_5, \ldots, x_m)(x_4 - b_4)^k \right.$$

$$\left. \left. \left. - \frac{\partial^{k-1} f}{\partial x_4^{k-1}}(s_1, s_2, s_3, a_4, x_5, \ldots, x_m)(x_4 - a_4)^k \right) ds_1 ds_2 ds_3 \right] \right\}$$

$$+ \left(\frac{n^3}{(n-1)! \prod\limits_{j=1}^{4} (b_j - a_j)} \right) \left(\int_{\prod\limits_{j=1}^{4} [a_j,b_j]} (x_4 - s_4)^{n-1} k(s_4, x_4) \right.$$

$$\left. \times \frac{\partial^n f}{\partial x_4^n}(s_1, s_2, s_3, s_4, x_5, \ldots, x_m) ds_1 ds_2 ds_3 ds_4 \right). \tag{29.14}$$

etc. proving the claim. $\qquad\square$

We make

Remark 29.8. For $i = 1, \ldots, m$ denote

$$A_i(x_i, x_{i+1}, \ldots, x_m) := \left(\frac{n^{i-1}}{\prod\limits_{j=1}^{i} (b_j - a_j)} \right) \sum_{k=1}^{n-1} \left(\frac{n-k}{k!} \right)$$

$$\times \left\{ \int_{\prod\limits_{j=1}^{i-1} [a_j, b_j]} \left(\frac{\partial^{k-1} f}{\partial x_i^{k-1}}(s_1, s_2, \ldots, s_{i-1}, b_i, x_{i+1}, \ldots, x_m)(x_i - b_i)^k \right.\right.$$

$$\left.\left. - \frac{\partial^{k-1} f}{\partial x_i^{k-1}}(s_1, s_2, \ldots, s_{i-1}, a_i, x_{i+1}, \ldots, x_m)(x_i - a_i)^k \right) ds_1 ds_2 \ldots ds_{i-1} \right\}.$$

$$(29.15)$$

When $n = 1$ then $A_i = 0$, $i = 1, \ldots, m$.

Also we denote

$$B_i(x_i, x_{i+1}, \ldots, x_m) := \left(\frac{n^{i-1}}{(n-1)! \prod\limits_{j=1}^{i} (b_j - a_j)} \right) \left(\int_{\prod\limits_{j=1}^{i} [a_j, b_j]} (x_i - s_i)^{n-1} k(s_i, x_i) \right.$$

$$\left. \times \frac{\partial^n f}{\partial x_i^n}(s_1, s_2, \ldots, s_i, x_{i+1}, \ldots, x_m) ds_1 \cdots ds_i \right).$$

$$(29.16)$$

That is

$$T_i = A_i + B_i, \quad i = 1, \ldots, m. \tag{29.17}$$

Thus by (29.1) we derive

$$E_f(x_1, \ldots, x_m) := f(x_1, \ldots, x_m)$$

$$- \frac{n^m}{\prod\limits_{j=1}^{m} (b_j - a_j)} \int_{\prod\limits_{j=1}^{m} [a_i, b_j]} f(s_1, \ldots, s_m) ds_1 \cdots ds_m - \sum_{i=1}^{m} A_i = \sum_{i=1}^{m} B_i. \tag{29.18}$$

Thus

$$|E_f(x_1, \ldots, x_m)| \leq \sum_{i=1}^{m} |B_i|. \tag{29.19}$$

Next we estimate E_f via some Ostrowski type inequalities, see Theorems 29.9 – 29.12.

Theorem 29.9. *Assume all as in Theorem 29.6. Then*

$$|E_f(x_1,\ldots,x_m)| \le \frac{\sum_{i=1}^{m}}{(n-1)!}\left\{\left\|\frac{\partial^n f}{\partial x_i^n}(\overbrace{\cdots}^{-i-},x_{i+1},\ldots,x_m)\right\|_\infty \frac{n^{i-1}}{(b_i-a_i)}\right.$$

$$\times\left[\left(\frac{b_i(b_i-x_i)^n - a_i(x_i-a_i)^n}{n}\right)\right.$$

$$+\sum_{\lambda=0}^{n-1}\binom{n-1}{\lambda}(-1)^{n-\lambda-1}\left\{\frac{(n+3)x_i^{n+1}}{(\lambda+2)(n-\lambda+1)}\right.$$

$$\left.\left.\left.-\left(\frac{x_i^\lambda a_i^{n-\lambda+1}}{(n-\lambda+1)} + \frac{x_i^{n-\lambda-1}b_i^{\lambda+2}}{(\lambda+2)}\right)\right\}\right]\right\}, \quad \forall(x_1,\ldots,x_m)\in\prod_{j=1}^{m}[a_j,b_j]. \quad (29.20)$$

Proof. We observe that

$$|B_i| \le \left(\frac{n^{i-1}}{(n-1)!\prod_{j=1}^{i}(b_j-a_j)}\right)\left(\int_{\prod_{j=1}^{i}[a_j,b_j]} |x_i-s_i|^{n-1}|k(s_i,x_i)|\right.$$

$$\left.\times\left|\frac{\partial^n f}{\partial x_i^n}(s_1,s_2,\ldots,s_i,x_{i+1},\ldots,x_m)\right|ds_1\cdots ds_i\right)$$

$$\le \frac{n^{i-1}}{(n-1)!(b_i-a_i)}\left\|\frac{\partial^n f}{\partial x_i^n}(\overbrace{\cdots}^{-i-},x_{i+1},\ldots,x_m)\right\|_\infty$$

$$\times\int_{a_i}^{b_i}|x_i-s_i|^{n-1}|k(s_i,x_i)|ds_i =: (*). \quad (29.21)$$

We find that

$$\int_{a_i}^{b_i}|x_i-s_i|^{n-1}|k(s_i,x_i)|ds_i$$

$$= \int_{a_i}^{x_i}|x_i-s_i|^{n-1}|k(s_i,x_i)|ds_i + \int_{x_i}^{b_i}|x_i-s_i|^{n-1}|k(s_i,x_i)|ds_i$$

$$= \int_{a_i}^{x_i}(x_i-s_i)^{n-1}(s_i-a_i)ds_i + \int_{x_i}^{b_i}(s_i-x_i)^{n-1}(b_i-s_i)ds_i$$

$$= \left[\left(\frac{b_i(b_i-x_i)^n - a_i(x_i-a_i)^n}{n}\right)\right.$$

$$+\sum_{\lambda=0}^{n-1}\binom{n-1}{\lambda}(-1)^{n-\lambda-1}\left\{\frac{(n+3)x_i^{n+1}}{(\lambda+2)(n-\lambda+1)}\right.$$

$$\left.\left.-\left(\frac{x_i^\lambda a_i^{n-\lambda+1}}{(n-\lambda+1)} + \frac{x_i^{n-\lambda-1}b_i^{\lambda+2}}{(\lambda+2)}\right)\right\}\right] =: \gamma_i. \quad (29.22)$$

Therefore we get

$$(*) \le \frac{n^{i-1}}{(n-1)!(b_i-a_i)}\left\|\frac{\partial^n f}{\partial x_i^n}(\overbrace{\cdots}^{-i-},x_{i+1},\ldots,x_m)\right\|_\infty \gamma_i. \quad (29.23)$$

Finally we find

$$|E_f(x_1,\ldots,x_m)| \le \sum_{i=1}^{m} \frac{n^{i-1}}{(n-1)!(b_i-a_i)} \left\| \frac{\partial^n f}{\partial x_i^n} (\overbrace{\cdots}^{-i-}, x_{i+1},\ldots,x_m) \right\|_{\infty} \gamma_i, \quad (29.24)$$

proving the claim. $\square$

We continue with

Theorem 29.10. *Assume all as in Theorem 29.6, and*

$$\frac{\partial^n f}{\partial x_i^n} \in L_{\infty}\left(\prod_{j=1}^{m}[a_j,b_j]\right), \quad i=1,\ldots,m.$$

Then

$$|E_f(x_1,x_2,\ldots,x_m)| \le \frac{\sum_{i=1}^{m}}{(n-1)!}\left\|\frac{\partial^n f}{\partial x_i^n}\right\|_{\infty}$$
$$\times \left\{ \frac{n^{i-1}}{(b_i-a_i)}\left[\left(\frac{b_i(b_i-x_i)^n - a_i(x_i-a_i)^n}{n}\right)\right.\right.$$
$$+ \sum_{\lambda=0}^{n-1}\binom{n-1}{\lambda}(-1)^{n-\lambda-1}\left\{\frac{(n+3)x_i^{n+1}}{(\lambda+2)(n-\lambda+1)} - \left(\frac{x_i^{\lambda}a_i^{n-\lambda+1}}{(n-\lambda+1)}\right)\right.$$
$$\left.\left.\left.+ \frac{x_i^{n-\lambda-1}b_i^{\lambda+2}}{(\lambda+2)}\right)\right\}\right]\right\}, \quad \forall(x_1,\ldots,x_m) \in \prod_{j=1}^{m}[a_i,b_j]. \quad (29.25)$$

Proof. From Theorem 29.9. $\square$

Next we give

Theorem 29.11. *Suppose all as in Theorem 29.6, and let $p,q > 1$ such that $\frac{1}{p} + \frac{1}{q} = 1$. Then*

$$|E_f(x_1,\ldots,x_m)| \le \frac{\sum_{i=1}^{m}}{(n-1)!}\left\{\left(\frac{n^{i-1}}{(b_i-a_i)\left(\prod_{j=1}^{i-1}(b_j-a_j)\right)^{1/p}}\right)\right.$$
$$\times \left\{\left((x_i-a_i)^{nq+1} + (b_i-x_i)^{nq+1}\right)B((n-1)q+1,q+1)\right\}^{1/q}$$
$$\times \left.\left\|\frac{\partial^n f}{\partial x_i^n}(\overbrace{\cdots}^{-i-}, x_{i+1},\ldots,x_m)\right\|_{p}\right\}, \quad \forall(x_1,\ldots,x_m) \in \prod_{j=1}^{m}[a_j,b_j]. \quad (29.26)$$

Proof. We notice that

$$|B_i| \le \left(\frac{n^{i-1}}{(n-1)! \prod_{j=1}^{i} (b_j - a_j)} \right) \left\| \frac{\partial^n f}{\partial x_i^n} (\overbrace{\cdots}^{-i-}, x_{i+1}, \ldots, x_m) \right\|_p$$

$$\times \left(\int_{a_i}^{b_i} |x_i - s_i|^{q(n-1)} |k(s_i, x_i)|^q ds_i \right)^{1/q} \left(\prod_{j=1}^{i-1} (b_j - a_j) \right)^{1/q}$$

$$\text{(by [242], p. 256)}$$

$$= \left(\frac{n^{i-1}}{(n-1)!(b_i - a_i)\left(\prod_{j=1}^{i-1} (b_j - a_j) \right)^{1/p}} \right)$$

$$\times \left\{ ((x_i - a_i)^{nq+1} + (b_i - x_i)^{nq+1}) B((n-1)q + 1, q + 1) \right\}^{1/q}$$

$$\times \left\| \frac{\partial^n f}{\partial x_i^n} (\overbrace{\cdots}^{-i-}, x_{i+1}, \ldots, x_m) \right\|_p , \tag{29.27}$$

proving the claim. $\qquad\square$

We also present

Theorem 29.12. *Suppose all as in Theorem 29.6. Then*

$$|E_f(x_1, \ldots, x_m)| \le \frac{\sum_{i=1}^{m}}{(n-1)!} \left\{ \frac{n^{i-1}}{\prod_{j=1}^{i} (b_j - a_j)} (\max(x_i - a_i, b_i - x_i))^n \right.$$

$$\left. \times \left\| \frac{\partial^n f}{\partial x_i^n} (\overbrace{\cdots}^{-i-}, x_{i+1}, \ldots, x_m) \right\|_1 \right\}, \quad \forall (x_1, \ldots, x_m) \in \prod_{j=1}^{m} [a_j, b_j]. \tag{29.28}$$

Proof. We observe that

$$|B_i| \le \left(\frac{n^{i-1}}{(n-1)! \prod_{j=1}^{i} (b_j - a_j)} \right) \left(\sup_{s_i \in [a_i, b_i]} |x_i - s_i|^{n-1} |k(s_i, x_i)| \right)$$

$$\times \left\| \frac{\partial^n f}{\partial x_i^n} (\overbrace{\cdots}^{-i-}, x_{i+1}, \ldots, x_m) \right\|_1$$

$$\leq \left(\frac{n^{i-1}}{(n-1)! \prod\limits_{j=1}^{i} (b_j - a_j)} \right) \left(\max(x_i - a_i, b_i - x_i) \right)^n$$

$$\times \left\| \frac{\partial^n f}{\partial x_i^n} (\overbrace{\cdots}^{-i-}, x_{i+1}, \ldots, x_m) \right\|_1, \tag{29.29}$$

proving the claim. $\qquad\square$

We give

Corollary 29.13. *Assume all as in Theorem 29.6. Then*

$$|E_f(x_1, \ldots, x_m)| \leq \min\{\text{R.H.S.}(29.20), \text{R.H.S.}(29.26), \text{R.H.S.}(29.28)\},$$

where $p, q > 1$: $\frac{1}{p} + \frac{1}{q} = 1$, $\forall (x_1, \ldots, x_m) \in \prod\limits_{i=1}^{m} [a_i, b_i]$.

Proof. By (29.20), (29.26) and (29.28). $\qquad\square$

We proceed with the comparison of integral means. For that we make

Remark 29.14. Let $[c_i, d_i] \subseteq [a_i, b_i]$, $i = 1, \ldots, m$. Let μ be a probability measure on

$$\left(\prod_{i=1}^{m} [c_i, d_i], \mathcal{P}\left(\prod_{i=1}^{m} [c_i, d_i] \right) \right),$$

$\mathcal{P}$ stands for the powerset. Then

$$\left| \int_{\prod\limits_{i=1}^{m} [c_i, d_i]} f(x_1, \ldots, x_m) d\mu - \frac{n^m}{\prod\limits_{j=1}^{m} (b_j - a_j)} \int_{\prod\limits_{j=1}^{m} [a_j, b_j]} f(s_1, \ldots, s_m) ds_1 \cdots ds_m \right.$$

$$\left. - \sum_{i=1}^{m} \int_{\prod\limits_{i=1}^{m} [c_i, d_i]} A_i(x_i, \ldots, x_m) d\mu \right|$$

$$= \left| \int_{\prod\limits_{i=1}^{m} [c_i, d_i]} E_f(x_1, \ldots, x_m) d\mu \right| \leq \int_{\prod\limits_{i=1}^{m} [c_i, d_i]} |E_f(x_1, \ldots, x_m)| d\mu. \tag{29.30}$$

We conclude

Theorem 29.15. *Suppose all as in Theorem 29.6, and $\frac{\partial^n f}{\partial x_i^n} \in L_\infty\left(\prod\limits_{j=1}^{m} [a_j, b_j] \right)$, $i = 1, \ldots, m$. Let $[c_i, d_i] \subseteq [a_i, b_i]$, $i = 1, \ldots, m$ and μ be a probability measure on*

$$\left(\prod_{i=1}^{m} [c_i, d_i], \mathcal{P}\left(\prod_{i=1}^{m} [c_i, d_i] \right) \right).$$

Then

$$\left| \int_{\prod\limits_{j=1}^{m}[c_j,d_j]} f(x_1,\dots,x_m)d\mu - \frac{n^m}{\prod\limits_{j=1}^{m}(b_j-a_j)} \int_{\prod\limits_{j=1}^{m}[a_j,b_j]} f(s_1,\dots,s_m)ds_1\cdots ds_m \right.$$

$$\left. - \sum_{i=1}^{m} \int_{\prod\limits_{i=1}^{m}[c_i,d_i]} A_i(x_i,\dots,x_m)d\mu \right|$$

$$\leq \frac{\sum\limits_{i=1}^{m}}{(n-1)!}\left\|\frac{\partial^n f}{\partial x_i^n}\right\|_\infty \left\{ \frac{n^{i-1}}{(b_i-a_i)}\left[\frac{1}{n}\left(b_i \int_{\prod\limits_{j=1}^{m}[c_j,d_j]} (b_i-x_i)^n d\mu\right.\right.\right.$$

$$\left. - a_i \int_{\prod\limits_{j=1}^{m}[c_j,d_j]} (x_i-a_i)^n d\mu\right)$$

$$+ \sum_{\lambda=0}^{n-1}\binom{n-1}{\lambda}(-1)^{n-\lambda-1}\left\{ \frac{(n+3)\int_{\prod\limits_{j=1}^{m}[c_j,d_j]} x_i^{n+1}d\mu}{(\lambda+2)(n-\lambda+1)} \right.$$

$$\left.\left.\left.\left. - \left(\frac{a_i^{n-\lambda+1}\int_{\prod\limits_{j=1}^{m}[c_j,d_j]} x_i^\lambda d\mu}{(n-\lambda+1)} + \frac{b_i^{\lambda+2}\int_{\prod\limits_{j=1}^{m}[c_j,d_j]} x_i^{n-\lambda-1}d\mu}{(\lambda+2)} \right) \right\}\right]\right\}\right\}. \tag{29.31}$$

Proof. We use (29.25) and (29.30). □

Similar results to (29.31) can be derived by integrating against μ the inequalities (29.26) and (29.28).

We continue with Grüss type inequalities

Remark 29.16. Let $f,g\colon \prod\limits_{i=1}^{m}[a_i,b_i]\to\mathbb{R}$ as in Theorem 29.6. Then by (29.1) we obtain

$$f(x_1,\dots,x_m) = \frac{n^m}{\prod\limits_{j=1}^{m}(b_j-a_j)} \int_{\prod\limits_{j=1}^{m}[a_j,b_j]} f(s_1,\dots,s_m)ds_1\cdots ds_m$$

$$+ \sum_{i=1}^{m} T_i^f(x_i,x_{i+1},\dots,x_m), \tag{29.32}$$

where T_i^f as in (29.2), and

$$g(x_1,\dots,x_m) = \frac{n^m}{\prod\limits_{j=1}^{m}(b_j-a_j)} \int_{\prod\limits_{j=1}^{m}[a_j,b_j]} g(s_1,\dots,s_m)ds_1\cdots ds_m$$

$$+ \sum_{i=1}^{m} T_i^g(x_i,x_{i+1},\dots,x_m),\forall(x_1,\dots,x_m)\in\prod_{i=1}^{m}[a_j,b_j]. \tag{29.33}$$

Therefore

$$
\begin{aligned}
&f(x_1,\ldots,x_m)g(x_1,\ldots,x_m)\\
&= \frac{n^m}{\prod\limits_{j=1}^{m}(b_j - a_j)} g(x_1,\ldots,x_m) \int_{\prod\limits_{j=1}^{m}[a_j,b_j]} f(s_1,\ldots,s_m)ds_1\cdots ds_m\\
&\quad + g(x_1,\ldots,x_m)\left(\sum_{i=1}^{m} T_i^f(x_i,\ldots,x_m)\right),
\end{aligned}
\tag{29.34}
$$

and

$$
\begin{aligned}
&f(x_1,\ldots,x_m)g(x_1,\ldots,x_m)\\
&= \frac{n^m}{\prod\limits_{j=1}^{m}(b_j - a_j)} f(x_1,\ldots,x_m) \int_{\prod\limits_{j=1}^{m}[a_j,b_j]} g(s_1,\ldots,s_m)ds_1\cdots ds_m\\
&\quad + f(x_1,\ldots,x_m)\left(\sum_{i=1}^{m} T_i^g(x_i,\ldots,x_m)\right).
\end{aligned}
\tag{29.35}
$$

Consequently after integration we get

$$
\begin{aligned}
&\int_{\prod\limits_{j=1}^{m}[a_j,b_j]} f(x_1,\ldots,x_m)g(x_1,\ldots,x_m)dx_1\cdots dx_m\\
&= \frac{n^m}{\prod\limits_{j=1}^{m}(b_j - a_j)} \left(\int_{\prod\limits_{j=1}^{m}[a_j,b_j]} g(x_1,\ldots,x_m)dx_1\cdots dx_m\right)\\
&\quad \times \left(\int_{\prod\limits_{j=1}^{m}[a_j,b_j]} f(x_1,\ldots,x_m)dx_1\cdots dx_m\right)\\
&\quad + \int_{\prod\limits_{j=1}^{m}[a_j,b_j]} \left[g(x_1,\ldots,x_m)\left(\sum_{i=1}^{m} T_i^f(x_i,\ldots,x_m)\right)\right]dx_1\cdots dx_m.
\end{aligned}
\tag{29.36}
$$

Similarly we have

$$
\begin{aligned}
&\int_{\prod\limits_{j=1}^{m}[a_j,b_j]} f(x_1,\ldots,x_m)g(x_1,\ldots,x_m)dx_1\cdots dx_m\\
&= \frac{n^m}{\prod\limits_{j=1}^{m}(b_j - a_j)} \left(\int_{\prod\limits_{j=1}^{m}[a_j,b_j]} f(x_1,\ldots,x_m)dx_1\cdots dx_m\right)\\
&\quad \times \left(\int_{\prod\limits_{j=1}^{m}[a_j,b_j]} g(x_1,\ldots,x_m)dx_1\cdots dx_m\right)\\
&\quad + \int_{\prod\limits_{j=1}^{m}[a_j,b_j]} \left[f(x_1,\ldots,x_m)\left(\sum_{i=1}^{m} T_i^g(x_i,\ldots,x_m)\right)\right]dx_1\cdots dx_m.
\end{aligned}
\tag{29.37}
$$

By (29.36) and (29.37) we obtain

$$\int_{\prod\limits_{j=1}^{m}[a_j,b_j]} f(x_1,\ldots,x_m)g(x_1,\ldots,x_m)dx_1\cdots dx_m$$

$$-\frac{n^m}{\prod\limits_{j=1}^{m}(b_j-a_j)}\left(\int_{\prod\limits_{j=1}^{m}[a_j,b_j]} f(x_1,\ldots,x_m)dx_1\cdots dx_m\right)$$

$$\times\left(\int_{\prod\limits_{j=1}^{m}[a_j,b_j]} g(x_1,\ldots,x_m)dx_1\cdots dx_m\right)$$

$$=\int_{\prod\limits_{j=1}^{m}[a_j,b_j]}\left[g(x_1,\ldots,x_m)\left(\sum_{i=1}^{m}T_i^f(x_i,\ldots,x_m)\right)\right]dx_1\cdots dx_m$$

$$=\int_{\prod\limits_{j=1}^{m}[a_j,b_j]}\left[f(x_1,\ldots,x_m)\left(\sum_{i=1}^{m}T_i^g(x_i,\ldots,x_m)\right)\right]dx_1\cdots dx_m. \quad (29.38)$$

We conclude that

$$\int_{\prod\limits_{j=1}^{m}[a_j,b_j]} fg\,d\overrightarrow{x}-\frac{n^m}{\prod\limits_{j=1}^{m}(b_j-a_j)}\left(\int_{\prod\limits_{j=1}^{m}[a_j,b_j]} f(\overrightarrow{x})d\overrightarrow{x}\right)$$

$$\times\left(\int_{\prod\limits_{j=1}^{m}[a_j,b_j]} g(\overrightarrow{x})d\overrightarrow{x}\right)=\frac{1}{2}\left[\int_{\prod\limits_{j=1}^{m}[a_j,b_j]}\left[g(\overrightarrow{x})\left(\sum_{i=1}^{m}T_i^f(x_i,\ldots,x_m)\right)\right.\right.$$

$$\left.\left.+f(\overrightarrow{x})\left(\sum_{i=1}^{m}T_i^g(x_i,\ldots,x_m)\right)\right]d\overrightarrow{x}\right]. \quad (29.39)$$

Remember by (29.17) for $i=1,\ldots,m$ that

$$T_i^f(x_i,\ldots,x_m)=A_i^f(x_i,\ldots,x_m)+B_i^f(x_i,\ldots,x_m), \quad (29.40)$$

and

$$T_i^g(x_i,\ldots,x_m)=A_i^g(x_i,\ldots,x_m)+B_i^g(x_i,\ldots,x_m). \quad (29.41)$$

We call and observe that

$$
\Delta_{(f,g)} := \int_{\prod_{j=1}^{m} [a_j,b_j]} f g \, d\overrightarrow{x} - \frac{n^m}{\prod_{j=1}^{m} (b_j - a_j)} \left(\int_{\prod_{j=1}^{m} [a_j,b_j]} f \, d\overrightarrow{x} \right)
$$

$$
\times \left(\int_{\prod_{j=1}^{m} [a_j,b_j]} g \, d\overrightarrow{x} \right) - \frac{1}{2} \left[\int_{\prod_{j=1}^{m} [a_j,b_j]} \left[g(\overrightarrow{x}) \left(\sum_{i=1}^{m} A_i^f (x_i, \ldots, x_m) \right) \right. \right.
$$

$$
\left. \left. + f(\overrightarrow{x}) \left(\sum_{i=1}^{m} A_i^g (x_i, \ldots, x_m) \right) \right] d\overrightarrow{x} \right]
$$

$$
= \frac{1}{2} \left[\int_{\prod_{j=1}^{m} [a_j,b_j]} \left[g(\overrightarrow{x}) \left(\sum_{i=1}^{m} B_i^f (x_i, \ldots, x_m) \right) \right. \right.
$$

$$
\left. \left. + f(\overrightarrow{x}) \left(\sum_{i=1}^{m} B_i^g (x_i, \ldots, x_m) \right) \right] d\overrightarrow{x} \right] =: \Gamma(f,g). \tag{29.42}
$$

Clearly we obtain that

$$
|\Delta_{(f,g)}| \le \frac{1}{2} \left[\|g\|_\infty \left\{ \sum_{i=1}^{m} \left(\prod_{j=1}^{i-1} (b_j - a_j) \right. \right. \right.
$$

$$
\times \int_{\prod_{j=i}^{m} [a_j,b_j]} |B_i^f (x_i, \ldots, x_m)| dx_i \cdots dx_m \left. \right) \right\}
$$

$$
+ \|f\|_\infty \left\{ \sum_{i=1}^{m} \left(\prod_{j=1}^{i-1} (b_j - a_j) \int_{\prod_{j=i}^{m} [a_j,b_j]} |B_i^g (x_i, \ldots, x_m)| dx_i \cdots dx_m \right) \right\} \right].
$$

$$
\tag{29.43}
$$

We state the established Grüss type result.

Theorem 29.17. *Let* $f, g \colon \prod_{i=1}^{m} [a_i, b_i] \to \mathbb{R}$ *as in Theorem 29.6. Then*

$$
\left| \int_{\prod_{j=1}^{m} [a_j,b_j]} f g \, d\overrightarrow{x} - \frac{n^m}{\prod_{j=1}^{m} (b_j - a_j)} \left(\int_{\prod_{j=1}^{m} [a_j,b_j]} f \, d\overrightarrow{x} \right) \right.
$$

$$
\times \left(\int_{\prod_{j=1}^{m} [a_j,b_j]} g \, d\overrightarrow{x} \right) - \frac{1}{2} \left[\int_{\prod_{j=1}^{m} [a_j,b_j]} \left[g(\overrightarrow{x}) \left(\sum_{i=1}^{m} A_i^f (x_i, \ldots, x_m) \right) \right. \right.
$$

$$
\left. \left. \left. + f(\overrightarrow{x}) \left(\sum_{i=1}^{m} A_i^g (x_i, \ldots, x_m) \right) \right] d\overrightarrow{x} \right] \right|
$$

$$\leq \frac{1}{2}\left[\|g\|_\infty\left\{\sum_{i=1}^m\left(\prod_{j=1}^{i-1}(b_j-a_j)\int_{\prod\limits_{j=i}^m[a_j,b_j]}|B_i^f(x_i,\ldots,x_m)|dx_i\cdots dx_m\right)\right\}\right.$$

$$\left.+\|f\|_\infty\left\{\sum_{i=1}^m\left(\prod_{j=1}^{i-1}(b_j-a_j)\int_{\prod\limits_{j=i}^m[a_j,b_j]}|B_i^g(x_i,\ldots,x_m)|dx_i\cdots dx_m\right)\right\}\right].(29.44)$$

We make

Remark 29.18. Let $p,q>1$: $\frac{1}{p}+\frac{1}{q}=1$. Suppose

$$B_i^f,B_i^q\in L_q\left(\prod_{j=i}^m[a_j,b_j]\right),\quad i=1,\ldots,m.$$

From (29.42) and Hölder's inequality we obtain

$$|\Gamma(f,g)|$$

$$\leq\frac{1}{2}\left[\sum_{i=1}^m\left(\int_{\prod\limits_{j=1}^m[a_j,b_j]}|g(\overrightarrow{x})||B_i^f(x_i,\ldots,x_m)|d\overrightarrow{x}\right.\right.$$

$$\left.\left.+\int_{\prod\limits_{j=1}^m[a_j,b_j]}|f(\overrightarrow{x})||B_i^g(x_i,\ldots,x_m)|d\overrightarrow{x}\right)\right]$$

$$\leq\frac{1}{2}\left\{\sum_{i=1}^m\left[\|g\|_p\left(\int_{\prod\limits_{j=1}^m[a_j,b_j]}|B_i^f(x_i,\ldots,x_m)|^qd\overrightarrow{x}\right)^{1/q}\right.\right.$$

$$\left.\left.+\|f\|_p\left(\int_{\prod\limits_{j=1}^m[a_j,b_j]}|B_i^g(x_i,\ldots,x_m)|^qd\overrightarrow{x}\right)^{1/q}\right]\right\}\qquad(29.45)$$

$$=\frac{1}{2}\left\{\sum_{i=1}^m\left[\left(\prod_{j=1}^{i-1}(b_j-a_j)\right)^{1/q}\left\{\|g\|_p\right.\right.\right.$$

$$\times\left(\int_{\prod\limits_{j=i}^m[a_j,b_j]}|B_i^f(x_i,\ldots,x_m)|^qdx_i\cdots dx_m\right)^{1/q}$$

$$\left.\left.\left.+\|f\|_p\left(\int_{\prod\limits_{j=i}^m[a_j,b_j]}|B_i^g(x_i,\ldots,x_m)|^qdx_i\cdots dx_m\right)^{1/q}\right\}\right]\right\}.\qquad(29.46)$$

We have established the following Grüss type result.

Theorem 29.19. *Let* $f,g\colon\prod_{i=1}^m[a_i,b_i]\to\mathbb{R}$ *as in Theorem 29.6. Let* $p,q>$
1: $\frac{1}{p}+\frac{1}{q}=1$. *Assume*

$$B_i^f,B_i^g\in L_q\left(\prod_{j=i}^m[a_j,b_j]\right),\quad i=1,\ldots,m.$$

Then

$$\left| \int_{\prod_{j=1}^{m} [a_j, b_j]} fg\, d\overrightarrow{x} - \frac{n^m}{\prod_{j=1}^{m} (b_j - a_j)} \left(\int_{\prod_{j=1}^{m} [a_j, b_j]} f\, d\overrightarrow{x} \right) \right.$$

$$\times \left(\int_{\prod_{j=1}^{m} [a_j, b_j]} g\, d\overrightarrow{x} \right) - \frac{1}{2} \left[\int_{\prod_{j=1}^{m} [a_j, b_j]} \left[g(\overrightarrow{x}) \left(\sum_{i=1}^{m} A_i^f(x_i, \ldots, x_m) \right) \right. \right.$$

$$\left. \left. + f(\overrightarrow{x}) \left(\sum_{i=1}^{m} A_i^g(x_i, \ldots, x_m) \right) \right] d\overrightarrow{x} \right] \Bigg| \leq \frac{1}{2} \left\{ \sum_{i=1}^{m} \left[\left(\prod_{j=1}^{i-1} (b_j - a_j) \right)^{1/q} \left\{ \|g\|_p \right. \right. \right.$$

$$\times \left(\int_{\prod_{j=i}^{m} [a_j, b_j]} |B_i^f(x_i, \ldots, x_m)|^q dx_i \cdots dx_m \right)^{1/q}$$

$$\left. \left. \left. + \|f\|_p \left(\int_{\prod_{j=i}^{m} [a_j, b_j]} |B_i^g(x_i, \ldots, x_m)|^q dx_i \cdots dx_m \right)^{1/q} \right\} \right] \right\}. \tag{29.47}$$

We derive the last Grüss type result.

Theorem 29.20. *Let* $f, g \colon \prod_{i=1}^{m} [a_i, b_i] \to \mathbb{R}$ *as in Theorem 29.6. Assume* $\frac{\partial^n f}{\partial x_i^n}$, $\frac{\partial^n g}{\partial x_i^n}$, $i = 1, \ldots, m$ *are continuous. Then*

$$\left| \int_{\prod_{j=1}^{m} [a_j, b_j]} fg\, d\overrightarrow{x} - \frac{n^m}{\prod_{j=1}^{m} (b_j - a_j)} \left(\int_{\prod_{j=1}^{m} [a_j, b_j]} f\, d\overrightarrow{x} \right) \right.$$

$$\times \left(\int_{\prod_{j=1}^{m} [a_j, b_j]} g\, d\overrightarrow{x} \right) - \frac{1}{2} \left[\int_{\prod_{j=1}^{m} [a_j, b_j]} \left[g(\overrightarrow{x}) \left(\sum_{i=1}^{m} A_i^f(x_i, \ldots, x_m) \right) \right. \right.$$

$$\left. \left. + f(\overrightarrow{x}) \left(\sum_{i=1}^{m} A_i^g(x_i, \ldots, x_m) \right) d\overrightarrow{x} \right] \right|$$

$$\leq \frac{1}{2(n-1)!} \left\{ \sum_{i=1}^{m} n^{i-1} (b_i - a_i)^n \left[\|g\|_1 \left\| \frac{\partial^n f}{\partial x_i^n} \right\|_\infty + \|f\|_1 \left\| \frac{\partial^n g}{\partial x_i^n} \right\|_\infty \right] \right\}. \tag{29.48}$$

Proof. Use of (29.45). $\qquad \square$

29.3 Applications

I) We treat the case of $n = 1$, $m = 2$. Let $[a_1, b_1] \times [a_2, b_2] \subseteq \mathbb{R}^2$.

General Assumptions 29.21. Let $f \colon [a_1, b_1] \times [a_2, b_2] \to \mathbb{R}$.

We assume

1) for $j = 1, 2$ we have that $f(s_1, x_2)$, $f(x_1, s_2)$ are absolutely continuous in $s_1 \in [a_1, b_1]$, $s_2 \in [a_2, b_2]$, for every $x_2 \in [a_2, b_2]$, $x_1 \in [a_1, b_1]$, respectively.

2) $\frac{\partial f}{\partial x_1}(s_1, x_2)$, $\frac{\partial f}{\partial x_2}(s_1, s_2)$ are continuous on $[a_1, b_1]$, $[a_1, b_1] \times [a_2, b_2]$, respectively, for every $x_2 \in [a_2, b_2]$.

3) f is continuous on $[a_1, b_1] \times [a_2, b_2]$.

We mention

Brief Assumptions 29.22. Let $f : [a_1, b_1] \times [a_2, b_2] \to \mathbb{R}$ with f, $\frac{\partial f}{\partial x_1}$, $\frac{\partial f}{\partial x_2}$ being continuous on $[a_1, b_1] \times [a_2, b_2]$.

We give

Corollary 29.23. *Let f as in General Assumptions 29.21 or Brief Assumptions 29.22, $(x_1, x_2) \in [a_1, b_1] \times [a_2, b_2]$. Then*

$$f(x_1, x_2) = \frac{1}{(b_1 - a_1)(b_2 - a_2)} \int_{[a_1,b_1] \times [a_2,b_2]} f(s_1, s_2) ds_1 ds_2 + T_1 + T_2, \quad (29.49)$$

where

$$T_1 = T_1(x_1, x_2) = \frac{1}{(b_1 - a_1)} \int_{a_1}^{b_1} K(s_1, x_1) \frac{\partial f}{\partial x_1}(s_1, x_2) ds_1$$

$$= B_1(x_1, x_2), \quad (29.50)$$

$$T_2 = T_2(x_2) = \frac{1}{(b_1 - a_1)(b_2 - a_2)} \int_{a_1}^{b_1} \int_{a_2}^{b_2} K(s_2, x_2) \frac{\partial f}{\partial x_2}(s_1, s_2) ds_1 ds_2$$

$$= B_2(x_2). \quad (29.51)$$

Proof. By Theorem 29.6. $\qquad\qquad\qquad\qquad\qquad\qquad\qquad\qquad\qquad\qquad\qquad \square$

We need

Remark 29.24. Denote here

$$E_f(x_1, x_2) = f(x_1, x_2) - \frac{1}{(b_1 - a_1)(b_2 - a_2)} \int_{a_1}^{b_1} \int_{a_2}^{b_2} f(s_1, s_2) ds_1 ds_2$$

$$= B_1 + B_2. \quad (29.52)$$

That is

$$|E_f(x_1, x_2)| \le |B_1| + |B_2|. \quad (29.53)$$

We give the following special Ostrowski type inequalities.

Corollary 29.25. *Suppose all as in Corollary 29.23. Then*

$$|E_f(x_1, x_2)| = \left| f(x_1, x_2) - \frac{1}{(b_1 - a_1)(b_2 - a_2)} \int_{a_1}^{b_1} \int_{a_2}^{b_2} f(s_1, s_2) ds_1 ds_2 \right|$$

$$\le \left\| \frac{\partial f}{\partial x_1}(\cdot, x_2) \right\|_{\infty, [a_1, b_1]} \frac{[a_1^2 + b_1^2 - x_1(a_1 + b_1)]}{(b_1 - a_1)}$$

$$+ \left\| \frac{\partial f}{\partial x_2}(\cdot, \cdot) \right\|_{\infty, [a_1, b_1] \times [a_2, b_2]} \frac{[a_2^2 + b_2^2 - x_2(a_2 + b_2)]}{(b_2 - a_2)},$$

$$\forall (x_1, x_2) \in [a_1, b_1] \times [a_2, b_2]. \quad (29.54)$$

Proof. By Theorem 29.9. $\qquad\qquad\qquad\qquad\qquad\qquad\qquad\qquad\qquad\qquad\qquad \square$

We continue with

Corollary 29.26. *Suppose all as in Corollary 29.23, and $\frac{\partial f}{\partial x_1}$, $\frac{\partial f}{\partial x_2} \in L_\infty([a_1, b_1] \times [a_2, b_2])$. Then*

$$|E_f(x_1, x_2)| \le \max\left\{\left\|\frac{\partial f}{\partial x_1}\right\|_\infty, \left\|\frac{\partial f}{\partial x_2}\right\|_\infty\right\}\left\{\left(\frac{a_1^2 + b_1^2 - x_1(a_1 + b_1)}{b_1 - a_1}\right)\right.$$

$$\left. + \left(\frac{a_2^2 + b_2^2 - x_2(a_2 + b_2)}{b_2 - a_2}\right)\right\}, \quad \forall(x_1, x_2) \in [a_1, b_1] \times [a_2, b_2]. \qquad (29.55)$$

Proof. By (29.54). $\qquad\qquad\square$

Next we have

Corollary 29.27. *Suppose all as in Corollary 29.23. Then*

$$\left| f(x_1, x_2) - \frac{1}{(b_1 - a_1)(b_2 - a_2)} \int_{a_1}^{b_1} \int_{a_2}^{b_2} f(s_1, s_2)ds_1 ds_2 \right|$$

$$\le \frac{1}{(b_1 - a_1)}\sqrt{\frac{(x_1 - a_1)^3 + (b_1 - x_1)^3}{3}}\left\|\frac{\partial f}{\partial x_1}(\cdot, x_2)\right\|_{2, [a_1, b_1]}$$

$$+ \frac{1}{(b_2 - a_2)\sqrt{b_1 - a_1}}\sqrt{\frac{(x_2 - a_2)^3 + (b_2 - x_2)^3}{3}}\left\|\frac{\partial f}{\partial x_2}(\cdot, \cdot)\right\|_{2, [a_1, b_1] \times [a_2, b_2]},$$

$$(29.56)$$

for all $(x_1, x_2) \in [a_1, b_1] \times [a_2, b_2]$.

Proof. By (29.26). $\qquad\qquad\square$

We present

Corollary 29.28. *Assume all as in Corollary 29.23. Then*

$$\left| f(x_1, x_2) - \frac{1}{(b_1 - a_1)(b_2 - a_2)} \int_{a_1}^{b_1} \int_{a_2}^{b_2} f(s_1, s_2)ds_1 ds_2 \right|$$

$$\le \frac{1}{b_1 - a_1}\max(x_1 - a_1, b_1 - x_1)\left\|\frac{\partial f}{\partial x_1}(\cdot, x_2)\right\|_{1, [a_1, b_1]}$$

$$+ \frac{1}{(b_1 - a_1)(b_2 - a_2)}\max(x_2 - a_2, b_2 - x_2)\left\|\frac{\partial f}{\partial x_2}(\cdot, \cdot)\right\|_{1, [a_1, b_1] \times [a_2, b_2]},$$

$$(29.57)$$

for all $(x_1, x_2) \in [a_1, b_1] \times [a_2, b_2]$.

Proof. By (29.28). $\qquad\qquad\square$

We finish Ostrowski type inequality applications for $n - 1$, $m - 2$ with

Corollary 29.29. *Assume all as in Corollary 29.23. Then*

$$\left| f(x_1, x_2) - \frac{1}{(b_1 - a_1)(b_2 - a_2)} \int_{a_1}^{b_1} \int_{a_2}^{b_2} f(s_1, s_2) ds_1 ds_2 \right|$$

$$\leq \ \min\{\text{R.H.S.}(29.54), \text{R.H.S.}(29.56), \text{R.H.S.}(29.57)\}, \tag{29.58}$$

for all $(x_1, x_2) \in [a_1, b_1] \times [a_2, b_2]$.

We proceed with the related comparison of means.

Corollary 29.30. *Suppose all as in Corollary 29.23, and* $\frac{\partial f}{\partial x_1}$, $\frac{\partial f}{\partial x_2} \in L_\infty\big([a_1, b_1] \times [a_2, b_2]\big)$. *Let* $[c_i, d_i] \subseteq [a_i, b_i]$, $i = 1, 2$ *and* μ *be a probability measure on* $\big([c_1, d_1] \times [c_2, d_2]\big)$, $\mathcal{P}\big([c_1, d_1] \times [c_2, d_2]\big)$. *Then*

$$\left| \int_{\prod_{j=1}^{2} [c_i, d_i]} f(x_1, x_2) d\mu(x_1, x_2) - \frac{1}{(b_1 - a_1)(b_2 - a_2)} \int_{\prod_{j=1}^{2} [a_j, b_j]} f(s_1, s_2) ds_1 ds_2 \right|$$

$$\leq \left\| \frac{\partial f}{\partial x_1} \right\|_\infty \left\{ \frac{1}{(b_1 - a_1)} \left[b_1 \left(b_1 - \int_{\prod_{j=1}^{2} [c_j, d_j]} x_1 d\mu \right) - a_1 \left(\int_{\prod_{j=1}^{2} [c_j, d_j]} x_1 d\mu - a_1 \right) \right] \right\}$$

$$+ \left\| \frac{\partial f}{\partial x_2} \right\|_\infty \left\{ \frac{1}{(b_2 - a_2)} \left[b_2 \left(b_2 - \int_{\prod_{j=1}^{2} [c_j, d_j]} x_2 d\mu \right) - a_2 \left(\int_{\prod_{j=1}^{2} [c_j, d_j]} x_2 d\mu - a_2 \right) \right] \right\}. \tag{29.59}$$

Proof. By (29.31). $\qquad\qquad\qquad\qquad\qquad\qquad\qquad\qquad\qquad\qquad\qquad\qquad\qquad \square$

We present the Grüss type inequalities for $n = 1$, $m = 2$.

Corollary 29.31. *Let* $f, g \colon [a_1, b_1] \times [a_2, b_2] \times \mathbb{R}$ *as in Corollary 29.23. Then*

$$\left| \int_{\prod_{j=1}^{2} [a_j, b_j]} f g \, d\vec{x} - \frac{1}{\prod_{j=1}^{2} (b_j - a_j)} \left(\int_{\prod_{j=1}^{2} [a_j, b_j]} f \, d\vec{x} \right) \right.$$

$$\left. \times \left(\int_{\prod_{j=1}^{2} [a_j, b_j]} g \, d\vec{x} \right) \right| \leq \frac{1}{2} \left[\|g\|_\infty \left\{ \int_{\prod_{j=1}^{2} [a_j, b_j]} |B_1^f(x_1, x_2)| dx_1 dx_2 \right. \right.$$

$$\left. + (b_1 - a_1) \int_{a_2}^{b_2} |B_2^f(x_2)| dx_2 \right\} + \|f\|_\infty \left\{ \int_{\prod_{j=1}^{2} [a_j, b_j]} |B_1^g(x_1, x_2)| dx_1 dx_2 \right.$$

$$\left. \left. + (b_1 - a_1) \int_{a_2}^{b_2} |B_2^g(x_2)| dx_2 \right\} \right]. \tag{29.60}$$

Proof. By (29.44). $\qquad\qquad\qquad\qquad\qquad\qquad\qquad\qquad\qquad\qquad\qquad\qquad\qquad \square$

We continue with

Corollary 29.32. *Let $f, g\colon [a_1, b_1] \times [a_2, b_2] \to \mathbb{R}$ as in Corollary 29.23. Suppose*

$$B_i^f, B_i^g \in L_2\left(\prod_{j=i}^{2}[a_j, b_j]\right), \quad i = 1, 2.$$

Then

$$\left| \int_{\prod_{j=1}^{2}[a_j,b_j]} fg\,d\overrightarrow{x} - \frac{1}{\prod_{j=1}^{2}(b_j - a_j)}\left(\int_{\prod_{j=1}^{2}[a_j,b_j]} f\,d\overrightarrow{x} \right) \right.$$

$$\times \left.\left(\int_{\prod_{j=1}^{2}[a_j,b_j]} g\,d\overrightarrow{x} \right) \right| \le \frac{1}{2}\left\{ \left[\|g\|_2\left(\int_{\prod_{j=1}^{2}[a_j,b_j]} (B_1^f(x_1,x_2))^2\,dx_1dx_2 \right)^{1/2} \right. \right.$$

$$\left. + \|f\|_2\left(\int_{\prod_{j=1}^{2}[a_j,b_j]} (B_1^g(x_1,x_2))^2\,dx_1dx_2 \right)^{1/2} \right]$$

$$\left. + \left[\sqrt{b_1 - a_1}\left\{ \|g\|_2\left(\int_{a_2}^{b_2} (B_2^f(x_2))^2\,dx_2 \right)^{1/2} + \|f\|_2\left(\int_{a_2}^{b_2} (B_2^g(x_2))^2\,dx_2 \right)^{1/2} \right\} \right] \right\}.$$

$$(29.61)$$

Proof. By (29.47). $\qquad\square$

We also give

Corollary 29.33. *Let $f, g\colon [a_1, b_1] \times [a_2, b_2] \to \mathbb{R}$ as in Corollary 29.23. Suppose $\frac{\partial f}{\partial x_i}, \frac{\partial g}{\partial x_i}, i = 1, 2$ are continuous. Then*

$$\left| \int_{\prod_{j=1}^{2}[a_j,b_j]} fg\,d\overrightarrow{x} - \frac{1}{\prod_{j=1}^{2}(b_j - a_j)}\left(\int_{\prod_{j=1}^{2}[a_j,b_j]} f\,d\overrightarrow{x} \right) \right.$$

$$\times \left.\left(\int_{\prod_{j=1}^{2}[a_j,b_j]} g\,d\overrightarrow{x} \right) \right| \le \frac{1}{2}\left\{ (b_1 - a_1)\left[\|g\|_1\left\|\frac{\partial f}{\partial x_1}\right\|_\infty + \|f\|_1\left\|\frac{\partial g}{\partial x_1}\right\|_\infty \right] \right.$$

$$\left. + (b_2 - a_2)\left[\|g\|_1\left\|\frac{\partial f}{\partial x_2}\right\|_\infty + \|f\|_1\left\|\frac{\partial g}{\partial x_2}\right\|_\infty \right] \right\}.$$

$$(29.62)$$

Proof. By (29.48). $\qquad\square$

II) Here we treat the case of $n = 2$, $m = 3$. Let $\prod_{j=1}^{3}[a_j, b_j] \subseteq \mathbb{R}^3$.

General Assumptions 29.34. Let $f\colon \prod_{j=1}^{3}[a_j, b_j] \to \mathbb{R}$.

We assume

1) that $\frac{\partial f}{\partial x_1}(s_1, x_2, x_3)$, $\frac{\partial f}{\partial x_2}(x_1, s_2, x_3)$, $\frac{\partial f}{\partial x_3}(x_1, x_2, s_3)$ are absolutely continuous in s_1, s_2, s_3, respectively, for every $(x_1, x_2, x_3) \in \prod_{j=1}^{3} [a_j, b_j]$.

2) We have that $\frac{\partial^2 f}{\partial x_1^2}(s_1, x_2, x_3)$, $\frac{\partial^2 f}{\partial x_2^2}(s_1, s_2, x_3)$, $\frac{\partial^2 f}{\partial x_3^2}(s_1, s_2, s_3)$ are continuous on $[a_1, b_1]$, $[a_1, b_1] \times [a_2, b_2]$, $\prod_{j=1}^{3} [a_j, b_j]$, respectively, for any $(x_2, x_3) \in \prod_{j=2}^{3} [a_j, b_j]$.

3) f is continuous on $\prod_{j=1}^{3} [a_j, b_j]$.

We mention

Brief Assumptions 29.35. Let $f \colon \prod_{j=1}^{3} [a_j, b_j] \to \mathbb{R}$ with $\frac{\partial^\ell f}{\partial x_j^\ell}$ for $\ell = 0, 1, 2$; $j = 1, 2, 3$, are continuous on $\prod_{j=1}^{3} [a_j, b_j]$.

We present

Corollary 29.36. *Let f as in General Assumptions 29.34 or Brief Assumptions 29.35, $(x_1, x_2, x_3) \in \prod_{j=1}^{3} [a_j, b_j]$. Then*

$$f(x_1, x_2, x_3) = \frac{8}{\prod_{j=1}^{3} (b_j - a_j)} \int_{\prod_{j=1}^{3} [a_j, b_j]} f(s_1, s_2, s_3) ds_1 ds_2 ds_3 + \sum_{i=1}^{3} T_i. \tag{29.63}$$

Here for $i = 1, 2, 3$ we put

$$T_1 = T_1(x_1, x_2, x_3) = \frac{1}{(b_1 - a_1)}$$

$$\times \left\{ \left(f(b_1, x_2, x_3)(x_1 - b_1) - f(a_1, x_2, x_3)(x_1 - a_1) \right) \right.$$

$$\left. + \int_{a_1}^{b_1} (x_1 - s_1) k(s_1, x_1) \frac{\partial^2 f}{\partial x_1^2}(s_1, x_2, x_3) ds_1 \right\}, \tag{29.64}$$

$$T_2 = T_2(x_2, x_3) = \frac{2}{(b_1 - a_1)(b_2 - a_2)} \left\{ \int_{a_1}^{b_1} \left(f(s_1, b_2, x_3)(x_2 - b_2) \right. \right.$$

$$\left. - f(s_1, a_2, x_3)(x_2 - a_2) \right) ds_1$$

$$\left. + \int_{\prod_{j=1}^{2} [a_j, b_j]} (x_2 - s_2) k(s_2, x_2) \frac{\partial^2 f}{\partial x_2^2}(s_1, s_2, x_3) ds_1 ds_2 \right\}, \tag{29.65}$$

$$T_3 = T_3(x_3) = \frac{4}{\prod\limits_{j=1}^{3}(b_j - a_j)}\left\{ \int_{\prod\limits_{j=1}^{2}[a_j,b_j]} \left(f(s_1,s_2,b_3)(x_3 - b_3) \right. \right.$$

$$- f(s_1,s_2,a_3)(x_3 - a_3))ds_1 ds_2$$

$$+ \int_{\prod\limits_{j=1}^{3}[a_j,b_j]} (x_3 - s_3)k(s_3,x_3)\frac{\partial^2 f}{\partial x_3^2}(s_1,s_2,s_3)ds_1 ds_2 ds_3 \right\}. \tag{29.66}$$

We need

Remark 29.37. We denote

$$A_1 = A_1(x_1,x_2,x_3) = \frac{f(b_1,x_2,x_3)(x_1 - b_1) - f(a_1,x_2,x_3)(x_1 - a_1)}{b_1 - a_1}, \tag{29.67}$$

$$B_1 = B_1(x_1,x_2,x_3)$$

$$= \frac{1}{(b_1 - a_1)}\int_{a_1}^{b_1} (x_1 - s_1)k(s_1,x_1)\frac{\partial^2 f}{\partial x_1^2}(s_1,x_2,x_3)ds_1, \tag{29.68}$$

also

$$A_2 = A_2(x_2,x_3) = \frac{2}{(b_1 - a_1)(b_2 - a_2)}\int_{a_1}^{b_1} \left(f(s_1,b_2,x_3)(x_2 - b_2) \right.$$

$$- f(s_1,a_2,x_3)(x_2 - a_2))ds_1, \tag{29.69}$$

$$B_2 = B_2(x_2,x_3) = \frac{2}{(b_1 - a_1)(b_2 - a_2)}\int_{\prod\limits_{j=1}^{2}[a_j,b_j]} (x_2 - s_2)k(s_2,x_2)$$

$$\times \frac{\partial^2 f}{\partial x_2^2}(s_1,s_2,x_3)ds_1 ds_2, \tag{29.70}$$

and

$$A_3 = A_3(x_3) = \frac{4}{\prod\limits_{j=1}^{3}(b_j - a_j)}\int_{\prod\limits_{j=1}^{2}[a_j,b_j]} \left(f(s_1,s_2,b_3)(x_3 - b_3) \right.$$

$$- f(s_1,s_2,a_3)(x_3 - a_3))ds_1 ds_2, \tag{29.71}$$

$$B_3 = B_3(x_3) = \frac{4}{\prod\limits_{j=1}^{3}(b_j - a_j)}\int_{\prod\limits_{j=1}^{3}[a_j,b_j]} (x_3 - s_3)K(s_3,x_3)$$

$$\times \frac{\partial^2 f}{\partial x_3^2}(s_1,s_2,s_3)ds_1 ds_2 ds_3. \tag{29.72}$$

Clearly here

$$T_i = A_i + B_i, \quad i = 1,2,3. \tag{29.73}$$

We also denote

$$E_f(x_1,x_2,x_3) := f(x_1,x_2,x_3) - \frac{8}{\prod\limits_{j=1}^{3}(b_j - a_j)}$$

$$\times \int_{\prod\limits_{j=1}^{3}[a_j,b_j]} f(s_1,s_2,s_3)ds_1 ds_2 ds_3 - \sum_{i-1}^{3} A_i = \sum_{i-1}^{3} B_i. \tag{29.74}$$

Therefore

$$|E_f(x_1, x_2, x_3)| \le \sum_{i=1}^{3} |B_i|. \tag{29.75}$$

We give

Corollary 29.38. *Assume all as in Corollary 29.36. Then*

$$|E_f(x_1, x_2, x_3)| \le \sum_{i=1}^{3} \left\{ \left\| \frac{\partial^2 f}{\partial x_i^2} (\overbrace{\cdots}^{-i-}, x_3) \right\|_\infty \frac{2^{i-1}}{(b_i - a_i)} \right.$$
$$\times \left[\left(\frac{b_i(b_i - x_i)^2 - a_i(x_i - a_i)^2}{2} \right) + \sum_{\lambda=0}^{1} (-1)^{1-\lambda} \left\{ \frac{5x_i^3}{(\lambda+2)(3-\lambda)} \right. \right.$$
$$\left. \left. \left. - \left(\frac{x_i^\lambda a_i^{3-\lambda}}{3-\lambda} + \frac{x_i^{1-\lambda} b_i^{\lambda+2}}{\lambda+2} \right) \right\} \right] \right\}, \quad \forall(x_1, x_2, x_3) \in \prod_{j=1}^{3} [a_j, b_j]. \tag{29.76}$$

Proof. By (29.20). $\qquad\square$

We continue with

Corollary 29.39. *Suppose all as in Corollary 29.36, and*

$$\frac{\partial^2 f}{\partial x_i^2} \in L_\infty \left(\prod_{j=1}^{3} [a_j, b_j] \right), \quad i = 1, 2, 3.$$

Then

$$|E_f(x_1, x_2, x_3)| \le \sum_{i=1}^{3} \left\| \frac{\partial^2 f}{\partial x_i^2} \right\|_\infty \left\{ \frac{2^{i-1}}{(b_i - a_i)} \right.$$
$$\times \left[\left(\frac{b_i(b_i - x_i)^2 - a_i(x_i - a_i)^2}{2} \right) + \sum_{\lambda=0}^{1} (-1)^{1-\lambda} \left\{ \frac{5x_i^3}{(\lambda+2)(3-\lambda)} \right. \right.$$
$$\left. \left. \left. - \left(\frac{x_i^\lambda a_i^{3-\lambda}}{3-\lambda} + \frac{x_i^{1-\lambda} b_i^{\lambda+2}}{\lambda+2} \right) \right\} \right] \right\}, \quad \forall(x_1, x_2, x_3) \in \prod_{j=1}^{3} [a_j b_j]. \tag{29.77}$$

Proof. By (29.25). $\qquad\square$

Next we give

Corollary 29.40. *Assume all as in Corollary 29.36. Then*

$$|E_f(x_1, x_2, x_3)| \le \sum_{i=1}^{3} \left\{ \left(\frac{2^{i-1}}{(b_i - a_i)\left(\prod_{j=1}^{i-1} (b_j - a_j) \right)^{1/2}} \right) \right.$$
$$\left. \times \left(\sqrt{\frac{(x_i - a_i)^5 + (b_i - x_i)^5}{30}} \right) \left\| \frac{\partial^2 f}{\partial x_i^2} (\overbrace{\cdots}^{-i-}, x_3) \right\|_2 \right\}, \tag{29.78}$$

for all $(x_1, x_2, x_3) \in \prod_{j=1}^{3} [a_j, b_j]$.

Proof. By (29.26). $\qquad\square$

We also present

Corollary 29.41. *Suppose all as in Corollary 29.36. Then*

$$|E_f(x_1, x_2, x_3)| \leq \sum_{i=1}^{3} \left\{ \frac{2^{i-1}}{\prod\limits_{j=1}^{i}(b_j - a_j)} \left(\max(x_i - a_i, b_i - x_i)\right)^2 \right.$$

$$\left. \times \left\| \frac{\partial^2 f}{\partial x_i^2}(\overbrace{\cdots}^{-i-}, x_3) \right\|_1 \right\}, \qquad \forall (x_1, x_2, x_3) \in \prod_{j=1}^{3}[a_j, b_j]. \qquad (29.79)$$

Proof. By (29.28). $\qquad \square$

We give

Corollary 29.42. *Assume all as in Corollary 29.36. Then*

$$|E_f(x_1, x_2, x_3)| \leq \min\{\text{R.H.S.}(29.76), \text{R.H.S.}(29.78), \text{R.H.S.}(29.79)\}, \quad (29.80)$$

for all $(x_1, x_2, x_3) \in \prod_{j=1}^{3}[a_j, b_j]$.

Proof. By Corollary 29.13. $\qquad \square$

Next we do the comparison of means.

Corollary 29.43. *Assume all as in Corollary 29.36, and*

$$\frac{\partial^2 f}{\partial x_i^2} \in L_\infty\left(\prod_{j=1}^{3}[a_j, b_j]\right), \quad i = 1, 2, 3.$$

Let $[c_i, d_i] \subseteq [a_i, b_i]$, $i = 1, 2, 3$ *and* μ *be a probability measure on*

$$\left(\prod_{i=1}^{3}[c_i, d_i], \ \mathcal{P}\left(\prod_{i=1}^{3}[c_i, d_i]\right)\right).$$

Then

$$\left| \int_{\prod\limits_{j=1}^{3}[c_j,d_j]} f(x_1, x_2, x_3)d\mu - \frac{8}{\prod\limits_{j=1}^{3}(b_j - a_j)} \int_{\prod\limits_{j=1}^{3}[a_j,b_j]} f(s_1, s_2, s_3)ds_1 ds_2 ds_3 \right.$$

$$\left. - \sum_{i=1}^{3} \int_{\prod\limits_{j=1}^{3}[c_j,d_j]} A_i(x_i, \cdot, x_3)d\mu \right|$$

$$\leq \sum_{i=1}^{3} \left\| \frac{\partial^2 f}{\partial x_i^2} \right\|_\infty \left\{ \frac{2^{i-1}}{(b_i - a_i)} \left[\frac{1}{2}\left(b_i \int_{\prod\limits_{j=1}^{3}[c_j,d_j]} (b_i - x_i)^2 d\mu \right. \right. \right.$$

$$\left. - a_i \int_{\prod\limits_{j=1}^{3}[c_j,d_j]} (x_i - a_i)^2 d\mu \right) + \sum_{\lambda=0}^{1}(-1)^{1-\lambda}\left\{ \frac{5\int_{\prod\limits_{j=1}^{3}[c_j,d_j]} x_i^3 d\mu}{(\lambda+2)(3-\lambda)} \right.$$

$$\left. \left. \left. - \left(\frac{a_i^{3-\lambda}\int_{\prod\limits_{j=1}^{3}[c_j,d_j]} x_i^\lambda d\mu}{(3-\lambda)} + \frac{b_i^{\lambda+2}\int_{\prod\limits_{j=1}^{3}[c_j,d_j]} x_i^{1-\lambda} d\mu}{(\lambda+2)} \right) \right\} \right] \right\}. \qquad (29.81)$$

Proof. By (29.31). $\qquad \square$

We continue with the Grüss type inequalities for the case of $n = 2$, $m = 3$.

Corollary 29.44. *Suppose all as in Corollary 29.36 for $f, g \colon \prod_{j=1}^{3}[a_j, b_j] \to \mathbb{R}$. Then*

$$
\left| \int_{\prod_{j=1}^{3}[a_j,b_j]} fg\, d\overrightarrow{x} - \frac{8}{\prod_{j=1}^{3}(b_j - a_j)} \left(\int_{\prod_{j=1}^{3}[a_j,b_j]} f\, d\overrightarrow{x} \right) \right.
$$

$$
\times \left(\int_{\prod_{j=1}^{3}[a_j,b_j]} g\, d\overrightarrow{x} \right) - \frac{1}{2} \left[\int_{\prod_{j=1}^{3}[a_j,b_j]} \left[g(\overrightarrow{x}) \left(\sum_{i=1}^{3} A_i^f(x_i, \cdot, x_3) \right) \right.\right.
$$

$$
\left.\left.\left. + f(\overrightarrow{x}) \left(\sum_{i=1}^{3} A_i^g(x_i, \cdot, x_3) \right) \right] d\overrightarrow{x} \right] \right|
$$

$$
\leq \frac{1}{2} \left[\|g\|_\infty \left\{ \sum_{i=1}^{3} \left(\prod_{j=1}^{i-1}(b_j - a_j) \int_{\prod_{j=i}^{3}[a_j,b_j]} |B_i^f(x_i, \cdot, x_3)|\, dx_i \cdots dx_3 \right) \right\} \right.
$$

$$
\left. + \|f\|_\infty \left\{ \sum_{i=1}^{3} \left(\prod_{j=1}^{i-1}(b_j - a_j) \int_{\prod_{j=i}^{3}[a_j,b_j]} |B_i^g(x_i, \cdot, x_3)|\, dx_i \cdots dx_3 \right) \right\} \right]. \quad (29.82)
$$

Proof.   By (29.44). $\square$

It follows

Corollary 29.45. *Let $f, g \colon \prod_{j=1}^{3}[a_j, b_j] \to \mathbb{R}$ as in Corollary 29.36. Suppose*

$$
B_i^f, B_i^g \in L_2 \left(\prod_{j=i}^{3}[a_j, b_j] \right), \quad i = 1, 2, 3.
$$

Then

$$
\left| \int_{\prod_{j=1}^{3}[a_j,b_j]} fg\, d\overrightarrow{x} - \frac{8}{\prod_{j=1}^{3}(b_j - a_j)} \left(\int_{\prod_{j=1}^{3}[a_j,b_j]} f\, d\overrightarrow{x} \right) \right.
$$

$$
\times \left(\int_{\prod_{j=1}^{3}[a_j,b_j]} g\, d\overrightarrow{x} \right) - \frac{1}{2} \left[\int_{\prod_{j=1}^{3}[a_j,b_j]} \left[g(\overrightarrow{x}) \left(\sum_{i=1}^{3} A_i^f(x_i, \cdot, x_3) \right) \right.\right.
$$

$$
\left.\left.\left. + f(\overrightarrow{x}) \left(\sum_{i=1}^{3} A_i^g(x_i, \cdot, x_3) \right) \right] d\overrightarrow{x} \right] \right| \leq \frac{1}{2} \left\{ \sum_{i=1}^{3} \left[\left(\prod_{j=1}^{i-1}(b_j - a_j) \right)^{1/2} \left\{ \|g\|_2 \right.\right.\right.
$$

$$
\times \left(\int_{\prod_{j=i}^{3}[a_j,b_j]} (B_i^f(x_i, \cdot, x_3))^2\, dx_i \cdots dx_3 \right)^{1/2}
$$

$$
\left.\left.\left. + \|f\|_2 \left(\int_{\prod_{j=i}^{3}[a_j,b_j]} (B_i^g(x_i, \cdot, x_3))^2\, dx_i \cdots dx_3 \right)^{1/2} \right\} \right] \right\}. \quad (29.83)
$$

Proof. By (29.47). $\qquad\square$

We also have

Corollary 29.46. *Let* $f, g\colon \prod_{j=1}^{3} [a_j, b_j] \to \mathbb{R}$ *as in Corollary 29.36. Suppose* $\frac{\partial^2 f}{\partial x_i^2}$, $\frac{\partial^2 g}{\partial x_i^2}$, $i = 1, 2, 3$ *are continuous. Then*

$$\left| \int_{\prod_{j=1}^{3} [a_j, b_j]} fg\, d\overrightarrow{x} - \frac{8}{\prod_{j=1}^{3} (b_j - a_j)} \left(\int_{\prod_{j=1}^{3} [a_j, b_j]} f\, d\overrightarrow{x} \right) \right.$$

$$\times \left(\int_{\prod_{j=1}^{3} [a_j, b_j]} g\, d\overrightarrow{x} \right) - \frac{1}{2} \left[\int_{\prod_{j=1}^{3} [a_j, b_j]} \left[g(\overrightarrow{x}) \left(\sum_{i=1}^{3} A_i^f (x_i, \cdot, x_3) \right) \right. \right.$$

$$\left. \left. \left. + f(\overrightarrow{x}) \left(\sum_{i=1}^{3} A_i^g (x_i, \cdot, x_3) \right) \right] d\overrightarrow{x} \right] \right|$$

$$\leq \frac{1}{2} \left\{ \sum_{i=1}^{3} 2^{i-1} (b_i - a_i)^2 \left[\|g\|_1 \left\| \frac{\partial^2 f}{\partial x_i^2} \right\|_\infty + \|f\|_1 \left\| \frac{\partial^2 f}{\partial x_i^2} \right\|_\infty \right] \right\}. \qquad (29.84)$$

Proof. By (29.48). $\qquad\square$

Finally we give a second inductive proof of Theorem 29.6.

Another **Proof of Theorem 29.6.**

We have that

$$f(x_1, \ldots, x_m, x_{m+1})$$

$$= \frac{n^m}{\prod_{j=1}^{m} (b_j - a_j)} \int_{\prod_{j=1}^{m} [a_j, b_j]} f(s_1, \ldots, s_m, x_{m+1}) ds_1 \cdots ds_m + \sum_{i=1}^{m} T_i, \quad (29.85)$$

where for $i, \ldots, m$, we put

$$T_i := T_i(x_i, \ldots, x_m, x_{m+1}) := \left(\frac{n^{i-1}}{\prod_{j=1}^{i} (b_j - a_j)} \right) \sum_{k=1}^{n-1} \left(\frac{(n-k)}{k!} \right)$$

$$\left\{ \int_{\prod_{j=1}^{i-1} [a_j, b_j]} \left(\frac{\partial^{k-1} f}{\partial x_i^{k-1}} (s_1, \ldots, s_{i-1}, b_i, x_{i+1}, \ldots, x_m, x_{m+1})(x_i - b_i)^k \right. \right.$$

$$\left. \left. - \frac{\partial^{k-1} f}{\partial x_i^{k-1}} (s_1, \ldots, s_{i-1}, a_i, x_{i+1}, \ldots, x_m, x_{m+1})(x_i - a_i)^k \right) ds_1 \cdots ds_{i-1} \right\}$$

$$+ \left(\frac{n^{i-1}}{(n-1)! \prod_{j=1}^{i} (b_j - a_j)} \right) \left(\int_{\prod_{j=1}^{i} [a_j, b_j]} (x_i - s_i)^{n-1} k(s_i, x_i) \right.$$

$$\left. \frac{\partial^n f}{\partial x_i^n} (s_1, \ldots, s_i, x_{i+1}, \ldots, x_m, x_{m+1}) ds_1 \ldots ds_i \right). \qquad (29.86)$$

But by Fink's identity, see (29.3), we have

$$
\begin{aligned}
&f(s_1, s_m, x_{m+1}) \\
&= \frac{n}{(b_{m+1} - a_{m+1})} \int_{a_{m+1}}^{b_{m+1}} f(s_1, \ldots, s_m, s_{m+1}) ds_{m+1} \\
&\quad + \frac{1}{(b_{m+1} - a_{m+1})} \sum_{k=1}^{n-1} \left(\frac{n-k}{k!} \right) \left(\frac{\partial^{k-1} f}{\partial x_{m+1}^{k-1}} (s_1, \ldots, s_m, b_{m+1})(x_{m+1} - b_{m+1})^k \right. \\
&\quad \left. - \frac{\partial^{k-1} f}{\partial x_{m+1}^{k-1}} (s_1, \ldots, s_m, a_{m+1})(x_{m+1} - a_{m+1})^k \right) \\
&\quad + \frac{1}{(n-1)!(b_{m+1} - a_{m+1})} \int_{a_{m+1}}^{b_{m+1}} (x_{m+1} - s_{m+1})^{n-1} \\
&\quad k\left(s_{m+1}, x_{m+1} \right) \frac{\partial^n f}{\partial x_{m+1}^n} (s_1, \ldots, s_m, s_{m+1}) ds_{m+1}.
\end{aligned}
\tag{29.87}
$$

Next we put (29.87) into (20.85). We obtain

$$
\begin{aligned}
&f(x_1, \ldots, x_{m+1}) \\
&= \frac{n^m}{\prod_{j=1}^{m}(b_j - a_j)} \int_{\prod_{j=1}^{m}[a_j, b_j]} \left\{ \frac{n}{(b_{m+1} - a_{m+1})} \int_{a_{m+1}}^{b_{m+1}} f(s_1, \ldots, s_m, s_{m+1}) ds_{m+1} \right. \\
&\quad + \frac{1}{(b_{m+1} - a_{m+1})} \sum_{k=1}^{n-1} \left(\frac{n-k}{k!} \right) \left(\frac{\partial^{k-1} f}{\partial x_{m+1}^{k-1}} (s_1, \ldots, s_m, b_{m+1})(x_{m+1} - b_{m+1})^k \right. \\
&\quad \left. - \frac{\partial^{k-1} f}{\partial x_{m+1}^{k-1}} (s_1, \ldots, s_m, a_{m+1})(x_{m+1} - a_{m+1})^k \right) \\
&\quad + \frac{1}{(n-1)!(b_{m+1} - a_{m+1})} \int_{a_{m+1}}^{b_{m+1}} (x_{m+1} - s_{m+1})^{n-1} \\
&\quad \left. k\left(s_{m+1}, x_{m+1} \right) \frac{\partial^n f}{\partial x_{m+1}^n} (s_1, \ldots, s_m, s_{m+1}) ds_{m+1} \right\} ds_1 \ldots ds_m + \sum_{i=1}^{m} T_i.
\end{aligned}
$$

Therefore

$$
\begin{aligned}
&f(x_1, \ldots, x_{m+1}) \\
&= \frac{n^{m+1}}{\prod_{j=1}^{m+1}(b_j - a_j)} \int_{\prod_{j=1}^{m+1}[a_j, b_j]} f(s_1, \ldots, s_{m+1}) ds_1 \ldots ds_{m+1} + \sum_{i=1}^{m+1} T_i,
\end{aligned}
$$

where

$$
T_{m+1} := T_{m+1}(x_{m+1}) := \frac{n^m}{\prod_{j=1}^{m+1}(b_j - a_j)} \sum_{k=1}^{n-1} \left(\frac{n-k}{k!} \right)
$$

$$\int_{\prod\limits_{j=1}^{m}[a_j,b_j]}\left(\frac{\partial^{k-1}f}{\partial x_{m+1}^{k-1}}(s_1,\ldots,s_m,b_{m+1})(x_{m+1}-b_{m+1})^k\right.$$

$$\left.-\frac{\partial^{k-1}f}{\partial x_{m+1}^{k-1}}(s_1,\ldots,s_m,a_{m+1})(x_{m+1}-a_{m+1})^k\right)ds_1\ldots ds_m$$

$$+\frac{n^m}{(n-1)!\prod\limits_{j=1}^{m+1}(b_j-a_j)}\int_{\prod\limits_{j=1}^{m+1}[a_j,b_j]}(x_{m+1}-s_{m+1})^{n-1}$$

$$k\left(s_{m+1},x_{m+1}\right)\frac{\partial^n f}{\partial x_{m+1}^n}(s_1,\ldots,s_m,s_{m+1})ds_1\ldots ds_{m+1}.$$

That is proving the claim. $\qquad\square$

Bibliography

[1] Abramowitz, M. and Stegun, I.A. (1965), *Handbook of Mathematical Functions with Formulae, Graphs and Mathematical Tables*, National Bureau of Standards, Applied Math. Series 55, 4th printing, Washington.

[2] Acosta G. and Durán, R.G. (2003) An Optimal Poincaré inequality in L_1 for convex domains, *Proc. A.M.S.*, Vol. **132** (1), 195–202.

[3] Adamski, W. (1991) An integral representation theorem for positive bilinear forms, *Glasnik Matematicki*, Ser III **26 (46)**, No. 1-2, 31–44.

[4] Agarwal, R.P. (1983) *An Integro-Differential Inequality, General Inequalities*, III (E.F. Beckenbach and W. Walter, eds.), Birkhäuser, Basel, 501-503.

[5] Agarwal, R.P. (2000) *Difference Equations and Inequalities. Theory, Methods and Applications*, Second Edition, Monographs and Textbooks in Pure and Applied Mathematics, 228. Marcel Dekker, Inc., New York.

[6] Agarwal R.P. and Pang, P.Y.H. (1995) *Opial Inequalities with Applications in Differential and Difference Equations*, Kluwer Academic Publishers, Dordrecht, Boston, London.

[7] Agarwal, R.P. and Wong, P.J.Y (1993) *Error Inequalities in Polynomial Interpolation and their Applications*, Mathematics and its Applications, 262, Kluwer Academic Publishers, Dordrecht, Boston, London.

[8] Aliprantis, C.D. and Burkinshaw, O. (1998) *Principles of Real Analysis*, third edition, Academic Press, Boston, New York.

[9] Aljinovic, A.A., Dedic, Lj., Matic, M. and Pecaric, J. (2005) On Weighted Euler harmonic Identities with Applications, *Mathematical Inequalities and Applications*, **8**, No. 2, 237–257.

[10] Aljinovic, A.A., Matic, M. and Pecaric, J. (2005) Improvements of some Ostrowski type inequalities, *J. of Computational Analysis and Applications*, **7**, No. 3, 289–304.

[11] Aljinovic, A.A. and Pecaric, J. (2005) The weighted Euler identity, *Mathematical Inequalities and Applications*, **8**, No. 2, 207–221.

[12] Aljinovic, A.A. and Pecaric, J. (2007) Generalizations of weighted Euler Identity and Ostrowski type inequalities, *Adv. Stud. Contemp. Math.* (Kyungshang) **14**, No. 1, 141–151.

[13] Aljinovic, A.A., Pecaric, J. and Vukelic, A. (2005) The extension of Montgomery identity via Fink identity with applications, *Journal of Inequalities and Applications*, Vol. **2005**, No. 1, 67–80.

[14] Alon, N. and Naor, A. (2006) Approximating the cut-norm via Grothendieck's inequality, *SIAM J. Comput.*, **35**, No. 4, 787–803 (electronic).

[15] Anastassiou, G.A. (1993) *Moments in Probability and Approximation Theory*, Pitman Research Notes in Math., 287, Longman Sci. and Tech., Harlow, U.K.

[16] Anastassiou, G.A. (1995) Ostrowski type inequalities, *Proc. AMS*, **123**, 3775–3781.

[17] Anastassiou, G.A. (1997) Multivariate Ostrowski type inequalities, *Acta Math. Hungarica*, **76**, No. 4, 267–278.

[18] Anastassiou, G.A. (1998) Opial type inequalities for linear differential operators, *Mathematical Inequalities and Applications*, **1**, No. 2, 193–200.

[19] Anastassiou, G.A. (1998) General fractional Opial type inequalities, *Acta Applicandae Mathematicae*, **54**, No. 3, 303–317.

[20] Anastassiou, G.A. (1999) Opial type inequalities involving fractional derivatives of functions, *Nonlinear Studies*, **6**, No. 2, 207–230.

[21] Anastassiou, G.A. (2001) *Quantitative Approximations*, Chapman and Hall/CRC, Boca Raton, New York.

[22] Anastassiou, G.A. (2001) Taylor integral remainders and moduli of smoothness, Y.Cho, J.K.Kim and S.Dragomir (eds.), *Inequalities Theory and Applications*, **1**, Nova Science Publ., New York, 1–31.

[23] Anastassiou, G.A. (2002) Multidimensional Ostrowski inequalities, revisited, *Acta Mathematica Hungarica*, **97**, No. 4, 339–353.

[24] Anastassiou, G.A. (2002) Univariate Ostrowski inequalities, revisited, *Monatshefte Math.*, **135**, 175–189.

[25] Anastassiou, G.A. (2002) Multivariate Montgomery identities and Ostrowski inequalities, *Numer. Funct. Anal. and Opt.*, **23**, No. 3-4, 247–263.

[26] Anastassiou, G.A. (2002) Integration by parts on the Multivariate domain, *Annals of University of Oradea*, fascicola mathematica, Tom **IX**, 5–12.

[27] Anastassiou, G.A. (2003) On Grüss type multivariate integral inequalities, *Mathematica Balkanica*, New Series Vol. **17**, Fasc.1-2, 1–13.

[28] Anastassiou, G.A. (2003) A new expansion formula, *Cubo Matematica Educational*, **5**, No. 1, 25–31.

[29] Anastassiou, G.A. (2006) Multivariate Euler type identity and Ostrowski type inequalities, *Proceedings of the International Conference on Numerical Analysis and Approximation theory*, NAAT 2006, Cluj-Napoca (Romania), July 5-8, 27–54.

[30] Anastassiou, G.A. (2006) Difference of general integral means, *Journal of Inequalities in Pure and Applied Mathematics*, **7**, No. 5, Article 185, pp. 13, http://jipam.vu.edu.au.

[31] Anastassiou, G.A. (2007) Opial type inequalities for semigroups, *Semigroup Forum*, **75**, 625–634.

[32] Anastassiou, G.A. (2007) On Hardy-Opial Type Inequalities, *J. of Math. Anal. and Approx. Th.*, **2** (**1**), 1–12.

[33] Anastassiou, G.A. (2007) Chebyshev-Grüss type and comparison of integral means inequalities for the Stieltjes integral, *Panamerican Mathematical Journal*, **17**, No. 3, 91–109.

[34] Anastassiou, G.A. (2007) Multivariate Chebyshev-Gr üss and comparison of integral means type inequalities via a multivariate Euler type identity, *Demonstratio Mathematica*, **40**, No. 3, 537–558.

[35] Anastassiou, G.A. (2007) High order Ostrowski type Inequalities, *Applied Math. Letters*, **20**, 616–621.

[36] Anastassiou, G.A. (2007) Opial type inequalities for Widder derivatives, *Panamer. Math. J.*, **17**, No. 4, 59–69.

[37] Anastassiou, G.A. (2007) Grüss type inequalities for the Stieltjes Integral, *J. Nonlinear Functional Analysis and Appls.*, **12**, No. 4, 583–593.

[38] Anastassiou, G.A. (2007) Multivariate Euler type identity and optimal multivariate Ostrowski type inequalities, *Advances in Nonlinear Variational Inequalities*, **10**, No. 2, 51–104.

[39] Anastassiou, G.A. (2007) Multivariate Fink type identity and multivariate Ostrowski, comparison of means and Grüss type inequalities, *Mathematical and Computer Modelling*, **46**, 351–374.

[40] Anastassiou, G.A. (2007) Optimal multivariate Ostrowski Euler type inequalities, *Studia Univ. "Babes-Bolyai", ser. Mathematica*, Vol. **LII**, No. 1, 25–61.

[41] Anastassiou, G.A. (2007) Chebyshev-Grüss type inequalities via Euler type and Fink identities, *Mathematical and Computer Modelling*, **45**, 1189–1200.

[42] Anastassiou, G.A. (2007) Ostrowski type inequalities over balls and shells via a Taylor-Widder formula, JIPAM, *J. Inequal. Pure Appl. Math.*, **8**, No. 4, Article 106, 13 pp.

[43] Anastassiou, G.A. (2008) Representations of functions, *Nonlinear Functional Analysis and Applications*, **13**, No. 4, 537-561.

[44] Anastassiou, G.A. (2008) Poincaré and Sobolev type inequalities for vector valued functions, *Computers and Mathematics with Applications*, **56**, 1102–1113.

[45] Anastassiou, G.A. (2008) Opial type inequalities for cosine and sine operator functions, *Semigroup Forum*, **76**, No. 1, 149–158.

[46] Anastassiou, G.A. (2008) Chebyshev-Grüss type inequalities on $\mathbb{R}^N$ over spherical shells and balls, *Applied Math. Letters*, **21**, 119–127.

[47] Anastassiou, G.A. (2008) Ostrowski type inequalities over spherical shells, *Serdica Math. J.*, **34**, 629–650.

[48] Anastassiou, G.A. (2008) Poincare type Inequalities for linear differential operators, *CUBO*, **10**, No. 3, 13–20.

[49] Anastassiou, G.A. (2008) Grothendieck type inequalities, *Applied Math. Letters*, **21**, 1286-1290.

[50] Anastassiou, G.A. (2009) Poincaré like inequalities for semigroups, cosine and sine Operator functions, *Semigroup Forum*, **78**, 54-67.

[51] Anastassiou, G.A. (2009) Distributional Taylor formula, *Nonlinear Analysis*, **70**, 3195-3202.

[52] Anastassiou, G.A. (2008) Opial type inequalities for vector valued functions, *Bulletin of Hellenic Mathematical Society*, **55**, 1-8.

[53] Anastassiou, G.A. (2009) *Fractional Differentiation Inequalities*, Springer Verlag, Berlin - New York, 693 pp.

[54] Anastassiou, G.A. (2009) Poincaré and Sobolev type inequalities for Widder derivatives, *Demonstratio Mathematica*, **42**, No. 2, 283-296.

[55] Anastassiou G.A. and Dragomir, S.S. (2001) On some estimates of the remainder in Taylor's formula, *J. of Math. Analysis and Appl.*, **263**, 246–263.

[56] Anastassiou G.A. and Goldstein, J. (2007) Ostrowski type inequalities over Euclidean domains, *Rend. Lincei Mat. Appl.*, **18**, 305–310.

[57] Anastassiou G.A. and Goldstein, J. (2008) Higher order Ostrowski type inequalities over Euclidean domains, *Journal of Math. Anal. Appl.*,**337**, 962–968.

[58] Anastassiou, G.A., Goldstein, G.R. and Goldstein, J.A. (2004) Multidimensional Opial inequalities for functions vanishing at an interior point, *Atti Accad. Lincei Cl. Fis. Mat. Natur. Rend. Math. Acc. Lincei*, s. 9, v. **15**, 5–15.

[59] Anastassiou, G.A., Goldstein, G.R. and Goldstein, J.A. (2006)Multidimensional weighted Opial inequalities, *Applicable Analysis*, **85**, No. 5, 579–591.

[60] Anastassiou G.A. and Papanicolaou, V. (2002) A new basic sharp integral inequality, *Revue Roumaine de Mathematiques Pure et Appliquees*, **47**, 399–402.

[61] Anastassiou G.A. and Pecaric, J. (1999) General weighted Opial inequalities for linear differential operators, *J. Mathematical Analysis and Applications*, **239**, No. 2, 402–418.

[62] Apostol, T.M. (1974) *Mathematical Analysis*, Second Edition, Addison-Wesley Publ. Co.

[63] Aumann, S., Brown, B.M. and Schmidt, K.M. (2008) A Hardy-Littlewood-type inequality for the p-Laplacian, *Bull. Lond. Math. Soc.*, **40**, No. 3, 525–531.

[64] Azar, L.E. (2008) On some extensions of Hardy-Hilbert's inequality and applications, *J. Inequal. Appl.*, Art. ID 546829, 14 pp.

[65] Bainov D. and Simeonov, P. (1992) *Integral Inequalities and Applications*, translated by R. A. M. Hoksbergen and V. Covachev, Mathematics and its Applications (East European Series), **57**, Kluwer Academic Publishers Group, Dordrecht.

[66] Bakry, D., Barthe, F., Cattiaux, P. and Guilin, A. (2008) A simple proof of the Poincaré inequality for a large class of probability measures including the log-concave case, *Electron. Commun. Probab.*, **13**, 60–66.

[67] Bao, G., Xing, Yuming and Wu, Congxin (2007) Two-weight Poincaré inequalities for the projection operator and A-harmonic tensors on Riemannian manifolds, *Illinois J. Math.*, **51**, No. 3, 831–842.

[68] Barnett, N.S., Cerone, P. and Dragomir, S.S. (2001) Ostrowski type inequalities for multiple integrals, pp. 245–281, *Chapter 5 from Ostrowski Type Inequalities and Applications in Numerical Integration*, edited by S. S. Dragomir and T. Rassias, on line: http://rgmia.vu.edu.au/monographs, Melbourne.

[69] Barnett N.S. and Dragomir, S.S. (1998) An inequality of Ostrowski's type for cumulative distribution functions, RGMIA (*Research Group in Mathematical Inequalities and Applications*), Vol. **1**, No. 1, pp. 3–11, on line: http://rgmia.vu.edu.au

[70] Barnett N.S. and Dragomir, S.S. (1998) An Ostrowski type inequality for double integrals and applications for cubature formulae, *RGMIA Research Report Collection*, **1**, 13–22.

[71] Barnett N.S. and Dragomir, S.S. (1998) An Ostrowski type inequality for a random variable whose probability density function belongs to $L_\infty[a, b]$, *RGMIA Research Group in Mathematical Inequalities and Applications*, Vol. **1**, No. 1, pp. 23–31, on line: http://rgmia.vu.edu.au

[72] Barnett N.S. and Dragomir, S.S. (2001) An Ostrowski type inequality for double integrals and applications for cubature formulae, *Soochow J. Math.*, **27**, No. 1, 1–10.

[73] Barnett, N.S. and Dragomir, S.S. (2007) On the weighted Ostrowski inequality. JIPAM, *J. Inequal. Pure Appl. Math.*, **8**, No. 4, Article 96, pp. 10.

[74] Bartle, R.G. (1976) *The Elements of Real Analysis*, Second Edition, Wiley, New York.

[75] Beesack, P.R. (1962) On an integral inequality of Z. Opial, *Trans. Amer. Math. Soc.*, **104**, 479–475.

[76] Bohner, M. and Matthews, T. (2008) Ostrowski inequalities on time scales, JIPAM, *J. Inequal. Pure Appl. Math.*, **9**, no. 1, Article 6, pp. 8.

[77] Burenkov, V.I. (1974) Sobolev's integral representation and Taylor's formula, (Russian), Studies in the theory of differentiable functions of several variables and its applications, V. *Trudy Mat. Inst. Steklov.*, **131**, 33–38.

[78] Burenkov, V.I. (1998) *Sobolev Space on Domains*, Teubner-Texte zur Mathematik [Teubner Texts in Mathematics], **137**. B. G. Teubner Verlagsgesellschaft mbtt, Stuttgart, pp. 312.

[79] Butzer, P.L. and Berens, H. (1967) *Semi-Groups of Operators and Approximation*, Springer-Verlag, New York.

[80] Butzer, P.L. and Tillmann, H.G. (1960) Approximation theorems for semigroups of bounded linear transformations, *Math. Annalen*, **140**, 256–262.

[81] Canavati, J.A. (1987) The Riemann-Liouville integral, *Nieuw Archief Voor Wiskunde*, **5**, No. 1, 53–75.

[82] Cerone, P. (2002) Difference between weighted integral means, *Demonstratio Mathematica*, Vol. **35**, No. 2, pp. 251–265.

[83] Cerone, P. (2003) Approximate multidimensional integration through dimension reduction via the Ostrowski functional, *Nonlinear Funct. Anal. Appl.*, **8**, No. 3, 313–333.

[84] Cerone, P. and Dragomir, S.S. (2000) Trapezoidal-type rules from an inequalities point of view, *Chapter 3 in Handbook of Analytical-Computational Methods in Applied Mathematics*, edited by George Anastassiou, Chapman and Hall/CRC, pp. 65–134.

[85] Cerone, P. and Dragomir, S.S. (2000) Midpoint type rules from an inequality point of view, *Chapter 4 in Handbook of Analytical-Computational Methods in Applied Mathematics*, edited by George Anastassiou, Chapman and Hall/CRC, pp. 135–200.

[86] Cerone, P. and Dragomir S.S (2004) and Ferdinand Österreicher, Bounds on extended f-divergences for a variety of classes, *Kybernetika* (Prague), **40**, No. 6, 745–756.

[87] Cerone, P. and Dragomir, S.S. (2008) *Advances in Inequalities for Special Functions*, edited volume, Nova Publ., New York.

[88] Chebyshev, P.L. (1882) Sur les expressions approximatives des integrales definies par les autres prises entre les mêmes limites, *Proc. Math. Soc. Charkov*, **2**, 93–98.

[89] Cheung, W.-S. and Zhao, C.-J. (2007) On a discrete Opial-type inequality, JIPAM, *J. Inequal. Pure Appl. Math.*, **8**, No. 4, Article 98, pp. 4.

[90] Cheung, Wing-Sum and Zhao, Chang-Jian (2007) On Opial-type integral inequalities, *J. Inequal. Appl.*, Art. ID 38347, pp. 15.

[91] Cheung, W.-S., Dandan, Zhao and Pecaric, J. (2007) Opial-type inequalities for differential operators, *Nonlinear Anal.*, **66**, No. 9, 2028–2039.

[92] Chua, S.-K. and Wheeden, R.L. (2006) A note on sharp 1-dimensional Poincaré inequalities, *Proc. AMS*, Vol. **134**, No. 8, 2309–2316.

[93] Chyan, D.-K., Shaw, S.-Y. and Piskarev, S. (1999) On Maximal regularity and semi-variation of Cosine operator functions, *J. London Math. Soc.*, (2) **59**, 1023–1032.

[94] Cianchi, A. (2008) Sharp Morrey-Sobolev inequalities and the distance from extremals, *Trans. Amer. Math. Soc.*, **360**, No. 8, 4335–4347.

[95] Cianchi, A. and Ferone, A. (2008) A strengthened version of the Hardy-Littlewood inequality, *J. Lond. Math. Soc.* (2) **77**, No. 3, 581–592.

[96] Civljak, A., Dedic, Lj. and Matic, M. (2007) On Ostrowski and Euler-Grüss type inequalities involving measures, *J. Math. Inequal.*, **1**, No. 1, 65–81.

[97] Cho, H.R. and Lee, J. (2005) Inequalities for the integral means of holomorphic functions in the strongly pseudoconvex domain, *Commun. Korean Math. Soc.*, **20**, No. 2, 339–350.

[98] Dedic, Lj., Matic, M. and Pecaric, J. (2000) On generalizations of Ostrowski inequality via some Euler- type identities, *Mathematical Inequalities and Applications*, Vol. **3**, No. 3, 337–353.

[99] Dedic, Lj., Matic, M., Pecaric, J. and Vukelic, A. (2002) On generalizations of Ostrowski inequality via Euler harmonic identities, *Journal of Inequalities and Applications*, Vol. **7**, No. 6, 787–805.

[100] Draghici, C. (2006) Inequalities for integral means over symmetric sets, *J. Math. Anal. Appl.*, **324**, No. 1, 543–554.

[101] Dragomir, S.S. (1999) New estimation of the remainder in Taylor's formula using Grüss' type inequalities and applications, *Math. Ineq. Appl.*, **2(2)**, 183–193.

[102] Dragomir, S.S. (2001) Ostrowski's inequality for mappings of bounded variation and applications, *Math. Ineq. and Appl.*, **4 (1)**, 59-66, and on-line in (http://rgmia.vu.edu.au); RGMIA Res. Rep. Coll., 2 (1) (1999), 103–110.

[103] Dragomir, S.S. (2000) *A Survey on Cauchy-Buniakowsky-Schwartz Type Discrete Inequalities*, RGMIA Monographs, Victoria University. (on-line http://rgmia.vu.edu.au/monographs/).

[104] Dragomir, S.S. (2000) *Semi-Inner Products and Applications*, RGMIA Monographs, Victoria University. (on-line http://rgmia.vu.edu.au/monographs/).

[105] Dragomir, S.S. (2000) *Some Gronwall Type Inequalities and Applications*, RGMIA Monographs, Victoria University. (on-line http://rgmia.vu.edu.au/monographs/).

[106] Dragomir, S.S. (2000) Some inequalities of Grüss type, *Indian J. Pure Appl. Math.*, **31**, 397–415.

[107] Dragomir, S.S. (2002) New estimates of the Čebyšev functional for Stieltjes integrals and applications, *RGMIA*, **5**, Supplement Article 27, electronic, http://rgmia.vu.edu.au

[108] Dragomir, S.S. (2002) Sharp bounds of Čebyšev functional for Stieltjes integrals and applications, *RGMIA*, **5**, Supplement Article 26, electronic, http://rgmia.vu.edu.au

[109] Dragomir, S.S. (2003) *Some Gronwall Type Inequalities and Applications*, Nova Science Publishers, Inc., Hauppauge, NY.

[110] Dragomir, S.S. (2003) A weighted Ostrowski type inequality for functions with values in Hilbert spaces and Applications, *J. Korean Math. Soc.*, **40**, 207–224.

[111] Dragomir, S.S. (2004) **Semi-Inner Products and Applications**, Nova Science Publishers, Inc., Hauppauge, NY.

[112] Dragomir, S.S. (2004) *Discrete Inequalities of the Cauchy–Bunyakovsky–Schwarz Type*, Nova Science Publishers, Inc., Hauppauge, NY.

[113] Dragomir, S.S. (2004) Inequalities of Grüss type for the Stieltjes integral and applications, *Kragujevac J. Math.*, **26**, 89–122.

[114] Dragomir, S.S. (Ed.) (2004), *Advances in inequalities of the Schwarz, Grüss and Bessel Type in Inner Product Spaces*, RGMIA Monographs, Victoria University. (on-line http://rgmia.vu.edu.au/monographs/).

[115] Dragomir, S.S. (Ed.) (2005) *Advances in Inequalities of the Schwarz, Triangle and Heisenberg Type in Inner Product Spaces*, RGMIA Monographs, Victoria University. (on-line http://rgmia.vu.edu.au/monographs/).

[116] Dragomir, S.S. (2005) *Advances in inequalities of the Schwarz, Grüss and Bessel Type in Inner Product Spaces*, Nova Science Publishers, Inc., Hauppauge, NY.

[117] Dragomir, S.S. (2007) *Advances in Inequalities of the Schwarz, Triangle and Heisenberg Type in Inner Product Spaces*, Nova Science Publishers, Inc., Hauppauge, NY.

[118] Dragomir, S.S. (2007) Grüss type discrete inequalities in inner product spaces, revisited, *Inequality theory and applications*, Vol. **5**, Nova Sci. Publ., NY, 61-69.

[119] Dragomir, S.S. (2007) Sharp Grüss-type inequalities for functions whose derivatives are of bounded variation, JIPAM, *J. Inequal. Pure Appl. Math.*, **8**, No. 4, Article 117, pp. 13.

[120] Dragomir, S.S. and Barnett, N.S. (1998) An Ostrowski type inequality for mappings whose second derivatives are bounded and applications, *RGMIA Research Report Collection*, Vol. **1**, No. 2, 69–77, on line http://rgmia.vu.edu.au

[121] Dragomir, S.S. and Barnett, N.S. (1999) An Ostrowski type inequality for mappings whose second derivatives are bounded and applications, *J. Indian Math. Soc. (S.S.)*, *66* (1-4), 237–245.

[122] Dragomir, S.S., Barnett, N.S. and Cerone, P. (1998) An Ostrowski type inequality for double integrals in terms of L_p -norms and applications in numerical integration, *RGMIA Research Report Collection*, Vol.1, No. 2, 79–87, on line http://rgmia.vu.edu.au

[123] Dragomir, S.S., Barnett, N.S. and Cerone, P. (1999) An n-dimension version of Ostrowski's inequality for mappings of the Holder Type, RGMIA Res. Rep. Collect, 2, 169–180.

[124] Dragomir, S.S., Barnett, N.S. and Cerone, P. (2000) An n-dimensional version of Ostrowski's inequality for mappings of the Hölder type, *Kyungpook Math. J.*, **40**, No. 1, 65–75.

[125] Dragomir, S.S., Cerone, P., Barnett, N.S. and Roumeliotis, J. (1999) An inequality of the Ostrowski type for double integrals and applications to cubature formulae, *RGMIA Research Report Collection*, **2**, 781–796.

[126] Dragomir, S.S., Cerone, P., Barnett, N.S. and Roumeliotis, J. (2000) An inequality of the Ostrowski type for double integrals and applications to cubature formulae, *Tamsui Oxford Journal of Mathematical Sciences*, **16**, 1–16.

[127] Dragomir, S.S. and Gosa, A.C. (2007) A generalization of an Ostrowski inequality in inner product spaces, in: *Inequality Theory and Applications*, Vol. **4**, 61–64, Nova Sci. Publ., New York.

[128] Dragomir, S.S. and Pearce, C.E.M. (2000) *Selected Topics on Hermite-Hadamard Inequalities and Applications*, RGMIA Monographs, Victoria University. (on-line http://rgmia.vu.edu.au/monographs/).

[129] Dragomir, S.S. and Pecaric, J. (1989) Refinements of some inequalities for isotonic functionals, *Anal. Num. Theor. Approx.*, **11**, 61–65.

[130] Dragomir, S.S., Pečarić, J.E. and Tepes, B. (2007) Note on integral version of the Grüss inequality for complex functions, in: *Inequality Theory and Applications*, Vol. **5**, 91–96, Nova Sci. Publ., New York.

[131] Dragomir, S.S. and Rassias, T.M. (2000) *Ostrowski Type Inequalities and Applications in Numerical Integration*, RGMIA Monographs, Victoria University. (on-line http://rgmia.vu.edu.au/monographs/).

[132] Dudley, R.M. (1989) *Real Analysis and Probability*, Wadsworth and Brooks/Cole, Pacific Grove, CA.

[133] J. Duoandikoetxea, J. (2001) A unified approach to several inequalities involving functions and derivatives, *Czechoslovak Mathematical Journal*, **51** (126), 363–376.

[134] Durán, A.L., Estrada, R. and Kanwal, R.P. (1996) Pre-asymptotic expansions, *J. of Math. Anal. and Appl.*, **202**, 470–484.

[135] Estrada, R. and Kanwal, R.P. (1990) A Distributional theory for asymptotic expansions, *Proceedings of the Royal Society of London*, Series A, Mathematical and Physical Sciences, Vol. **428**, No. 1875, 399–430.

[136] Estrada, R. and Kanwal, R.P. (1992) The asymptotic expansion of certain multi-dimensional generalized functions, *J. of Math. Analysis and Appl.*, **163**, 264–283.

[137] Estrada, R. and Kanwal, R.P. (1993) Taylor expansions for distributions, *Mathematical Methods in the Applied Sciences*, Vol. **16**, 297–304.

[138] Estrada, R. and Kanwal, R.P. (2002) *A Distributional Approach to Asymptotics, Theory and Applications*, 2nd Edition, Birkhäuser, Boston, Basel, Berlin.

[139] Evans, L.C. (1998) *Partial Differential Equations*, Graduate Studies in Mathematics, vol. **19**, American Mathematical Society, Providence, RI.

[140] Fattorini, H.O. (1968) Ordinary differential equations in linear topological spaces, *Int. J. Diff. Equations*, **5**, 72–105.

[141] Fink, A.M. (1992) Bounds on the deviation of a function from its averages, *Czechoslovak Mathematical Journal*, **42** (117), No. 2, 289–310.

[142] Fleming, W. (1977) *Functions of Several Variables*, Springer-Verlag, New York, Berlin, Undergraduate Texts in Mathematics, 2nd edition.

[143] Flores-Franulič, A. and Romá n-Flores, H. (2007) A Chebyshev type inequality for fuzzy integrals, *Appl. Math. Comput.*, **190**, No. 2, 1178–1184.

[144] Gao, Peng (2008) Hardy-type inequalities via auxiliary sequences, *J. Math. Anal. Appl.*, **343**, No. 1, 48–57.

[145] Gavrea, I. (2006) On Chebyshev type inequalities involving functions whose derivatives belong to L_p spaces via isotonic functionals, JIPAM, *J. Inequal. Pure Appl. Math.*, **7**, No. 4, Article 121, 6 pp. (electronic).

[146] Gel'fand, I.M. and Shilov, G.E. (1964) *Generalized Functions*, Vol. **I**, Academic Press, New York, London.

[147] Goldstein, J.A. (1985) *Semigroups of Linear Operators and Applications*, Oxford Univ. Press, Oxford.

[148] Goyal, S.P. (2007) Manita Bhagtani and Kantesh Gupta, Integral mean inequalities for fractional calculus operators of analytic multivalent function, *Bull. Pure Appl. Math.*, **1**, No. 2, 214–226.

[149] Grothendieck, A. (1956) Résumé de la th éorie métrique des produits tensoriels topologiques, *Bol. Soc. Matem. São Paolo*, **8**, 1–79.

[150] Grüss, G. (1935) Über das Maximum des absoluten Betrages von

$$\left[\left(\frac{1}{b-a} \right) \int_a^b f(x)g(x)dx - \left(\frac{1}{(b-a)^2} \int_a^b f(x)dx \int_a^b g(x)dx \right) \right],$$

Math. Z., **39**, pp. 215–226.

[151] Güney, H. Ö. and Owa, S. (2007) Integral means inequalities for fractional derivatives of a unified subclass of prestarlike functions with negative coefficients, *J. Inequal. Appl.*, Art. ID 97135, 9 pp.

[152] Guyker, J. (2006) An inequality for Chebyshev connection coefficients, JIPAM, *J. Inequal. Pure Appl. Math.*, **7**, No. 2, Article 67, 11 pp. (electronic).

[153] Haagerup, U. (1987) A new upper bound for the complex Grothendieck constant, *Israel J. Math.*, **60**, 199–224.

[154] Haluska, J. and Hutnik, O. (2007) Some inequalities involving integral means, *Tatra Mt. Math. Publ.*, **35**, 131–146.

[155] Han, J. (2008) A class of improved Sobolev-Hardy inequality on Heisenberg groups, *Southeast Asian Bull. Math.*, **32**, No. 3, 437–444.

[156] Hardy, G.H., Littlewood, J.E. and Polya, G. (1988) *Inequalities*, Reprint of the 1952 edition, Cambridge Mathematical Library, Cambridge University Press, Cambridge.

[157] Harjulehto, P. and Hurri-Syrjanen, R. (2008) On a (q,p)-Poincaré inequality, *J. Math. Anal. Appl.*, **337**, No. 1, 61–68.

[158] Hewitt, E. and Stromberg, K. (1965) *Real and Abstract Analysis*, Springer-Verlag, New York, Berlin.

[159] Hille, E. and Phillips, R.S. (1957) *Functional Analysis and Semigroups*, revised edition, pp. XII a. 808, Amer. Math. Soc. Colloq. Publ., Vol. **31**, Amer. Math. Soc., Providence, RI.

[160] Hognas, G. (1977) Characterization of weak convergence of signed measures on [0, 1], *Math. Scand.*, **41**, 175–184.

[161] Johnson, J. (1985) An elementary characterization of weak convergence of measures, *The American Math. Monthly*, **92**, No. 2, 136–137.

[162] Junge, M. (2005) Embedding of the operator space OH and the logarithmic little Grothendieck inequality, *Invent. Math.*, **161**, No. 2, 225–286.

[163] Karlin, S. and Studden, W.J. (1966) *Tchebycheff Systems: with Applications in Analysis and Statistics*, Interscience, New York.

[164] Katznelson, Y. (1976) *An introduction to Harmonic Analysis*, Dover, New York.

[165] Keith, S. and Zhong, Xiao (2008) The Poincaré inequality is an open ended condition, *Ann. of Math.*, **(2) 167**, No. 2, 575–599.

[166] Kemperman, J.H.B. (1965) On the sharpness of Tchebycheff type inequalities, *Indagationes Math.*, **27**, 554–601.

[167] Kingman, J. and Taylor, S. (1966) *Introduction to Measure and Probability*, Cambridge University Press, Cambridge, NY.

[168] Korovkin, P.P. (1986) *Inequalities*, Translated from the Russian by Sergei Vrubel, Reprint of the 1975 edition, Little Mathematics Library, Mir, Moscow, distributed by Imported Publications, Chicago, IL.

[169] Kreider, D., Kuller, R., Ostberg, D. and Perkins, F. (1966) *An Introduction to Linear Analysis*, Addison-Wesley Publishing Company, Inc., Reading, Mass., USA.

[170] Krivine, I.L. (1977) Sur la constante de Grothendieck, *Comptes Rendus Acad. Sci. Paris*, **284**, 445–446.

[171] Krylov, V.I. (1962) *Approximate Calculation of Integrals*, Macmillan, New York, London.

[172] Kwong, M.K. (2007) On an Opial inequality with a boundary condition, JIPAM, *J. Inequal. Pure Appl. Math.*, **8**, No. 1, Article 4, 6 pp. (electronic).

[173] Lakshmikantham, V. and Leela, S. (1969) *Differential and Integral Inequalities: Theory and Applications*, vol. **I**: Ordinary Differential Equations, Mathematics in Science and Engineering, Vol. 55-I, Academic Press, New York - London.

[174] Lakshmikantham, V. and Leela, S. (1969) *Differential and Integral Inequalities: Theory and Applications*, vol. **II**: Functional, Partial, Abstract and Complex Differential Equations, Mathematics in Science and Engineering, Vol. 55-II, Academic Press, New York - London.

[175] Lehrbäck, J. (2008) Pointwise Hardy inequalities and uniformly fat sets, *Proc. Amer. Math. Soc.*, **136**, No. 6, 2193–2200.

[176] Levinson, N. (1964) On an inequality of Opial and Beesack, *Proc. Amer. Math. Soc.*, **15**, 565–566.

[177] Lian, W.-C., Yu, S.-L., Wong, F.-H. and Lin, S.-W. (2005) Nonlinear type of Opial's inequalities on time scales, *Int. J. Differ. Equ. Appl.*, **10**, No. 1, 101–111.

[178] Lieb, E.H. and Loss, M. (2001) *Analysis*, 2nd Edition, American Math. Soc., Providence, R. I.

[179] Lin, C.T. and Yang, G.S. (1983) A generalized Opial's inequality in two variables, *Tamkang J. Math.*, **15**, 115–122.

[180] Liu, W. and Dong, J. (2008) On new Ostrowki type inequalities, *Demonstratio Math.*, **41**, No. 2, 317–322.

[181] Liu, Zheng (2007) Notes on a Grüss type inequality and its application, *Vietnam J. Math.*, **35**, No. 2, 121–127.

[182] Lust-Piquard, F. and Xu, Q. (2007) The little Grothendieck theorem and Khintchine inequalities for symmetric spaces of measurable operators, *J. Funct. Anal.*, **244**, No. 2, 488–503.

[183] Mao, Y.-H. (2008) General Sobolev type inequalities for symmetric forms, *J. Math. Anal. Appl.*, **338**, No. 2, 1092–1099.

[184] Markov, A. (1884) *On Certain Applications of Algebraic Continued Fractions*, Thesis, St. Petersburg.

[185] Matic, M., Pecaric, J. and Ujevic, N. (2001) Weighted version of multivariate Ostrowski type inequalities, *Rocky Mountain J. Math.*, **31**, No. 2, 511–538.

[186] Mitrinovic, D.S., Pecaric, J.E. and Fink, A.M. (1991) *Inequalities Involving Functions and Their Integrals and Derivatives*, Mathematics and its Applications (East European Series), **53**, Kluwer Academic Publishers Group, Dordrecht.

[187] Mitrinovic, D.S., Pecaric, J.E. and Fink, A.M. (1993) *Classical and New Inequalities in Analysis*, Kluwer Academic, Dordrecht.

[188] Mitrinovic, D.S., Pecaric, J.E. and Fink, A.M. (1994) *Inequalities for Functions and their Integrals and Derivatives*, Kluwer Academic Publishers, Dordrecht.

[189] Mond, B., Pecaric, J. and Peric, I. (2006) On reverse integral mean inequalities, *Houston J. Math.*, **32**, No. 1, 167–181.

[190] Mulholland, H. and Rogers, C. (1958) Representation theorems for distribution functions, *Proc. London Math. Soc.*, **8**, 177–223.

[191] Nagy, B. (1974) On cosine operator functions in Banach spaces, *Acta Scientiarum Mathematicarum Szeged*, **36**, 281–289.

[192] Nagy, B. (1977) Approximation theorems for cosine operator functions, *Acta Mathematica Academiae Scientiarum Hungaricae*, **29** (1-2), 69–76.

[193] Necaev, I.D. (1973) Integral inequalities with gradients and derivatives, *Soviet. Math. Dokl.*, **22**, 1184–1187.

[194] Olech, C. (1960) A simple proof of a certain result of Z. Opial, *Ann. Polon. Math.*, **8**, 61–63.

[195] Opial, Z. (1960) Sur une inegalité, *Ann. Polon. Math.*, **8**, 29–32.

[196] Ostrowski, A. (1938) Über die Absolutabweichung einer differentiebaren Funktion von ihrem Integralmittelwert, *Comment. Math. Helv.*, **10**, 226–227.

[197] Pachpatte, B.G. (2000) On an inequality of Ostrowski type in three independent variables, *J. Math. Anal. Appl.*, **249**, No. 2, 583–591.

[198] Pachpatte, B.G. (2001) An inequality of Ostrowski type in n independent variables, *Facta Univ. Ser. Math. Inform.*, No. **16**, 21–24.

[199] Pachpatte, B.G. (2002) On multivariate Ostrowski type inequalities, JIPAM, *J. Inequal. Pure. Appl. Math.*, **3**, No. 4, Article 58, 5 pp. (electronic).

[200] Pachpatte, B.G. (2002) On Grüss type inequalities for double integrals, *J. Math. Anal. Appl.*, **267**, 454–459.

[201] Pachpatte, B.G. (2003) New weighted multivariate Grüss type inequalities, electronic Journal: *Journal of Inequalities in Pure and Applied Mathematics*, http://jipam.vu.edu.au/, Volume 4, Issue 5, Article 108.

[202] Pachpatte, B.G. (2005) Inequalities similar to Opial's inequality involving higher order derivatives, *Tamkang J. Math.*, **36**, No. 2, 111–117.

[203] Pachpatte, B.G. (2005) A note on Opial type finite difference inequalities, *Tamsui Oxf. J. Math. Sci.*, **21**, No. 1, 33–39.

[204] Pachpatte, B.G. (2005) *Mathematical Inequalities*, North Holland Mathematical Library, **67**, Elsevier B. V., Amsterdam.

[205] Pachpatte, B.G. (2006) *Integral and Finite Difference Inequalities and Applications*, North-Holland Mathematics Studies, 205, Elsevier Science B. V., Amsterdam.

[206] Pachpatte, B.G. (2006) On Chebyshev-Grüss type inequalities via Pecaric's extension of the Mongomery identity, *Journal of Inequalities in Pure and Applied Mathematics*, http://jipam.vu.edu.au/, Vol. **7**, Issue 1, Article 11.

[207] Pachpatte, B.G. (2007) Some new Ostrowski and Gr üss type inequalities, *Tamkang J. Math.*, **38**, No. 2, 111–120.

[208] Pachpatte, B.G. (2007) New inequalities of Ostrowski-Grüss type, *Fasc. Math.*, No. **38**, 97–104.

[209] Pachpatte, B.G. (2007) On a new generalization of Ostrowski type inequality, *Tamkang J. Math.*, **38**, No. 4, 335–339.

[210] Pachpatte, B.G. (2007) New generalization of certain Ostrowski type inequalities, *Tamsui Oxf. J. Math. Sci.*, **23**, No. 4, 393–403.

[211] Pachpatte, B.G. (2007) New discrete Ostrowski-Grüss like inequalities, *Facta. Univ. Ser. Math. Inform.*, **22**, No. 1, 15–20.

[212] Pachpatte, B.G. (2008) A note on Grüss type inequalities via Cauchy's mean value theorem, *Math. Inequal. Appl.*, **11**, No. 1, 75–80.

[213] Pecaric, J.E. (1980) On the Cebyshev inequality, *Bul. Sti. Tehn. Inst. Politehn. "Traian Vuia" Timisoara*, **25 (39)**, No. 1, 5–9.

[214] Pecaric, J. and Peric, I. (2006) A multidimensional generalization of the Lupas-Ostrowski inequality, *Acta Sci. Math. (Szeged)*, **72**, 65–72.

[215] Pecaric, J.E., Proschan, F. and Tong, Y.L. (1992) *Convex Functions, Partial Orderings and Statistical Applications*, Mathematics in Science and Engineering, **187**, Academic Press, Inc., Boston, MA.

[216] Pecaric, J. and Vukelic, A. (2006) Milovanovic-Pecaric-Fink inequality for difference of two integral means, *Taiwanese J. Math.*, **10**, No. 4, 933–947.

[217] Peralta, A.M. and Palacios, A.R. (2001) Grothendieck's inequalities revisited, in: *Recent Progress in Functional Analysis*, (Valencia, 2000), 409–423, North-Holland Math. Stud., **189**, North-Holland, Amsterdam.

[218] Plotnikova, E.A. (2008) Integral representations and the generalized Poincaré inequality on Carnot groups, (Russian) *Sibirsk. Mat. Zh.*, **49**, No. 2, 420–436.

[219] Polya, G. and Szego, G. (1951) *Isoperimetric Inequalities in Mathematical Physics*, Annals of Mathematics Studies, No. **27**, Princeton University Press, Princeton, N. J.

[220] Rafig, A. and Zafar, F. (2007) New bounds for the first inequality of Ostrowski-Grüss type and applications in numerical integration, *Nonlinear Funct. Anal. Appl.*, **12**, No. 1, 75–85.

[221] Richter, H. (1957) Parameterfreie Abschätzung und Realisierung von Erwartungswerten, *Blätter der Deutschen Gesellschaft für Versicherungsmathematik*, **3**, 147–161.

[222] Riesz, F. (1911) Sur certaines systèmes singuliers d'équations integrales, *Ann. Sci. Ecole Norm. Sup.*, **28**, 33–62.

[223] Rogers, L.J. (1888) An extension of a certain theorem in inequalities, *Messenger of Math.*, **17**, 145–150.

[224] Royden, H.L. (1968) *Real Analysis*, Second edition, Macmillan, New York, London.

[225] Sababheh, M. (2007) Two-sided probabilistic versions of Hardy's inequality, *J. Fourier Anal. Appl.*, **13**, No. 5, 577–587.

[226] Schumaker, L. (1981) *Spline Functions Basic Theory*, Wiley, New York.

[227] Schwartz, L. (1967) *Analyse Mathematique*, Paris.

[228] Selberg, H.L. (1940) Zwei Ungleichungen zur Ergänzung des Tchebycheffschen Lemmas, *Skand. Aktuarietidskrift*, **23**, 121–125.

[229] Shen, Z. and Zhao, P. (2008) Uniform Sobolev inequalities and absolute continuity of periodic operators, *Trans. Amer. Math. Soc.*, **360**, No. 4, 1741–1758.

[230] Shilov, G. (1974) *Elementary Functional Analysis*, The MIT Press Cambridge, Massachusetts.

[231] Sofo, A. (2002) Integral inequalities of the Ostrowski type, JIPAM, *J. Inequal. Pure Appl. Math.*, **3**, No. 2, Article 21, 39 pp. (electronic).

[232] Sova, M. (1966) *Cosine Operator Functions*, Rozprawy Matematyczne, **XLIX**, Warszawa.

[233] Spiegel, M.R. (1963) *Advances Calculus*, Schaum's Outline Series, McGraw-Hill Book Co., New York.

[234] Stankovic, B. (1996) Taylor Expansion for generalized functions, *J. of Math. Analysis and Appl.*, **203**, 31–37.

[235] Stević, S. (2007) Area type inequalities and integral means of harmonic functions on the unit ball, *J. Math. Soc. Japan*, **59**, No. 2, 583–601.

[236] Sulaiman, W.T. (2008) On a reverse of Hardy-Hilbert's integral inequality in its general form, *Int. J. Math. Anal. (Ruse)*, **2**, No. 1-4, 67–74.

[237] Troy, W.C. (2001) On the Opial-Olech-Beesack inequalities, USA-Chile Workshop on Nonlinear Analysis, *Electron. J. Diff. Eqns.*, Conf., **06**, 297–301. http://ejde.math.swt.edu or http://ejde.math.unt.edu

[238] Tseng, K.-L., Hwang, S.R. and Dragomir, S.S. (2008) Generalizations of weighted Ostrowski type inequalities for mappings of bounded variation and their applications, *Comput. Math. Appl.*, **55**, No. 8, 1785–1793.

[239] Watanabe, K., Kametaka, Y., Nagai, A., Takemura, K. and Yamagishi, H. (2008) The best constant of Sobolev inequality on a bounded interval, *J. Math. Anal. Appl.*, **340**, No. 1, 699–706.

[240] Wen, J.J. and Gao, C.B. (2008) The best constants of Hardy type inequalities for $p = -1$, *J. Math. Res. Exposition*, **28**, No. 2, 316–322.

[241] Wen, J.J. and Wang, W.-l. (2007) Chebyshev type inequalities involving permanents and their applications, *Linear Algebra Appl.*, **422**, No. 1, 295–303.

[242] Whittaker, E.T. and Watson, G.N. (1927) *A Course in Modern Analysis*, Cambridge University Press.

[243] Widder, D.V. (1928) A Generalization of Taylor's Series, *Transactions of AMS*, **30**, No. 1, 126–154.

[244] Willett, D. (1968) The existence - uniqueness theorems for an nth order linear ordinary differential equation, *Amer. Math. Monthly*, **75**, 174–178.

[245] Wong, F.-H., Lian, W.-C., Yu, S.-L. and Yeh, C.-C. (2008) Some generalizations of Opial's inequalities on time scales, *Taiwanese J. Math.*, **12**, No. 2, 463–471.

[246] Xu, Q. (2006) Operator-space Grothendieck inequalities for noncommutative L_p−space, *Duke Math. J.*, **131**, No. 3, 525–574.

[247] Yang, G.S. (1982) Inequality of Opial-type in two variables, *Tamkang J. Math.*, **13**, 255–259.

[248] Yeh, C.-C. (2008) Ostrowski's inequality on time scales, *Appl. Math. Lett.*, **21**, No. 4, 404–409.

[249] Zhang, Q.S. (2008) A uniform Sobolev inequality for Ricci flow with surgeries and applications, *C. R. Math. Acad. Sci. Paris* **346**, No. 9-10, 549–552.

[250] Zhao, C.-J. and Cheung, W.-S. (2008) On multivariate Grüss inequalities, *J. Inequal. Appl.*, Art. ID 249438, 8 pp.

[251] Ziemian, B. (1988) *Taylor Formula for Distributions*, Dissertationes Math. (Rozprawy Mat.), **264**, 56 pp.

[252] Ziemian, B. (1989) The Melin transformation and multidimensional generalized Taylor expansions of singular functions, *J. Fac. Sci. Univ. Tokyo Sect. IA Math.*, **36**, No. 2, 263–295.

[253] Ziemian, B. (1990) Generalized Taylor expansions and theory of resurgent functions of Jean E'calle, in: *Generalized Functions and Convergence* (Katowice, 1988), 285-295, World Sci. Publ., Teaneck, NJ.

List of Symbols

Index